AF410813

Steroid Compounds with Potential Biological Activity

Steroid Compounds with Potential Biological Activity

Editors

Marina Savić
Erzsébet Mernyák
Jovana Ajdukovic
Suzana Jovanović-Šanta

Basel • Beijing • Wuhan • Barcelona • Belgrade • Novi Sad • Cluj • Manchester

Editors

Marina Savić
Department of Chemistry,
Biochemistry and
Environmental Protection
University of Novi Sad
Novi Sad
Serbia

Erzsébet Mernyák
Department of Organic
Chemistry
University of Szeged
Szeged
Hungary

Jovana Ajdukovic
Department of Chemistry,
Biochemistry and
Environmental Protection
University of Novi Sad
Novi Sad
Serbia

Suzana Jovanović-Šanta
Department of Chemistry,
Biochemistry and
Environmental Protection
University of Novi Sad
Novi Sad
Serbia

Editorial Office
MDPI
St. Alban-Anlage 66
4052 Basel, Switzerland

This is a reprint of articles from the Special Issue published online in the open access journal *Molecules* (ISSN 1420-3049) (available at: www.mdpi.com/journal/molecules/special_issues/ Steroid_Compounds_Biological_Activity).

For citation purposes, cite each article independently as indicated on the article page online and as indicated below:

Lastname, A.A.; Lastname, B.B. Article Title. *Journal Name* **Year**, *Volume Number*, Page Range.

ISBN 978-3-0365-9617-4 (Hbk)
ISBN 978-3-0365-9616-7 (PDF)
doi.org/10.3390/books978-3-0365-9616-7

Contents

About the Editors

Marina Savić

Dr. Marina Savić is an Associate Professor at the Department of Chemistry, Biochemistry and Environmental Protection, Faculty of Sciences at the University of Novi Sad in Serbia. She has been engaged in research at the Department since 2004, and the focus of her research is the synthesis and modification of the steroid skeleton in the androstane series, with the aim of obtaining new biologically active derivatives, more precisely, compounds with potential anticancer activity. She is the co-author of 27 papers. She is an active reviewer of numerous international journals.

Erzsébet Mernyák

Dr. Erzsebet Mernyak is an Associate Professor at the University of Szeged in Hungary. She is the holder of a large number of national awards for scientific excellence. She is the co-author of more than 70 papers. She is a member of the Isoprenoid Society, the Hungarian Chemical Society and the Steroidal and Terpenoidal Chemistry Working Group of the Hungarian Academy of Sciences. She is an active reviewer of numerous international journals.

Jovana Ajdukovic

Dr. Jovana Ajduković works as an Associate Professor at the Department of Chemistry, Faculty of Sciences, University of Novi Sad in Serbia. The field of her research/teaching is organic and medicinal chemistry, particularly the synthesis and chemical transformations of steroid hormones and their biological evaluation. She is the author of a university textbook and co-author of 30 papers. She is an active reviewer of numerous international journals. In 2018, she was awarded with the DAAD Scholarship "Research Stays for University Academics and Scientists" for a short research stay, by The German Academic Exchange Service.

Suzana Jovanović-Šanta

Dr. Suzana Jovanović-Šanta is Full Professor at the University of Novi Sad Faculty of Sciences, Department of Chemistry, Biochemistry and Environmental Protection. She is teaching biochemistry-related courses on graduate, master and PhD studies of biochemistry and chemistry. Her research area encompasses the study of the biological activity of modified steroids and other compounds in vitro on cell cultures and animal tissues and in vivo on animal models as well as the testing of antiestrogenic, anticancer and antioxidant activity and the measuring of the effect on steroid-converting enzymes. She has published more than 50 research articles. During her scientific work, she has been project manager in three and researcher in three international research projects as well as eight national research projects. She is an active reviewer of numerous international journals. She is Vice President of the Serbian Chemical Society and Vice President of the Serbian Biochemical Society.

Preface

The Reprint of the Special Issue of *Molecules,* titled "Steroid Compounds with Potential Biological Activity", represents an effort by both authors and editors to provide insight into the biological potential of various compounds based on steroid structure.

This reprint is mainly intended for researchers and university teachers who, as part of their scientific research or teaching process, deal with the synthesis, isolation, characterization and testing of the biological activity of various compounds with a steroid nucleus. This topic is constantly evolving since steroids show excellent bioactivity, lipophilic character and good bioavailability, with less toxic effects compared to many commercial drugs.

We earnestly hope that this reprint could become a resource for newcomers to the fields of medicinal chemistry, organic chemistry, biochemistry, biology, and computer chemistry, as well as for experienced researchers and scientists in academia.

We sincerely thank the authors of the manuscripts, the reviewers who selflessly contributed to the quality of each individual manuscript with their expertise and useful suggestions, as well as the editorial team of the journal *Molecules* for technical assistance and support.

Marina Savić, Erzsébet Mernyák, Jovana Ajdukovic, and Suzana Jovanović-Šanta
Editors

Article

Novel 4-Azapregnene Derivatives as Potential Anticancer Agents: Synthesis, Antiproliferative Activity and Molecular Docking Studies

Vanessa Brito [1], Adriana Oliveira Santos [1], Gilberto Alves [1], Paulo Almeida [1] and Samuel Silvestre [1,2,*]

[1] Health Sciences Research Centre (CICS-UBI), Universidade da Beira Interior, Av. Infante D. Henrique, 6200-506 Covilhã, Portugal
[2] Center for Neuroscience and Cell Biology (CNC), University of Coimbra, 3004-517 Coimbra, Portugal
* Correspondence: sms@ubi.pt

Abstract: A series of novel 21E-arylidene-4-azapregn-5-ene steroids has been successfully designed, synthesized and structurally characterized, and their antiproliferative activity was evaluated in four different cell lines. Within this group, the 21E-(pyridin-3-yl)methylidene derivative exhibited significant cytotoxic activity in hormone-dependent cells LNCaP (IC$_{50}$ = 10.20 µM) and T47-D cells (IC$_{50}$ = 1.33 µM). In PC-3 androgen-independent cells, the steroid 21E-p-nitrophenylidene-4-azapregn-5-ene was the most potent of this series (IC$_{50}$ = 3.29 µM). Considering these results, the 21E-(pyridin-3-yl)methylidene derivative was chosen for further biological studies on T47-D and LNCaP cells, and it was shown that this azasteroid seems to lead T47-D cells to apoptotic death. Finally, molecular docking studies were performed to explore the affinity of these 4-azapregnene derivatives to several steroid targets, namely 5α-reductase type 2, estrogen receptor α, androgen receptor and CYP17A1. In general, compounds presented higher affinity to 5α-reductase type 2 and estrogen receptor α.

Keywords: 4-azasteroids; 4-azapregnenes; aldol condensation; molecular docking; antiproliferative activity

Citation: Brito, V.; Santos, A.O.; Alves, G.; Almeida, P.; Silvestre, S. Novel 4-Azapregnene Derivatives as Potential Anticancer Agents: Synthesis, Antiproliferative Activity and Molecular Docking Studies. *Molecules* **2022**, *27*, 6126. https://doi.org/10.3390/molecules27186126

Academic Editors: Marina Savić, Erzsébet Mernyák, Jovana Ajdukovic and Suzana Jovanović-Šanta

Received: 25 August 2022
Accepted: 15 September 2022
Published: 19 September 2022

Publisher's Note: MDPI stays neutral with regard to jurisdictional claims in published maps and institutional affiliations.

1. Introduction

Natural, semi-synthetic and synthetic steroidal derivatives are a rich source of potential drug candidates, being a current focus of investigation [1–3]. In addition, steroids exhibit several biological advantages, such as a low toxicity, less vulnerability to multidrug resistance and high bioavailability [4–6]. Therefore, a large number of modified steroids have been described, principally, as potential antitumor agents [7,8]. For example, some 17(E)-picolinylidene androstane derivatives were reported as potential inhibitors of prostate and breast cancer cell growth. Interesting antiproliferative activity against prostate PC-3 cells was observed and this effect is correlated with the cytotoxic effect observed for abiraterone, a drug with clinical use in prostate cancer (PCa) treatment [9].

Among other explored chemical modifications, the insertion of a 16-arylidene group into the steroid structure has also been associated with significant cytotoxic effects in several cell lines, being considered a relevant pharmacophore for anticancer activity [10–13]. Moreover, several pregnane derivatives have emerged as potential antitumor agents, showing relevant results in this context [5,6,14–18]. Therefore, considering this, Banday et al. reported the synthesis of novel benzylidene pregnenolone derivatives and their antiproliferative effects against a panel of human cancer cell lines. This research showed interesting results, with some of these steroids presenting very potent effects especially against colon (HCT-15) and breast (MCF-7) cancer cells [16].

Other semi-synthetic steroidal derivatives widely explored over the years are 4-azasteroids, such as finasteride and dutasteride, which are clinically used to treat benign prostatic hyperplasia (BPH), inhibiting the conversion of testosterone to 5α-dihydrotestosterone

(DHT) by 5α-reductase (5AR) enzymes [7]. In addition, 5AR inhibitors have been evaluated for their potential use as chemopreventive agents against PCa, since all isoforms of this enzyme are present in PCa at increased levels. Moreover, different 4-azasteroids are also being explored as anticancer agents. For example, a research study reported the synthesis of 4-azasteroidal purine nucleoside analogs and their evaluation as antitumor agents in MCF-7 and PC-3 cell lines, and some of these derivatives revealed potent antiproliferative activity against PCa cells [19]. Recently, we also described the synthesis, biological evaluation and in silico studies of novel 16-arylidene-4-azaandrostenes with antiproliferative effects, and interesting results were observed in prostatic cancer cell lines. In addition, a significant selectivity towards cancer cell lines was found for all azaandrostenes, presenting low cytotoxicity in non-cancerous human fibroblasts. Moreover, the molecular docking studies showed that these 4-azaandrostene derivatives can interact with 5AR type 2, as well as with other common targets of steroidal drugs [20].

Based on the aforementioned studies and intending to develop compounds with higher potency and selectivity, the present work foucuses on the preparation of new 21-arylideneazapregnene derivatives and the biological evaluation of their effects in several cell lines, such as prostatic (LNCaP, PC-3) and breast (T47-D) cancer cells, and on non-cancerous dermal fibroblasts. The structure of the synthesized steroids was validated through adequate structural characterization techniques. The cytotoxic effects evaluation of the novel arylidenes was performed by the 3-(4,5-dimethylthiazol-2-yl)-2,5-diphenyltetrazolium bromide (MTT) assay. Moreover, the antiproliferative effects in T47-D and LNCaP cells for the most relevant steroids were further explored by different cellular and molecular biology techniques. Furthermore, in silico molecular docking simulations against important targets of steroidal molecules were performed [8,9,21–23].

2. Results and Discussion

2.1. Chemistry

The synthesis of 4-azapregn-5-ene-3,20-dione derivatives has been carried out as depicted in Scheme 1. Progesterone was treated with sodium periodate and potassium permanganate to form **1**, by an oxidative cleavage reaction, with an excellent yield of 89%. Then, 4-azapregn-4-ene-3,20-dione (**2**) was obtained from an azacyclization reaction of **1**. Distinct procedures for this reaction with different catalysts or reagents and energy sources are described [24,25]. Within these methods, the use of acetic acid and ammonium acetate proved to be a practical approach and allowed for the preparation of the desired product with a high yield, but the removal of acetic acid revealed to be a difficult task. To overcome this issue, extraction with a larger portion of dichloromethane (DCM), followed by washing the organic layer with brine three times, to ensure the total removal of the acetic acid present in the reactional mixture, was required. Infrared (IR) and nuclear magnetic resonance (NMR) spectroscopy data of **1** and **2** were similar to the described in the literature [25].

The changes on the side chain of steroids are relevant since these modifications frequently lead to more effective receptor binding or increased bioavailability [15]. Taking into account the cytotoxicity against tumor cells associated with structural analogs, such as 16*E*-arylidene-4-azaandrostene derivatives, it was decided to synthesize novel 4-azapregnene derivatives with modifications on the side chain by introducing an arylidene group [17,20,26–28]. Particularly, the 4-azaandrostene derivatives reported by us presented some interesting cytotoxicity results, including a relevant selectivity toward cancer cell lines since, generally, low cytotoxicity was detected in non-cancerous human fibroblasts. Furthermore, these steroids induced a reduction of viability in LNCaP cells comparable to the observed with finasteride [20]. Similarly, in the present work, an aldol condensation of 4-azapregn-4-ene-3,20-dione with several aldehydes at room temperature in alkaline medium afforded the corresponding arylidene derivatives **3a–g** in very acceptable yields, as shown in Table 1 [11,13]. The ^{1}H NMR spectra revealed, typically, the signals of the vinylic protons, being 21-H and 21a-H at ≈7.47 and ≈6.70 ppm, respectively. An observed coupling constant of 16.0 Hz unequivocally indicated the *E* geometry for the double bond, which is in agreement with the literature [17,18,29].

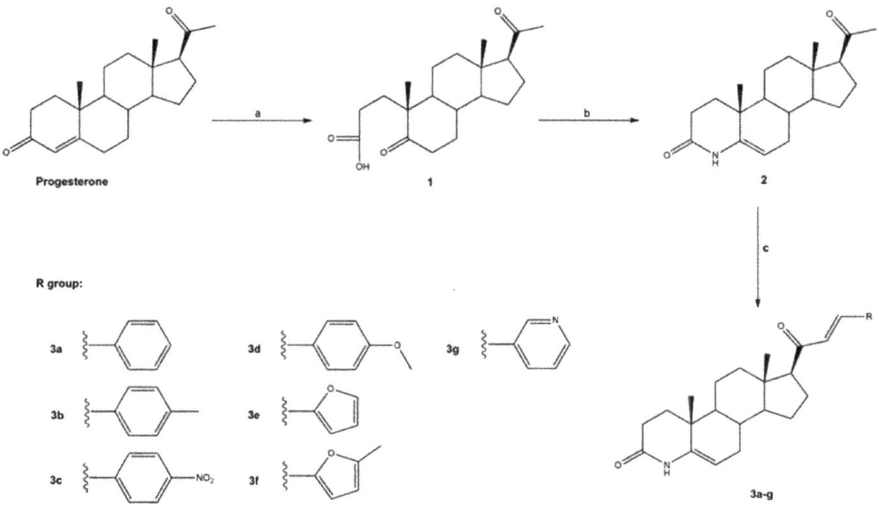

Scheme 1. Synthesis of the steroids **3a–g** from progesterone. Reagents, reaction conditions and yields: (a) $NaIO_4$, $KMnO_4$, Na_2CO_3, *i*-PrOH, reflux 3 h, 89%; (b) CH_3COONH_4, CH_3COOH, reflux 4 h, 98%; (c) KOH, aldehyde, EtOH, r.t., 12–24 h, 46–84%.

Table 1. New 21-arylidene-4-azapregn-5-ene derivatives synthesized, and respective overall yields (%) [1].

Aldehyde	Arylidenesteroid	Overall Yield (%)
Benzaldehyde	**3a**	40
p-tolualdehyde	**3b**	65
4-nitrobenzaldehyde	**3c**	58
4-methoxybenzaldehyde	**3d**	48
2-furaldehyde	**3e**	54
5-methyl-2-furaldehyde	**3f**	73
Pyridine-3-carboxaldehyde	**3g**	55

[1] The overall yields are the global yield after the three synthetic steps from progesterone.

2.2. Biology

2.2.1. Screening of Cell Proliferation Effects

The effects of steroids **3a–g** and synthetic precursors (progesterone, **1** and **2**) on the proliferation of LNCaP, PC-3, T47-D and normal human dermal fibroblasts (NHDF) cells were examined by the MTT proliferation assay. These cell lines were used as models of androgen-dependent and androgen-independent prostate cancer, hormone-responsive breast cancer and non-cancerous cells, respectively, to evaluate the selective cytotoxicity of these compounds. Due to the structural similarities between the prepared azasteroidal derivatives and finasteride, this drug was also included in the study, as well as testosterone and DHT, allowing to compare the effect on cell proliferation in relation to the novel arylidenes herein described. Additionally, 5-fluorouracil (5-FU), a clinically used antitumor agent, was also included in the assay as a positive control. Firstly, cells were exposed to all compounds at 30 µM, during 72 h, in a screening assay, similarly to which was performed by other authors [30–32]. Following this preliminary evaluation, concentration–response studies were performed for the most active derivatives and 5-FU in all cell lines, and the half-maximal inhibitory concentration (IC_{50}) of the tested compounds was determined.

Screening and IC_{50} results are shown in Table 2. The IC_{50} determination was performed for arylidenes **3a**, **3c** and **3g** in all cell lines, and **3b** in T47-D and PC-3 cells. Interestingly, the steroid **3g** was the most potent in two of the three tumor cell lines, with an IC_{50} of 10.20 µM in LNCaP cells and 1.33 µM in T47-D cells. In PC-3 cell line, derivatives **3c** and **3g** presented very close IC_{50} values, 3.29 µM and 3.64 µM, respectively. Considering these results, azasteroids **3c** and **3g** seemed promising candidates for more advanced studies, despite the relevant cytotoxicity of **3g** against the non-cancerous cells.

Table 2. In vitro antiproliferative activities of the tested compounds against LNCaP, PC-3, T47-D and NHDF cell lines, after 72 h of exposure. MTT screening results at 30 µM (data presented as average percentage of negative control $\pm$ SD) and determined IC_{50} values (µM) with the respective R squared (R^2) [1].

Compound		LNCaP	PC-3	T47-D	NHDF
Progesterone	Screening (30 µM)	64.3 ± 16.1	42.9 ± 5.4	62.9 ± 6.8	96.6 ± 14.3
	IC_{50} (µM); R^2	–	–	–	–
Testosterone	Screening (30 µM)	99.2 ± 4.2	89.5 ± 9.0	72.0 ± 4.0	94.2 ± 7.3
	IC_{50} (µM); R^2	–	–	–	–
DHT	Screening (30 µM)	138.9 ± 3.9	93.1 ± 4.9	90.7 ± 9.4	100.5 ± 7.4
	IC_{50} (µM); R^2	–	–	–	–
1	Screening (30 µM)	96.4 ± 16.1	102.1 ± 5.2	130.8 ± 20.8	100.0 ± 6.0
	IC_{50} (µM); R^2	–	–	–	–
2	Screening (30 µM)	58.3 ± 13.6	96.7 ± 9.3	137.0 ± 18.0	106.1 ± 6.0
	IC_{50} (µM); R^2	–	–	–	–
3a	Screening (30 µM)	6.5 ± 5.3	1.4 ± 1.0	6.5 ± 5.4	25.1 ± 7.4
	IC_{50} (µM); R^2	27.47; 0.93	12.60; 0.96	16.94; 0.97	63.99; 0.91
3b	Screening (30 µM)	62.7 ± 9.0	25.9 ± 3.8	45.0 ± 11.6	83.0 ± 3.6
	IC_{50} (µM); R^2	–	25.94; 0.98	44.99; 0.98	–
3c	Screening (30 µM)	5.2 ± 2.6	0.7 ± 0.5	6.7 ± 4.6	20.3 ± 2.8
	IC_{50} (µM); R^2	16.26; 0.97	3.29; 0.99	6.56; 0.98	6.94; 0.93
3d	Screening (30 µM)	88.0 ± 15.5	58.2 ± 6.0	73.9 ± 14.6	52.4 ± 2.8
	IC_{50} (µM); R^2	–	–	–	–
3e	Screening (30 µM)	85.3 ± 6.7	82.7 ± 6.5	92.6 ± 7.4	111.6 ± 13.9
	IC_{50} (µM); R^2	–	–	–	–
3f	Screening (30 µM)	88.9 ± 17.7	108.1 ± 7.8	104.3 ± 11.3	89.7 ± 9.7
	IC_{50} (µM); R^2	–	–	–	–
3g	Screening (30 µM)	0.7 ± 1.0	0.1 ± 0.1	1.0 ± 0.4	23.8 ± 3.2
	IC_{50} (µM)	10.20; 0.90	3.64; 0.98	1.33; 0.98	3.09; 0.90
Finasteride	Screening (30 µM)	88.5 ± 2.4	78.8 ± 8.7	103.5 ± 17.8	62.8 ± 5.9
	IC_{50} (µM); R^2	–	–	–	–
5-FU	Screening (30 µM)	18.6 ± 2.6	10.8 ± 3.1	42.3 ± 6.4	33.6 ± 8.2
	IC_{50} (µM); R^2	1.50; 0.98	3.30; 0.99	0.70; 0.99	29.44; 0.97

[1] The cells were treated with the following distinct concentrations: 0.01, 0.1, 1, 10, 50 and 100 µM for 72 h. The antiproliferative effect was accessed by the MTT assay and the IC_{50} values were determined by sigmoidal fitting. Data shown are representative of at least two independent experiments. DHT = 5α-dihydrotestosterone; 5-FU = 5-fluorouracil.

2.2.2. Characterization of the Cytotoxic Effect of **3g** on T47-D Cells

The 21*E*-(pyridin-3-yl)methylidene derivative **3g**, was the most potent of this series of steroids on T47-D cells, with a IC_{50} of 1.33 µM. In order to better evaluate the cytotoxic effect of this compound, a fluorescence microscopy study by staining cells with Hoechst 3342—a nucleus marker—and PI—a cell death marker—was performed. For this, after 24, 48 and 72 h of treatment with **3g** at 20 µM, cells were observed through optical microscopy and then stained for fluorescence microscopy observation. Cells exposed to **3g** had a more frequent PI staining than positive and negative controls (Figure 1), as evidenced by the quantification of a higher ratio of PI/Hoechst 3342 staining (Figure 2). These results indicate that cells incubated with **3g** are dying with increased frequency from an early time point after treatment when compared to negative and positive controls.

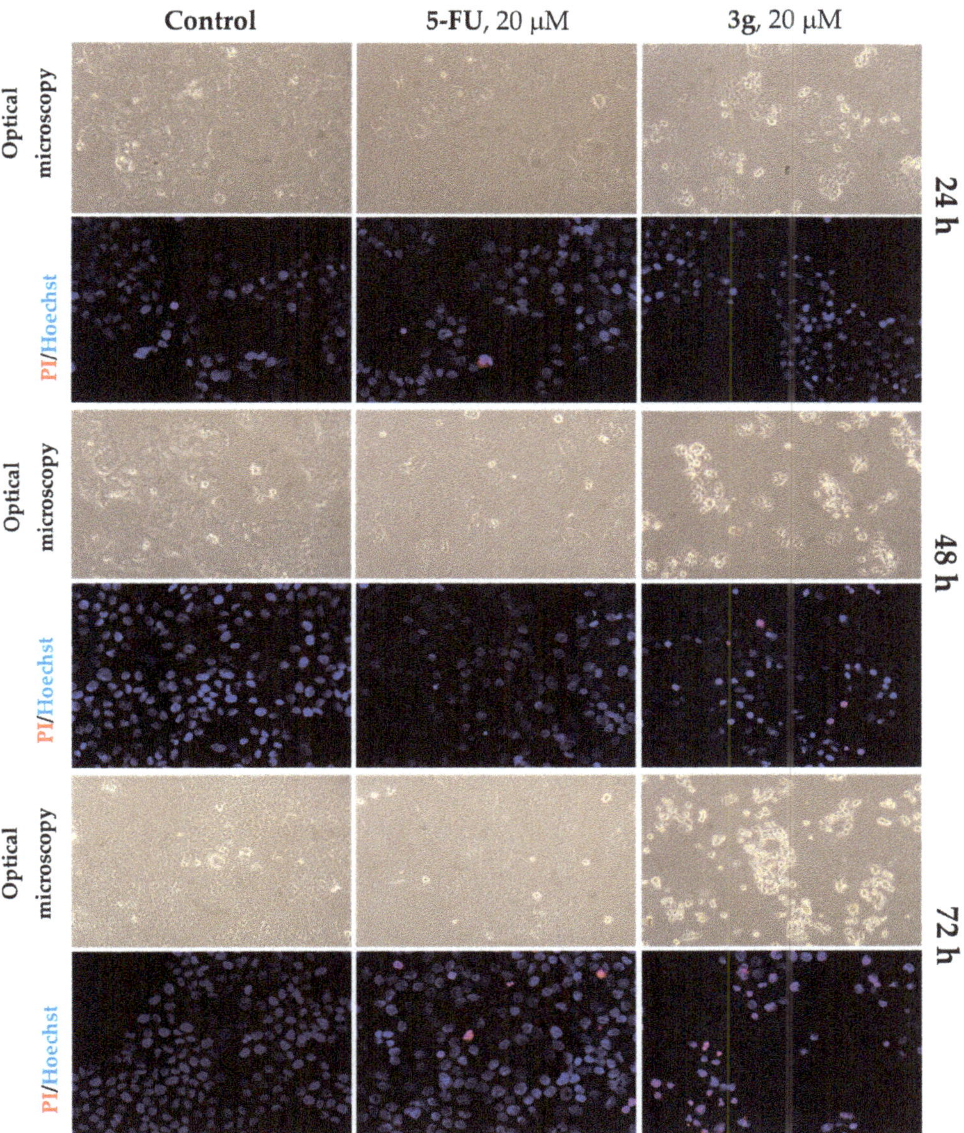

Figure 1. PI/Hoechst 3342 staining results. T47-D cells were incubated for 24, 48 and 72 h with 5-fluorouracil (5-FU) and **3g** at 20 µM. Untreated cells were considered the control of the experiment. Cells were visualized in an Optical microscope Olympus CKX41 coupled to a digital camera (Olympus SP-500UZ) and in an Axio Imager A1 microscope (fluorescence). The resulting images were treated in ImageJ software.

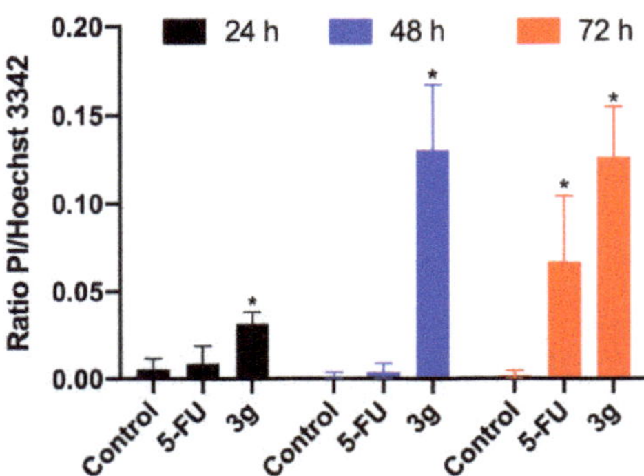

Figure 2. Ratio of PI/Hoechst 3342 staining results. T47-D cells were incubated during 24, 48 and 72 h with 5-fluorouracil (5-FU) and **3g** at 20 μM. Untreated cells were considered the control of the experiment. Cells were visualized in an Axio Imager A1 microscope, and the resulting images were treated in ImageJ. Data are expressed as mean ± SD and represent at least two experiments. * $p < 0.05$ compared to control.

Several morphological modifications occurred over time when cells were exposed to **3g** and, also, to 5-FU, being more evident when cells were treated with the azasteroid (Figure 1). These were mainly cell rounding and the decrease of the cells' area and volume, which were quantified with ImageJ software—Figure 3A. In this context, the **3g**-incubated cultures showed a significant decrease in the nuclear area after all incubation times. The decrease in cell nuclei area due to DNA condensation/loss is an indicator of apoptosis, and this correlation was validated by Eidet et al. (2014) [33]. The nearest neighbor analysis (NND) of T47-D cells when treated with the testing compound is shown in Figure 3B. The NND is measured to assess the cell nuclei distribution, and is defined as the distance between the centroid of each nucleus and its closest one. This analysis was employed to evaluate if the treatment (5-FU and **3g**) causes changes in cell distribution. Moreover, cells treated for 48 and 72 h with 5-FU and **3g** displayed different distances, which means a more aleatory placement of the cells. The literature points out that these results are in agreement with the occurrence of apoptosis, which seems to cause a more unequal cell spacing [33].

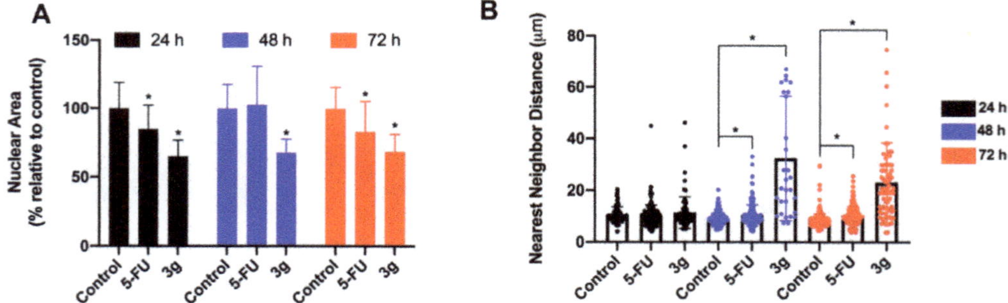

Figure 3. (**A**) T47D nuclear area of untreated cells (control) and cells treated with 5-fluorouracil (5-FU) and **3g** at 20 μM, incubated 24, 48 and 72 h; (**B**) nearest neighbor analysis of T47-D cells with the same conditions referred previously. Data are mean ± SD (**A**) or individual points ± SD (**B**) and represent at least two experiments. * $p < 0.05$ compared to control.

Lastly, nuclei with a morphology coincidental with apoptotic cells are indicated with red arrows, and they are presenting a circular and pycnotic form (Figure 4).

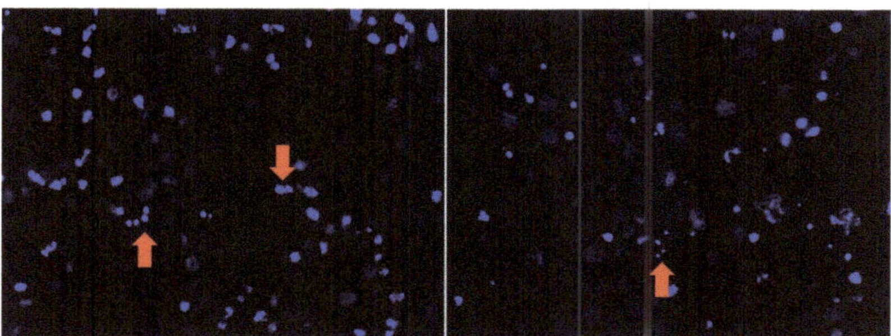

Figure 4. T47-D cells stained with Hoechst 3342, after treatment with **3g** during 72 h. Pycnotic cells with circular morphology are indicative of apoptosis.

To explore the possible mechanism of apoptotic death in T47-D cells, the caspase-9 activity was measured using a Caspase-Glo kit from Promega looking to consolidate caspase-dependent apoptosis. It is known that caspase-9 has an important role in the intrinsic or mitochondrial pathway of the apoptosis death mechanism [34]. In this experiment, cells were exposed for 48 h to doxorubicin (DOX) at 10 µM, as the positive control, and to **3g** at 10 and 20 µM, and the results obtained are shown in Figure 5. After 48 h of treatment, significant elevation in caspase-9 activity was observed in cells treated with DOX and with both concentrations of **3g**, which is a relevant indication that this steroid triggers the apoptotic mechanism of cell death.

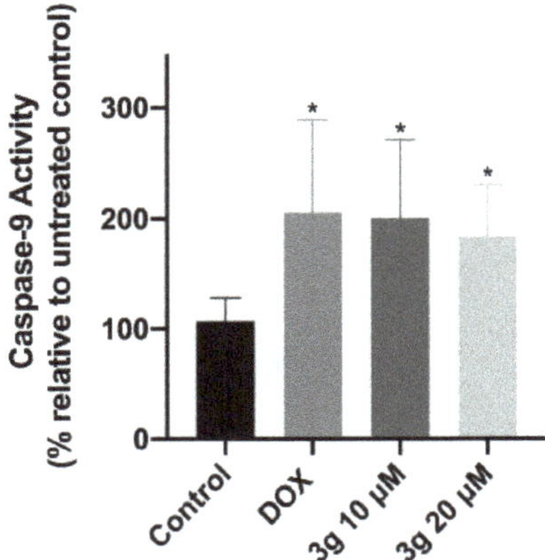

Figure 5. T47D caspase-9 activity results. Cells were treated for 48 h with doxorubicin (DOX) at 10 µM, as the positive control, and **3g** at 10 and 20 µM, and the results are relative to untreated cells (control). The activity of caspase-9 was evaluated using a Caspase-Glo kit from Promega, as described in Material and Methods. Data are expressed as mean ± SD (2 independent assays, n = 4 in each). * $p < 0.05$ compared to control.

2.2.3. Characterization of the Cytotoxic Effect of **3g** on LNCaP Cells

In our previous work, 16*E*-arylidene-4-azaandrost-5-ene derivatives were synthesized, their antiproliferative effects were evaluated and further studies were performed with the most promising steroid in LNCaP cells [20]. Due to the similarity of 16*E*-arylidene-4-azaandrostenes and these novel 21*E*-arylidene-4-azapregnene derivatives, identical studies in this cell line were performed, namely a flow cytometry assay and microscopic cell morphology observation. To assess the viability of LNCaP cells when exposed to steroid **3g**, a flow cytometry assay after PI staining was performed. Cells were treated with **3g** and

5-FU during 24 and 72 h, and then a microscopic cell observation was performed to assess the possible morphological alterations and cell density changes triggered by incubation with **3g** and compare with controls (Figure 6). A cell density reduction and the preservation of healthy cells morphology, when treated with finasteride, was observed. In contrast, cells treated with **3g**, after 24 and 72 h of treatment, appeared as detached, forming clusters and presenting a round form.

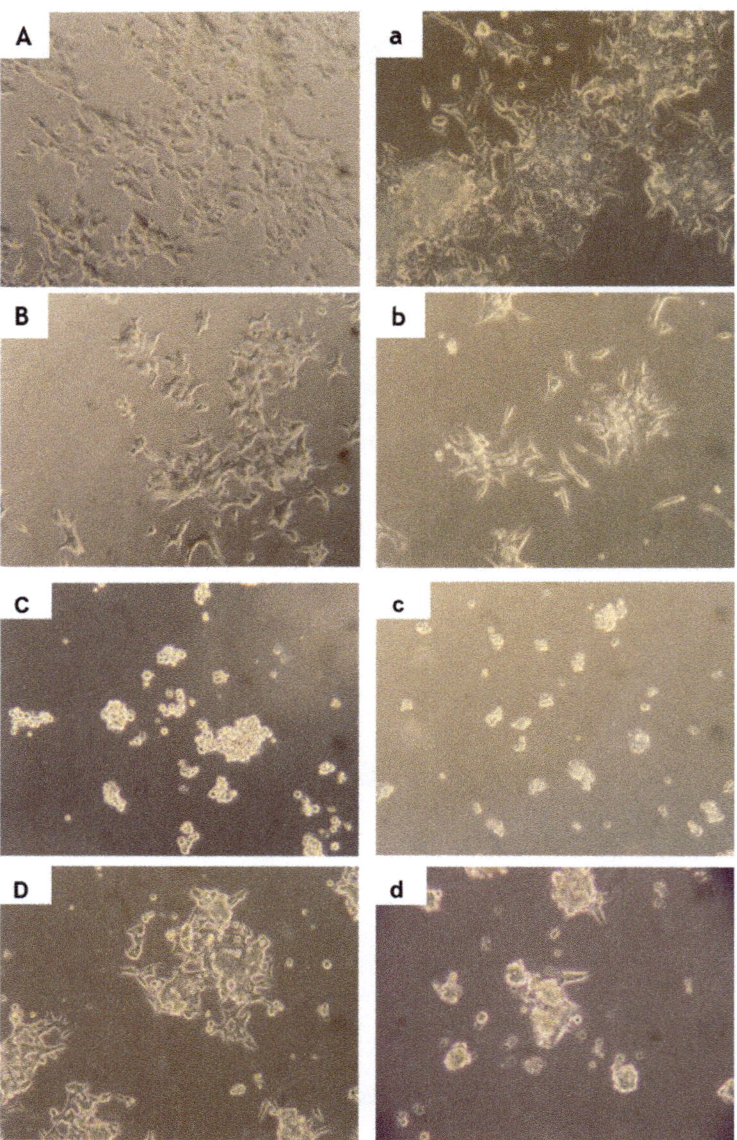

Figure 6. Photograph shots after incubation of LNCaP cells with the compounds (24 and 72 h, zoom: 100×). (**A**) Control after 24 h, (**a**) control after 72 h; (**B**) cells after 24 h of incubation with finasteride, (**b**) cells after 72 h of incubation with finasteride; (**C**) cells after 24 h of incubation with **3g**, (**c**) cells after 72 h of incubation with **3g**; and (**D**) cells after 24 h of incubation with 5-fluorouracil (5-FU), (**d**) cells after 72 h of incubation with 5-FU.

Cell viability of LNCaP cells when exposed to steroid **3g**, finasteride and 5-FU was more accurately evaluated by flow cytometry quantification of PI permeant (dead) vs. non-permeant cells (live cells) (Figure 7) [35]. Derivative **3g** induced a drastic reduction in the number of living cells after 24 h of incubation. After 72 h, the number of viable cells was almost absent. On the other hand, it can be noted that after 24 and 72 h of incubation a relevant proportion of cells is in R3, a region that, despite being less well defined, can be associated with partial PI permeability or increased autofluorescence (early apoptotic cells) and/or cell debris with degraded DNA (advanced cell death stage) [35]. This study suggests that the mechanism of cell death by **3g** is different in relation to finasteride or 5-FU.

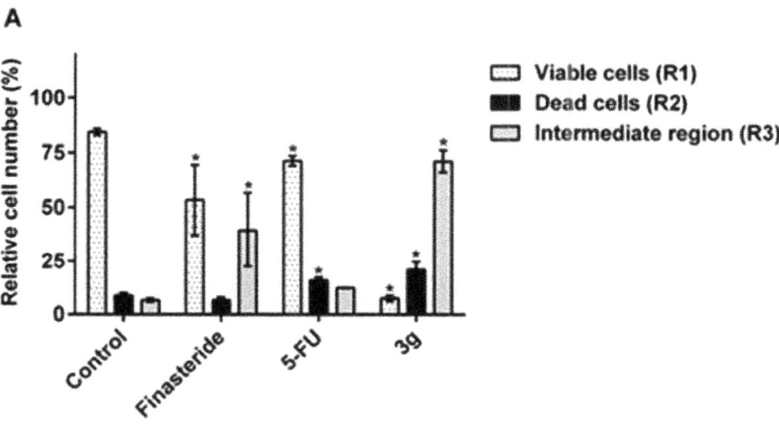

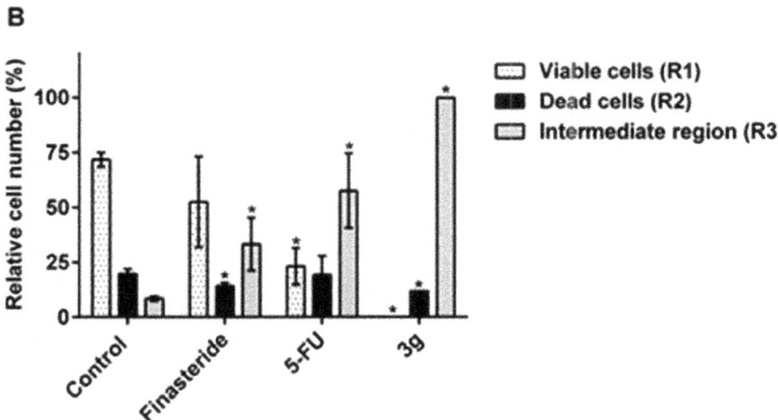

Figure 7. Distribution of the events in the regions R1, R2 and R3 (relative cell number): (**A**) after 24 and (**B**) 72 h of treatment with finasteride, 5-fluorouracil (5-FU) and **3g**. Data are expressed as mean ± SD of the percentage of cells in the different regions (n = 4–6) and are representative of at least two independent experiments. * $p < 0.05$ vs. the control (ANOVA).

2.3. Molecular Docking

An in silico study was performed using molecular docking simulations. The principal aim of this study was to assess the affinity and the existence of potential interactions between these new arylidenes and several proteins that are known targets of steroidal drugs currently used in the treatment of BHP, PCa and breast cancer: 5AR type 2, AR, ERα and CYP17A1. The structures of these targets were chosen based on several important criteria, specifically: the existence of high-resolution X-ray crystal structure, available structure in a complex with a steroidal drug or other similar ligands, and, finally, the protein have to be a target of clinically-approved steroid-based anticancer drugs in the treatment of breast or prostate cancers. On this set, 5AR type 2, an important target of 4-azasteroids used in the symptomatic treatment of BPH, is the only exception. Therefore, this study was performed for each 4-azapregnene **3a–g** against the mentioned proteins. Three-dimensional structural coordinates of protein receptors were retrieved from the protein data bank (PDB), and molecular docking simulations were executed using AutoDockTools. To validate the docking method, simulations were carried out between crystallized ligands/drugs with the respective proteins and all control re-docking simulations were able to reproduce the ligand–protein interaction geometries presented in the respective crystal structures with a root-mean-square distance (RMSD) ≤ 2.0 Å. The results of re-docking in all simulations exhibit a RMSD equal to or lower than 1.0 Å (Table 3). Based on the control docking simulations, predicted binding energies < -11.00 kcal.mol^{-1} were considered to be significant in the cases of 5AR type 2, AR and CYP17A1, and <-9.00 kcal.mol^{-1} in the case of ERα.

Table 3. Results of re-docking using GA of protein–ligand complexes, PDB accession codes and resolution [1].

Protein–Ligand Complex	PDB Code	Resolution (Å)	Number of Clusters	Energy (kcal.mol^{-1}) and RMSD (Å)			
				Lowest Energy	RMSD of Lowest Energy	Average Energy	Average RMSD
5AR type 2 + Adduct of Finasteride/NADPH	7BW1	2.8	1	−11.90	0.78	−6.89	1.80
ERα + 17β-estradiol	1A52	2.8	1	−9.28	0.51	−9.28	0.53
AR + DHT	2AMA	1.9	1	−11.20	0.79	−11.20	0.79
CYP17A1 + Abiraterone	3RUK	2.6	1	−11.78	0.23	−11.78	0.29

[1] 5AR = 5α-reductase; NADPH = nicotinamide adenine dinucleotide phosphate; Erα = estrogen receptor α; AR = androgen receptor; DHT = 5α-dihydrotestosterone; CYP17A1 = cytochrome P450 family 17 subfamily A member 1.

It is quite evident that very strong binding energies were predicted for all seven steroids for the active site of 5AR type 2, ERα and CYP17A1, while no significant binding was observed for the majority of these derivatives for the active site of AR (Table 4). These results show the possibility that novel synthesized compounds can potentially interact with the 5AR type 2 and possibly act as 5ARIs, which can be explained by the fact these are 4-azasteroids. In fact, all novel derivatives have similar affinity energy values to the control finasteride. Moreover, in the molecular docking with ERα, all steroids showed higher affinity when compared with 17β-estradiol. Concerning CYP17A1, the affinity energy values obtained to these steroidal derivatives are relatively similar to abiraterone. Although these values are higher than the reference (except for **3d**), a potential interaction between these novel steroids and CYP17A1 cannot be discarded. The principal interactions between the macromolecules and the best-scored derivatives in molecular docking simulations were also analyzed. Through the analysis of principal interactions, it is possible to conclude that the amide group of these derivatives could be essential in establishing polar interactions with the different amino acids of studied proteins. Additionally, the important interactions between proteins (specific amino acids) and crystallized ligands were present in docking simulations for best-scored compounds.

Table 4. Predicted binding energies of steroids **3a–g** calculated from molecular docking against known protein targets of steroidal molecules: 5AR type 2, ERα, AR and CYP17A1. Binding energies of ligands (in bold) present in the X-ray crystal structures were calculated by re-docking [1].

Compound	5AR Type 2	ERα	AR	CYP17A1
	Autodock binding energies (kcal.mol^{-1})			
3a	−11.79	−10.87	−9.38	−10.24
3b	−12.02	−11.44	−8.88	−11.17
3c	−11.70	−10.43	−5.10	−9.32
3d	−11.63	−11.05	−7.32	−11.99
3e	−11.07	−10.78	−9.70	−10.91
3f	−11.24	−11.43	−9.98	−10.95
3g	−11.71	−10.67	−10.35	−10.93
Finasteride	**−11.90**	−	−	−
17β-estradiol	−	**−9.28**	−	−
DHT	−	−	**−11.20**	−
Abiraterone	−	−	−	**−11.78**

[1] 5AR = 5α-reductase; Erα = estrogen receptor α; AR = androgen receptor; CYP17A1 = cytochrome P450 family 17 subfamily A member 1; DHT = 5α-dihydrotestosterone.

3. Materials and Methods

3.1. Chemistry

3.1.1. General Considerations

Reagents and solvents were acquired from standard sources and were purified and/or dried whenever crucial using standard procedures before use. Finasterida Tetrafarma^{TM} 5 mg was purchased from *Tetrafarma – Produtos Farmacêuticos*, Lda, Portugal and the active steroid was extracted from the tablets [36,37]. Finasteride was extracted with high purity, based on the NMR spectra acquired. The reactions were performed under heating and magnetic stirring using Heidolph plates. TLC analysis was carried out using 0.20 mm Al-backed silica-gel plates (Macherey-Nagel 60 F254, Duren, Germany), and after elution, the plates were visualized under UV radiation (254 nm) in a CN-15.LC UV chamber. Then a revelation step with an ethanol/sulfuric acid (95:5) mixture, followed by heating at 120 °C was performed. For the isolation and purification of product **1**, a column chromatography, using silica gel (0.063–0.200 mm or 0.040–0.063 mm) acquired from Merck (New Jersey, USA), was used. The eluents used are indicated as a v/v proportion in the experimental procedure. A Büchi R-215 rotavapor system was used for the evaporation of solvents. Attenuated total reflectance (ATR) IR spectra were collected on a Thermoscientific Nicolet iS10: smart iTR, equipped with a diamond ATR crystal. For IR data acquisition, each solid sample was placed onto the crystal and the spectrum was recorded. An air spectrum was used as a reference in absorbance calculations. The sample spectra were collected at room temperature in the 4000–600 cm^{-1} range by averaging 16 scans at a spectral resolution of 2 cm^{-1}. ^{1}H- and ^{13}C-NMR spectra were recorded on a Bruker Avance 400 MHz spectrometer (^{1}H NMR at 400.13 MHz and ^{13}C NMR at 100.62 MHz), and were processed with the software TOPSPIN (v. 3.1) (Bruker, Fitchburg, WI, USA). Chloroform (CDCl$_3$) was the solvent used in these experiments. Chemical shifts are reported in parts per million (δ) relative to TMS or solvent as an internal standard. Coupling constants (*J* values) are reported in hertz (Hz), and splitting multiplicities are described as s = singlet, d = doublet, t = triplet, combinations of above or m = multiplet. Spectra are available on Supplementary Material (S1). ESI-TOF mass spectrometry was performed by the microanalysis service on a QSTAR XL instrument.

3.1.2. Synthesis

5-Oxo-A-nor-3,5-secopregnan-3-oic acid (**1**)

To a solution of progesterone (943.41 mg, 3 mmol) dissolved in isopropanol (25 mL) was added a warmed solution of sodium carbonate (510 mg, 4.8 mmol) in water (3 mL). This mixture was carried to reflux, and a heated solution of sodium periodate (5.37 g,

25.1 mmol) and potassium permanganate (70 mg, 0.4 mmol) in water (6 mL) was added dropwise over 1 h. Then, the reflux was maintained for an additional 3 h, and after this period the reaction was cooled to 30 °C. The solids formed were removed by filtration with celite, and washed with water. The combined filtrates were concentrated under reduced pressure to eliminate isopropanol. The aqueous residue was cooled and acidified with the concentrated hydrochloric acid aqueous solution until precipitate formation. The product was extracted with ethyl acetate (3 × 80 mL), washed with brine and dried with anhydrous sodium sulfate. After removal of solvent under reduced pressure, the residue was purified by column chromatography (silica gel, ethyl acetate/petroleum ether 40–60 °C, 1:1) to afford **1** as a white solid (983.30 mg, 89%) [24]; mp 168–169 °C.

4-Azapregn-4-ene-3,20-dione (**2**)

A mixture of **1** (476.8 mg, 1.4 mmol) and ammonium acetate (647.4 mg, 8.4 mmol) in glacial acetic acid (12 mL) was refluxed for 4 h. At the end of the reaction, the mixture was cooled followed by water addition (100 mL). The product was extracted with DCM (3 × 100 mL). The organic phase was washed with brine and dried with anhydrous sodium sulfate. The solvent was removed under reduced pressure to give **2** as an orange solid (433.3 mg, 98%) [25]; mp 152–155 °C.

General procedure for the preparation of 4-azapregn-5-ene-3,20-dione derivatives **3a–g** by aldol condensation

A mixture of ethanolic solution (5 mL) of compound **2** (94.64 mg, 0.3 mmol), aldehyde (0.36 mmol) and aqueous solution of potassium hydroxide (100 µL, 50% *w*/*w*) was stirred for 12–24 h at room temperature. The reaction mixture was worked up by first adding water to induce the precipitation (10 mL), and then filtered and washed with water to obtain **3a–g** [15].

21*E*-(Phenylmethylidene)-4-azapregn-5-ene-3,20-dione (**3a**)

Pallid yellow powder (55.9 mg, 46%); mp 220–223 °C; IR (cm^{-1}): 3191, 3060, 2937, 2873, 1693, 1674, 1600, 978, 837. ^{1}H NMR (CDCl$_3$, 400 MHz) δ: 7.72 (1H, s, -NH), 7.49 (3H, m, H$_{Ar}$+21a-CH), 7.40–7.27 (3H, m, H$_{Ar}$), 6.71 (1H, d, *J* = 16.0 Hz, 21-CH), 4.79 (1H, s, 6-H), 1.02 (3H, s, 19-CH$_3$), 0.60 (3H, s, 18-CH$_3$). ^{13}C NMR (CDCl$_3$, 101 MHz) δ: 200.11, 169.61, 141.68, 139.88, 134.74, 130.35, 128.92, 128.28, 126.67, 103.31, 63.51, 61.88, 56.89, 47.93, 44.98, 38.78, 34.26, 31.52, 29.65, 28.39, 24.58, 22.75, 20.98, 18.72, 13.53. HRMS (ESI-TOF) *m/z*: [M + H]$^+$ Calcd for C$_{27}$H$_{34}$NO$_2$ 404.2584; Found 404.2589.

21*E*-[(4-Methylphenyl)methylidene]-4-azapregn-5-ene-3,20-dione (**3b**)

Pallid orange solid (92.1 mg, 74%); mp 191–193 °C; IR (cm^{-1}): 3192, 3059, 2936, 2871, 1675, 1664, 1597, 990, 837, 809. ^{1}H NMR (CDCl$_3$, 400 MHz) δ: 7.75 (1H, s, -NH), 7.47 (1H, d, *J* = 15.9 Hz, 21a-CH), 7.39 (2H, d, *J* = 7.8 Hz, H$_{Ar}$), 7.13 (2H, d, *J* = 7.7 Hz, H$_{Ar}$), 6.67 (1H, d, *J* = 15.9 Hz, 21-CH), 4.79 (1H, s, 6-H), 2.31 (3H, s, 7'-CH$_3$), 1.02 (3H, s, 19-CH$_3$), 0.59 (3H, s, 18-CH$_3$). ^{13}C NMR (CDCl$_3$, 101 MHz) δ: 200.16, 169.65, 140.85, 139.87, 131.98, 129.66, 128.30, 125.77, 103.36, 63.51, 61.78, 56.89, 47.94, 44.95, 44.01, 38.76, 34.26, 31.51, 29.65, 28.38, 24.58, 24.36, 22.76, 20.97, 18.72, 13.51. HRMS (ESI-TOF) *m/z*: [M + H]$^+$ Calcd for C$_{28}$H$_{36}$NO$_2$ 418.2741; Found 418.2735.

21*E*-[(4-Nitrophenyl)methylidene]-4-azapregn-5-ene-3,20-dione (**3c**)

Pallid orange solid (89.6 mg, 67%); mp 250–251 °C; IR (cm^{-1}): 3193, 3075, 2936, 2868, 1681, 1650, 1592, 1514, 1341, 982, 834, 801. ^{1}H NMR (CDCl$_3$, 400 MHz) δ: 8.18 (2H, d, *J* = 7.7 Hz, H$_{Ar}$), 7.63 (2H, d, *J* = 7.6 Hz, H$_{Ar}$), 7.52–7.43 (2H, m, -NH+21a-CH), 6.79 (1H, d, *J* = 16.0 Hz, 21-CH), 4.76 (1H, s, 6-H), 1.03 (3H, s, 19-CH$_3$), 0.61 (3H, s, 18-CH$_3$). ^{13}C NMR (CDCl$_3$, 101 MHz) δ: 199.83, 169.89, 148.49, 143.48, 138.55, 128.80, 124.18, 123.73, 121.23, 103.03, 62.42, 56.87, 47.99, 45,18, 38.83, 31.77, 31.54, 29.60, 25.94, 24.53, 20.98, 18.73, 14.82, 13.61. HRMS (ESI- TOF) *m/z*: [M + H]$^+$ Calcd for C$_{27}$H$_{33}$N$_2$O$_4$ 449.2435; Found 449.2438.

21*E*-[(4-Methoxyphenyl)methylidene]-4-azapregn-5-ene-3,20-dione (**3d**)

Yellow solid (70.9 mg, 55%); mp 280–282 °C; IR (cm^{-1}): 3188, 3067, 2953, 2871, 2855, 1683, 1651, 1593, 1514, 1108, 986, 833, 805. ^{1}H NMR (CDCl$_3$, 400 MHz) δ: 7.65 (1H, s, -NH), 7.52–7.39 (3H, m, H$_{Ar}$+21a-CH), 6.84 (2H, d, *J* = 8.5 Hz, H$_{Ar}$), 6.60 (1H, d, *J* = 15.9 Hz, 21-CH), 4.78 (1H, s, 6-H), 3.78 (3H, s, 7'-CH$_3$), 1.02 (3H, s, 19-CH$_3$), 0.59 (3H, s, 18-CH$_3$).

^{13}C NMR (CDCl$_3$, 101 MHz) δ: 200.04, 169.57, 161.49, 141.46, 139.89, 129.98, 127.41, 124.53, 114.38, 103.29, 61.77, 56.89, 55.41, 47.95, 44.93, 44.01, 38.76, 34.28, 31.52, 29.65, 28.40, 24.58, 22.78, 20.94, 18.72, 13.50. HRMS (ESI-TOF) *m/z*: [M + H]$^+$ Calcd for C$_{28}$H$_{36}$NO$_3$ 434.2690; Found 434.2692.

21*E*-[(Furan-2-yl)methylidene]-4-azapregn-5-ene-3,20-dione (**3e**)

Pallid yellow solid (73.0 mg, 62%); mp 260–261 °C; IR (cm^{-1}): 3193, 3122, 2937, 2870, 1684, 1662, 1603, 1551, 1014, 974, 815. ^{1}H NMR (CDCl$_3$, 400 MHz) δ: 7.61 (1H, s, -NH), 7.42 (1H, s, H$_{Ar}$), 7.26 (1H, d, *J* = 15.6 Hz, 21a-CH), 6.60 (2H, d, *J* = 15.6 Hz, H$_{Ar}$+21-CH), 6.42 (1H, dd, *J* = 3.2, 1.8 Hz, H$_{Ar}$), 4.77 (1H, s, 6-H), 1.02 (3H, s, 19-CH$_3$), 0.59 (3H, s, 18-CH$_3$). ^{13}C NMR (CDCl$_3$, 101 MHz) δ: 199.74, 169.45, 151.38, 144.63, 139.95, 127.76, 123.90, 115.75, 112.52, 103.15, 62.31, 56.84, 47.92, 44.96, 38.69, 34.28, 31.77, 31.54, 29.64, 28.44, 24.55, 22.67, 20.95, 18.73, 13.48. HRMS (ESI-TOF) *m/z*: [M + H]$^+$ Calcd for C$_{25}$H$_{32}$NO$_3$ 394.2377; Found 394.2382.

21*E*-[(5-Methylfuran-2-yl)methylidene]-4-azapregn-5-ene-3,20-dione (**3f**)

Orange solid (102.6 mg, 84%); mp 200–203 °C; IR (cm^{-1}): 3200, 3119, 2937, 2870, 1665, 1626, 1606, 1569, 1019, 969, 784.^{1}H NMR (CDCl$_3$, 400 MHz) δ: 7.69 (1H, s, -NH), 7.19 (1H, d, *J* = 15.4, 21a-CH), 6.51 (2H, d, *J* = 15.7 Hz, H$_{Ar}$+21-CH), 6.03 (1H, s, H$_{Ar}$), 4.79 (1H, s, 6-H), 2.29 (3H, s, 5′-CH$_3$), 1.02 (3H, s, 19-CH$_3$), 0.58 (3H, s, 18-CH$_3$). ^{13}C NMR (CDCl$_3$, 101 MHz) δ: 199.79, 169.50, 155.53, 149.99, 139.95, 127.96, 122.24, 117.67, 109.16, 103.21, 63.51, 62.15, 56.86, 47.96, 44.94, 44.01, 38.69, 34.27, 31.55, 29.66, 28.42, 24.57, 22.75, 20.95, 18.73, 13.99. HRMS (ESI-TOF) *m/z*: [M + H]+ Calcd for C$_{26}$H$_{34}$NO$_3$ 408.2533; Found 408.2530.

21*E*-[(Pyridin-3-yl)methylidene]-4-azapregn-5-ene-3,20-dione (**3g**)

Beige solid (76.0 mg, 63%); mp 275–277 °C; IR (cm^{-1}): 3196, 3089, 2971, 2839, 2864, 1698, 1681, 1647, 1606, 1586, 1477, 998, 811. ^{1}H NMR (CDCl$_3$, 400 MHz) δ: 8.72 (1H, s, H$_{Ar}$), 8.54 (1H, dd, *J* = 4.9, 1.6 Hz, H$_{Ar}$), 7.79 (1H, d, *J* = 7.8 Hz, H$_{Ar}$), 7.71 (1H, s, -NH), 7.47 (1H, d, *J* = 15.9 Hz, 21a-CH), 7.28 (1H, m, H$_{Ar}$), 6.67 (1H, d, *J* = 15.9 Hz, 21-CH), 4.78 (1H, s, 6-H), 1.07 (3H, s, 19-CH$_3$), 0.61 (3H, s, 18-CH$_3$). ^{13}C NMR (CDCl$_3$, 101 MHz) δ: 199.60, 169.44, 150.85, 149.71, 139.96, 137.79, 134.60, 130.59, 128.34, 123.79, 103.06, 62.13, 59.31, 56.87, 50.66, 47.88, 45.08, 38.81, 34.27, 31.76, 29.62, 28.43, 24.54, 20.96, 18.73, 13.57. HRMS (ESI-TOF) *m/z*: [M + H]$^+$ Calcd for C$_{26}$H$_{33}$N$_2$O$_2$ 405.2537; Found 405.2534.

3.2. Biology

3.2.1. Cell Culture

Cell lines, namely LNCaP, PC-3, T47-D and NHDF were obtained from American Type Culture Collection (ATCC; Manassas, VA, USA) and cultured in 75 cm^2 culture flasks at 37 °C in a humidified air incubator with 5% CO$_2$. LNCaP, PC-3 and T47-D cells, which were used in passages 20th to 27th, 25th to 29th and 10th to 15th, respectively, were cultured in RPMI 1640 medium (Sigma-Aldrich, Inc., St. Louis, MO, USA) with 10% fetal bovine serum (FBS; Sigma-Aldrich, Inc. St. Louis, MO, USA) and 1% of the antibiotic mixture of 10,000 IU/mL penicillin G and 100 mg/mL of streptomycin (Sp, Sigma-Aldrich, Inc. St. Louis, MO, USA). Finally, NHDF cells (Normal Human Dermal Fibroblasts) were cultured in RPMI 1640 medium supplemented with 10% FBS, 2 mM *L*-glutamine, 10 mM HEPES, 1 mM sodium pyruvate and 1% of an antibiotic/antimycotic mixture (10,000 U/mL penicillin G, 100 mg/mL streptomycin and 25 µg/mL amphotericin B) (Ab; Sigma-Aldrich, Inc. St. Louis, MO, USA), and these cells were used in passages 10 to 12. For all cell types, the medium was renewed every 2–3 days until cells reach nearly the confluence state. Then, they were detached gently by trypsinization (trypsin-EDTA solution: 0.125 g/L of trypsin and 0.02 g/L of EDTA).

3.2.2. Preparation of Compounds Solutions

All the stock solutions were prepared by dissolving compounds in dimethyl sulfoxide (DMSO; Sigma-Aldrich, Inc., St. Louis, MO, USA) at 10 mM and stored at 4 °C. From these solutions, the several diluted compounds solutions in different concentrations were freshly prepared in a complete culture medium before each experiment. The DMSO concentration in final solutions did not interfere with the cell viability (<1%).

3.2.3. MTT Cell Proliferation Assay [38]

As previously referred, cells were trypsinized after reaching a near confluence state and counted by the trypan blue exclusion assay in a Neubauer chamber. Then, 100 μL of cell suspension/well with an initial density of 2×10^4 cells/mL were seeded in 96-well culture plates (Nunc, Apogent, Denmark) and left to adhere for 48 h. After cells adherence, the medium was replaced by the distinct solutions of the compounds in the study (30 μM for preliminary studies and 0.01, 0.1, 1, 10, 50 and 100 μM for concentration–response studies) in the appropriate medium for approximately 72 h. 5-FU and finasteride were used as positive controls and untreated cells were used as the negative control. Each experiment was performed in quadruplicate and independently performed at least two times. The in vitro antiproliferative effects were evaluated by the MTT assay (Sigma-Aldrich, Inc. St. Louis, MO, USA). After the incubation period, the medium was removed and 100 μL of phosphate buffer saline (NaCl 137 mM, KCl 2.7 mM, Na_2HPO_4 10 mM and KH_2PO_4 1.8 mM in deionized water and pH adjusted to 7.4) were used to wash the cells. Then, 100 μL of the MTT solution (5 mg/mL) was prepared in the appropriate serum-free medium and was added to each well, followed by incubation for 4 h at 37 °C. Hereafter, the MTT containing medium was removed and the formazan crystals were dissolved in DMSO. Then the absorbance was measured at 570 nm using a microplate spectrophotometer BIO-RAD xMarkTM. Cell viability values were expressed as percentages relative to the absorbance determined in the cells used as negative controls.

3.2.4. Fluorescence Microscopy Assay

To access the effects of the compound **3g** in T47-D cells, fluorescence microscopy assays with Hoechst 3342 and PI staining were performed. T47-D cells were seeded at 2×10^4 cells/mL in a 24-well culture plate containing circular coverslips of 10 mm diameter, in a complete culture medium. After 48 h, cells were incubated during 24, 48 and 72 h, with 5-FU at 20 μM, as the positive control, steroid **3g** at 20 μM, and untreated cells were used as a negative control. Then, cells were incubated with 20 μL/per well of a solution of PI in PBS at 1 mg/mL for 25 min at 37 °C. The medium was removed, and the cells were fixed with formalin 4% for 15 min at room temperature. Cells were washed three times with PBS and then they were incubated with 30 μL/per coverslip of Hoechst 33,342 (1:500) for 10 min. After incubation, cells were washed three times with PBS. Coverslips were mounted on a drop of permanent mounting medium (Dako) on a microscope slide and visualized in a Zeiss Axio Imager A1 microscope with the 40 × objective. The PI/Hoechst 3342 ratio was calculated by dividing PI-positive cells by Hoechst 3342-positive cells and, then the values were normalized.

3.2.5. Cell Nuclear Morphology and Distribution Analysis with Imagej

The nuclear measurements were achieved by converting 16-bit photomicrographs of Hoechst 3342-stained nuclei, in different conditions, into 8-bit images and then these images were autothresholded to binary photos using the default method "Make binary" function of ImageJ v 1.49. Cell nuclei that are touching were separated and fragments were discarded based on the area through the "Analyze Particle" function. This function also provides several information pretended as nuclear area, circumference and form factor [33]. In addition, the "Nearest Neighbor Distance" was determined. This function allows measuring the distance between each cell nucleus and the nearest ones.

3.2.6. Flow Cytometry Assay

Similar to the previous study, the analysis of cell viability was accomplished by flow cytometry after staining dead cells with propidium iodide (PI; Invitrogen, Carlsbad, USA) [20]. Succinctly, 1 mL of a cell suspension was seeded in 12-well culture plates (initial cell density of 5×10^4 cells/mL of LNCaP cells for 24 h assay and 2×10^4 cells/mL of the same cell line for 72 h assay) in a complete culture medium. After 48 h, the cells were treated with finasteride and 5-FU, as positive controls, and compound **3g** at a concentration of 50 µM. Untreated cells were used as a negative control. At the pretended time point, the supernatant was collected and pooled with the cells harvested by trypsin treatment (each well was also washed with 400 µL of PBS before trypsin treatment). The resulting cell suspension was kept on ice and pelleted by centrifugation and resuspended with 400 µL of complete medium. Afterward, 5 µL of PI (1 mg/mL) was transferred to a FACS tube and 395 µL of the cell suspension was added to this tube. A total of 20,000 events (very small events excluded) were acquired using a FACSCalibur flow cytometer, using the FCS, SSC and FL3 (PI) channels. Acquisition and analysis were executed with CellQuestTM Pro (v. 5.1) software. Briefly, a region was created on the SSC/FCS contour plot to exclude events of very small size and complexity (considered not relevant debris). Then, in the FCS/FL3 contour plot gated on the previous region, three regions were created, one corresponding to viable cells, R1, another to dead cells, R2, and another corresponding to an intermediate subpopulation of cells, R3, which may include both large debris resulting from apoptotic death (with very low DNA content) and cells with partial permeability to PI (plots available on Supplementary material, S2). The percentages of each region were calculated for a total of R1 + R2 + R3. The experiment was performed in two dependent days, each in duplicate or triplicate wells.

3.2.7. Caspase-9 Activity Assay

Promega Caspase-Glo 9 assay (Promega, Madison, WI, USA) was employed to evaluate the caspase-9 activity. T47-D cells were seeded in a 96 multiwell plate at a density of 4×10^4 cells/mL. After 48 h, cells were treated with the positive control, DOX, at 10 µM and with 10 and 20 µM of **3g**, during 48 h. The assay was performed at the end of treatment, following the instructions provided by the manufacturer. Data are representative of at least two experiments (n = 4 in each).

3.2.8. Statistical Data Analysis

The data are expressed as a mean $\pm$ standard deviation (SD) and differences between groups were considered statistically significant when $p < 0.05$. The IC$_{50}$ values were determined through sigmoidal fitting analysis with variable slope and considering a 95% confidence level. All data shown are representative of at least two independent experiments.

3.3. Molecular Docking

3.3.1. Preparation of Proteins

The three-dimensional structural coordinates for 5α-reductase type 2 (5AR PDB code: 7BW1), estrogen receptor-α (ERα PDB code: 1A52), androgen receptor (AR PDB code: 2AMA), 17α-hydroxylase-17,20-lyase (CYP17A1 PDB code: 3RUK) and aromatase (PDB code: 3EQM) were downloaded from Protein Data Bank (www.rcsb.org, accessed on 12 August 2022). The coordinates of the ligands co-crystallized and water molecules were deleted using the software Chimera (v. 1.10.1), and histidine charges were defined to match the physiologic environment and the final structures were saved in PDB format. Then, non-polar hydrogens were merged in AutoDockTools (v. 1.5.6) from The Scripps Research Institute [39]. Kollman and Gasteiger partial charges were added. Finally, the prepared structures were converted from the PDB format to PDBQT for posterior employment in the docking study.

3.3.2. Preparation of Ligands

Chem3D (v. 12.0) software (by Cambridge ChemBioOffice 2010) was used to build ligands. Then, geometry optimization and energy minimization (MMFF94 force field: 500 steps of conjugate gradient energy minimization followed by 500 steps of steepest descent energy minimization with a convergence setting of 10×10^{-7}) were executed with

Avogadro (v. 1.0.1). The final structures were saved in a PDB file format. The ligands were completely prepared, choosing torsions and the structures were converted from PDB format to PDBQT, in AutoDockTools software.

3.3.3. Grid Map Calculations

AutoGrid4 was used to calculate Autodock grid maps for each macromolecule, based on the active site coordinates of the crystal structure. The size of all grid boxes was $40 \times 40 \times 40$ with 0.375 Å of spacing. Maps were calculated for each atom type in each ligand along with an electrostatic and desolvation map using a dielectric value of -0.1465.

3.3.4. Molecular Docking Simulations

Molecular docking studies were performed using the Lamarckian genetic algorithm and empirical free energy scoring function [39]. The maximum number of energy evaluations was 2,500,000, and the GA population size was 150. A total of 15 hybrid GA-LS runs were performed for each simulation. The results of these simulations were visualized in PyMol (The PyMol Molecular Graphics System v. 1.3, Schrödinger, LLC – www.pymol.org), built for educational use. All docking simulations conducted to validate the method, using the ligands present in crystal structures, were able to reproduce the ligand–protein interaction geometries. For the docking process to be considered successful, the RMSD value between ligand conformations (docked ligand and crystalized ligand) was less than 2.0 Å.

4. Conclusions

A new series of 4-azapregn-5-ene-3,20-diones bearing aromatic or heteroaromatic substituents at C21 were prepared by aldol condensation in good overall yields. The MTT cell proliferation assay showed some interesting cytotoxic activity of some of these compounds in tumor cell lines. In fact, 4-azapregnene derivatives **3a**, **3c** and **3g** presented relevant antiproliferative effects in tumor cells. The pyridinyl derivative, **3g**, had the lowest determined IC_{50} in LNCaP (10.20 µM) and T47-D cells (1.33 µM). The most cytotoxic steroid on PC-3 cells was **3c** (IC_{50} of 3.29 µM), but **3g** also had a very close IC_{50} value (3.64 µM). Therefore, considering these results, steroid **3g** was selected for further biological studies in T47-D cells. The assessment of cytotoxic effect was performed through fluorescence microscopy after PI/Hoechst 3342 staining and caspase-9 activity measurement. All the results indicate that this compound possibly led cells to death by triggering the apoptotic pathway. Moreover, from a flow cytometry assay with PI staining and microscopic observation, a drastic effect on cell viability can be observed when LNCaP cells were exposed to steroid **3g**, which is a very interesting result to explore in a further investigation. Molecular docking studies indicated that these novel 4-azapregnene derivatives can potentially interact with 5AR type 2, similarly to finasteride. Additionally, these simulations also showed the possibility of these novel derivatives interacting with the other tested targets of steroidal drugs, with the exception of AR. In conclusion, derivatives **3a**, **3c** and **3g** presented a remarkable cytotoxic effect on tumor cell lines.

Supplementary Materials: The following supporting information can be downloaded at: https://www.mdpi.com/article/10.3390/molecules27186126/s1. NMR and IR spectra of synthesized compounds (S1). Counter plots obtained from flow cytometry after PI staining are also available for control, finasteride and steroid **3g** (S2).

Author Contributions: V.B. and S.S. carried out the literature, conceived and designed the experiments. The methodology was carried out by V.B. and A.O.S. contributed to the characterization of cytotoxic activities. All the authors read and approved the final manuscript. Conceptualization, S.S. and V.B.; methodology, V.B. and A.O.S.; investigation, V.B.; funding and resources, V.B., A.O.S., G.A., P.A. and S.S.; writing–original draft preparation, V.B. and S.S.; supervision, G.A., P.A. and S.S. All authors have read and agreed to the published version of the manuscript.

Funding: This work is supported by FEDER funds through the POCI—COMPETE 2020—Operational Programme Competitiveness and Internationalization in Axis I—Strengthening Research, Technological Development and Innovation (Project No. 007491) and National Funds by the FCT—Foundation for Science and Technology (Project UID/Multi/00709). The NMR spectrometers are part of the Portuguese NMR Network (PTNMR) and are partially supported by the Infrastructure Project No. 022161 (co-financed by FEDER through COMPETE 2020, POCI and PORL and FCT through PIDDAC). V.B. also acknowledges the grants BID/ICI-FC/Santander Universidades-UBI/2016 and SFRH/BD/131059/2017 (FCT). It is also acknowledged founding from C4-Cloud Computing Competences Center project (CENTRO-01-0145-FEDER-000019).

Institutional Review Board Statement: Not applicable.

Informed Consent Statement: Not applicable.

Data Availability Statement: Not applicable.

Acknowledgments: All authors acknowledge the previously referred funding support, including the PhD grant SFRH/BD/131059/2017 (V.B.). Thanks are also to CICS-UBI and University of Beira Interior.

Conflicts of Interest: The authors declare no conflict of interest.

Sample Availability: Samples of the compounds are not available from the authors.

References

1. Ke, S. Recent Progress of Novel Steroid Derivatives and Their Potential Biological Properties. *Mini-Reviews Med. Chem.* **2018**, *18*, 1–31. [CrossRef] [PubMed]
2. Shaikh, S.; Verma, H.; Yadav, N.; Jauhari, M.; Bullangowda, J. Applications of Steroid in Clinical Practice: A Review. *ISRN Anesthesiol.* **2012**, *2012*, 1–11. [CrossRef]
3. Brito, V.; Alves, G.; Almeida, P.; Silvestre, S. Highlights on Steroidal Arylidene Derivatives as a Source of Pharmacologically Active Compounds: A Review. *Molecules* **2021**, *26*, 2032. [CrossRef]
4. Roy, J.; Deroy, P.; Poirier, D. 2β-(N-Substituted Piperazino)-5α-Androstane-3α,17β-Diols: Parallel Solid-Phase Synthesis and Antiproliferative Activity on Human. *J. Comb. Chem.* **2007**, *9*, 347–358. [CrossRef]
5. Banday, A.H.; Mir, B.P.; Lone, I.H.; Suri, K.A.; Kumar, H.M.S. Studies on Novel D-Ring Substituted Steroidal Pyrazolines as Potential Anticancer Agents. *Steroids* **2010**, *75*, 805–809. [CrossRef]
6. Fan, N.; Tang, J.; Li, H.; Li, X.; Luo, B.; Gao, J. Synthesis and Cytotoxic Activity of Some Novel Steroidal C-17 Pyrazolinyl Derivatives. *Eur. J. Med. Chem.* **2013**, *69*, 182–190. [CrossRef]
7. Salvador, J.A.R.; Pinto, R.M.A.; Silvestre, S.M. Steroidal 5α-Reductase and 17α-Hydroxylase/17,20-Lyase (CYP17) Inhibitors Useful in the Treatment of Prostatic Diseases. *J. Steroid Biochem. Mol. Biol.* **2013**, *137*, 199–222. [CrossRef]
8. Cortés-Benítez, F.; Cabeza, M.; Teresa, M.; Apan, R.; Bratoeff, E. Synthesis of 17b-N-Arylcarbamoylandrost-4-En-3-One Derivatives and Their Antiproliferative Effect on Human Androgen-Sensitive LNCaP Cell Line. *Eur. J. Med. Chem.* **2016**, *4*, 737–746. [CrossRef]
9. Ajdukovic, J.J.; Djurendic, E.A.; Petri, E.T.; Klisuric, O.R.; Celic, A.S.; Sakač, M.N.; Jakimov, D.S.; Penov Gaši, K.M. 17(E)-Picolinylidene Androstane Derivatives as Potential Inhibitors of Prostate Cancer Cell Growth: Antiproliferative Activity and Molecular Docking Studies. *Bioorg. Med. Chem.* **2013**, *21*, 7257–7266. [CrossRef]
10. Bansal, R.; Guleria, S. Synthesis of 16E-[3-Methoxy-4-(2-Aminoethoxy)Benzylidene]Androstene Derivatives as Potent Cytotoxic Agents. *Steroids* **2008**, *73*, 1391–1399. [CrossRef]
11. Bansal, R.; Thota, S.; Karkra, N.; Minu, M.; Zimmer, C.; Hartmann, R.W. Synthesis and Aromatase Inhibitory Activity of Some New 16E-Arylidenosteroids. *Bioorg. Chem.* **2012**, *45*, 36–40. [CrossRef] [PubMed]
12. Chattopadhaya, R.; Jindal, D.P.; Minu, M.; Gupta, R. Synthesis and Cytotoxic Studies of Hydroximino Derivatives of Some 16E-Arylidenosteroids. *Arzneim. Forshung - Drug Res.* **2004**, *556*, 551–556. [CrossRef]
13. Dubey, S.; Piplani, P.; Jindal, D.P. Synthesis and in Vitro Antineoplastic Evaluation of Certain 16-(4-Substituted Benzylidene) Derivatives of Androst-5-Ene. *Chem. Biodivers.* **2004**, *1*, 1529–1536. [CrossRef] [PubMed]
14. Choudhary, M.I.; Alam, M.S.; Yousuf, S.; Wu, Y.; Lin, A.; Shaheen, F. Pregnenolone Derivatives as Potential Anticancer Agents. *Steroids* **2011**, *76*, 1554–1559. [CrossRef]
15. Banday, A.H.; Shameem, S.A.; Jeelani, S. Steroidal Pyrazolines and Pyrazoles as Potential 5α-Reductase Inhibitors: Synthesis and Biological Evaluation. *Steroids* **2014**, *92*, 13–19. [CrossRef]
16. Banday, A.H.; Akram, S.M.M.; Shameem, S.A. Benzylidine Pregnenolones and Their Oximes as Potential Anticancer Agents: Synthesis and Biological Evaluation. *Steroids* **2014**, *84*, 64–69. [CrossRef]
17. Fan, N.; Han, Y.; Li, Y.; Gao, J.; Tang, J. Synthesis of Novel 4'-Acylamino Modified 21E-Benzylidene Steroidal Derivatives and Their Cytotoxic Activities. *Steroids* **2017**, *123*, 20–26. [CrossRef]

18. Shan, L.; Liu, H.; Huang, K.; Dai, G.; Cao, C.; Dong, R. Synthesis of 3b, 7a, 11a-Trihydroxy-Pregn-21-Benzylidene-5-En-20-One Derivatives and Their Cytotoxic Activities. *Bioorg. Med. Chem. Lett.* **2009**, *19*, 6637–6639. [CrossRef]
19. Huang, L.-H.; Xu, H.-D.; Yang, Z.-Y.; Zheng, Y.-F.; Liu, H.-M. Synthesis and Anticancer Activity of Novel C6-Piperazine Substituted Purine Steroid-Nucleosides Analogues. *Steroids* **2014**, *82*, 1–6. [CrossRef]
20. Brito, V.; Santos, A.O.; Almeida, P.; Silvestre, S. Novel 4-Azaandrostenes as Prostate Cancer Cell Growth Inhibitors: Synthesis, Antiproliferative Effects and Molecular Docking Studies. *Comptes Rendus Chim.* **2018**, *22*, 73–83. [CrossRef]
21. Xiao, Q.; Wang, L.; Fan, H.; Wei, Z. Structure of Human Steroid 5α-Reductase 2 with the Anti-Androgen Drug Finasteride. *Nat. Commun.* **2020**, *11*, 5430. [CrossRef] [PubMed]
22. Kumar, R.; Malla, P.; Verma, A.; Kumar, M. Design of Potent Human Steroid 5α-Reductase Inhibitors: 3D-QSAR CoMFA, CoMSIA and Docking Studies. *Med. Chem. Res.* **2013**, *22*, 4568–4580. [CrossRef]
23. Yao, Z.; Xu, Y.; Zhang, M.; Jiang, S.; Nicklaus, M.C.; Liao, C. Discovery of a Novel Hybrid from Finasteride and Epristeride as 5a-Reductase Inhibitor. *Bioorganic Med. Chem. Lett.* **2011**, *21*, 475–478. [CrossRef] [PubMed]
24. Borthakur, M.; Boruah, R.C. A Microwave Promoted and Lewis Acid Catalysed Solventless Approach to 4-Azasteroids. *Steroids* **2008**, *73*, 637–641. [CrossRef] [PubMed]
25. Morzycki, J.W.; Lotowski, Z.; Wilczewska, A.Z.; Stuart, J.D. Synthesis of 4,17-Diazasteroid Inhibitors of Human 5α-Reductase. *Bioorg. Med. Chem.* **1996**, *4*, 1209–1215. [CrossRef]
26. Bansal, R.; Acharya, P.C. Synthesis and Antileukemic Activity of 16E-[4-(2-Carboxy)Ethoxybenzylidene]-Androstene Amides. *Steroids* **2012**, *77*, 552–557. [CrossRef]
27. Fan, N.-J.; Bai, Y.-B.; Zhang, F.-Y.; Luo, B.; Tang, J.-J.; Zhang, Q.-Z.; Gao, J.-M. Synthesis and Cytotoxicity of Some Novel 21E-Benzylidene Steroidal Derivatives. *Steroids* **2013**, *78*, 874–879. [CrossRef]
28. Chen, S.R.; Wu, H.; Zhao, H.Y.; Zhang, Y.M.; Li, P.Q.; Zhao, L.M. Synthesis and Antiproliferative Activity of Novel 4-Azasteroidal-17-Hydrazone Derivatives. *J. Chem. Res.* **2019**, *43*, 130–134. [CrossRef]
29. Iványi, Z.; Szabó, N.; Huber, J.; Wölfling, J.; Zupkó, I.; Szécsi, M.; Wittmann, T.; Schneider, G. Synthesis of D-Ring-Substituted (5′R)- and (5′S)-17b-Pyrazolinylandrostene Epimers and Comparison of Their Potential Anticancer Activities. *Steroids* **2012**, *77*, 566–574. [CrossRef]
30. Figueiredo, J.; Serrano, J.L.; Cavalheiro, E.; Keurulainen, L.; Yli-Kauhaluoma, J.; Moreira, V.M.; Ferreira, S.; Domingues, F.C.; Silvestre, S.; Almeida, P. Trisubstituted Barbiturates and Thiobarbiturates: Synthesis and Biological Evaluation as Xanthine Oxidase Inhibitors, Antioxidants, Antibacterial and Anti-Proliferative Agents. *Eur. J. Med. Chem.* **2018**, *143*, 829–842. [CrossRef]
31. Baji, Á.; Kiss, T.; Wölfling, J.; Kovács, D.; Igaz, N.; Gopisetty, M.K.; Kiricsi, M.; Frank, É. Multicomponent Access to Androstano-Arylpyrimidines under Microwave Conditions and Evaluation of Their Anti-Cancer Activity in Vitro. *J. Steroid Biochem. Mol. Biol.* **2017**, *172*, 79–88. [CrossRef] [PubMed]
32. de la Guardia, C.; Stephens, D.E.; Dang, H.T.; Quijada, M.; Larionov, O.V.; Lleonart, R. Antiviral Activity of Novel Quinoline Derivatives against Dengue Virus Serotype 2. *Molecules* **2018**, *23*, 672. [CrossRef] [PubMed]
33. Eidet, J.R.; Pasovic, L.; Maria, R.; Jackson, C.J.; Utheim, T.P. Objective Assessment of Changes in Nuclear Morphology and Cell Distribution Following Induction of Apoptosis. *Diagn. Pathol.* **2014**, *9*, 1–9. [CrossRef] [PubMed]
34. Li, P.; Zhou, L.; Zhao, T.; Liu, X.; Zhang, P. Caspase-9: Structure, Mechanisms and Clinical Application. *Oncotarget* **2017**, *8*, 23996–24008. [CrossRef]
35. Santos, A.; Sarmento-Ribeiro, A.B.; Pedroso De Lima, M.C.; Simões, S.; Moreira, J.N. Simultaneous Evaluation of Viability and Bcl-2 in Small-Cell Lung Cancer. *Cytom. Part A* **2008**, *73*, 1165–1172. [CrossRef]
36. Armarego, W.L.F. Chapter 3 - Purification of Organic Chemicals. In *Purification of Laboratory Chemicals (Eighth Edition)*; Armarego, W.L.F., Ed.; Butterworth-Heinemann: Oxford, UK, 2017; pp. 95–634. ISBN 978-0-12-805457-4.
37. Trapani, G.; Dazzi, L.; Pisu, M.G.; Reho, A.; Seu, E.; Biggio, G. A Rapid Method for Obtaining Finasteride, a 5α-Reductase Inhibitor, from Commercial Tablets. *Brain Res. Protoc.* **2002**, *9*, 130–134. [CrossRef]
38. Santos, D.; Medeiros-Silva, J.; Cegonho, S.; Alves, E.; Ramilo-Gomes, F.; Santos, A.O.; Silvestre, S.; Cruz, C. Cell Proliferation Effects of Calix[4]Arene Derivatives. *Tetrahedron* **2015**, *71*, 7593–7599. [CrossRef]
39. Morris, G.M.; Goodsell, D.S.; Halliday, R.S.; Huey, R.; Hart, W.E.; Belew, R.K.; Olson, A.J.; Al, M.E.T. Automated Docking Using a Lamarckian Genetic Algorithm and an Empirical Binding Free Energy Function. *J. Comput. Chem.* **1998**, *19*, 1639–1662. [CrossRef]

Article

The Isolation, Structure Elucidation and Bioactivity Study of Chilensosides A, A$_1$, B, C, and D, Holostane Triterpene Di-, Tri- and Tetrasulfated Pentaosides from the Sea Cucumber *Paracaudina chilensis* (Caudinidae, Molpadida)

Alexandra S. Silchenko *, Sergey A. Avilov, Pelageya V. Andrijaschenko, Roman S. Popov, Ekaterina A. Chingizova, Boris B. Grebnev, Anton B. Rasin and Vladimir I. Kalinin

G.B. Elyakov Pacific Institute of Bioorganic Chemistry, Far Eastern Branch of the Russian Academy of Sciences, Pr. 100-letya Vladivostoka 159, 690022 Vladivostok, Russia
* Correspondence: silchenko_als@piboc.dvo.ru; Tel./Fax: +7-(423)2-31-40-50

Abstract: Five new triterpene (4,4,14-trimethylsterol) di-, tri- and tetrasulfated pentaosides, chilensosides A (**1**), A$_1$ (**2**), B (**3**), C (**4**), and D (**5**) were isolated from the Far-Eastern sea cucumber *Paracaudina chilensis*. The structures were established on the basis of extensive analysis of 1D and 2D NMR spectra and confirmed by HR-ESI-MS data. The structural variability of the glycosides concerned the pentasaccharide chains. Their architecture was characterized by the upper semi-chain consisting of three sugar units and the bottom semi-chain of two sugars. Carbohydrate chains of compounds **2–5** differed in the quantity and positions of sulfate groups. The interesting structural features of the glycosides were: the presence of two sulfate groups at C-4 and C-6 of the same glucose residue in the upper semi-chain of **1, 2, 4**, and **5** and the sulfation at C-3 of terminal glucose residue in the bottom semi-chain of **4** that makes its further elongation impossible. Chilensoside D (**5**) was the sixth tetrasulfated glycoside found in sea cucumbers. The architecture of the sugar chains of chilensosides A–D (**1–5**), the positions of sulfation, the quantity of sulfate groups, as well as the aglycone structures, demonstrate their similarity to the glycosides of the representatives of the order Dendrochirotida, confirming the phylogenetic closeness of the orders Molpadida and Dendrochirotida. The cytotoxic activities of the compounds **1–5** against human erythrocytes and some cancer cell lines are presented. Disulfated chilensosides A$_1$ (**2**) and B (**3**) and trisulfated chilensoside C (**4**) showed significant cytotoxic activity against human cancer cells.

Keywords: *Paracaudina chilensis*; Molpadida; triterpene glycosides; chilensosides; sea cucumber; cytotoxic activity

check for updates

Citation: Silchenko, A.S.; Avilov, S.A.; Andrijaschenko, P.V.; Popov, R.S.; Chingizova, E.A.; Grebnev, B.B.; Rasin, A.B.; Kalinin, V.I. The Isolation, Structure Elucidation and Bioactivity Study of Chilensosides A, A$_1$, B, C, and D, Holostane Triterpene Di- and Tetrasulfated Pentaosides from the Sea Cucumber *Paracaudina chilensis* (Caudinidae, Molpadida). *Molecules* **2022**, *27*, 7655. https://doi.org/10.3390/molecules27217655

Academic Editors: Marina Savić, Erzsébet Mernyák, Jovana Ajdukovic and Suzana Jovanović-Šanta

Received: 4 October 2022
Accepted: 2 November 2022
Published: 7 November 2022

Publisher's Note: MDPI stays neutral with regard to jurisdictional claims in published maps and institutional affiliations.

1. Introduction

Despite triterpene glycosides from sea cucumbers having a rather long history of investigations, there are some systematic groups, including the order Molpadida, comprising the studied species *Paracaudina chilensis*, which are poorly studied or unexplored chemically. The majority of recent research concerning the sea cucumber triterpene glycosides deals with the structure elucidation of the compounds isolated from representatives of the orders Dendrochirotida, Synallactida, and Holothuriida [1–9]. The use of mass-spectrometry-based metabolomics for the solving of diverse chemical and biological issues concerning secondary metabolites has become very popular and has provided some significant results in the exploration of triterpene glycoside chemical diversity, their content and composition in different body parts [10–13], and their chemotaxonomy [14,15]. The application of this approach in combination with molecular phylogenetic analysis allowed to clarify the evolution of the Holothuroidea taxons [16]. Different investigations have also confirmed the defensive role of glycosides [17,18]. The biosynthetic studies of triterpene glycosides are very difficult to conduct. They began in the 1970s through the introduction of radioactively labeled precursors to the organisms-producers and showed some contradictory results [19–21]. However, it has been established that precursors of triterpenoids can be either lanosterol or parkeol depending on the intranuclear double bond position in the biosynthesizing aglycones [22–25]. Oxydosqualenecyclases (OSCs) are the enzymes processing the cyclization of 2,3-oxidosqualene, making diverse triterpene alcohols. This stage is the branchpoint of steroids and triterpenoids biosynthesis. In the process of steroidogenesis, the triterpene precursors are enzymatically demethylated at positions C-4, C-14, double bond positions are changed and side chains are modified. The sea cucumbers are characterized by the presence of steroids with uncommon chemical features ($\Delta^{9(11)}$- and Δ^7-derivatives) instead of Δ^5-sterols characteristic for other animals. This is explained by the presence of the membranolytic triterpene glycosides targeting the Δ^5-steroids. Therefore, to protect their own membranes from the action of these toxins, the steroid composition has been evolutionary changed. It was supposed that Δ^7-steroids are formed as result of modification of dietary Δ^5-sterols, while $\Delta^{9(11)}$-steroid compounds are biosynthesized from parkeol, formed *de novo* by OSCs in the sea cucumbers. Noticeably, some sea cucumber species contain 14α-methylated and 4α,14α-dimethylated $\Delta^{9(11)}$-sterols [23]. The recent genetic studies of *Apostichopus japonicus* explained this phenomenon by the absence of the gene for 14-sterol-demethylase in its genome [26]. There is little molecular/genetic research on genes and corresponding enzymes participating in triterpenoid biosynthesis in sea cucumbers. They only concerned transcriptomic analysis of tissues of *Holothuria scabra* and *Stichopus horrens* and have led to identification of the sequences corresponding to some genes of the mevalonate pathway [27,28]. The decoding of sequences of OSCs from *Apostichopus japonicus* followed by the expression of the genes integrated into the yeast genome allowed for the identification of parkeol and 9βH-lanosta-7,24-dien-3β-ol as the products [26].

Finally, there exists a large number of investigations devoted to the biological activity, including anticancer activity, of sea cucumber glycosides, which are valuable sources of new drug candidates [29–35].

Only preliminary research on the glycosidic composition of *Paracaudina ransonetii* (=*Paracaudina chilensis*) has been previously published [36]. The taxonomic status of the order Molpadida raises questions for biologists dealing with the systematics of Holothuroidea. Some support the idea of closeness of Molpadida and Dendrochirotida, while others consider molpadiids as being near to Aspidochirotida [37]. From this viewpoint, investigations on the glycosides of representatives of the order Molpadida are relevant for the searching for new structural variants, which broaden our knowledge concerning chemical biodiversity and chemotaxonomy. Glycosides have been successfully used as chemotaxonomic markers of different sea cucumber systematic groups [38–41]. Therefore, the analysis of chemical peculiarities of the glycosides of *P. chilensis* can help to resolve this dilemma.

New triterpene glycosides, chilensosides A (**1**), A$_1$ (**2**), B (**3**), C (**4**) and D (**5**), were isolated from the Far Eastern sea cucumber *Paracaudina chilensis*. The chemical structures of **1**–**5** were established by the analyses of the ^{1}H, ^{13}C NMR, 1D TOCSY and 2D NMR (^{1}H, ^{1}H-COSY, HMBC, HSQC, ROESY) spectra as well as HR-ESI mass spectra. All the original spectra are presented in Figures S1–S40 in the Supplementary Materials. The hemolytic activity against human erythrocytes and cytotoxic activities against human neuroblastoma SH-SY5Y, adenocarcinoma HeLa, colorectal adenocarcinoma DLD-1, leukemia promyeloblast HL-60 and monocytic THP-1 cells were examined.

2. Results and Discussion

2.1. Structural Elucidation of the Glycosides

The crude glycosidic fraction of the sea cucumber *Paracaudina chilensis* was obtained as a result of hydrophobic chromatography of the concentrated ethanolic extract on a Polychrom-1 column (powdered Teflon, Biolar, Latvia). Its subsequent separation by chromatography on Si gel columns with the stepped gradient of the system of eluents

CHCl3/EtOH/H2O used in ratios (100:100:17), (100:125:25), and (100:150:50) gave the fractions I–III. Each of the obtained fractions was additionally purified on a Si gel column with the solvent system CHCl3/EtOH/H2O (100:125:25), which resulted in the isolation of five subfractions I.0, I.1, II, III.1 and III.2. The individual compounds **1–5** (Figure 1) were isolated through HPLC of these subfractions on the silica-based column Supelcosil LC-Si (4.6 × 150 mm), and reversed-phase columns Supelco Discovery HS F5-5 (10 × 250 mm) and Diasfer 110 C-8 (4.6 × 250 mm).

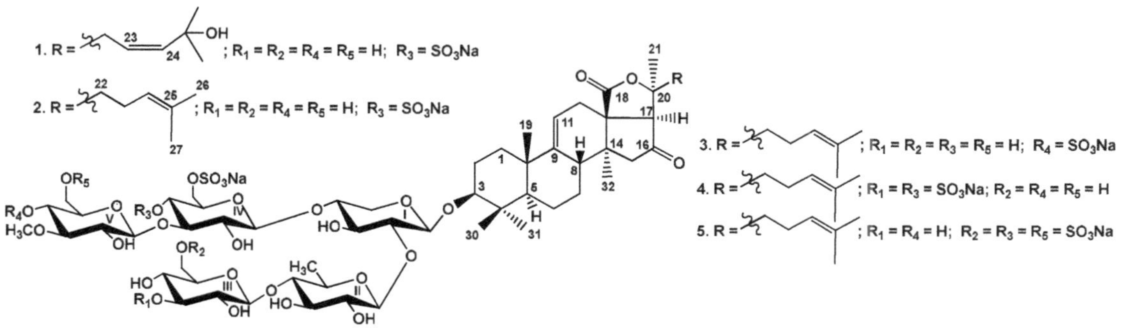

Figure 1. Chemical structures of glycosides isolated from *Paracaudina chilensis*: **1**—chilensoside A; **2**—chilensoside A$_1$; **3**—chilensoside B; **4**—chilensoside C; **5**—chilensoside D.

The configurations of the monosaccharide residues in the glycosides **1–5** were assigned as *D* based on the biogenetic analogies with the monosaccharides from all other known sea cucumber triterpene glycosides.

The molecular formula of chilensoside A (**1**) was determined to be $C_{60}H_{92}O_{34}S_2Na_2$ from the $[M_{2Na}-Na]^-$ ion peak at *m/z* 1443.4800 (calc. 1443.4815), and $[M_{2Na}-2Na]^{2-}$ ion peak at *m/z* 710.2466 (calc. 710.2461) in the (−) HR-ESI-MS (Figure S8). The ^{13}C NMR spectrum of the aglycone part of chilensoside A (**1**) demonstrated the signals of quaternary oxygen-bearing carbons at δ_C 176.8 (C-18) and 82.8 (C-20), corresponding to 18(20)-lactone and the signals of olefinic carbons at δ_C 151.1 (C-9), 111.2 (C-11) (Table 1, Figures S1–S7), indicating the presence of 9(11)-double bond, typical of many sea cucumber glycosides. An additional deshielded signal at δ_C 214.6 was assigned to C-16 oxo-group in the holostane nucleus, confirmed by the singlet signal of H-17 at δ_H 2.89 with a corresponding carbon signal at δ_C 61.2 (C-17). The protons of the side chain H-22/H-23/H-24 formed the isolated spin system deduced by the COSY spectrum, indicating the presence of an additional 23Z(24)-double bond (δ_{H-23} 5.90 (dd, *J* = 6.3; 11.8 Hz), δ_{H-24} 5.90 (d, *J* = 11.8 Hz)). The presence of the signal of quaternary oxygen-bearing carbon at δ_C 81.3 along with the coincidence of the signals of 26, 27-methyl groups to each other (δ_C 24.7 (C-26, C-27), δ_H 1.42 (s, H-26, H-27)) indicated the presence of hydroxyl at C-25. The side chain structure was confirmed by the HMBC correlations H-24/C-25; H-23/C-22; H-26(27)/C-24 (Table 1). The same aglycone was found only once earlier in cladoloside A$_5$ from the sea cucumber *Cladolabes schmeltzii* [42].

Table 1. ^{13}C and ^1H NMR chemical shifts and HMBC and ROESY correlations of aglycone moiety of chilensoside A (**1**).

Position	δ_C Mult. [a]	δ_H Mult. (*J* in Hz) [b]	HMBC	ROESY
1	36.0 CH_2	1.72 m		H-11
		1.31 m		H-3, H-5, H-11
2	26.6 CH_2	2.06 m		
		1.84 m		H-30
3	88.6 CH	3.13 dd (4.5; 11.3)		H-5, H-31, H1-Xyl1
4	39.5 C			
5	52.7 CH	0.79 brd (11.2)	C: 10, 19	H-1, H-3, H-31
6	20.9 CH_2	1.57 m		H-31
		1.38 m		H-30
7	28.2 CH_2	1.56 m		
		1.17 m		H-32
8	38.6 CH	3.11 m		H-19
9	151.1 C			
10	39.6 C			
11	111.2 CH	5.26 brs		H-1
12	31.8 CH_2	2.61 d (16.7)		H-17, H-32
		2.45 dd (6.7; 16.7)		H-21
13	55.8 C			
14	42.0 C			
15	52.0 CH_2	2.41 d (15.2)	C: 13, 16	H-32
		2.15 d (15.2)		
16	214.6 C			
17	61.1 CH	2.89 s	C: 12, 16, 18	H-12, H-21, H-32
18	176.8 C			
19	21.9 CH_3	1.26 s	C: 1, 5, 9, 10	H-1, H-2, H-8, H-30
20	82.8 C			
21	27.0 CH_3	1.44 s	C: 17, 20, 22	H-12, H-17, H-22
22	41.9 CH_2	2.55 dd (4.8; 13.8)		H-24
		2.33 m		H-24
23	123.9 CH	5.90 dd (6.3; 11.8)	C: 22	
24	140.1 CH	5.90 d (11.8)	C: 25	H-22, H-26, H-27
25	81.3 C			
26	24.7 CH_3	1.42 s	C: 24, 27	H-24
27	24.7 CH_3	1.42 s	C: 24, 26	H-24
30	16.6 CH_3	0.96 s	C: 3, 4, 5, 31	H-2, H-6, H-19, H-31
31	27.9 CH_3	1.13 s	C: 3, 4, 5, 30	H-3, H-5, H-6, H-30, H-1 Xyl1
32	20.5 CH_3	0.88 s	C: 8, 13, 14, 15	H-7, H-12, H-15, H-17

[a] Recorded at 176.04 MHz in C_5D_5N/D_2O (4/1). [b] Recorded at 700.13 MHz in C_5D_5N/D_2O (4/1). The original spectra of **1** are provided in Figures S1–S7.

The ^1H and ^{13}C NMR spectra of the carbohydrate chain of chilensosides A (**1**) (Table 2, Figures S1–S7) and A_1 (**2**) (Table S1, Figures S9–S15) were coincident to each other, indicating the identity of sugar moieties of **1**, **2**. The spectra of **1** demonstrated five characteristic doublets of anomeric protons at δ_H 4.66–5.20 (*J* = 7.1–8.1 Hz) and the signals of anomeric carbons at δ_C 102.2–104.7, indicating the presence of a pentasaccharide chain and β-configurations of glycosidic bonds. The coherent analysis of the ^1H, ^1H-COSY, 1D TOCSY, HSQC and ROESY spectra of **1** indicated the presence of one xylose (Xyl1), one quinovose (Qui2), two glucose (Glc3 and Glc4), and 3-*O*-methylglucose (MeGlc5) residues. The ROE- and HMBC correlations showed the positions of glycosidic linkages (Table 2, Figures S1–S7), which indicated that the carbohydrate chain of **1** was branched by C-4 Xyl1 having bottom semi-chain composed of two sugar units and the upper semi-chain from three.

Table 2. ^{13}C and ^{1}H NMR chemical shifts and HMBC and ROESY correlations of carbohydrate moiety of chilensoside A (**1**).

Atom	δ_C Mult. [a,b c]	δ_H Mult. (*J* in Hz) [d]	HMBC	ROESY
Xyl1 (1 → C-3)				
1	104.7 CH	4.66 d (7.2)	C: 3	H-3; H-3, 5 Xyl1
2	**82.0** CH	3.93 t (8.6)		H-1 Qui2
3	75.1 CH	4.13 t (8.6)	C: 4 Xyl1	H-1, 5 Xyl1
4	**78.3** CH	4.11 m		H-1 Glc4
5	63.5 CH$_2$	4.32 m	C: 3 Xyl1	
		3.61 m		H-1 Xyl1
Qui2 (1 → 2Xyl1)				
1	104.4 CH	5.02 d (7.1)	C: 2 Xyl1	H-2 Xyl1; H-3, 5 Qui2
2	75.5 CH	3.88 t (9.1)	C: 1, 3 Qui2	H-4 Qui2
3	75.1 CH	4.02 t (9.1)	C: 2, 4 Qui2	H-1, 5 Qui2
4	**86.3** CH	3.51 t (9.1)	C: 3 Qui2, 1 Glc3	H-1 Glc3; H-2 Qui2
5	71.4 CH	3.70 dd (6.1; 9.2)		H-1, 3 Qui2
6	17.9 CH$_3$	1.61 d (6.1)	C: 4, 5 Qui2	H-4 Qui2
Glc3 (1 → 4Qui2)				
1	104.6 CH	4.80 d (8.1)	C: 4 Qui2	H-4 Qui2; H-3, 5 Glc3
2	74.4 CH	3.86 t (8.7)	C: 1, 3 Glc3	
3	77.2 CH	4.13 m	C: 4 Glc3	H-1, 5 Glc3
4	70.9 CH	3.92 m	C: 5 Glc3	
5	77.6 CH	3.92 m	C: 6 Glc3	H-1, 3 Glc3
6	61.8 CH$_2$	4.39 d (11.2)		H-4 Glc3
		4.06 m	C: 5 Glc3	
Glc4 (1 → 4Xyl1)				
1	102.2 CH	4.87 d (7.8)	C: 4 Xyl1	H-4 Xyl1; H-3, 5 Glc4
2	73.8 CH	3.93 t (8.9)	C: 1 Glc4	
3	**82.9** CH	4.37 t (8.9)	C: 2, 4 Glc4; 1 MeGlc5	H-1 MeGlc5; H-1 Glc4
4	*75.6* CH	4.78 t (8.9)	C: 3, 5, 6 Glc4	
5	74.5 CH	4.27 t (8.9)		H-1 Glc4
6	*68.2* CH$_2$	5.49 m		
		4.71 dd (8.9; 11.2)		
MeGlc5 (1 → 3Glc4)				
1	104.4 CH	5.20 d (7.8)	C: 3 Glc4	H-3 Glc4; H-3,5 MeGlc5
2	74.3 CH	3.99 t (8.9)	C: 1, 3 MeGlc5	
3	86.9 CH	3.65 t (8.9)	C: 2, 4 MeGlc5; OMe	H-1, 5 Me Glc5; OMe
4	70.0 CH	3.91 t (8.9)	C: 3, 5 MeGlc5	
5	77.4 CH	3.86 m		H-1 MeGlc5
6	62.0 CH$_2$	4.34 d (12.3)		H-4 MeGlc5
		4.09 dd (6.7; 12.3)	C: 5 MeGlc5	
OMe	60.5 CH$_3$	3.76 s	C: 3 MeGlc5	

[a] Recorded at 176.04 MHz in C_5D_5N/D_2O (4/1). [b] Bold = interglycosidic positions. [c] Italic = sulfate position. [d] Recorded at 700.13 MHz in C_5D_5N/D_2O (4/1). Multiplicity by 1D TOCSY. The original spectra of **1** are provided in Figures S1–S7.

Noticeably, such architecture of carbohydrate chains is not common for the holothuroid glycosides, but similar sugar moieties have been found in some glycosides of recently studied species of sea cucumbers: *Thyonidium kurilensis* [43] and *Psolus chitonoides* [6] (order Dendrochirotida).

The availability of two sulfate groups in the sugar moiety of **1** was deduced on the basis of shifting effects observed in its ^{13}C NMR spectrum. These were the signals of two hydroxy methylene groups of glucopyranose residues at δ_C 61.8 (C-6 Glc3) and 62.0 (C-6 MeGlc5), indicating the absence of sulfate groups in these positions and one signal at δ_C 68.2 (deshielded due to α-shifting effect of sulfate group) corresponding to sulfated at C-6 Glc4 residue. Additional shifting effects of the sulfate group became evident when the ^{13}C NMR spectrum of **1** was compared with the spectrum of the carbohydrate part of kuriloside A$_1$ [43]. The signals of all monosaccharides in the spectra of these glycosides were close to each other, with the exception of the signals of glucose residue in the upper semi-chain (Glc4). The signals of C-3 Glc4 and C-5 Glc4 were shielded in the spectrum of **1** (to δ_C 82.9 and 74.5, correspondingly) in comparison with the same signals in the spectrum of kuriloside A$_1$ (δ_C 86.9 and 75.7, correspondingly) due to the β-shifting effect of the sulfate group, which was attached to C-4 Glc4 of **1**. This was confirmed by an α-shifting effect: the signal of C-4 Glc4 in the spectrum of chilensoside A (**1**) was deshielded (δ_C 75.6) when compared with the signal of C-4 Glc4 in the spectrum of kuriloside A$_1$ (δ_C 69.6). Therefore, two sulfate groups were attached to one monosaccharide unit (Glc4) in the sugar chain of **1**. Such a structural feature was also recently found in psolusoside P from *Psolus fabricii* [44]. However, chilensoside A (**1**) is a new combination of some unusual structural features: aglycone side chain structure, carbohydrate chain architecture and the positions of sulfate groups.

The (−)ESI-MS/MS of **1** (Figure S8) demonstrated the fragmentation of $[M_{2Na}-Na]^-$ ion at *m/z* 1443.5 with ion peaks observed at *m/z* 1179.5 $[M_{2Na}-Na-Glc-SO_3Na+2H]^-$, 1135.5 $[M_{2Na}-Na-Glc-Qui+H]^-$, 1010.4 $[M_{2Na}-Na-MeGlc-2HSO_4Na]^-$, 417.1 $[M_{2Na}-Na-MeGlc-Glc(OSO_3Na)_2-Agl]^-$, and 255.0 $[M_{2Na}-Na-MeGlc-Glc(OSO_3Na)_2-Glc-Agl]^-$, corroborating the sequence of monosaccharides and the aglycone structure of **1**.

These data indicate that chilensoside A (**1**) is 3β-*O*-{β-D-glucopyranosyl-(1 → 4)-β-D-quinovopyranosyl-(1 → 2)-[3-*O*-methyl-β-D-glucopyranosyl-(1 → 3)-4,6-*O*-sodium disulfate-β-D-glucopyranosyl-(1 → 4)]-β-D-xylopyranosyl}-16-oxo,25-hydroxyholosta-9(11), 23*Z*(24)-diene.

The aglycones of chilensosides A$_1$ (**2**), B (**3**), C (**4**) and D (**5**) (Tables 3 and S2–S4, Figures S9–S14, S17–S22, S25–S30 and S33–S38) were identical to each other and to those of cladoloside A$_4$ [42] and psolusoside D$_1$ [45]. This holostane aglycone has the same polycyclic system as **1** and differs in the side chain structure with a 24(25)-double bond.

The molecular formula of chilensoside A$_1$ (**2**) was determined to be $C_{60}H_{92}O_{33}S_2Na_2$ from the $[M_{2Na}-Na]^-$ ion peak at *m/z* 1427.4928 (calc. 1427.4865), and $[M_{2Na}-2Na]^{2-}$ ion peak at *m/z* 702.2510 (calc. 702.2487) in the (−)HR-ESI-MS (Figure S16). The (−)ESI-MS/MS of **2** (Figure S16) demonstrated the fragmentation of $[M_{2Na}-Na]^-$ ion at *m/z* 1427.5, *m/z*: 1120.5 $[M_{2Na}-Na-Glc-Qui+H]^-$, 915.4 $[M_{2Na}-Na-Glc-Qui-2SO_3Na+3H]^-$, 667.1 $[M_{2Na}-Na-Glc-Qui-Agl]^-$, 417.1 $[M_{2Na}-Na-MeGlc-Glc(OSO_3Na)_2-Agl]^-$.

All these data indicate that chilensoside A$_1$ (**2**) is 3β-*O*-{β-D-glucopyranosyl-(1 → 4)-β-D-quinovopyranosyl-(1 → 2)-[3-*O*-methyl-β-D-glucopyranosyl-(1 → 3)-4,6-*O*-sodium disulfate-β-D-glucopyranosyl-(1 → 4)]-β-D-xylopyranosyl}-16-oxoholosta-9(11),24(25)-diene.

The molecular formula of chilensoside B (**3**) was determined to be $C_{60}H_{92}O_{33}S_2Na_2$ from the $[M_{2Na}-Na]^-$ ion peak at *m/z* 1427.4881 (calc. 1427.4865), and $[M_{2Na}-2Na]^{2-}$ ion peak at *m/z* 702.2499 (calc. 702.2487) in the (−)HR-ESI-MS (Figure S24). The ^1H and ^{13}C NMR spectra of the carbohydrate chain of chilensoside B (**3**) (Table 4, Figures S17–S23) demonstrated five characteristic doublets of anomeric protons at δ_H 4.66–5.18 (*J* = 7.1–8.1 Hz) and five signals of anomeric carbons at δ_C 102.3–104.7, indicating the presence of a pentasaccharide chain and β-configurations of glycosidic bonds. The extensive analysis of the ^1H, ^1H-COSY, 1D TOCSY, HSQC, ROESY and HMBC spectra (Table 4, Figures S17–S23) of **3** indicated the same monosaccharide composition, positions of glycosidic linkages, and architecture established for the glycosides **1**, **2**. The differences in the chemical shifts of carbon signals of chilensosides A (**1**) and B (**3**) were attributed to the diverse positions of sulfate groups. The signal of C-4 Glc4 in the ^{13}C NMR spectrum of **3** was shielded to δ_C 68.9 instead of δ_C 75.6 in **1** due to the absence of a sulfate group in this position of **3**. Additionally, the signal of C-3 Glc4 was deshielded to 85.9 due to the glycosylation effect and the absence of the β-shifting effect of sulfate group. The signal of C-6 Glc4 at δ_C 67.2 was characteristic for the sulfated hydroxy methylene group of the glucopyranose unit. Therefore, the glucose residue attached to C-4 Xyl1 of the carbohydrate chain of **3** bears one sulfate group at C-6. The comparison of the signals assigned to carbons of the 3-*O*-methylglucose unit of the compounds **3** (Table 4) and **1** (Table 2) showed that the signal of C-4 MeGlc5 of **3** was deshielded by 6.1 ppm (to δ_C 76.1) and the signals of C-3 and

C-5 MeGlc5 were shielded by 1.7 and 1.0 ppm, corresponding to the shifting effects of the sulfate group attached to C-4 MeGlc5 of chilensoside B (**3**). Thus, the glycoside **3** is a new disulfated pentaoside having sulfate groups at C-6 Glc4 and C-4 MeGlc5. The compound with identical positions of sulfates but differing in the terminal xylose residue in the bottom semi-chain was chitonoidoside H, found recently in the sea cucumber *Psolus chitonoides* [6].

Table 3. ^{13}C and ^{1}H NMR chemical shifts, HMBC and ROESY correlations of aglycone moiety of chilensoside A_1 (**2**).

Position	δ_C Mult. [a]	δ_H Mult. (*J* in Hz) [b]	HMBC	ROESY
1	36.0 CH$_2$	1.72 m		H-11
		1.31 m		
2	26.7 CH$_2$	2.07 m		
		1.85 m		
3	88.6 CH	3.13 dd (5.0; 12.3)		H-5, H-31, H1-Xyl1
4	39.5 C			
5	52.7 CH	0.79 brd (11.2)		H-1, H-3, H-7, H-31
6	20.8 CH$_2$	1.58 m		
		1.39 m		H-19, H-30
7	28.2 CH$_2$	1.58 m		H-15
		1.16 m		
8	38.6 CH	3.13 m		H-19
9	151.0 C			
10	39.6 C			
11	111.2 CH	5.27 m		H-1
12	32.0 CH$_2$	2.64 brd (15.5)	C: 13, 18	H-32
		2.48 brd (15.5)		H-21
13	55.9 C			
14	42.0 C			
15	51.9 CH$_2$	2.39 d (16.4)	C: 13, 16	H-7, H-32
		2.09 m		
16	214.6 C			
17	61.3 CH	2.88 s	C: 16, 18, 21	H-12, H-21, H-32
18	176.7 C			
19	21.9 CH$_3$	1.26 s	C: 1, 5, 9, 10	H-1, H-2, H-8, H-30
20	83.4 C			
21	26.6 CH$_3$	1.47 s	C: 17, 20, 22	H-12, H-17, H-23
22	38.6 CH$_2$	1.80 m		
		1.58 m		
23	22.9 CH$_2$	2.27 m		H-21
		2.01 m		
24	123.8 CH	5.03 m	C: 27	H-26
25	132.2 C			
26	25.4 CH$_3$	1.56 s	C: 24, 25, 27	H-24
27	17.5 CH$_3$	1.52 s	C: 24, 25, 26	H-23
30	16.5 CH$_3$	0.95 s	C: 3, 4, 5, 31	H-2, H-6, H-19, H-31
31	27.9 CH$_3$	1.13 s	C: 3, 4, 5, 30	H-3, H-5, H-6, H-30, H-1 Xyl1
32	20.5 CH$_3$	0.89 s	C: 8, 13, 14, 15	H-7, H-12, H-15, H-17

[a] Recorded at 125.67 MHz in C_5D_5N/D_2O (4/1). [b] Recorded at 500.12 MHz in C_5D_5N/D_2O (4/1). The original spectra of **2** are provided Figures S9–S15.

Table 4. ^{13}C and ^{1}H NMR chemical shifts and HMBC and ROESY correlations of carbohydrate moiety of chilensoside B (**3**).

Atom	δ_C Mult. [a][b][c]	δ_H Mult. (*J* in Hz) [d]	HMBC	ROESY
Xyl1 (1 → C-3)				
1	104.7 CH	4.66 d (7.9)	C: 3	H-3; H-3, 5 Xyl1
2	**82.0** CH	3.95 t (7.9)	C: 1 Qui2; 1, 3 Xyl1	H-1 Qui2
3	75.0 CH	4.15 t (7.9)	C: 4 Xyl1	H-1, 5 Xyl1
4	**78.2** CH	4.14 m	C: 3 Xyl1	H-1 Glc4
5	63.4 CH$_2$	4.37 dd (5.3; 11.8)	C: 1, 3 Xyl1	
		3.62 m		H-1 Xyl1
Qui2 (1 → 2Xyl1)				
1	104.5 CH	5.02 d (7.1)	C: 2 Xyl1	H-2 Xyl1; H-3, 5 Qui2
2	75.6 CH	3.88 t (8.9)	C: 1, 3 Qui2	H-4 Qui2
3	75.0 CH	4.00 t (8.9)	C: 2, 4 Qui2	H-1, 5 Qui2
4	**86.2** CH	3.52 t (8.9)	C: 1 Glc3; 3, 5 Qui2	H-1 Glc3; H-2 Qui2
5	71.4 CH	3.69 m		H-1, 3 Qui2
6	17.8 CH$_3$	1.62 d (5.1)	C: 4, 5 Qui2	H-4 Qui2
Glc3 (1 → 4Qui2)				
1	104.4 CH	4.81 d (7.8)	C: 4 Qui2	H-4 Qui2; H-3, 5 Glc3
2	74.4 CH	3.87 t (8.6)	C: 1, 3 Glc3	
3	77.2 CH	4.13 t (8.6)	C: 2, 4 Glc3	H-1, 5 Glc3
4	70.8 CH	3.93 m	C: 3 Glc3	
5	77.6 CH	3.92 m		H-1, 3 Glc3
6	61.8 CH$_2$	4.39 d (11.2)		
		4.06 dd (4.3; 11.2)	C: 5 Glc3	
Glc4 (1 → 4Xyl1)				
1	102.3 CH	4.89 d (8.1)	C: 4 Xyl1	H-4 Xyl1; H-3, 5 Glc4
2	73.2 CH	3.83 t (9.3)	C: 1, 3 Glc4	
3	**85.9** CH	4.17 t (9.3)	C: 1 MeGlc5; 2, 4 Glc4	H-1 MeGlc5; H-1 Glc4
4	68.9 CH	3.86 t (9.3)	C: 3, 5, 6 Glc4	H-6 Glc4
5	75.1 CH	4.05 t (9.3)		H-1 Glc4
6	*67.2* CH$_2$	4.95 d (11.2)		
		4.65 brd (11.2)	C: 5 Glc4	
MeGlc5 (1 → 3Glc4)				
1	104.3 CH	5.18 d (7.5)	C: 3 Glc4	H-3 Glc4; H-3,5 MeGlc5
2	74.0 CH	3.86 t (9.3)	C: 1, 3 MeGlc5	H-4 MeGlc5
3	85.2 CH	3.71 t (9.3)	C: 2, 4 MeGlc5; OMe	H-1 Me Glc5; OMe
4	*76.1* CH	4.88 t (9.3)	C: 3, 5, 6 MeGlc5	H-2, 6 MeGlc5
5	76.4 CH	3.85 t (9.3)		H-1 MeGlc5
6	61.7 CH$_2$	4.50 d (11.2)		
		4.33 dd (5.6; 11.2)		
OMe	60.6 CH$_3$	3.93 s	C: 3 MeGlc5	

[a] Recorded at 176.04 MHz in C_5D_5N/D_2O (4/1). [b] Bold = interglycosidic positions. [c] Italic = sulfate position. [d] Recorded at 700.13 MHz in C_5D_5N/D_2O (4/1). Multiplicity by 1D TOCSY. The original spectra of **3** are provided in Figures S17–S23.

The (−)ESI-MS/MS of **3** (Figure S24) demonstrated the fragmentation of $[M_{2Na}-Na]^-$ ion at *m/z* 1427.5, resulting in the ion peaks appearance at *m/z* 1307.5 $[M_{2Na}-Na-NaHSO_4]^-$, 1149.5 $[M_{2Na}-Na-MeGlcOSO_3Na+H]^-$, 987.4 $[M_{2Na}-Na-MeGlcOSO_3Na-Glc+H]^-$, 841.4 $[M_{2Na}-Na-MeGlcOSO_3Na-Glc-Qui+H]^-$, 667.1 $[M_{2Na}-Na-Agl-Glc-Qui-H]^-$. The fragmentation of $[M_{2Na}-2Na]^{2-}$ ion at *m/z* 702.2 led to the ion peak at *m/z* 621.7 $[M_{2Na}-2Na-Glc]^{2-}$, and 548.2 $[M_{2Na}-2Na-Glc-Qui]^{2-}$, confirming the structure of **3**.

These data indicate that chilensoside B (**3**) is 3β-O-{β-D-glucopyranosyl-(1 → 4)-β-D-quinovopyranosyl-(1 → 2)-[4-*O*-sodium sulfate-3-*O*-methyl-β-D-glucopyranosyl-(1 → 3)-6-*O*-sodium sulfate-β-D-glucopyranosyl-(1 → 4)]-β-D-xylopyranosyl}-16-oxoholosta-9(11),24(25)-diene.

The molecular formula of chilensoside C (**4**) was determined to be $C_{60}H_{91}O_{36}S_3Na_3$ from the $[M_{3Na}-Na]^-$ ion peak at *m/z* 1529.4300 (calc. 1529.4253), $[M_{3Na}-2Na]^{2-}$ ion peak at *m/z* 753.2206 (calc. 753.2180) and $[M_{3Na}-3Na]^{3-}$ ion peak at *m/z* 494.4839 (calc. 494.4823) in the (−)HR-ESI-MS (Figure S32).

The ^{1}H and ^{13}C NMR spectra of the carbohydrate chain of chilensoside C (**4**) (Table 5, Figures S25–S31) demonstrated five characteristic doublets of anomeric protons at δ_H 4.65–5.21 (*J* = 6.5–8.5 Hz) and five signals of anomeric carbons at δ_C 102.4–104.7, indicating the presence of a pentasaccharide chain and *β*-configurations of glycosidic bonds.

Table 5. ^{13}C and ^{1}H NMR chemical shifts and HMBC and ROESY correlations of carbohydrate moiety of chilensoside C (**4**).

Atom	δ_C Mult. [a,b,c]	δ_H Mult. (*J* in Hz) [d]	HMBC	ROESY
Xyl1 (1 → C-3)				
1	104.7 CH	4.65 d (6.5)	C: 3	H-3; H-3, 5 Xyl1
2	**82.3** CH	3.89 t (8.4)	C: 1 Qui2; 1, 3 Xyl1	H-1 Qui2
3	75.0 CH	4.08 m	C: 4 Xyl1	H-1 Xyl1
4	**78.9** CH	4.07 m		H-1 Glc4
5	63.4 CH$_2$	4.31 m	C: 3 Xyl1	
		3.60 dd (9.3; 11.2)		H-1, 3 Xyl1
Qui2 (1 → 2Xyl1)				
1	104.6 CH	4.94 d (7.5)	C: 2 Xyl1	H-2 Xyl1; H-3, 5 Qui2
2	75.4 CH	3.89 t (9.3)	C: 1, 3 Qui2	H-4 Qui2
3	74.8 CH	4.04 t (9.3)	C: 2, 4 Qui2	H-1, 5 Qui2
4	**86.1** CH	3.51 t (9.3)	C: 1 Glc3; 3, 5 Qui2	H-1 Glc3; H-2 Qui2
5	71.5 CH	3.68 dd (6.5; 9.3)		H-1, 3 Qui2
6	17.7 CH$_3$	1.60 d (6.5)	C: 4, 5 Qui2	H-4 Qui2
Glc3 (1 → 4Qui2)				
1	104.4 CH	4.82 d (8.5)	C: 4 Qui2	H-4 Qui2; H-3, 5 Glc3
2	73.1 CH	3.90 t (8.5)	C: 3 Glc3	
3	*84.3* CH	5.03 t (8.5)	C: 2, 4 Glc3	H-1, 5 Glc3
4	69.8 CH	3.93 m	C: 3, 5 Glc3	
5	77.1 CH	3.93 m		H-1, 3 Glc3
6	61.5 CH$_2$	4.35 brd (11.7)		
		3.99 d (11.7)	C: 5 Glc3	
Glc4 (1 → 4Xyl1)				
1	102.4 CH	4.85 d (8.5)	C: 4 Xyl1	H-4 Xyl1; H-3, 5 Glc4
2	73.7 CH	3.93 t (8.5)	C: 1, 3 Glc4	H-4 Glc4
3	**82.9** CH	4.37 t (8.5)	C: 1 MeGlc5; 2, 4 Glc4	H-1 MeGlc5; H-1, 5 Glc4
4	*75.6* CH	4.78 t (8.5)	C: 3, 5, 6 Glc4	H-2 Glc4
5	74.3 CH	4.28 t (8.5)	C: 4 Glc4	H-1, 3 Glc4
6	*68.3* CH$_2$	5.50 brd (9.0)		
		4.70 brd (9.9)	C: 5 Glc4	
MeGlc5 (1 → 3Glc4)				
1	104.2 CH	5.21 d (7.7)	C: 3 Glc4	H-3 Glc4; H-3,5 MeGlc5
2	74.6 CH	4.00 t (8.6)	C: 1, 3 MeGlc5	
3	86.9 CH	3.65 t (8.6)	C: 2, 4 MeGlc5; OMe	H-1, 5 Me Glc5; OMe
4	70.0 CH	3.91 t (8.6)	C: 3, 5, 6 MeGlc5	
5	77.4 CH	3.87 t (8.6)		H-1, 3 MeGlc5
6	62.0 CH$_2$	4.35 brd (11.5)		H-4 MeGlc5
		4.09 dd (5.7; 11.5)	C: 5 MeGlc5	
OMe	60.3 CH$_3$	3.76 s	C: 3 MeGlc5	

[a] Recorded at 176.04 MHz in C$_5$D$_5$N/D$_2$O (4/1). [b] Bold = interglycosidic positions. [c] Italic = sulfate position. [d] Recorded at 700.13 MHz in C$_5$D$_5$N/D$_2$O (4/1). Multiplicity by 1D TOCSY. The original spectra of **4** are provided in Figures S25–S31.

The extensive analysis of the ^{1}H, ^{1}H-COSY, 1D TOCSY, HSQC, ROESY, and HMBC spectra of **4** indicated the same monosaccharide composition, glycosidic bond locations and architecture of carbohydrate chains as in the previously discussed glycosides **1**−**3**. Differences were found in the quantity of sulfate groups, which was also confirmed by MS data, where three-charged ions were registered, indicating the presence of three sulfate groups.

The comparison of the ^{13}C NMR spectra of sugar moieties of **4** and **1** showed the coincidence of all the signals except the signals of glucose residue in the bottom semi-chain. The signal of C-3 Glc3 was deshielded to δ_C 84.3 in the spectrum of **4**, which could be explained by the α-shifting effect of the sulfate group as well as by the glycosylation effect. However, the latter was excluded due to the absence of the ROE- and HMBC correlations of H-3 Glc3 with any protons or carbons of neighboring monosaccharide residues (Table 5). Moreover, the signals of C-2 Glc3 and C-4 Glc3 in the spectrum of **4** were shielded to δ_C 73.1 and 69.8, respectively, in comparison with the corresponding signals in the spectrum of **1** due to β-shifting effect of sulfate group at C-3 Glc3. Therefore, the third sulfate group in chilensoside C (**4**) was unique for the glycosides position at C-3 Glc3 instead of the characteristic glycosidic bond position in the glycosides with normal (consisting of three monosaccharide units) bottom semi-chain. Such a location of the sulfate group makes further elongation of the carbohydrate chain of **4** impossible. The rest of the sulfate groups were attached to C-4 Glc4 and C-6 Glc4 in chilensoside C (**4**), by the same manner as in chilensosides A (**1**), and A$_1$ (**2**).

The (−)ESI-MS/MS of **4** (Figure S32) demonstrated the fragmentation of $[M_{3Na}-Na]^-$ ion at m/z 1529.5, which resulted in the ion peaks at m/z 1015.4 $[M_{3Na}-Na-GlcOSO_3Na-Qui-SO_3Na]^-$, 987.4 $[M_{3Na}-Na-MeGlc-Glc(OSO_3Na)_2]^-$, 605.2 $[M_{3Na}-Na-MeGlc-NaHSO_4]^-$. The fragmentation of $[M_{3Na}-2Na]^{2-}$ ion at m/z 753.2 led to the presence of the ion peaks at m/z 702.2 $[M_{3Na}-2Na-SO_3Na]^{2-}$, 605.2 $[M_{3Na}-2Na-MeGlc-NaHSO_4]^{2-}$.

These data indicate that chilensoside C (**4**) is 3β-O-{3-O-sodium sulfate-β-D-glucopyranosyl-(1 → 4)-β-D-quinovopyranosyl-(1 → 2)-[3-O-methyl-β-D-glucopyranosyl-(1 → 3)-4,6-O-sodium disulfate-β-D-glucopyranosyl-(1 → 4)]-β-D-xylopyranosyl}-16-oxoholosta-9(11),24(25)-diene.

The molecular formula of chilensoside D (**5**) was determined to be $C_{60}H_{90}O_{39}S_4Na_4$ from the $[M_{4Na}-Na]^-$ ion peak at m/z 1631.3667 (calc. 1631.3641), $[M_{4Na}-2Na]^{2-}$ ion peak at m/z 804.1886 (calc. 804.1874), $[M_{4Na}-3Na]^{3-}$ ion peak at m/z 528.4631 (calc. 528.4619) and $[M_{4Na}-4Na]^{4-}$ ion peak at m/z 390.6005 (calc. 390.5991) in the (−) HR-ESI-MS (Figure S40). Chilensoside D (**5**), analogously to compounds **1–4**, has a pentasaccharide branched by C-4 Xyl1 chain consisting of xylose, quinovose, two glucose and 3-O-methylglucose residues deduced from thorough analysis of its 1D and 2D NMR spectra (Table 6, Figures S33–S39). The availability of four-charged ion peaks in the ESI-MS spectra of **5** indicated four sulfate groups are present in its carbohydrate chain. The analysis of 1D TOCSY spectrum corresponding to Glc3 showed strongly deshielded signals of protons of the hydroxy methylene group at δ_H 4.61 (m) and 5.00 (d, J = 11.9 Hz), which were assigned to the corresponding carbon signal at δ_C 67.6. These data indicate that the glucose residue in the bottom semi-chain was sulfated by C-6. The glucose unit (Glc4) attached to C-4 Xyl1 in chilensoside D (**5**) had two sulfate groups at C-4 Glc4 and C-6 Glc4, deduced from the deshielding of its signals to δ_C 75.1 and 68.5, respectively. The fourth sulfate group was positioned at C-6 MeGlc5 because of the deshielding of the signals of hydroxy methylene group to δ_C 67.0 and δ_H 4.99 (brd, J = 11.9 Hz); 4.78 (dd, J = 5.1; 11.9 Hz). Therefore, chilensoside D (**5**) is a new, sixth tetrasulfated glycoside found in sea cucumbers [7,44].

Table 6. ^{13}C and ^{1}H NMR chemical shifts and HMBC and ROESY correlations of carbohydrate moiety of chilensoside D (**5**).

Atom	δ_C Mult. [a,b,c]	δ_H Mult. (J in Hz) [d]	HMBC	ROESY
Xyl1 (1 → C-3)				
1	104.7 CH	4.63 d (8.0)	C: 3	H-3; H-5 Xyl1
2	**82.7** CH	3.74 t (8.0)	C: 1 Qui2; 1, 3 Xyl1	H-1 Qui2
3	75.5 CH	4.01 t (8.0)	C: 2, 4 Xyl1	
4	**80.7** CH	3.97 t (8.0)		H-1 Glc4
5	63.5 CH$_2$	4.40 dd (5.3; 11.5)		
		3.64 dd (8.0; 11.5)		H-1 Xyl1
Qui2 (1 → 2Xyl1)				
1	104.5 CH	4.77 d (8.6)	C: 2 Xyl1	H-2 Xyl1; H-3, 5 Qui2
2	75.2 CH	3.91 t (8.6)	C: 1, 3 Qui2	H-4 Qui2
3	74.4 CH	4.04 t (8.6)	C: 4 Qui2	H-1 Qui2
4	**86.1** CH	3.29 t (8.6)	C: 1 Glc3; 3 Qui2	H-1 Glc3; H-2 Qui2
5	71.6 CH	3.62 t (8.6)		H-1 Qui2
6	17.8 CH$_3$	1.55 d (6.1)	C: 4, 5 Qui2	H-4 Qui2
Glc3 (1 → 4Qui2)				
1	104.6 CH	4.64 d (7.9)	C: 4 Qui2	H-4 Qui2; H-3 Glc3
2	73.9 CH	3.75 t (8.6)	C: 1, 3 Glc3	
3	76.8 CH	4.10 t (8.6)	C: 2, 4 Glc3	H-1 Glc3
4	70.6 CH	3.81 t (8.6)	C: 3, 5, 6 Glc3	
5	75.1 CH	4.07 m		H-1 Glc3
6	67.6 CH$_2$	5.00 d (11.9)		
		4.61 m		
Glc4 (1 → 4Xyl1)				
1	103.4 CH	4.80 d (7.4)	C: 4 Xyl1	H-4 Xyl1; H-3, 5 Glc4
2	73.3 CH	3.95 t (8.6)	C: 1, 3 Glc4	
3	**79.4** CH	4.50 t (8.6)	C: 1 MeGlc5; 2, 4 Glc4	H-1 MeGlc5; H-1, 5 Glc4
4	*75.1* CH	4.78 m		
5	73.8 CH	4.28 t (8.6)		H-1, 3 Glc4
6	*68.5* CH$_2$	5.46 dd (7.5; 12.5)		
		4.66 m		
MeGlc5 (1 → 3Glc4)				
1	101.8 CH	5.33 d (7.6)	C: 3 Glc4	H-3 Glc4; H-3,5 MeGlc5
2	73.5 CH	4.00 t (7.6)	C: 1, 3 MeGlc5	
3	86.4 CH	3.62 t (7.6)	C: 4 MeGlc5; OMe	H-1, 5 Me Glc5; OMe
4	69.5 CH	4.01 t (7.6)	C: 5 MeGlc5	
5	75.7 CH	4.01 t (7.6)		H-1, 3 MeGlc5
6	*67.0* CH$_2$	4.99 brd (11.9)		
		4.78 dd (5.1; 11.9)		
OMe	60.4 CH$_3$	3.75 s	C: 3 MeGlc5	

[a] Recorded at 125.67 MHz in C$_5$D$_5$N/D$_2$O (4/1). [b] Bold = interglycosidic positions. [c] Italic = sulfate position. [d] Recorded at 500.12 MHz in C$_5$D$_5$N/D$_2$O (4/1). Multiplicity by 1D TOCSY. The original spectra of **5** are provided in Figures S33–S39.

The (−) ESI-MS/MS of chilensoside D (**5**) (Figure S40) demonstrated the fragmentation of [M$_{4Na}$−Na]$^-$ ion at *m/z* 1631.5, leading to the presence of the ion peak at *m/* 987.4 [M$_{4Na}$−Na−MeGlcOSO$_3$Na−Glc(OSO$_3$Na)$_2$]$^-$, of [M$_{4Na}$−2Na]$^{2-}$ ion at *m/z* 804.2 leading to the ion peak at *m/z* 753.2 [M$_{4Na}$−2Na−SO$_3$Na+H]$^{2-}$, and of [M$_{4Na}$−3Na]$^{3-}$ ion at *m/z* 528.5 leading to the ion peak at *m/z* 494.3 [M$_{4Na}$−3Na−SO$_3$Na+H]$^{3-}$.

These data indicate that chilensoside D (**5**) is 3β-*O*-{6-*O*-sodium sulfate-β-D-glucopyranosyl-(1 → 4)-β-D-quinovopyranosyl-(1 → 2)-[6-*O*-sodium sulfate-3-*O*-methyl-β-D-glucopyranosyl-(1 → 3)-4,6-*O*-sodium disulfate-β-D-glucopyranosyl-(1 → 4)]-β-D-xylopyranosyl}-16-oxoholosta-9(11),24(25)-diene.

The structural peculiarities of the glycosides of *P. chilensis* showed similarity to the compounds of the representatives of the order Dendrochirotida, i.e., sea cucumbers of the species *Thyonidium kurilensis* and *Psolus chitonoides* (the same architecture of the carbohydrate chains), *Psolus fabricii* (attachment of sulfates to C-4 Glc4 and C-6 Glc4) and *Cladolabes schmeltzii* (the same aglycones). All these data significantly support the phylogenetic closeness of the order Molpadida to the order Dendrochirotida, rather than to the order Aspidochirotida (in accordance with the system of Pawson and Fell). This order is absent in the last revision of the system of the class Holothuroidea, and the families, which were part of it, are now included in the orders Holothuriida, Persiculida and Synallactida [46]. The obtained structural data are in good agreement with the phylogenetic study of Holothuroidea using a multi-gene approach, which showed poor support of Molpadida as a sister group to Synallactida but demonstrated the close relationship of Molpadida to Dendrochirotida [46].

2.2. Bioactivity of the Glycosides

Cytotoxic activity of chilensosides A–D (**1–5**) against human cells, including erythrocytes and cancer cell lines SH-SY5Y, HeLa, DLD-1, HL-60, and THP-1, was studied. The earlier tested chitonoidoside L [7] was used as the positive control (Table 7).

Table 7. The cytotoxic activities of glycosides **1–5**, and chitonoidoside L (positive control) against human erythrocytes, and SH-SY5Y, HeLa, DLD-1, HL-60, THP-1 human cell lines.

Glycosides	ED_{50}, μM, Erythrocytes	Cytotoxicity, IC_{50} μM				
		SH-SY5Y	HeLa	DLD-1	HL-60	THP-1
Chilensoside A (**1**)	4.85 ± 0.10	>100.00	>100.00	>100.00	84.90 ± 2.96	75.29 ± 2.12
Chilensoside A_1 (**2**)	1.45 ± 0.12	38.78 ± 0.08	21.80 ± 0.22	29.78 ± 2.34	22.78 ± 0.60	30.26 ± 1.60
Chilensoside B (**3**)	0.96 ± 0.01	30.73 ± 2.65	20.34 ± 0.32	45.08 ± 1.63	5.68 ± 0.05	19.57 ± 0.75
Chilensoside C (**4**)	1.18 ± 0.05	26.43 ± 0.91	17.79 ± 0.56	35.44 ± 1.76	14.21 ± 1.23	18.59 ± 0.79
Chilensoside D (**5**)	10.26 ± 0.53	>100.00	>100.00	>100.00	64.74 ± 3.63	56.23 ± 2.74
Chitonoidoside L	1.24 ± 0.02	7.58 ± 0.13	10.88 ± 0.04	11.48 ± 0.25	6.96 ± 0.44	9.00 ± 0.75

The less active compounds in the series were chilensosides A (**1**) and D (**5**). The first of these substances has a hydroxyl group in the aglycone side chain, which is the cause of the decrease in its membranolytic activity [3]. In fact, its structural analog chilensoside A_1 (**2**), with the same carbohydrate chain and aglycone without the OH-group, demonstrated high hemolytic and cytotoxic effects against all tested cell lines. Chilensoside D (**5**) is a tetrasulfated glycoside that itself is not the cause of activity depletion, because it is known that tetrasulfated hexaosides from *P. chitonoides*, chitonoidosides K and L, were significantly active [7]. The combination of sulfate group positions in **5**, especially at C-6 Glc3 and C-6 MeGlc5, probably negatively affected the activity. Disulfated chilensosides A_1 (**2**), B (**3**) and trisulfated chilensoside C (**4**) displayed similar cytotoxicity. The differing sulfate groups in these glycosides were attached to C-4 or C-3 of glucopyranose units while C-6 positions of terminal sugar residues were free from sulfation.

The differential sensitivity of the cell lines in relation to the cytotoxic action of sea cucumber glycosides depended both on the glycoside's chemical structures and the composition of cellular membranes [47]. In the current tests, erythrocytes were, as usual, more sensitive than cancer cells to the action of the glycosides, but leukemia cells (promyeloblast HL-60 and monocytic THP-1) displayed increased sensitivity compared to the other cancer cells.

Therefore, three of the five glycosides isolated from *P. chilensis* demonstrated high hemolytic and moderate cytotoxic activities against cancer cells. These data, along with the previous investigations of highly polar tri- and tetrasulfated glycosides [7], indicate the possible potential of these water-soluble compounds to be used as anticancer drugs.

3. Materials and Methods

3.1. General Experimental Procedures

Specific rotation, PerkinElmer 343 Polarimeter (PerkinElmer, Waltham, MA, USA); NMR, Bruker AMX 500 (Bruker BioSpin GmbH, Rheinstetten, Germany) (500.12/125.67 MHz (^{1}H/^{13}C) spectrometer; Bruker AVANCE III-700 spectrometer at 700.13 MHz/176.04 MHz (1H/13C); ESI MS (positive and negative ion modes), Agilent 6510 Q-TOF apparatus (Agilent Technology, Santa Clara, CA, USA), sample concentration 0.01 mg/mL; HPLC, Agilent 1260 Infinity II with a differential refractometer (Agilent Technology, Santa Clara, CA, USA); columns Supelcosil LC-Si (4.6 × 150 mm, 5 μm) and Discovery HS F5-5 (10 × 250 mm, 5 μm) (Supelco, Bellefonte, PA, USA), Diasfer 110 C-8 (4.6 × 250 mm, 5 μm) (Biochemmack, Moscow, Russia).

3.2. Animals and Cells

Specimens of the sea cucumber *Paracaudina chilensis* (family Cuadinidae; order Molpadida) were collected in the Troitsa bay, Japan sea in August 2019 by scuba diving from 2–5 m depth. The animals were taxonomically determined by Boris B. Grebnev. Voucher specimens are kept in G.B. Elyakov PIBOC FEB RAS, Vladivostok, Russia.

Human erythrocytes were purchased from the Station of Blood Transfusion in Vladivostok. The cells of human adenocarcinoma line HeLa were provided by the N.N. Blokhin National Medicinal Research Center of Oncology of the Ministry of Health Care of the Russian Federation, (Moscow, Russia). The cells of human colorectal adenocarcinoma line DLD-1 CCL-221™, human promyeloblast cell line HL-60 CCL-240, human monocytic THP-1 TIB-202™ cells and human neuroblastoma line SH-SY5Y CRL-2266™ were received from ATCC (Manassas, VA, USA). HeLa cell line was cultured in DMEM (Gibco Dulbecco's Modified Eagle Medium) with 1% penicillin/streptomycin sulfate (Biolot, St. Petersburg, Russia) and 10% fetal bovine serum (FBS) (Biolot, St. Petersburg, Russia). The cells of DLD-1, HL-60, and THP-1 lines were cultured in RPMI medium with 1% penicillin/streptomycin (Biolot, St. Petersburg, Russia) and 10% fetal bovine serum (FBS) (Biolot, St. Petersburg, Russia). All the cells were incubated at 37 °C in a humidified atmosphere with 5% (v/v) CO_2. SH-SY5Y were cultured MEM (Minimum Essential Medium) with 1% penicillin/streptomycin sulfate (Biolot, St. Petersburg, Russia) and with fetal bovine serum (Biolot, St. Petersburg, Russia) to a final concentration of 10%.

This study was conducted according to the guidelines of the Declaration of Helsinki and was approved by the Ethics Committee of the Pacific Institute of Bioorganic Chemistry (Protocol No. 0037.12.03.2021).

3.3. Extraction and Isolation

The sea cucumbers (36 specimens) were kept in EtOH at +4 °C. Then, they were minced by cutting into pieces and extracted with refluxing EtOH (2 L vol.) for 4 hrs. The extract was concentrated to dryness in vacuum, dissolved in H2O, and chromatographed on a Polychrom-1 column (powdered Teflon, Biolar, Latvia). We first eluted the inorganic salts and impurities with H2O and then the glycosides with 50% EtOH to give 1300 mg of crude glycoside fraction. This was subjected to column chromatography on Si gel using the stepwise gradient of solvent systems CHCl3/EtOH/H2O: 100:100:17 → 100:125:25 → 100:150:50 as mobile phase as the first stage of purification. Three fractions, I (377 mg), II (378 mg) and III (338 mg), were obtained. Each of them was subsequently rechromatographed on an Si gel column using the solvent system CHCl3/EtOH/H2O (100:125:25) as mobile phase, resulting in the isolation of subfractions: I.0 (22 mg), I.1 (120 mg), II (286 mg), III.1 (66 mg) and III.2 (177 mg). HPLC of the subfraction I.0 on silica-based column Supelcosil LC-Si (4.6 × 150 mm, 5 μm) with $CHCl_3$/MeOH/H_2O (55/30/4) as mobile phase resulted in the isolation of three fractions (I.0.1–I.0.3). The subsequent HPLC of fraction I.0.3 on Supelco Discovery HS F5-5 (10 × 250 mm) column with MeOH/H_2O/NH_4OAc (1 M water solution), ratio (50/48.5/1.5), as mobile phase led to the isolation of 5.5 mg of chilensoside A (**1**). HPLC of the subfraction I.1 on the silica-based column Supelcosil LC-Si (4.6 × 150 mm,

5 µm) in the same conditions used for the subtraction I.0 resulted in the isolation of three fractions (I.1.1–I.1.3). The HPLC of the fraction I.1.3 on Supelco Discovery HS F5-5 (10 × 250 mm) column with $MeOH/H_2O/NH_4OAc$ (1 M water solution) (65/33.5/1.5) as mobile phase led to isolation of individual chilensosides A_1 (**2**) (2.2 mg) and B (**3**) (4.8 mg) as well as the subfraction I.1.3.1. The repeated HPLC of the latter on the same column with $MeOH/H_2O/NH_4OAc$ (1 M water solution) (55/43.5/1.5) as mobile phase resulted in the obtaining of 3.8 mg of chilensoside C (**4**). The HPLC of the subfraction III.1 on silica-based column Supelcosil LC-Si (4.6 × 150 mm, 5 µm) with $CHCl_3/MeOH/H_2O$ (55/25/3) as mobile phase followed by HPLC of the obtained fraction on Diasfer 110 C-8 (4.6 × 250 mm) column with $MeOH/H_2O/NH_4OAc$ (1 M water solution) (49/49/2) as mobile phase gave three fractions III.1.1–III.1.3. The rechromatography of III.1.3 on Diasfer 110 C-8 (4.6 × 250 mm) column with $MeOH/H_2O/NH_4OAc$ (1 M water solution) (50/48/2) led to isolation of 2.2 mg of chilensoside D (**5**).

3.3.1. Chilensoside A (**1**)

Colorless powder; $[\alpha]_D^{20} -48°$ (*c* 0.1, 50% MeOH). NMR: See Tables 1 and 2, Figures S1–S8. (−)HR-ESI-MS *m/z*: 1443.4800 (calc. 1443.4815) $[M_{2Na}-Na]^-$, 710.2466 (calc. 710.2461) $[M_{2Na}-2Na]^{2-}$. (−)ESI-MS/MS *m/z*: 1179.5 $[M_{2Na}-Na-C_6H_{11}O_5\ (Glc)-SO_3Na+2H]^-$, 1135.5 $[M_{2Na}-Na-C_6H_{11}O_5\ (Glc)-C_6H_{10}O_4\ (Qui)+H]^-$, 1010.4 $[M_{2Na}-Na-C_7H_{13}O_6\ (MeGlc)-2HSO_4Na]^-$, 417.1 $[M_{2Na}-Na-C_7H_{13}O_6\ (MeGlc)-C_6H_8O_{11}S_2Na_2\ (Glc(OSO_3Na)_2)-C_{30}H_{43}O_4\ (Agl)]^-$, and 255.0 $[M_{2Na}-Na-C_7H_{13}O_6\ (MeGlc)-C_6H_8O_{11}S_2Na_2\ (Glc(OSO_3Na)_2)-C_6H_{11}O_5\ (Glc)-C_{30}H_{43}O_4\ (Agl)]^-$.

3.3.2. Chilensoside A_1 (**2**)

Colorless powder; $[\alpha]_D^{20} -36°$ (*c* 0.1, 50% MeOH). NMR: See Tables 3 and S1, Figures S9–S16. (−)HR-ESI-MS *m/z*: 1427.4928 (calc. 1427.4865) $[M_{2Na}-Na]^-$, 702.2510 (calc. 702.2487) $[M_{2Na}-2Na]^{2-}$; (−)ESI-MS/MS *m/z*: 1119.5 $[M_{2Na}-Na-C_6H_{11}O_5\ (Glc)-C_6H_{10}O_4\ (Qui)+H]^-$, 915.4 $[M_{2Na}-Na-C_6H_{11}O_5\ (Glc)-C_6H_{10}O_4\ (Qui)-2SO_3Na+3H]^-$, 667.1 $[M_{2Na}-Na-C_6H_{11}O_5\ (Glc)-C_6H_{10}O_4\ (Qui)-C_{30}H_{43}O_3\ (Agl)]^-$, 417.1 $[M_{2Na}-Na-C_7H_{13}O_6\ (MeGlc)-C_6H_8O_{11}S_2Na_2\ (Glc(OSO_3Na)_2)-C_{30}H_{43}O_3\ (Agl)]^-$.

3.3.3. Chilensoside B (**3**)

Colorless powder; $[\alpha]_D^{20} -53°$ (*c* 0.1, 50% MeOH). NMR: See Tables 4 and S2, Figures S17–S24. (−)HR-ESI-MS *m/z*: 1427.4881 (calc. 1427.4865) $[M_{2Na}-Na]^-$, 702.2499 (calc. 702.2487) $[M_{2Na}-2Na]^{2-}$; (−)ESI-MS/MS *m/z*: 1307.5 $[M_{2Na}-Na-NaHSO_4]^-$, 1149.5 $[M_{2Na}-Na-C_7H_{12}O_8SNa\ (MeGlcOSO_3Na)+H]^-$, 987.4 $[M_{2Na}-Na-C_7H_{12}O_8SNa\ (MeGlcOSO_3Na)-C_6H_{11}O_5\ (Glc)+H]^-$, 841.4 $[M_{2Na}-Na-C_7H_{12}O_8SNa\ (MeGlcOSO_3Na)-C_6H_{11}O_5\ (Glc)-C_6H_{10}O_4\ (Qui)+H]^-$, 667.1 $[M_{2Na}-Na-C_{30}H_{43}O_3\ (Agl)-C_6H_{11}O_5\ (Glc)-C_6H_{10}O_4\ (Qui)-H]^-$, 621.7 $[M_{2Na}-2Na-C_6H_{11}O_5\ (Glc)]^{2-}$, and 548.2 $[M_{2Na}-2Na-C_6H_{11}O_5\ (Glc)-C_6H_{10}O_4\ (Qui)]^{2-}$.

3.3.4. Chilensoside C (**4**)

Colorless powder; $[\alpha]_D^{20} -56°$ (*c* 0.1, 50% MeOH). NMR: See Tables 5 and S3, Figures S25–S32. (−)HR-ESI-MS *m/z*: 1529.4300 (calc. 1529.4253) $[M_{3Na}-Na]^-$, 753.2206 (calc. 753.2180) $[M_{3Na}-2Na]^{2-}$, 494.4839 (calc. 494.4823) $[M_{3Na}-3Na]^{3-}$; (−)ESI-MS/MS *m/z*: 1015.4 $[M_{3Na}-Na-C_6H_{10}O_8SNa\ (GlcOSO_3Na)-C_6H_{10}O_4\ (Qui)-SO_3Na]^-$, 987.4 $[M_{3Na}-Na-C_7H_{13}O_5\ (MeGlc)-C_6H_8O_{11}S_2Na_2\ (Glc(OSO_3Na)_2)]^-$, 605.2 $[M_{3Na}-Na-C_7H_{13}O_5\ (MeGlc)-NaHSO_4]^-$, 702.2 $[M_{3Na}-2Na-SO_3Na]^{2-}$, 605.2 $[M_{3Na}-2Na-C_7H_{13}O_5\ (MeGlc)-NaHSO_4]^{2-}$.

3.3.5. Chilensoside D (**5**)

Colorless powder; $[\alpha]_D^{20} -61°$ (*c* 0.1, 50% MeOH). NMR: See Tables S4 and 6, Figures S33–S39. (−) HR-ESI-MS *m/z*: 1631.3667 (calc. 1631.3641) $[M_{4Na}-Na]^-$, 804.1886 (calc. 804.1874) $[M_{4Na}-2Na]^{2-}$, 528.4631 (calc. 528.4619) $[M_{4Na}-3Na]^{3-}$, 390.6005 (calc.

390.5991) $[M_{4Na}-4Na]^{4-}$ (Figure S40); $(-)$ ESI-MS/MS *m/z* 987.4 $[M_{4Na}-Na-C_7H_{12}O_8SNa$ (MeGlcOSO$_3$Na)$-C_6H_7O_{11}S_2Na_2$ (Glc(OSO$_3$Na)$_2$)]$^-$, 753.2 $[M_{4Na}-2Na-SO_3Na+H]^{2-}$, 494.3 $[M_{4Na}-3Na-SO_3Na+H]^{3-}$.

3.4. Cytotoxic Activity (MTT Assay) (for SH-SY5Y, HeLa and DLD-1 Cells)

All the studied substances (including chitonoidoside L used as positive control) were tested in concentrations between 0.1 μM to 100 μM using 2-fold dilution in d-H2O. The cell suspension (180 μL) and solutions (20 μL) of tested compounds in different concentrations were injected in wells of 96-well plates (SH-SY5Y, 1×10^4 cells/well, HeLa and DLD-1, $6 \times 10^3/200$ μL) and incubated at 37 °C for 24 h in atmosphere with 5% CO_2. Then, 100 μL of fresh medium was added instead of the tested substances in the same volume of medium. After that, 10 μL of MTT (3-(4,5-dimethylthiazol-2-yl)-2,5-diphenyltetrazolium bromide) (Sigma-Aldrich, St. Louis, MO, USA) stock solution (5 mg/mL) was added to each well, and the microplate was incubated for 4 h. Next, each well was additionally incubated for 18 h with 100 μL of SDS-HCl solution (1 g SDS/10 mL d-H2O/17 μL 6 N HCl). Multiskan FC microplate photometer (Thermo Fisher Scientific, Waltham, MA, USA) was used to measure the absorbance of the converted dye formazan at 570 nm. Cytotoxic activity of the tested compounds was calculated as the concentration that caused 50% cell metabolic activity inhibition (IC50). The experiments were carried out in triplicate, $p < 0.05$.

3.5. Cytotoxic Activity (MTS Assay) (for HL-60 and THP-1 Cells)

The cells of HL-60 line ($10 \times 10^3/200$ μL) and THP-1 ($6 \times 10^3/200$ μL) were placed in 96-well plates at 37 °C for 24 h in a 5% CO_2 incubator. The cells were treated with tested substances and chitonoidoside L as positive control at concentrations from 0 to 100 μM for an additional 24 h incubation. Then, the cells were incubated with 10 μL MTS ([3-(4,5-dimethylthiazol-2-yl)-5-(3-carboxymethoxyphenyl)-2-(4-sulfophenyl)-2H-tetrazolium) for 4 h, and the absorbance in each well was measured at 490/630 nm with plate reader PHERA star FS (BMG Labtech, Ortenberg, Germany). The experiments were carried out in triplicate and the mean absorbance values were calculated. The results were presented as the percentage of inhibition that produced a reduction in absorbance after tested compounds treatment compared to the non-treated cells (negative control), $p < 0.01$.

3.6. Hemolytic Activity

Erythrocytes were isolated from human blood (AB(IV) Rh+) by centrifugation with phosphate-buffered saline (PBS) (pH 7.4) at 4 °C for 5 min by 450 g on centrifuge LABOFUGE 400R (Heraeus, Hanau, Germany) three times. Then, the residue of erythrocytes was re-suspended in ice cold phosphate saline buffer (pH 7.4) to a final optical density of 1.5 at 700 nm, and kept on ice. For the hemolytic assay, 180 μL of erythrocyte suspension was mixed with 20 μL of test compound solution (including chitonoidoside L used as positive control) in V-bottom 96-well plates. After 1 h of incubation at 37 °C, plates were exposed to centrifugation for 10 min at 900 g on laboratory centrifuge LMC-3000 (Biosan, Riga, Latvia). Then, 100 μL of supernatant was carefully selected and transferred in new flat-plates, respectively. Lysis of erythrocytes was determined by measuring the concentration of hemoglobin in the supernatant with microplate photometer Multiskan FC (Thermo Fisher Scientific, Waltham, MA, USA), $\lambda = 570$ nm. The effective dose causing 50% hemolysis of erythrocytes (ED50) was calculated using the computer program SigmaPlot 10.0. All the experiments were carried out in triplicate, $p < 0.01$.

4. Conclusions

As a result of investigation of the glycosidic composition of the sea cucumber *Paracaudina chilensis*, the structures of five new glycosides, chilensosides A–D (**1**–**5**), were established and their cytotoxic activities were studied. Two different aglycones were found and one of them was a part of four compounds. Four diverse carbohydrate chains were detected in the studied glycosides. They differed in the quantity of sulfate groups: two in chilensosides of groups A (**1**, **2**) and B (**3**), three in chilensoside C (**4**), and four in chilensoside D (**5**). The positions of sulfation were also variable: two sulfates were attached to C-4 and C-6 of Glc4 residue in the glycosides **1**, **2**; additional third sulfate group bonded C-3 Glc3 in chilensoside C (**4**); two sulfates bonded different monosaccharide residues, C-6 Glc4 and C-4 Meglc5, in chilensoside B (**3**); and, finally, two positions of sulfation at C-6 Glc3 and C-6 MeGlc5, additional to those observed in **1**, **2**, were detected in chilensoside D (**5**).

Such diversity in sulfate group quantity and positions indicates the high enzymatic activity of sulfatases. They have low specificity to attach a sulfate group to different positions of the same or several monosaccharide residues in glycosides **1**–**5**. Especially interesting was the observation that the sulfatase could compete with the glycosidase, bonding the sulfate group to C-3 Glc3 in chilensoside C (**4**) instead of potential glycosylation of this position. Additionally, it is interesting to note that only the aglycones with intranuclear 9(11)-double were found. This indicates that one oxidosqualene cyclase (OSC), forming the parkeol (precursor of the glycosides with 9(11)-double bond), is expressed in *P. chilensis*.

The structures of the glycosides of *P. chilensis* were similar to those found in some representatives of the order Dendochirotida, confirming phylogenetic closeness of the order Molpadida to the order Dendrochirotida.

Rather high hemolytic and cytotoxic activity of three out of five isolated glycosides along with the previous investigations of highly polar tri- and tetrasulfated glycosides indicate the possible potential of these water-soluble compounds to be used as anticancer drugs.

Supplementary Materials: The following supporting information can be downloaded at: https://www.mdpi.com/article/10.3390/molecules27217655/s1, The original spectral data (Figures S1–S40 and Tables S1–S4).

Author Contributions: Conceptualization, A.S.S. and V.I.K.; methodology, A.S.S., S.A.A. and V.I.K.; investigation, A.S.S., S.A.A., R.S.P., A.B.R., E.A.C., P.V.A. and B.B.G.; writing—original draft preparation, A.S.S. and V.I.K.; review and editing, A.S.S. and V.I.K. All authors have read and agreed to the published version of the manuscript.

Funding: This research received no external funding.

Institutional Review Board Statement: The study was conducted according to the guidelines of the Declaration of Helsinki, and approved by the Ethics Committee of the Pacific Institute of Bioorganic Chemistry (Protocol No. 0037.12.03.2021).

Informed Consent Statement: Informed consent was obtained from all subjects involved in the study.

Acknowledgments: The study was carried out on the equipment of the Collective Facilities Center "The Far Eastern Center for Structural Molecular Research (NMR/MS) PIBOC FEB RAS".

Conflicts of Interest: The authors declare no conflict of interest.

Sample Availability: Samples of the compounds, chilensosides A–D (**1**–**5**), are available from the authors.

References

1. Kalinin, V.I.; Silchenko, A.S.; Avilov, S.A.; Stonik, V.A. Progress in the studies of triterpene glycosides from sea cucumbers (Holothuroidea, Echinodermata) between 2017 and 2021. *Nat. Prod. Commun.* **2021**, *16*, 10. [CrossRef]
2. Kalinin, V.I.; Silchenko, A.S.; Avilov, S.A.; Stonik, V.A. Non-holostane aglycones of sea cucumber triterpene glycosides. Structure, biosynthesis, evolution. *Steroids* **2019**, *147*, 42–51. [CrossRef] [PubMed]
3. Zelepuga, E.A.; Silchenko, A.S.; Avilov, S.A.; Kalinin, V.I. Structure-activity relationships of holothuroid's triterpene glycosides and some in silico insights obtained by molecular dynamics study on the mechanisms of their membranolytic action. *Mar. Drugs* **2021**, *19*, 604. [CrossRef] [PubMed]
4. Hawas, U.W.; Abou El-Kassem, L.T.; Shaher, F.M.; Ghandourah, M.; Al-Farawati, R. Sulfated triterpene glycosides from the Saudi Red Sea cucumber *Holothuria atra* with antioxidant and cytotoxic activities. *Thalass. Int. J. Mar. Sci.* **2021**, *37*, 817–824. [CrossRef]
5. Silchenko, A.S.; Kalinovsky, A.I.; Avilov, S.A.; Andrijaschenko, P.V.; Popov, R.S.; Dmitrenok, P.S.; Chingizova, E.A.; Kalinin, V.I. Unusual structures and cytotoxicities of chitonoidosides A, A$_1$, B, C, D, and E, six triterpene glycosides from the Far Eastern sea cucumber *Psolus chitonoides*. *Mar. Drugs* **2021**, *19*, 449. [CrossRef] [PubMed]

6. Silchenko, A.S.; Kalinovsky, A.I.; Avilov, S.A.; Andrijaschenko, P.V.; Popov, R.S.; Chingizova, E.A.; Kalinin, V.I.; Dmitrenok, P.S. Triterpene glycosides from the Far Eastern sea cucumber *Psolus chitonoides*: Chemical structures and cytotoxicities of chitonoidosides E_1, F, G, and H. *Mar. Drugs* **2021**, *19*, 696. [CrossRef]

7. Silchenko, A.S.; Avilov, S.A.; Andrijaschenko, P.V.; Popov, R.S.; Chingizova, E.A.; Dmitrenok, P.S.; Kalinovsky, A.I.; Rasin, A.B.; Kalinin, V.I. Structures and biologic activity of chitonoidosides I, J, K, K_1 and L–triterpene di-, tri- and tetrasulfated hexaosides from the sea cucumber *Psolus chitonoides*. *Mar. Drugs* **2022**, *20*, 369. [CrossRef]

8. Mondol, M.A.M.; Shin, H.J.; Rahman, M.A. Sea cucumber glycosides: Chemical structures, producing species and important biological properties. *Mar. Drugs* **2017**, *15*, 317. [CrossRef]

9. Vien, L.T.; Hanh, T.T.H.; Quang, T.H.; Thanh, Q.N.V.; Thao, D.T.; Cuong, N.X.; Nam, N.H.; Thung, D.C.; Kiem, P.V. Triterpene tetraglycosides from *Stichopus herrmanni* Semper, 1868. *Nat. Prod. Commun.* **2022**, *17*, 5. [CrossRef]

10. Van Dyck, S.; Gerbaux, P.; Flammang, P. Elucidation of molecular diversity and body distribution of saponins in sea cucumber *Holothuria forskali* (Echinodermata) by mass spectrometry. *Comp. Biochem. Physiol.* **2009**, *152B*, 124–134. [CrossRef]

11. Van Dyck, S.; Flammang, P.; Meriaux, C.; Bonnel, D.; Salzet, M.; Fourmier, I.; Wisztorski, M. Localization of secondary metabolites in marine invertebrates: Contribution of MALDI MSI for the study of saponins in Cuvierian tubules of *H. forskali*. *PLoS ONE* **2010**, *5*, 11. [CrossRef] [PubMed]

12. Van Dyck, S.; Gerbaux, P.; Flammang, P. Quialitative and quantitative saponin contents in five sea cucumbers from Indian Ocean. *Mar. Drugs* **2010**, *8*, 173–189. [CrossRef] [PubMed]

13. Bahrami, Y.; Zhang, W.; Franco, C.M.M. Distribution of saponins in the sea cucumber Holothuria lessoni; the body wall versus the viscera, and their biological activities. *Mar. Drugs* **2018**, *16*, 423. [CrossRef] [PubMed]

14. Bondoc, K.G.V.; Lee, H.; Gruz, L.J.; Lebrila, C.B.; Juinio-Menez, M.A. Chemical fingerprinting and phylogenetic mapping of saponin congeners from three tropical holothurian sea cucumbers. *Comp. Biochem. Physiol.* **2013**, *166B*, 182–193. [CrossRef]

15. Caulier, G.; Mezali, K.; Soualili, D.L.; Decroo, C.; Demeyer, M.; Eeckhaut, I.; Cerbauz, P.; Flammang, P. Chemical characterization of saponins contained in the body wall and the Cuvierian tubules of the sea cucumber *Holothuria (Platyperona) sanctori* (Delle Chiaje, 1823). *Biochem. Syst. Ecol.* **2016**, *68*, 119–127. [CrossRef]

16. Bryne, M.; Rowe, F.; Uthike, S. Molecular, phylogeny and evolution in the family Stichopodidae (Aspidochirotida: Holothurioidea) based on COI and 16S mitochondrial DNA. *Mol. Phylogenet. Evol.* **2010**, *56*, 1068–1081. [CrossRef]

17. Kamyab, E.; Rohde, S.; Kellerman, M.Y.; Schupp, P.J. Chemical defense mechanisms and ecological implications of Indo-Pacific holothurians. *Molecules* **2020**, *25*, 4008. [CrossRef]

18. Van Dyck, S.; Caulier, G.; Todeso, M.; Gebraux, P.; Fournier, I.; Wisztorsky, M.; Flammang, P. The triterpene glycosides of *Holohuria forskali*: Usefulness and efficiency as a chemical defense mechanism against predatory fish. *J. Exp. Biol.* **2011**, *214*, 1347–1356. [CrossRef]

19. Elyakov, G.B.; Stonik, V.; Levina, E.V.; Levin, V.S. Glycosides of marine invertebrates-III. Biosynthesis of stichoposides from acetate. *Comp. Biochem. Physiol.* **1975**, *52*, 321–323. [CrossRef]

20. Kelecom, A.; Daloze, D.; Tursch, B. Chemical studies of marine invertebrates-XXI. Six triterpene genins artifacts from thelothurins A and B, toxic saponins of the sea cucumber *Thelenota ananas* Jaeger (Echinodermata). Biosynthesis of the Thelothurins. *Tetrahedron* **1976**, *32*, 2353–2359. [CrossRef]

21. Sheikh, Y.M.; Djerassi, C. Bioconversion of lanosterol into holotoxingonin, a triterpenoid from the sea cucumber *Stichopus californicus*. *J. Chem. Soc. Chem. Commun.* **1976**, *24*, 1057–1058. [CrossRef]

22. Cordeiro, M.L.; Djerassi, C. Biosynthetic studies of marine lipids. 25. Biosynthesis of $\Delta^{9(11)}$ sterols and saponins in sea cucumbers. *J. Org. Chem.* **1990**, *55*, 2806–2813. [CrossRef]

23. Makarieva, T.N.; Stonik, V.A.; Kapustina, I.I.; Boguslavsky, V.M.; Dmitrenok, A.S.; Kalinin, V.I.; Cordeiro, M.L.; Djerassi, C. Biosynthetic studies of marine lipids. 42. Biosynthesis of steroid and triterpenoid metabolites in the sea cucumber *Eupentacta fraudatrix*. *Steroids* **1993**, *58*, 508–517. [CrossRef]

24. Kerr, R.G.; Chen, Z. In vivo and in vitro biosynthesis of saponin in sea cucumbers. *J. Nat. Prod.* **1995**, *58*, 172–176. [CrossRef] [PubMed]

25. Claereboudt, E.J.S.; Gualier, G.; Decroo, C.; Colson, E.; Gerbaux, P.; Claereboudt, M.R.; Schaller, H.; Flammang, P.; Deleu, M.; Eeckhaut, I. Triterpenoids in echinoderms: Fundamental differences in diversity and biosynthetic pathways. *Mar. Drugs* **2019**, *17*, 352. [CrossRef]

26. Li, Y.; Wang, R.; Xun, X.; Wang, J.; Bao, L.; Thimmappa, R.; Ding, J.; Jiang, J.; Zhang, L.; Li, T.; et al. Sea cucumber genome provides insights into saponin biosynthesis and aestivation regulation. *Cell Discov.* **2018**, *4*, 29. [CrossRef]

27. Mitu, S.A.; Bose, U.; Suwansa-ard, S.; Turner, L.H.; Zhao, M.; Elizur, A.; Osbourne, S.M.; Shaw, P.N.; Cummins, S.F. Evidence for a saponin biosynthesis pathway in the body wall of the commercially significant sea cucumber *Holothuria scabra*. *Mar. Drugs* **2017**, *15*, 349. [CrossRef]

28. Liu, H.; Kong, X.; Chen, J.; Zhang, H. De novo sequencing and transcriptome analysis of *Stichpous horrens* to reveal genes related to biosynthesis of triterpenoids. *Aquaculture* **2018**, *491*, 358–367. [CrossRef]

29. Kim, S.K.; Himaya, S.W.A. Triterpene glycosides from sea cucumbers and their biological activities. *Adv. Food Nutr. Res.* **2012**, *65*, 297–317. [CrossRef]

30. Khotimchenko, Y. Pharmacological potential of sea cucumbers. *Int. J. Mol. Sci.* **2020**, *19*, 1342. [CrossRef]

31. Park, J.-I.; Bae, H.-R.; Kim, C.G.; Stonik, V.A.; Kwak, J.Y. Relationships between chemical structures and functions of triterpene glycosides isolated from sea cucumbers. *Front. Chem.* **2014**, *2*, 77. [CrossRef] [PubMed]

32. Zhao, Y.-C.; Xue, C.-H.; Zhang, T.T.; Wang, Y.-M. Saponins from sea cucumber and their biological activities. *Agric. Food Chem.* **2018**, *66*, 7222–7237. [CrossRef] [PubMed]

33. Gomes, A.R.; Freitas, A.C.; Duarte, A.C.; Rocha-Santos, T.A.P. Echinoderms: A review of bioactive compounds with potential health effects. In *Studies in Natural Products Chemistry*; Rahman, A.U., Ed.; Elsevier, B.V.: Amsterdam, The Netherlands, 2016; Volume 49, pp. 1–54.

34. Chludil, H.D.; Murray, A.P.; Seldes, A.M.; Maier, M.S. Biologically active triterpene glycosides from sea cucumbers (Holothurioidea, Echinodermata). In *Studies in Natural Products Chemistry*; Rahman, A.U., Ed.; Elsevier Science, B.V.: Amsterdam, The Netherlands, 2003; Volume 28, pp. 587–616.

35. Maier, M.S. Biological activities of sulfated glycosides from Echinoderms. In *Studies in Natural Product Chemistry (Bioactive Natural Products)*; Rahman, A.U., Ed.; Elsevier Science Publisher: Amsterdam, The Netherlands, 2008; Volume 35, pp. 311–354.

36. Kalinin, V.I.; Malyutin, A.N.; Stonik, V.A. Caudinoside A − A new triterpene glycoside from the holothurian *Paracaudina ransonetii*. *Chem. Nat. Compd.* **1986**, *22*, 355–356. [CrossRef]

37. Smirnov, A.V. System of the class Holothuroidea. *Paleontol. J.* **2012**, *46*, 793–832. [CrossRef]

38. Kalinin, V.I.; Silchenko, A.S.; Avilov, S.A.; Stonik, V.A.; Smirnov, A.V. Sea cucumbers triterpene glycosides, the recent progress in structural elucidation and chemotaxonomy. *Phytochem. Rev.* **2005**, *4*, 221–236. [CrossRef]

39. Avilov, S.A.; Kalinin, V.I.; Smirnov, A.V. Use of triterpene glycosides for resolving taxonomic problems in the sea cucumber genus *Cucumaria* (Holothorioidea, Echinodermata). *Biochem. Syst. Ecol.* **2004**, *32*, 715–733.

40. Honey-Escandon, M.; Arreguin-Espinosa, R.; Solis-Martin, F.A.; Samyn, Y. Biological and taxonomic perspective of triterpenoid glycosides of sea cucumbers of the family Holothuriidae (Echinodermata, Holothuroidea). *Comp. Biochem. Physiol.* **2015**, *180B*, 16–39. [CrossRef]

41. Omran, N.E.; Salem, ·H.K.; Eissa, S.H.; Kabbash, A.M.; Kandeil, ·M.A.; Salem, M.A. Chemotaxonomic study of the most abundant Egyptian sea-cucumbers using ultra-performance liquid chromatography (UPLC) coupled to high-resolution mass spectrometry (HRMS). *Chemoecology* **2020**, *30*, 35–48. [CrossRef]

42. Silchenko, A.S.; Kalinovsky, A.I.; Avilov, S.A.; Andryjaschenko, P.V.; Dmitrenok, P.S.; Yurchenko, E.A.; Dolmatov, I.Y.; Savchenko, A.M.; Kalinin, V.I. Triterpene glycosides from the sea cucumber *Cladolabes schmeltzii* II. Structure and biological action of cladolosides A_1–A_6. *Nat. Prod. Commun.* **2014**, *9*, 1421–1428.

43. Silchenko, A.S.; Kalinovsky, A.I.; Avilov, S.A.; Andrijaschenko, P.V.; Popov, R.S.; Dmitrenok, P.S.; Chingizova, E.A.; Kalinin, V.I. Kurilosides A_1, A_2, C_1, D, E and F−triterpene glycosides from the Far Eastern sea cucumber *Thyonidium* (=*Duasmodactyla*) *kurilensis* (Levin): Structures with unusual non-holostane aglycones and cytotoxicities. *Mar. Drugs* **2020**, *18*, 551. [CrossRef]

44. Silchenko, A.S.; Kalinovsky, A.I.; Avilov, S.A.; Kalinin, V.I.; Andrijaschenko, P.V.; Dmitrenok, P.S.; Popov, R.S.; Chingizova, E.A. Structures and bioactivities of psolusosides B_1, B_2, J, K, L, M, N, O, P, and Q from the sea cucumber *Psolus fabricii*. The first finding of tetrasulfated marine low molecular weight metabolites. *Mar. Drugs* **2019**, *17*, 631. [CrossRef] [PubMed]

45. Silchenko, A.S.; Avilov, S.A.; Kalinovsky, A.I.; Kalinin, V.I.; Andrijaschenko, P.V.; Dmitrenok, P.S. Psolusosides C_1, C_2 and D_1, novel triterpene hexaosides from the sea cucumber *Psolus fabricii* (Psolidae, Dendrochirotida). *Nat. Prod. Commun.* **2018**, *13*, 1623–1628.

46. Miller, A.K.; Kerr, A.M.; Paulay, G.; Reich, M.; Wilson, N.G.; Carvajal, J.I.; Rouse, G.W. Molecular phylogeny of extant Holothuroidea (Echinodermata). *Mol. Phylogenet. Evol.* **2017**, *111*, 110–131. [CrossRef] [PubMed]

47. Aminin, D.L.; Menchinskaya, E.S.; Pislyagin, E.A.; Silchenko, A.S.; Avilov, S.A.; Kalinin, V.I. Sea cucumber triterpene glycosides as anticancer agents. In *Studies in Natural Product Chemistry*; Rahman, A.U., Ed.; Elsevier, B.V.: Amsterdam, The Netherlands, 2016; Volume 49, pp. 55–105.

Article

A Study on the Chemistry and Biological Activity of 26-Sulfur Analogs of Diosgenin: Synthesis of 26-Thiodiosgenin *S*-Mono- and Dioxides, and Their Alkyl Derivatives

Aneta M. Tomkiel [1,*], Dorota Czajkowska-Szczykowska [1], Ewa Olchowik-Grabarek [2], Lucie Rárová [3], Szymon Sękowski [2] and Jacek W. Morzycki [1,*]

[1] Laboratory of Natural Products, Department of Organic Chemistry, Faculty of Chemistry, University of Bialystok, K. Ciołkowskiego 1K, 15-245 Białystok, Poland
[2] Laboratory of Molecular Biophysics, Department of Microbiology and Biotechnology, Faculty of Biology, University of Bialystok, K. Ciołkowskiego 1 J, 15-245 Białystok, Poland
[3] Department of Experimental Biology, Faculty of Science, Palacký University, Šlechtitelů 27, CZ-78371 Olomouc, Czech Republic
* Correspondence: a.tomkiel@uwb.edu.pl (A.M.T.); morzycki@uwb.edu.pl (J.W.M.); Tel.: +48-85-738-80-44 (A.M.T.); +48-85-738-82-60 (J.W.M.)

Abstract: A chemoselective procedure for MCPBA oxidation of 26-thiodiosgenin to corresponding sulfoxides and sulfone was elaborated. An unusual equilibration of sulfoxides in solution was observed. Moreover, α-alkylation of sulfoxide and sulfone was investigated. Finally, the biological activity of obtained compounds was examined.

Keywords: 26-thiodiosgenin; thiol oxidation; sulfoxides; sulfones; sapogenins; steroids; antimicrobial activity; cytotoxicity

Citation: Tomkiel, A.M.; Czajkowska-Szczykowska, D.; Olchowik-Grabarek, E.; Rárová, L.; Sękowski, S.; Morzycki, J.W. A Study on the Chemistry and Biological Activity of 26-Sulfur Analogs of Diosgenin: Synthesis of 26-Thiodiosgenin *S*-Mono- and Dioxides, and Their Alkyl Derivatives. *Molecules* **2023**, *28*, 189. https://doi.org/10.3390/molecules28010189

Academic Editor: Carla Boga

Received: 30 November 2022
Revised: 12 December 2022
Accepted: 20 December 2022
Published: 26 December 2022

1. Introduction

Saponins are a class of chemical compounds abundant in various plant species [1]. More specifically, they are amphiphilic glycosides producing the soap-like foam when shaken in aqueous solutions. The lipophilic triterpene or steroid aglycones (sapogenins) are combined in these compounds with one or more hydrophilic sugar moieties. The steroidal sapogenins usually contain 27 carbon atoms and the oxidized side chain, which forms a spiroacetal system characteristic for steroidal spirostanes, e.g., diosgenin (**1**) (Figure 1). These compounds have received considerable attention as precursors for synthesizing sex hormones and various steroidal drugs.

There is also a group of naturally occurring compounds based on a C_{27} cholestane skeleton, which are essentially nitrogen analogs of spirostane sapogenins, e.g., solasodine (**2**) (Figure 1). These steroidal alkaloids (spirosolanes) are a class of secondary metabolites isolated from plants (mostly of the family *Solanaceae*), amphibians, and marine invertebrates [2–5]. Evidence accumulated in the last two decades demonstrates that steroidal alkaloids of the spirosolane group show a wide range of bioactivities, including anticancer, antimicrobial, anti-inflammatory, antinociceptive, etc., suggesting their great potential for pharmaceutical application. Several comprehensive review articles on the alkaloid bioactivity, especially anticancer activity, and the mechanism of their biological action have recently appeared [6–8].

The replacement of the F-ring oxygen atom in spirostane sapogenins with a different heteroatom severely affects the chemical properties of a steroid and may result in useful alterations to its biological activity. The potential of heterosteroids as novel drugs encourages organic chemists to undertake studies in this field. The sulfur analogs of steroidal sapogenins do not occur in nature but 26-thiodiosgenin (**3**), the diosgenin F-ring thia-counterpart, was described in the literature many years ago [9]. However, efficient syntheses of this compound have been reported only recently. In particular, the one-pot Wang synthesis [10], which involves the treatment of a solution of diosgenin in dichloromethane with hydrogen sulfide gas in the presence of $BF_3 \cdot Et_2O$ as a catalyst, was the most advantageous for us. It has been shown that replacing the F-ring oxygen atom of diosgenin with sulfur increases compound cytotoxicity against different cancer cell lines, especially in the case of glycosyl derivatives. For example, IC_{50} of a natural saponin, dioscine, against the lung cancer cell line A549 IC_{50} was found 4.02 µM versus 3.72 µM for 26-thiodioscine [11,12]. Though 26-thiodiosgenin (**3**) is now readily available from diosgenin, its chemistry has not been explored yet. The *Se*-analog of diosgenin (**4**), in which selenium atom replaces the F-ring spiroketal oxygen, has also been described, but we have found the literature procedure difficult to reproduce [12,13], i.e., the product we obtained proved to be a mixture of stereoisomers. The development in partial and total syntheses of thiasteroids has been reviewed, but steroid analogs containing sulfur in the side chain were not included in this article [14].

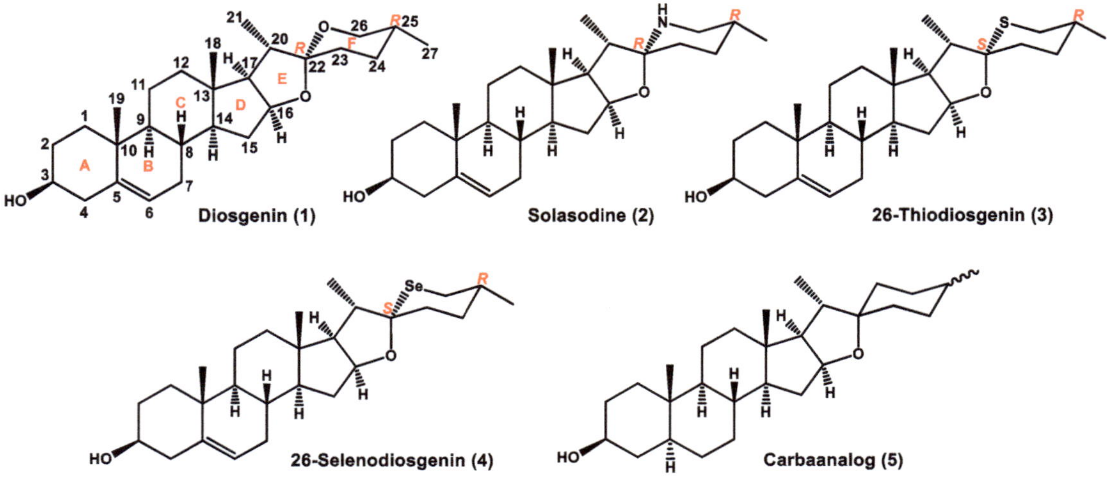

Figure 1. Diosgenin and its aza-, thia-, selena-, and carba-analogs.

We have recently described a simple synthesis of carbaanalogs of steroidal sapogenins along with their biological activity [15]. Unfortunately, the carbaanalog (**5**) was obtained as an inseparable mixture of *cis/trans* isomers (note that C22 and C25 are not stereogenic centers in this compound).

2. Results and Discussion

2.1. Chemistry

The presence of a soft sulfur atom in 26-thiodiosgenin (**3**) alters its chemical, physical, and biological properties, making it, among others, very susceptible to oxidation. In fact, diosgenin (**2**) can also be oxidized with different reagents; its oxidation sites are the C5-C6 double bond, the 3β-hydroxyl group, and the carbon atoms C20 or C23. However, the introduction of sulfur in place of oxygen changes the reactivity of compound and directs the oxidation of **3** to the sulfur atom. It is well known that the oxidation of sulfides is a two-step process. In the first step, only one oxygen atom is transferred from the oxidizing agent to sulfur, forming sulfoxide. Since a new stereogenic center is generated at sulfur during this step (provided that the starting sulfide is not symmetrical), two diastereomeric sulfoxides can be formed. Further oxidation of both sulfoxides leads to a single sulfone. The oxidation of sulfoxides to sulfones is relatively easy, and sometimes it is difficult to stop the oxidation of sulfides at an intermediate stage. The simplest method of sulfoxide and sulfone synthesis is the sulfide oxidation with halogen derivatives and metal-mediated oxidative systems [16]. However, due to growing concerns about chemical pollution and environmental protection, there is a tendency to use hydrogen peroxide or peroxy acids as atom-efficient and environmentally benign oxidizing agents [16,17]. The oxidation of organic sulfides with peroxycarboxylic acids or with hydrogen peroxide, which is usually activated by conventional acidic catalysts, often leads to various side reactions such as over-oxidation (sulfoxides to sulfones), epoxidation (if a double bond is present) and Baeyer–Villiger oxidation (if an oxo group is present). After testing several methods of the chemoselective sulfide oxidation, we turned our attention to the method employing hydrogen peroxide and *N*-hydroxysuccinimide (NHS) [18]. According to the literature, sulfoxides can be obtained with this reagent in acetone under reflux conditions, without over-oxidation to sulfones, and the method is compatible with the presence of sensitive groups, including alkenes and hydroxyl groups. Nonetheless, a complex mixture of products was formed when 26-thiodiosgenin (**3**) was subjected to the described reaction conditions. In turn, when compound **3** was treated with this oxidizing system under milder conditions (50 °C, 10 h), the corresponding sulfone **8** was obtained as the only product in 70% yield (Scheme 1). The formation of intermediate sulfoxides **6** and **7** was not observed, even after lowering the reaction temperature to 30 °C and reducing the reaction time. The yield of sulfone **8** then dropped to 68%.

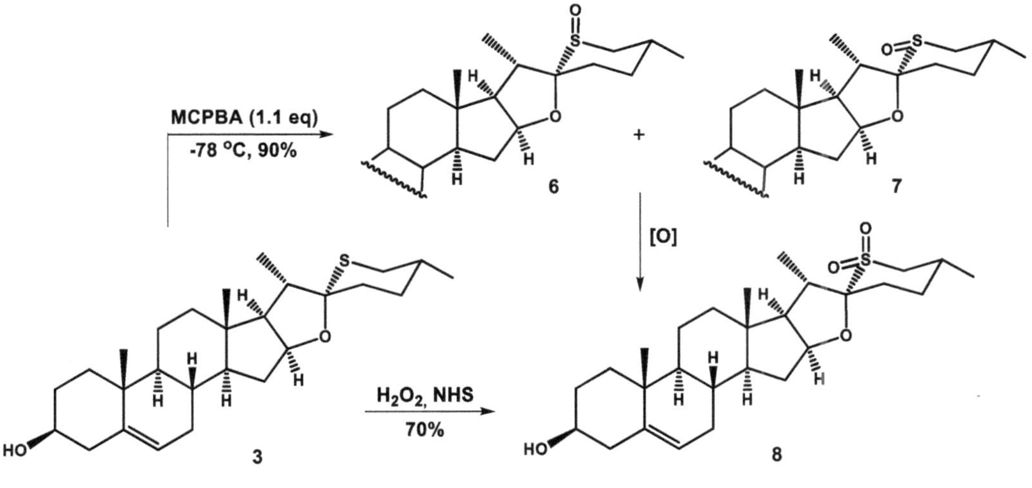

Scheme 1. Oxidation of 26-thiodiosgenin (**3**).

Another reagent recommended for oxidation of sulfides to sulfoxides is m-chloroperoxybenzoic acid (MCPBA) [19]. The reaction was carried out with 1.1 equiv. of this reagent at −78 °C (dry ice/acetone bath). The progress of the reaction was monitored by TLC, which showed a single spot of a very polar product (more polar than sulfone **8**). After two hours, the reaction came to completion and was quenched with dimethylsulfide.

Although the product appeared to be single by TLC, a detailed analysis of its spectra (Figure 2) revealed that it was a mixture of sulfoxides, one of which was by far predominant. In the ¹H NMR spectrum, a tiny signal at about 3 ppm derived from protons of the C26 methylene group of the minor sulfoxide (the C26 methylene protons of the major one appeared as a doublet at δ 2.78 and triplet at δ 2.57 ppm). Furthermore, the LC-MS analysis unambiguously confirmed the presence of trace amounts of the second sulfoxide. Unfortunately, the separation of these isomers was practically impossible, even by crystallization of crude product.

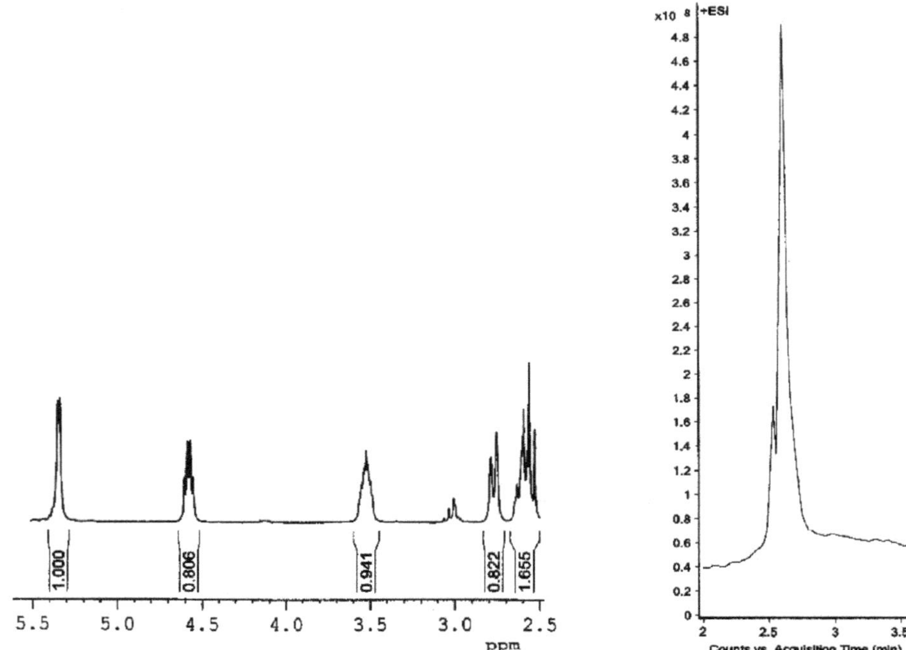

Figure 2. ¹H NMR spectrum and LC-MS chromatogram of crude product of 26-thiodiosgenin (**3**) oxidation with MCPBA.

The reaction of 26-thiodiosgenin (**3**) with 2.2 equiv. of MCPBA at −40 °C afforded sulfone **8** as the major product, in addition to 10% yield of sulfoxides **6/7** and 2% yield of the over-oxidized product 5,6-epoxysulfone (Scheme 2).

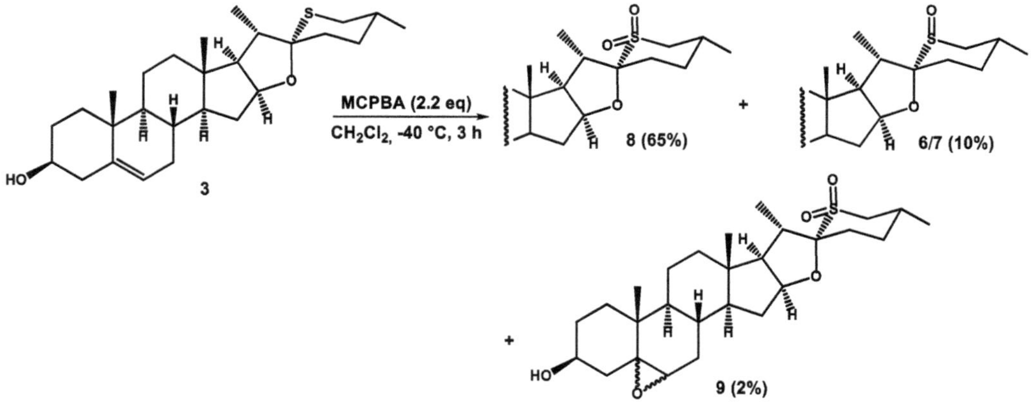

Scheme 2. Oxidation of 26-thiodiosgenin (**3**) with 2.2 equiv. of MCPBA at −40 °C.

The low temperature (−78 °C) MCPBA oxidation of 26-thiodiosgenin 4-nitrobenzoate (**3a**) was also carried out with the expectation of obtaining the well-crystallizing and easily separable products. Indeed, it turned out that the separation of isomeric sulfoxides was possible not only by HPLC but also by a silica gel (230–400 mesh) gravity flow column chromatography. The less polar minor product was obtained in 9% yield, while the major one in 71% yield (Scheme 3).

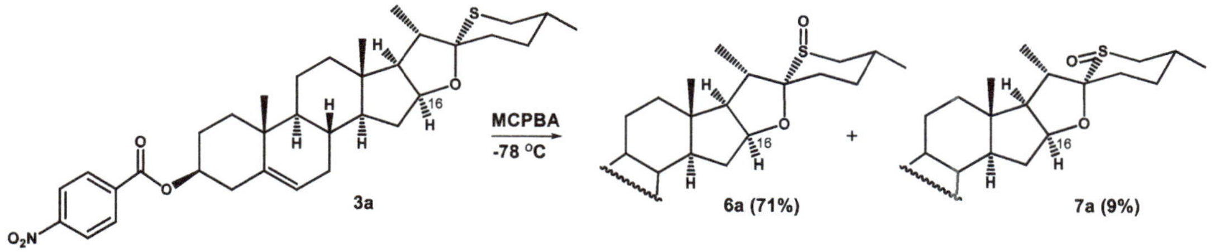

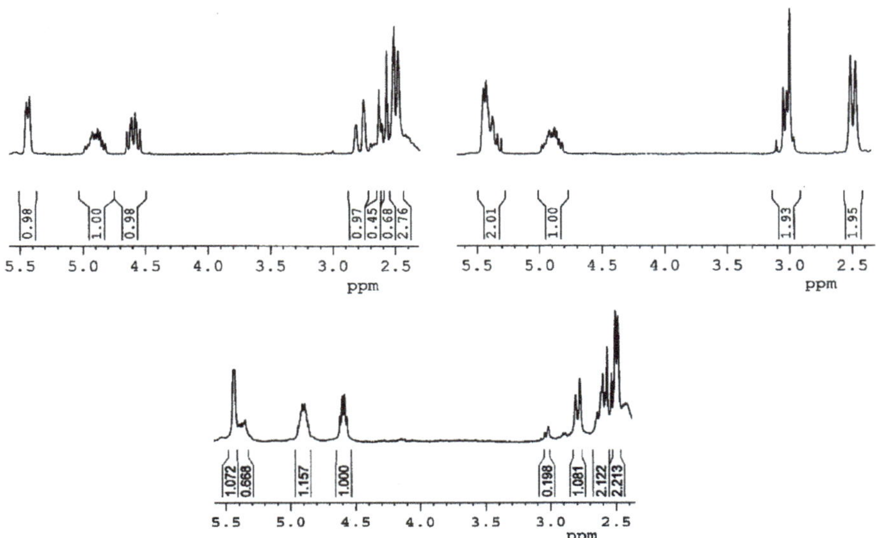

Scheme 3. Oxidation of 26-thiodiosgenin 3β-4-nitrobenzoate (**3a**) with MCPBA at −78 °C.

A detailed analysis of ^{1}H NMR spectra of sulfoxides (Figure 3) can unequivocally ascribe the configuration at the sulfur atom. The diagnostic for the configuration assignment proved to be the H-16α signal. Cone shielding anisotropy generated by the sulfinyl group is similar to that of the carbonyl group. The inspection of Dreiding models, as well as molecular modeling employing the MM+ calculations (Figure 4), show that the α proton at C16 is in close proximity to the sulfinyl group (deshielding zone) in the equatorial sulfoxide **7** (configuration *R* at sulfur), while the effect of the axial *S*-oxide **6** (configuration *S* at sulfur) on this proton is negligible. As can be seen from Table 1, the H-16α signal appears at δ 4.64 in 26-thiodiosgenin (**3**), at δ 4.55–4.58 for axial sulfoxides **6** (or **6a**), and it is strongly deshielded (to δ 5.37–5.38) for equatorial sulfoxides **7** (or **7a**).

Figure 3. ^{1}H NMR spectra of pure oxidation products of **3a** with MCPBA at −78 °C: **6a** (**upper left**), **7a** (**upper right**), and the crude reaction mixture (**below**).

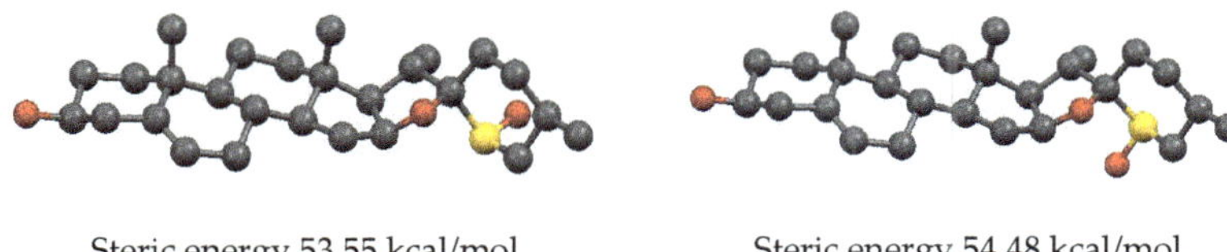

Steric energy 53.55 kcal/mol   Steric energy 54.48 kcal/mol

Figure 4. A ball and stick presentation of the MM+ optimized structures of 26-thiodiosgenin S-oxides: axial **6** (**Left**) and equatorial **7** (**Right**).

Table 1. Chemical shifts (δ, ppm) of H-16α, H-26α and H-26β in the ^{1}H NMR spectra of 26-thiodiosgenin (**3**), sulfoxides **6**, **7**, and sulfone **8**.

Compound * Proton	3	3a	3b	6	6a	6b	7	7a	8	8b
H-16α	4.64	4.65	4.64	4.58	4.55	4.58	~5.37	5.38	5.10	5.10
H-26α (t)	2.55	2.55	2.55	2.57	2.53	2.57	~3.03	~3.00	3.16	3.16
H-26β (d)	~2.30	2.30	2.30	2.78	2.74	2.78	~3.03	~3.00	2.55	2.63

* Compound numbers without additional indication refer to free C3-alcohols; **a** and **b** apply to their derivatives, C3-*p*-nitrobenzoates and C3-TBDMS-ethers, respectively.

The sulfoxides are usually configurationally stable. However, in our case, the sulfoxides are stable in solid state only, but in solution (dichloromethane) a slow equilibration between the isomeric sulfoxides **6** and **7** occurs. The solution of **6a** was allowed to stand under argon at room temperature and the progress of isomerization was monitored by TLC. The equilibrium was reached within 14 days, and then the ratio of **6a**:**7a** was 3:2. The same isomeric ratio was obtained when the equatorial sulfoxide **7a** was subjected to equilibration. This result is consistent with molecular mechanics calculations which showed a slightly lower steric energy for the axial sulfoxide (Figure 4). The equilibration between sulfoxides can be reached much faster (10 min) if a catalytic amount of *p*-TsOH is added. However, when the reaction mixture was allowed to stand for a longer time, further isomerization processes occurred (presumably at C20 and C22); as a result, up to 8 isomers can be formed. The tentative mechanism of the sulfoxide equilibration is shown in Scheme 4.

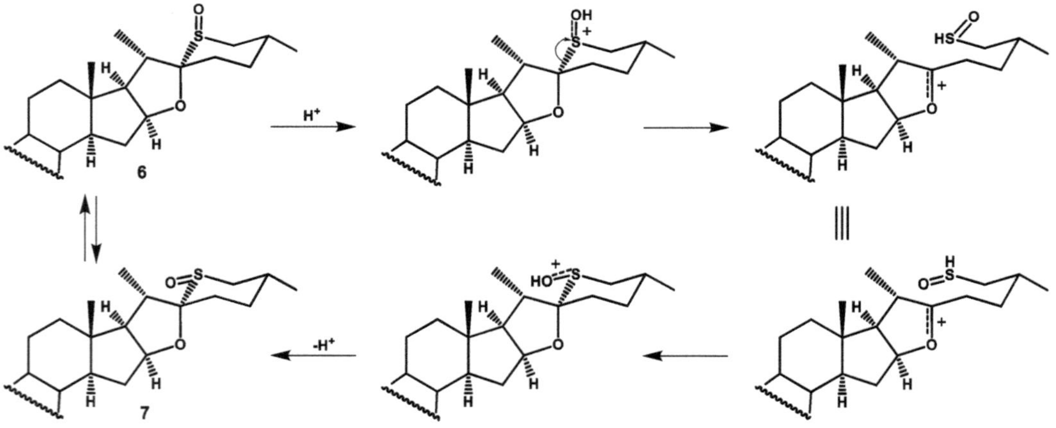

Scheme 4. The tentative mechanism of the sulfoxide isomerization.

This unusual behavior of sulfoxides **6** and **7** is caused by an easy cleavage of the C22-S bond under acidic conditions. The cleavage leads to the sulfenic acid and the relatively stable oxocarbenium ion. The reverse reaction recovers the sulfoxide, but the configuration at the sulfur atom may be inversed. It should be noted that if the sulfoxide is treated with acid for a longer time, the C20 proton may be abstracted from the oxocarbenium ion to give the C20–C22 double bond, and then an isomerization at these carbon atoms may also occur.

On the other hand, the C22-S bond in sulfoxides **6**/**7** and sulfone **8** should be resistant to basic conditions. If so, an α-alkylation should be possible. Consequently, the starting 26-thiodiosgenin (**3**) was converted to *tert*-butyldimethylsilyl ether **3b** with TBDMS-Cl and imidazole. Then **3b** was oxidized with 1.1 equiv. of MCPBA at −78 °C affording sulfoxide **6b**. After increasing the reaction temperature to −40 °C and raising the amount of oxidant to 2.2 equiv., the corresponding sulfone **8b** was also prepared. The obtained sulfoxide **6b** and sulfone **8b** were deprotonated with n-butyllithium and then treated with 2.5 equiv.

of methyl iodide (Scheme 5). The reaction of sulfoxide **6b** led to the formation of the dimethylated product **10b** in 91% yield. The α-methylation of sulfone **8b** also proceeded smoothly, but only the mono substituted product **11b** was obtained. The difference in the reaction course is probably due to a larger steric hindrance at C26 in the sulfone than in the corresponding sulfoxide. The α-methylation of sulfone **8b** proved to be highly stereoselective, providing only one product **11b** in 82% yield. The S configuration at the newly formed stereogenic center at C26 was concluded on the basis of the ^{1}H NMR signal of H-26, which appeared at δ 3.14 as a doublet of quartets with coupling constants $J = 6.9$ and 11.1 Hz. The latter comes from the coupling of two axial protons at C26 (alpha) and C25 (beta). It implies that the new methyl group at C26 assumed the equatorial position. The reaction of sulfone **8b** with ethyl iodide also provided the α-substituted product, albeit in slightly lower yield (76%). Attempts of α-alkylation of 26-thiodiosgenin 3-TBDMS-ether (**3b**) failed, probably due to a low acidity of the α-proton in this compound. This result could be expected given the lower acidity of dimethylsulfide (pK$_a$ = 45.0) compared with dimethylsulfoxide (pK$_a$ = 35.1) and dimethylsulfone (pK$_a$ = 31.1).

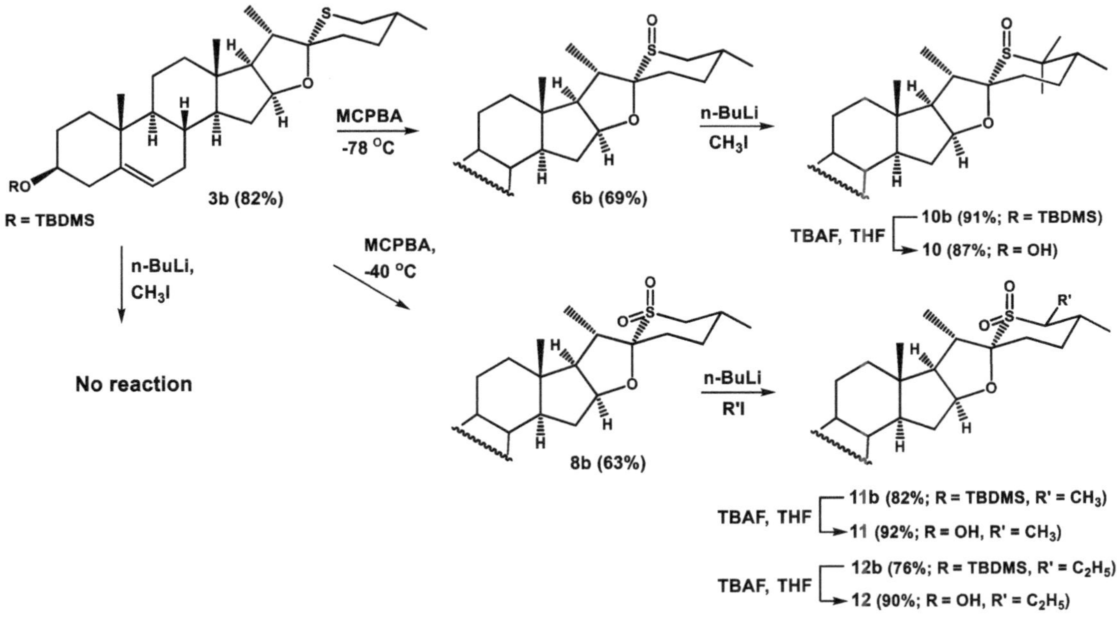

Scheme 5. Alkylation of sulfoxide **6b** and sulfone **8b**.

2.2. Biology

2.2.1. Antimicrobial Activity of Tested Compounds

Diosgenin (**1**) belongs to the group of steroidal compounds known as sapogenins, which are obtained from their glycoside forms, i.e., saponins [20]. It demonstrates various biological activities, i.a. anticancer, antioxidant, and anti-inflammatory activity [21]. In order to verify if newly synthesized sulfur analogs of diosgenin (**1**) possess antimicrobial activity, the MIC (minimal inhibition concentration) and MBC (minimal bactericidal concentration) values were estimated, using two bacterial strains: Gram-positive *Staphylococcus aureus* 8325-4 and Gram-negative *Escherichia coli* 35218. The obtained MIC and MBC values are shown in Table 2.

Table 2. Antibacterial activity of diosgenin (**1**) and its sulfur analogs.

Compound	*S. aureus* 8325-4			*E. coli* ATCC 35218		
	MIC [µg/mL]	MIC [µM]	MBC [µg/mL]	MIC [µg/mL]	MIC [µM]	MBC [µg/mL]
Diosgenin (**1**)	3.9	9.4	15.615	250	603	500
26-Thiodiosgenin (**3**)	1.95	4.5	3.9	62.5	145	125
Sulfoxide **6**	3.9	8.7	7.8	125	280	250
Sulfone **8**	3.9	8.4	7.8	125	270	250
α-Methylsulfone **11**	3.9	8.2	7.8	125	262	250
Carbaanalog **5**	7.8	18.9	31.25	500	1207	1000

MIC—minimal inhibition concentration; MBC—minimal bactericidal concentration.

Diosgenin (**1**) has a strong antimicrobial activity against *S. aureus* (Gram-positive bacteria), and its growth is inhibited at the concentration of 3.9 µg/mL, whereas four-times higher concentration (15.615 µg/mL) was estimated as MBC for these bacteria. Much weaker activity was detected for diosgenin (**1**) against Gram-negative *E. coli* where MIC and MBC were 250 µg/mL and 500 µg/mL, respectively. This effect may be due to the characteristic structure of the outer membrane of Gram-negative bacteria, which contains unique component—lipopolysaccharide [22]. Exchanging the oxygen atom with a sulfur atom in 26-thiodiosgenin (**3**) performed via chemical modification of the diosgenin F-ring results in an increase of antibacterial activity for both the *S. aureus* and *E. coli*. The obtained MIC and MBC values were: 1.95 µg/mL and 3.9 µg/mL for *S. aureus*, and 62.5 µg/mL and 125 µg/mL for *E. coli*. This may be a consequence of the presence of a sulfur atom in the structure of these compounds. It is well know that sulfur functional groups are found in many pharmaceuticals, including penicillin, prevacid (lansoprazole), seroquel (quetiapine), dapsone, or sulfamethoxazole [23]. From the chemical point of view, they belong to different classes of sulfur compounds, e.g., cyclic sulfides, sulfoxides, sulfones, or sulfonamides. The oxidation of 26-thiodiosgenin (**3**) to sulfoxide or sulfone as well as their α-methylation decreased antibacterial activity of compounds **6, 8, 11** compared with **3**. However, their antimicrobial activity was still stronger than that of diosgenin (**1**) (see Table 2). The weakest activity has been detected for carbaanalog **5** with the MIC values 7.8 µg/mL and 500 µg/mL for *S. aureus* and *E. coli*, respectively. This is in line with our previous findings for not methylated 5- and 6-membered F-ring carbaanalogs recently described [15]. The latter compound showed antimicrobial activity against *S. aureus* 8325-4 with the MIC = 4 µg/mL and against *E. coli* with the MIC = 512 µg/mL. The carbaanalog **5** demonstrated slightly lower antimicrobial potential against *S. aureus* 8325-4 (MIC = 7.8 µg/mL), probably due to the presence of an additional 27-methyl group. In the case of *E. coli* the activity was similar (MIC = 500 µg/mL for **5** and MIC = 512 µg/mL for its non-methylated analog). The determined MIC and MBC values allowed the conclusion that 26-thiodiosgenin (**3**) demonstrated the strongest antibacterial activity against *S. aureus* 8325-4 and *E. coli* ATCC 35218 among all tested compounds. Both its S-oxidation and α-methylation of the oxidized derivatives decreased their antimicrobial potential.

2.2.2. The Interaction of Studied Compounds with Proteins of Bacterial Cell Membranes

The results described above clearly showed that studied compounds exhibit antibacterial activity against *S. aureus* and *E. coli*. This feature may be due to the interaction between the molecule and specific component of the cell membrane, i.e., a protein or a phospholipid. In order to check if prepared thia-steroids possess the ability to interact with bacterial membranes proteins, we have analyzed fluorescence changes of tryptophan residues of membrane proteins as the marker of such interactions.

Based on fluorescence analysis, the Stern-Volmer plots were drawn (Figure 5) using equation given below (Equation (1)) [24].

$$\frac{F_0}{F} = 1 + K_{SV}[Q] \tag{1}$$

where: F_0 and F are the Trp fluorescence without and in the presence of quencher (studied compounds), K_{SV}—Stern-Volmer constant, $[Q]$—concentration of the quencher.

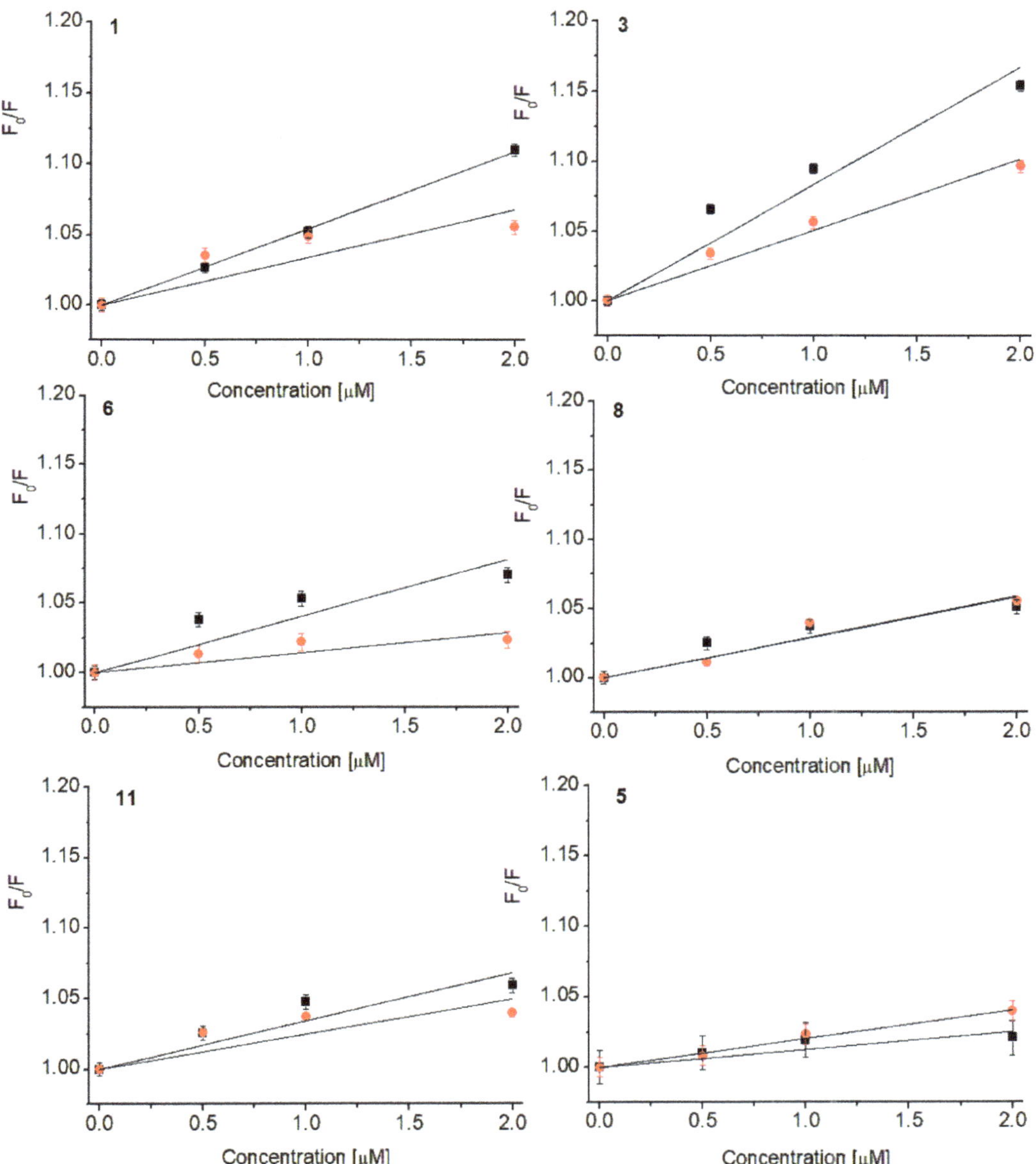

Figure 5. Stern-Volmer plots of Trp fluorescence quenching in *S. aureus* 8325-4 (■) and *E. coli* 35218 (●) membrane proteins. Values demonstrated as mean ± SD, concentrations below MIC.

Referring to Figure 5 it can be concluded that all compounds have an affinity for Trp^{214} residues in a hydrophobic pocket of Gram-positive (*S. aureus*) and Gram-negative (*E. coli*) bacterial membrane proteins, but the observed effect was visibly stronger for *S. aureus*. Based on the Equation (1), the Stern-Volmer constants have been calculated and are presented below (Table 3).

Table 3. Stern-Volmer (K_{SV}) and quenching (k_q) constants of compounds interactions with bacterial membrane proteins.

Compound	*S. aureus* 8325-4		*E. coli* ATCC 35218	
	$K_{SV}\ 10^4$ $[\mathbf{M^{-1}}]$	$k_q\ 10^{12}$ $[\mathbf{M^{-1}s^{-1}}]$	$K_{SV}\ 10^4$ $[\mathbf{M^{-1}}]$	$k_q\ 10^{12}$ $[\mathbf{M^{-1}s^{-1}}]$
Diosgenin (**1**)	5.441 ± 0.131	10.93 ± 0.27	3.384 ± 0.165	6.76 ± 0.35
26-Thiodiosgenin (**3**)	8.322 ± 0.142	16.62 ± 0.23	5.091 ± 0.152	10.02 ± 0.28
Sulfoxide **6**	4.051 ± 0.170	8.14 ± 0.35	1.456 ± 0.201	2.92 ± 0.36
Sulfone **8**	2.915 ± 0.175	5.82 ± 0.31	2.962 ± 0.105	5.92 ± 0.16
α-Methylsulfone **11**	3.442 ± 0.168	6.84 ± 0.34	2.470 ± 0.101	4.94 ± 0.17
Carbaanalog **5**	1.258 ± 0.189	2.52 ± 0.32	2.021 ± 0.198	4.04 ± 0.45

According to calculated K_{SV} constants, it can be inferred that diosgenin (**1**), as well as its analogs, possess higher affinity to Trp[214] residues localized in *S. aureus* membrane proteins in comparison with *E. coli*. These results are in good agreement with obtained MIC and MBC values. Thiodiosgenin (**3**) ($K_{SV} = (8.322 \pm 0.142) \times 10^4$ M^{-1} for *S. aureus* and $K_{SV} = (5.091 \pm 0.152) \times 10^4$ M^{-1} for *E. coli*) demonstrated higher ability to interact with membrane proteins of both used bacteria strains in comparison with diosgenin (**1**) ($K_{SV} = (5.441 \pm 0.131) \times 10^4$ M^{-1} and $K_{SV} = (3.384 \pm 0.165) \times 10^4$ M^{-1} for *S. aureus* and *E. coli*, respectively). The oxidation of thiodiosgenin (**3**) and further α-methylation of the corresponding sulfone decreased the affinity of these compounds to tryptophan residues localized in the membrane proteins of *S. aureus* and *E. coli* to the values lower than that of diosgenin (**1**) and thiodiosgenin (**3**). Despite different affinity of studied compounds to Trp[214] residue of the bacterial membrane proteins, it can be concluded that their antibacterial activity is closely related to it. The lowest K_{SV} in the interaction with *S. aureus* was obtained for carbaanalog **5** ($K_{SV} = (1.258 \pm 0.189) \times 10^4$ M^{-1}). It indicated the weakest affinity of this compound to the hydrophobic pockets of tryptophan in staphylococcal membrane proteins.

In order to check whether studied compounds form complexes with bacterial cells proteins, the quenching constants (k_q) have been calculated using equation below (Equation (2)) and are presented in Table 3.

$$k_q = \frac{K_{SV}}{\tau_0} \tag{2}$$

where: k_q—quenching constant, K_{SV}—Stern-Volmer constant, τ_0—fluorescence lifetime of fluorophore molecules.

The calculated k_q values for all tested compounds were greater than the one for the maximum scatter collision (2×10^{10} M^{-1}s^{-1}), thus it can be deduced that investigated molecules formed complexes with *S. aureus* and *E. coli* membrane proteins.

2.2.3. Cytotoxicity Study

The obtained sulfur analogs of diosgenin (**3, 6–12**) were briefly tested for cytotoxicity against three cancer lines (MCF7, K562, and HeLa) and normal human retina cells RPE-1. All tested compounds have not been shown to be toxic to normal RPE-1 cells, as well as towards MCF7 and K562 (Table 4). However, sulfoxides **7** and **10**, as well as alkylated sulfones **11** and **12**, exhibited a moderate toxicity against human cervical carcinoma cells (HeLa) with IC$_{50}$ 34.3, 32.2, 92.3 and 44.1 µM, respectively. Interestingly, the carbaanalog **5** proved also slightly cytotoxic against HeLa cancer cell line (IC$_{50}$ 38.2 µM).

Table 4. Cytotoxicity (IC_{50}) o diosgenin analogs against cancer and healthy cell lines after 72 h.

	IC_{50} (µM)			
Analogs	**MCF7**	**K562**	**HeLa**	**RPE-1**
Diosgenin (**1**)	>50	>50	>50	>50
26-Thiodiosgenin (**3**)	>50	>50	>50	>50
Sulfoxide **6**	>50	>50	>50	>50
Sulfoxide **7**	>50	>50	34.3 ± 3.0	>50
Sulfone **8**	>50	>50	>50	>50
Epoxysulfone **9**	>50	>50	>50	>50
Dimethylsulfoxide **10**	>50	>50	32.2 ± 1.7	>50
26-Methylsulfone **11**	>50	>50	92.3	>50
26-Ethylsulfone **12**	>50	>50	44.1 ± 0.1	>50
Carbaanalog **5**	>50	>50	38.2 ± 0.8	>50

3. Materials and Methods

3.1. Chemistry

3.1.1. General

The reagents were purchased from Merck, Alfa Aesar, or Acros. All solvents were freshly distilled prior to use. The dry solvents were prepared by distillation over the following drying agents: DMF (4 Å molecular sieves), THF (Na/benzophenone), CH_2Cl_2 (CaH_2).

The reactions were monitored by TLC on silica gel plates 60 F254 (Merck, Darmstadt, Germany) and spots were visualized either by UV Hand Lamp (Type: NU-4; 254 nm/365 nm, 2x4W, Herolab GmbH Laborgeräte, Wiesloch, Germany) or by charring with molybdophosphoric acid/cerium(IV) sulfate in H_2SO_4. The reaction products were isolated by chromatographic methods using JT Baker silica gel (J.T. Baker, Phillipsburg, NJ, USA), pore size 40 Å (70–230 mesh) (column chromatography) or 60 Å (230–400 mesh) (gravity flow column chromatography and dry flash chromatography).

^{1}H and ^{13}C NMR (400 and 100 MHz, respectively) spectra of all compounds were recorded using Bruker Avance II spectrometer (Bruker, Fällanden, Switzerland) in CDCl₃ and referenced to TMS (0.0 ppm) and CDCl₃ (77.0 ppm), respectively. Only selected signals in the ^{1}H NMR spectra are reported. The original ^{1}H and ^{13}C NMR spectra are contained in the Supplementary Materials. Infrared spectra were recorded using Attenuated Total Reflectance (ATR) as solid samples with Nicolet 6700 FT-IR spectrometer (Thermo Fisher Scientific, Waltham, MA, USA). Mass spectra were obtained at Accurate-Mass Q-TOF LC/MS 6530 spectrometer (Agilent, Santa Clara, NJ, USA) with electrospray ionization (ESI). Melting points were determined on a Kofler bench (Boetius type, Nagema, VEB Wägetechnik Rapido Radebeul, Dresden, Germany) melting point apparatus.

26-Thiodiosgenin (**3**) [10] and carbaanalog (**5**) [15] were prepared according to literature procedures.

3.1.2. Synthesis of 26-Thiodiosgenin 3β-4-Nitrobenzoate (3a)

26-Thiodiosgenin (**3**) (200 mg, 0.47 mmol) was dissolved in a mixture of dry pyridine (25 mL) and dichloromethane (10 mL), and then 4-nitrobenzoyl chloride (106 mg, 0.56 mmol) was added. The reaction mixture was stirred at room temperature. After completion of the reaction (7 days), it was poured into water, and extracted with dichloromethane (3 × 100 mL). The combined organic extracts were dried over anhydrous Na_2SO_4, and the solvent was evaporated in vacuo. The crude product was purified by dry flash chromatography with a hexane/ethyl acetate (99:1) mixture to afford ester **3a** (245 mg, 91%).

Compound **3a**: colorless crystals (hexane/ethyl acetate); mp 257–258 °C; IR (ATR) v_{max} 1601, 1520, 1449, 1337, 1270, 1106, 1013 cm^{-1}; ^{1}H NMR (CDCl₃, 400 MHz) δ 8.28 (2H, d, J = 8.7 Hz, H-Ar), 8.21 (2H, d, J = 8.7 Hz, H-Ar), 5.44 (1H, m, H-6), 4.90 (1H, m, H-3α), 4.65 (1H, m, H-16), 2.55 (1H, t, J = 12.6 Hz, H-26α), 2.30 (1H, d, J = 12.6 Hz, H-26β), 1.10 (3H, s, H-19), 1.03 (3H, d, J = 6.9 Hz, H-21), 0.94 (3H, d, J = 6.5 Hz, H-27), 0.83 (3H, s, H-18);

^{13}C NMR (CDCl$_3$, 100 MHz) δ 164.0 (C), 150.4 (C), 139.3 (C), 136.2 (C), 130.6 (2 × CH), 123.4 (2 × CH), 122.9 (CH), 97.5 (C), 81.6 (CH), 75.7 (CH), 62.8 (CH), 56.5 (CH), 49.9 (CH), 44.4 (CH), 40.3 (C), 39.7 (CH$_2$), 38.5 (CH$_2$), 38.0 (CH$_2$), 36.9 (C), 36.7 (CH$_2$), 33.3 (CH), 32.1 (CH$_2$), 32.0 (CH$_2$), 31.7 (CH$_2$), 31.4 (CH), 31.4 (CH$_2$), 27.7 (CH$_2$), 22.5 (CH$_3$), 20.8 (CH$_2$), 19.4 (CH$_3$), 16.5 (CH$_3$), 16.2 (CH$_3$); HRMS *m/z* 580.3104 (calcd for C$_{34}$H$_{46}$NO$_5$S$^+$, 580.3091).

3.1.3. Preparation of 26-Thiodiosgenin 3β-t-Butyldimethylsilyl Ether (3b)

Imidazole (48 mg, 0.7 mmol) and *t*-butyldimethylsilyl chloride (106 mg, 0.7 mmol) were added to a solution of 26-thiodiosgenin (**3**) (200 mg, 0.47 mmol) in dry dimethylformamide (20 mL). The reaction mixture was stirred for 24 h at room temperature, then poured into water, and extracted with diethyl ether (3 × 100 mL). The combined organic extracts were dried over anhydrous Na$_2$SO$_4$, and concentrated under reduced pressure. The crude product was purified by dry flash chromatography with hexane/ethyl acetate (97:3) elution to afford ether **3b** (207 mg, 82%).

Compound **3b**: colorless crystals (hexane/ethyl acetate); mp 235–237 °C; IR (ATR) ν_{max} 1458, 1373, 1248, 1092, 1014 cm^{-1}; ^{1}H NMR (CDCl$_3$, 400 MHz) δ 5.32 (1H, m, H-6), 4.64 (1H, dd, *J* = 15.4 Hz, *J* = 7.4 Hz, H-16), 3.49 (1H, m, H-3α), 2.55 (1H, dd, *J* = 12.7 Hz, *J* = 11.7 Hz, H-26α), 2.30 (1H, d, *J* = 12.7 Hz, H-26β), 1.027 (3H, d, *J* = 6.8 Hz, H-21), 1.026 (3H, s, H-19), 0.94 (3H, d, *J* = 6.5 Hz, H-27), 0.90 (9H, s, *t*-Bu-Si), 0.81 (3H, s, H-18), 0.07 (6H, s, (CH$_3$)$_2$Si); ^{13}C NMR (CDCl$_3$, 100 MHz) δ 141.6 (C), 120.9 (CH), 97.5 (C), 81.7 (CH), 72.6 (CH), 62.8 (CH), 56.7 (CH), 50.1 (CH), 44.4 (CH), 42.8 (CH$_2$), 40.3 (C), 39.8 (CH$_2$), 38.5 (CH$_2$), 37.3 (CH$_2$), 36.7 (C), 33.3 (CH), 32.10 (CH$_2$), 32.09 (CH$_2$), 32.06 (CH$_2$), 31.7 (CH$_2$), 31.5 (CH), 31.4 (CH$_2$), 25.9 (3 × CH$_3$), 22.5 (CH$_3$), 20.8 (CH$_2$), 19.4 (CH$_3$), 18.3 (C), 16.5 (CH$_3$), 16.2 (CH$_3$), −4.6 (2 × CH$_3$); HRMS *m/z* 545.3829 (calcd for C$_{33}$H$_{57}$O$_2$SSi$^+$, 545.3843).

3.1.4. General Procedure for S-Oxides Formation

Steroidal sulfide (**3**, **3a** or **3b**) (0.23, mmol) was dissolved in dry dichloromethane (20 mL), cooled to −78 °C (dry ice/acetone cooling bath), and *m*-chloroperoxybenzoic acid (56 mg, 0.25 mmol) was added. The solution was stirred at −78 °C for 2–3 h and monitored by TLC. When the reaction was completed, dimethyl sulfide (0.1 mL, 1.3 mmol) was added. The reaction mixture was poured into saturated aqueous solution of NaHCO$_3$, and extracted with dichloromethane (3 × 100 mL). The combined organic extracts were dried over anhydrous Na$_2$SO$_4$, and evaporated to dryness in vacuo. The residue was subjected to gravity flow column chromatography on silica gel 60 Å (230–400 mesh), which resulted in separation of compounds **6/7** (as a mixture), **6a** and **7a**, or **8b**.

Compound **6a**: was eluted with benzene/ethyl acetate (1:1) mixture in 71% yield. Colorless crystals (benzene/ethyl acetate); mp 150–151 °C; IR (ATR) ν_{max} 1600, 1521, 1452, 1345, 1270, 1168, 1107, 1017 cm^{-1}; ^{1}H NMR (CDCl$_3$, 400 MHz) δ 8.24 (2H, d, *J* = 9.0 Hz, H-Ar), 8.17 (2H, d, *J* = 9.0 Hz, H-Ar), 5.39 (1H, m, H-6), 4.84 (1H, m, H-3α), 4.55 (1H, m, H-16), 2.74 (1H, d, *J* = 13.3 Hz, H-26β), 2.53 (1H, t, *J* = 13.3 Hz, H-26α), 1.32 (3H, d, *J* = 7.2 Hz, H-21), 1.06 (3H, s, H-19), 0.94 (3H, d, *J* = 6.7 Hz, H-27), 0.80 (3H, s, H-18); ^{13}C NMR (CDCl$_3$, 100 MHz) δ 163.9 (C), 150.3 (C), 139.3 (C), 136.0 (C), 130.5 (2 × CH), 123.3 (2 × CH), 122.5 (CH), 99.6 (C), 84.2 (CH), 75.5 (CH), 63.6 (CH), 56.2 (CH), 49.7 (CH), 49.0 (CH$_2$), 43.7 (CH), 40.8 (C), 39.1 (CH$_2$), 37.9 (CH$_2$), 36.8 (CH$_2$), 36.6 (C), 32.5 (CH$_2$), 31.8 (CH$_2$), 31.2 (CH), 29.7 (CH$_2$), 27.65 (CH$_2$), 27.60 (CH$_2$), 21.3 (CH$_3$), 20.6 (CH$_2$), 19.9 (CH), 19.2 (CH$_3$), 16.7 (CH$_3$), 16.1 (CH$_3$); HRMS *m/z* 596.3020 (calcd for C$_{34}$H$_{46}$NO$_6$S$^+$, 596.3040).

Compound **7a**: was eluted with benzene/ethyl acetate (1:1) mixture in 9% yield. Colorless crystals (benzene/ethyl acetate); mp 175–176 °C; IR (ATR) ν_{max} 1522, 1453, 1342, 1268, 1163, 1108, 1038 cm^{-1}; ^{1}H NMR (CDCl$_3$, 400 MHz) δ 8.28 (2H, d, *J* = 9.0 Hz, H-Ar), 8.20 (2H, d, *J* = 9.0 Hz, H-Ar), 5.43 (1H, m, H-6), 5.38 (1H, m, H-16), 4.89 (1H, m, H-3α), 3.06-2.98 (2H, m, H-26), 1.46 (3H, d, *J* = 7.0 Hz, H-21), 1.09 (3H, s, H-19), 1.05 (3H, d, *J* = 6.6 Hz, H-27), 0.82 (3H, s, H-18); ^{13}C NMR (CDCl$_3$, 100 MHz) δ 164.0 (C), 150.4 (C), 139.2 (C), 136.2 (C), 130.6 (2 × CH), 123.4 (2 × CH), 122.8 (CH), 104.4 (C), 87.3 (CH), 75.6 (CH), 63.0 (CH), 56.0 (CH), 52.4 (CH$_2$), 49.8 (CH), 43.9 (CH), 40.7 (C), 39.4 (CH$_2$), 38.0 (CH$_2$), 36.8

(CH$_2$), 36.7 (C), 33.6 (CH$_2$), 33.2 (CH$_2$), 32.0 (CH$_2$), 31.3 (CH), 30.0 (CH$_2$), 29.7 (CH), 27.7 (CH$_2$), 21.5 (CH$_3$), 20.8 (CH$_2$), 19.3 (CH$_3$), 16.4 (2 x CH$_3$); HRMS *m/z* 596.3023 (calcd for C$_{34}$H$_{46}$NO$_6$S$^+$, 596.3040).

Compound **6b**: was eluted with hexane/ethyl acetate (7:3) mixture in 69% yield. Colorless crystals (hexane/ethyl acetate); mp 205–207 °C; IR (ATR) ν_{max} 1571, 1455, 1376, 1248, 1083, 1031 cm^{-1}; ^{1}H NMR (CDCl$_3$, 400 MHz) δ 5.32 (1H, m, H-6), 4.58 (1H, m, H-16), 3.49 (1H, m, H-3α), 2.78 (1H, d, J = 13.0 Hz, H-26β), 2.57 (1H, t, J = 13.0 Hz, H-26α), 1.36 (3H, d, J = 7.3 Hz, H-21), 1.03 (3H, s, H-19), 0.98 (3H, d, J = 6.7 Hz, H-27), 0.90 (9H, s, *t*-Bu-Si), 0.82 (3H, s,H-18), 0.07 (6H, s, (CH$_3$)$_2$Si); ^{13}C NMR (CDCl$_3$, 100 MHz) δ 141.7 (C), 120.6 (CH), 99.7 (C), 84.4 (CH), 72.5 (CH), 63.8 (CH), 56.5 (CH), 50.0 (CH), 49.1 (CH$_2$), 43.8 (CH), 42.7 (CH$_2$), 40.9 (C), 39.3 (CH$_2$), 37.3 (CH$_2$), 36.7 (C), 32.6 (CH$_2$), 31.99 (CH$_2$), 31.97 (CH$_2$), 31.4 (CH), 29.8 (CH$_2$), 27.8 (CH$_2$), 25.9 (3 × CH$_3$), 21.4 (CH$_3$), 20.7 (CH$_2$), 20.0 (CH), 19.4 (CH$_3$), 18.2 (C), 16.8 (CH$_3$), 16.2 (CH$_3$), −4.6 (2 × CH$_3$); HRMS *m/z* 561.3784 (calcd for C$_{33}$H$_{57}$O$_3$SSi$^+$, 561.3792).

The mixture of compounds **6** and **7** proved inseparable. These compounds were obtained in their pure forms by removing *p*-nitrobenzoyl groups from separated compounds **6a** and **7a**, what is described below.

3.1.5. General Procedure for Removing the 3β-4-Nitrobenzoyl Group

To a solution of steroidal sulfoxide (**6a** or **7a**) (50 mg) (0.08 mmol) in dry methanol (10 mL) NaOH (80 mg, 2 mmol) was added. The reaction was stirred for 24 h. Then, the solvent was evaporated in vacuo, the crude product was dissolved in dichloromethane and washed by water. The organic extract was dried over anhydrous Na$_2$SO$_4$, and concentrated under reduced pressure. The crude product (**6** or **7**) was purified by column chromatography on silica gel with hexane/ethyl acetate (3:7) mixture elution.

Compound **6**: 89% yield; colorless crystals (hexane/ethyl acetate); mp 143–146 °C; IR (ATR) ν_{max} 3331, 1449, 1377, 1348, 1164, 1139, 1052, 1017 cm^{-1}; ^{1}H NMR (CDCl$_3$, 400 MHz) δ 5.35 (1H, m, H-6), 4.58 (1H, m, H-16), 3.53 (1H, m, H-3α), 2.78 (1H, d, J = 13.0 Hz, H-26β), 2.57 (1H, t, J = 13.0 Hz, H-26α), 1.36 (3H, d, J = 7.3 Hz, H-21), 1.04 (3H, s, H-19), 0.98 (3H, d, J = 6.7 Hz, H-27), 0.83 (3H, s, H-18); ^{13}C NMR (CDCl$_3$, 100 MHz) δ 140.9 (C), 121.1 (CH), 99.7 (C), 84.4 (CH), 71.6 (CH), 63.8 (CH), 56.5 (CH), 50.0 (CH), 49.2 (CH$_2$), 43.9 (CH), 42.2 (CH$_2$), 40.9 (C), 39.3 (CH$_2$), 37.2 (CH$_2$), 36.6 (C), 32.6 (CH$_2$), 31.9 (CH$_2$), 31.6 (CH$_2$), 31.4 (CH), 29.8 (CH$_2$), 27.8 (CH$_2$), 21.4 (CH$_3$), 20.7 (CH$_2$), 20.0 (CH), 19.4 (CH$_3$), 16.8 (CH$_3$), 16.3 (CH$_3$); HRMS *m/z* 447.2921 (calcd for C$_{27}$H$_{43}$O$_3$S$^+$, 447.2927).

Compound **7**: 87% yield; colorless crystals (hexane/ethyl acetate); mp 165–166 °C; IR (ATR) ν_{max} 3420, 1455, 1377, 1341, 1156, 1138, 1074, 1033, 1011 cm^{-1}; ^{1}H NMR (CDCl$_3$, 400 MHz) δ 5.40–5.34 (2H, m, H-6, H-16), 3.53 (1H, m, H-3α), 3.07–2.98 (2H, m, H-26), 1.46 (3H, d, J = 8.8 Hz, H-21), 1.05 (3H, d, J = 6.6 Hz, H-27), 1.03 (3H, s, H-19), 0.82 (3H, s, H-18); ^{13}C NMR (CDCl$_3$, 100 MHz) δ 140.9 (C), 121.3 (CH), 104.5 (C), 87.4 (CH), 71.7 (CH), 63.1 (CH), 56.2 (CH), 52.6 (CH$_2$), 50.1 (CH), 44.1 (CH), 42.3 (CH$_2$), 40.7 (C), 39.6 (CH$_2$), 37.3 (CH$_2$), 36.7 (C), 33.7 (CH$_2$), 33.3 (CH$_2$), 32.0 (CH$_2$), 31.7 (CH$_2$), 31.5 (CH), 30.1 (CH$_2$), 29.7 (CH), 21.5 (CH$_3$), 20.9 (CH$_2$), 19.4 (CH$_3$), 16.4 (CH$_3$), 16.3 (CH$_3$); HRMS *m/z* 447.2984 (calcd for C$_{27}$H$_{43}$O$_3$S$^+$, 447.2927).

3.1.6. Procedures for Preparation of Sulfones

Procedure 1 (with H$_2$O$_2$ and NHS)

30% hydrogen peroxide (45 µL, 0.4 mmol) and NHS (*N*-hydroxysuccinimide) (23 mg, 0.2 mmol) were added to a solution of 26-thiodiosgenin (**3**) (40 mg, 0.1 mmol) in acetone (5 mL). The mixture was heated at 50 °C for 10 h. Then, the reaction mixture was poured into aqueous solution of NaHSO$_3$, and extracted with ethyl acetate (3 × 50 mL). The combined organic extracts were dried over anhydrous Na$_2$SO$_4$, and evaporated to dryness in vacuo. The crude product was purified by column chromatography on silica gel with hexane/ethyl acetate (3:1) elution to afford sulfone (**8**) (230 mg, 70%).

Compound **8**: was eluted with hexane/ethyl acetate (3:1) mixture in 70% yield. Colorless crystals (hexane/ethyl acetate); mp 205–207 °C; IR (ATR) ν_{max} 3573, 3431, 1464, 1291, 1261, 1050, 1021 cm^{-1}; ^{1}H NMR (CDCl$_3$, 400 MHz) δ 5.36 (1H, m, H-6), 5.10 (1H, m, H-16), 3.54 (1H, m, H-3α), 3.16 (1H, t, J = 13.2 Hz, H-26α), 2.55 (1H, d, J = 13.2 Hz, H-26β), 1.42 (3H, d, J = 7.3 Hz, H-21), 1.04 (3H, s, H-19), 1.02 (3H, d, J = 6.7 Hz, H-27), 0.80 (3H, s, H-18); ^{13}C NMR (CDCl$_3$, 100 MHz) δ 140.8 (C), 121.2 (CH), 103.3 (C), 85.5 (CH), 71.6 (CH), 63.9 (CH), 56.3 (CH), 55.2 (CH$_2$), 49.9 (CH), 42.9 (CH), 42.2 (CH$_2$), 40.8 (C), 39.3 (CH$_2$), 37.2 (CH$_2$), 36.6 (C), 35.2 (CH$_2$), 32.6 (CH$_2$), 31.9 (CH$_2$), 31.6 (CH$_2$), 31.4 (CH), 30.9 (CH), 29.1 (CH$_2$), 21.1 (CH$_3$), 20.8 (CH$_2$), 19.4 (CH$_3$), 16.2 (CH$_3$), 16.0 (CH$_3$); HRMS m/z 463.2856 (calcd for C$_{27}$H$_{43}$O$_4$S$^+$, 463.2877).

Procedure 2 (with MCPBA)

Steroidal sulfide (**3** or **3b**) (0.23 mmol) was dissolved in dry dichloromethane (20 mL) cooled to −40 °C (dry ice/acetonitrile cooling bath) and m-chloroperoxybenzoic acid (112 mg, 0.50 mmol) was added. The solution was stirred at −40 °C for 2–3 h and monitored by TLC. When the reaction was completed, dimethyl sulfide (0.2 mL, 2.6 mmol) was added. Then, the reaction mixture was poured into saturated aqueous solution of NaHCO$_3$, and extracted with dichloromethane (3 × 100 mL). The combined organic extracts were dried over anhydrous Na$_2$SO$_4$, and evaporated to dryness in vacuo. The residue was chromatographed on a silica gel column to afford pure compounds **8** or **8b** from **3** and **3b**, respectively.

In the reaction of **3**, in addition to **8** (65%), small amounts of 5,6-epoxides **9** (2%) and sulfoxides **6/7** (10%) were isolated. Compounds **6/7** and **8** were described above. Epoxides **9** were eluted with hexane/ethyl acetate (6:4) as a mixture of 5,6α and 5,6β epimers in the ratio of 7:3. Only major product (5,6α-epoxide) is described below.

5,6α-Epoxide **9**: IR (ATR) ν_{max} 3698, 3626, 3589, 3523, 1572, 1450, 1290, 1126, 1055 cm^{-1}; ^{1}H NMR (CDCl$_3$, 400 MHz) δ 5.08 (1H, m, H-16), 3.93 (1H, m, H-3α), 3.15 (1H, t, J = 12.7 Hz, H-26α), 2.92 (1H, d, J = 4.4 Hz, H-6), 2.64 (1H, d, J = 12.7 Hz, H-26β), 1.40 (3H, d, J = 7.3 Hz, H-21), 1.09 (3H, s, H-19), 1.01 (3H, d, J = 6.8 Hz, H-27), 0.74 (3H, s, H-18); ^{13}C NMR (CDCl$_3$, 100 MHz) δ 103.3 (C), 85.3 (CH), 68.7 (CH), 65.6 (C), 63.7 (CH), 58.9 (CH), 56.5 (CH), 55.2 (CH$_2$), 42.9 (CH), 42.3 (CH), 40.8 (C), 39.7 (CH$_2$), 38.9 (CH$_2$), 35.2 (CH$_2$), 35.0 (C), 32.4 (CH$_2$), 32.3 (CH$_2$), 31.0 (CH$_2$), 30.9 (CH), 29.5 (CH), 29.2 (CH$_2$), 28.9 (CH$_2$), 21.1 (CH$_3$), 20.4 (CH$_2$), 16.2 (CH$_3$), 16.03 (CH$_3$), 15.95 (CH$_3$); HRMS m/z 479.2822 (calcd for C$_{27}$H$_{43}$O$_5$S$^+$, 479.2826).

Compound **8b**: was eluted with hexane/ethyl acetate (3:1) in 63% yield. Colorless crystals (hexane/ethyl acetate); mp 228–230 °C; IR (ATR) ν_{max} 1454, 1374, 1284, 1249, 1081 cm^{-1}; ^{1}H NMR (CDCl$_3$, 400 MHz) δ 5.31 (1H, m, H-6), 5.10 (1H, m, H-16), 3.48 (1H, m, H-3α), 3.16 (1H, t, J = 13.2 Hz, H-26a), 2.63 (1H, dd, J = 13.2 Hz, J = 1.6 Hz, H-26b), 1.41 (3H, d, J = 7.3 Hz, H-21), 1.02 (3H, s, H-19), 1.01 (3H, d, J = 6.6 Hz, H-27), 0.89 (9H, s, t-Bu-Si), 0.79 (3H, s, H-18), 0.06 (6H, s, (CH$_3$)$_2$Si); ^{13}C NMR (CDCl$_3$, 100 MHz) δ 141.6 (C), 120.7 (CH), 103.3 (C), 85.5 (CH), 72.4 (CH), 63.8 (CH), 56.4 (CH), 55.2 (CH$_2$), 49.9 (CH), 42.9 (CH), 42.7 (CH$_2$), 40.7 (C), 39.3 (CH$_2$), 37.3 (CH$_2$), 36.6 (C), 35.1 (CH$_2$), 32.6 (CH$_2$), 32.0 (CH$_2$), 31.9 (CH$_2$), 31.4 (CH), 30.9 (CH), 29.1 (CH$_2$), 25.9 (3 × CH$_3$), 21.1 (CH$_3$), 20.7 (CH$_2$), 19.4 (CH$_3$), 18.2 (C), 16.2 (CH$_3$), 16.0 (CH$_3$), −4.6 (2 × CH$_3$); HRMS m/z 577.3755 (calcd for C$_{33}$H$_{57}$O$_4$SSi$^+$, 577.3741).

3.1.7. General Procedure for α-Alkylation

n-BuLi (0.08 mL, 2.5 M in hexane, 0.2 mmol) was added to a solution of steroidal sulfoxide or sulfone (**6b** or **8b**) (0.08 mmol) in dry THF (10 mL), and the mixture was stirred at room temperature for 15 min. After this time, alkyl iodide (CH$_3$I or C$_2$H$_5$I) (0.2 mmol) was added and stirring was continued for 45–90 min. Then, the reaction mixture was poured into water, and extracted with dichloromethane (3 × 50 mL). The combined organic extracts were dried over anhydrous Na$_2$SO$_4$, and evaporated to dryness in vacuo. The residue was subjected to chromatography on a silica gel column to afford pure compounds **10b**, **11b**, or **12b**.

Compound **10b**: was eluted with hexane/ethyl acetate (3:1) mixture in 91% yield. Colorless crystals (hexane/ethyl acetate); mp 198–200 °C; IR (ATR) ν_{max} 1426, 1359, 1219, 1091 cm^{-1}; ^{1}H NMR (CDCl$_3$, 400 MHz) δ 5.31 (1H, m, H-6), 4.65 (1H, m, H-16), 3.48 (1H, m, H-3α), 1.32 (3H, d, J = 7.4 Hz, H-21), 1.31 (3H, s, CH$_3$-C26), 1.23 (3H, s, CH$_3$-C26), 1.02 (3H, s, H-19), 0.90 (3H, d, J = 6.9 Hz, H-27), 0.89 (9H, s, t-Bu-Si), 0.81 (3H, s, H-18), 0.06 (6H, s, (CH$_3$)$_2$Si); ^{13}C NMR (CDCl$_3$, 100 MHz) δ 141.7 (C), 120.6 (CH), 103.2 (C), 84.0 (CH), 72.5 (CH), 63.5 (CH), 57.5 (C), 56.5 (CH), 50.1 (CH), 46.0 (CH), 42.8 (CH$_2$), 40.8 (C), 39.5 (CH$_2$), 37.3 (CH$_2$), 36.7 (C), 32.4 (CH$_2$), 32.0 (2 × CH$_2$), 31.3 (CH), 30.7 (CH), 28.5 (CH$_2$), 27.4 (CH$_2$), 26.2 (CH$_3$), 25.9 (3 × CH$_3$), 20.8 (CH$_2$), 19.4 (CH$_3$), 18.2 (C), 16.7 (CH$_3$), 16.2 (CH$_3$), 15.9 (CH$_3$), 15.6 (CH$_3$), −4.6 (2 x CH$_3$); HRMS m/z 589.4089 (calcd for C$_{35}$H$_{61}$O$_3$SSi$^+$, 589.4105).

Compound **11b**: was eluted with hexane/ethyl acetate (93:7) mixture in 82% yield. Colorless crystals (hexane/ethyl acetate); mp 198–200 °C; IR (ATR) ν_{max} 1459, 1375, 1290, 1251, 1123, 1081 cm^{-1}; ^{1}H NMR (CDCl$_3$, 400 MHz) δ 5.31 (1H, m, H-6), 5.07 (1H, m, H-16), 3.49 (1H, m, H-3α), 3.14 (1H, dk, J = 11.1 Hz, J = 6.9 Hz, H-26α), 1.42 (3H, d, J = 7.4 Hz, H-21), 1.29 (3H, d, J = 6.9 Hz, CH$_3$-C26), 1.02 (3H, s, H-19), 0.98 (3H, d, J = 6.6 Hz, H-27), 0.89 (9H, s, t-Bu-Si), 0.80 (3H, s, H-18), 0.06 (6H, s, (CH$_3$)$_2$Si); ^{13}C NMR (CDCl$_3$, 100 MHz) δ 141.6 (C), 120.7 (CH), 103.5 (C), 85.5 (CH), 72.5 (CH), 63.8 (CH), 57.9 (CH), 56.4 (CH), 50.0 (CH), 43.4 (CH), 42.8 (CH$_2$), 40.7 (C), 39.3 (CH$_2$), 37.3 (CH$_2$), 36.7 (C), 35.9 (CH), 34.8 (CH$_2$), 32.6 (CH$_2$), 32.03 (CH$_2$), 31.98 (CH$_2$), 31.5 (CH), 29.8 (CH$_2$), 25.9 (3 × CH$_3$), 20.7 (CH$_2$), 19.4 (CH$_3$), 18.9 (CH$_3$), 18.2 (C), 16.2 (CH$_3$), 16.0 (CH$_3$), 7.3 (CH$_3$), −4.6 (2 × CH$_3$); HRMS m/z 591.3928 (calcd for C$_{34}$H$_{59}$O$_4$SSi$^+$, 591.3898).

Compound **12b**: was eluted with hexane/ethyl acetate (99:1) mixture in 76% yield. Colorless crystals (hexane/ethyl acetate); mp 215–217 °C; IR (ATR) ν_{max} 1457, 1375, 1286, 1254, 1082 cm^{-1}; ^{1}H NMR (CDCl$_3$, 400 MHz) δ 5.31 (1H, m, H-6), 5.07 (1H, m, H-16), 3.49 (1H, m, H-3α), 2.99 (1H, dt, J = 11.4 Hz, J = 4.6 Hz, H-26α), 1.41 (3H, d, J = 7.3 Hz, H-21), 1.13 (3H, t, J = 7.6 Hz, CH$_3$CH$_2$-C26), 1.02 (3H, s, H-19), 1.01 (3H, d, J = 6.5 Hz, H-27), 0.89 (9H, s, t-Bu-Si), 0.79 (3H, s, H-18), 0.06 (6H, s, (CH$_3$)$_2$Si); ^{13}C NMR (CDCl$_3$, 100 MHz) δ 141.6 (C), 120.7 (CH), 103.7 (C), 85.4 (CH), 72.5 (CH), 63.8 (CH), 62.7 (CH), 56.4 (CH), 50.0 (CH), 43.4 (CH), 42.8 (CH$_2$), 40.7 (C), 39.4 (CH$_2$), 37.3 (CH$_2$), 36.7 (C), 34.8 (CH$_2$), 33.8 (CH), 32.6 (CH$_2$), 32.03 (CH$_2$), 31.98 (CH$_2$), 31.4 (CH), 29.9 (CH$_2$), 25.9 (3 × CH$_3$), 20.8 (CH$_2$), 19.4 (CH$_3$), 19.0 (CH$_3$), 18.2 (C), 16.6 (CH$_2$), 16.2 (CH$_3$), 16.0 (CH$_3$), 11.8 (CH$_3$), −4.6 (2 × CH$_3$); HRMS m/z 605.4031 (calcd for C$_{35}$H$_{61}$O$_4$SSi$^+$, 605.4054).

3.1.8. General Procedure for the Removal of the TBDMS Group

Steroidal 3β-t-butyldimethylsilyl ether (**10b**, **11b**, or **12b**) (0.08 mmol) was dissolved in dry THF (5 mL) and then tetra-n-butylammonium fluoride (0.8 mL, 1 M in THF, 0.8 mmol) was added dropwise. After stirring at room temperature for 2 h, the reaction mixture was poured into water and extracted with diethyl ether (3 × 50 mL). The combined organic layers were dried over Na$_2$SO$_4$, and evaporated in vacuo. The residue was purified by column chromatography on silica gel afforded compound **10**, **11**, or **12**.

Compound **10**: was eluted with hexane/ethyl acetate (7:13) mixture as a white amorphous solid in 87% yield. IR (ATR) ν_{max} 3338, 1445, 1346, 1250, 1139, 1062, 1014 cm^{-1}; ^{1}H NMR (CDCl$_3$, 400 MHz) δ 5.35 (1H, m, H-6), 4.66 (1H, m, H-16), 3.53 (1H, m, H-3α), 1.33 (3H, d, J = 7.4 Hz, H-21), 1.32 (3H, s, CH$_3$-C26), 1.23 (3H, s, CH$_3$-C26), 1.04 (3H, s, H-19), 0.91 (3H, d, J = 6.9 Hz, H-27), 0.82 (3H, s, H-18); ^{13}C NMR (CDCl$_3$, 100 MHz) δ 141.0 (C), 121.2 (CH), 103.2 (C), 84.0 (CH), 71.7 (CH), 63.6 (CH), 57.5 (C), 56.6 (CH), 50.1 (CH), 46.1 (CH), 42.3 (CH$_2$), 40.8 (C), 39.6 (CH$_2$), 37.2 (CH$_2$), 36.7 (C), 32.5 (CH$_2$), 31.7 (2 × CH$_2$), 31.4 (CH), 30.7 (CH), 28.6 (CH$_2$), 27.4 (CH$_2$), 26.2 (CH$_3$), 20.8 (CH$_2$), 19.4 (CH$_3$), 16.7 (CH$_3$), 16.2 (CH$_3$), 15.9 (CH$_3$), 15.6 (CH$_3$); HRMS m/z 475.3253 (calcd for C$_{29}$H$_{47}$O$_3$S$^+$, 475.3240).

Compound **11**: was eluted with hexane/ethyl acetate (3:1) mixture in 92% yield. Colorless crystals (hexane/ethyl acetate); mp 225–227 °C; IR (ATR) ν_{max} 3514, 3355, 1452, 1378, 1348, 1290, 1279, 1117, 1049 cm^{-1}; ^{1}H NMR (^{1}H NMR (CDCl$_3$, 400 MHz) δ 5.35 (1H, m, H-6), 5.07 (1H, m, H-16), 3.53 (1H, m, H-3α), 3.14 (1H, dq, J = 11.1 Hz, J = 7.0 Hz, H-26α), 1.43 (3H, d, J = 7.4 Hz, H-21), 1.29 (3H, d, J = 7.0 Hz, CH$_3$-C26), 1.03 (3H, s, H-19), 0.98

(3H, d, J = 6.6 Hz, H-27), 0.80 (3H, s, H-18); ^{13}C NMR (CDCl$_3$, 100 MHz) δ 140.9 (C), 121.2 (CH), 103.6 (C), 85.5 (CH), 71.7 (CH), 63.9 (CH), 57.9 (CH), 56.4 (CH), 49.9 (CH), 43.5 (CH), 42.3 (CH$_2$), 40.8 (C), 39.4 (CH$_2$), 37.2 (CH$_2$), 36.6 (C), 35.9 (CH), 34.9 (CH$_2$), 32.6 (CH$_2$), 32.0 (CH$_2$), 31.6 (CH$_2$), 31.5 (CH), 29.8 (CH$_2$), 20.8 (CH$_2$), 19.4 (CH$_3$), 18.9 (CH$_3$), 16.2 (CH$_3$), 16.0 (CH$_3$), 7.3 (CH$_3$); HRMS m/z 477.3039 (calcd for C$_{28}$H$_{45}$O$_4$S$^+$, 477.3033).

Compound **12**: was eluted with hexane/ethyl acetate (3:1) mixture in 90% yield. Colorless crystals (hexane/ethyl acetate); mp 210–212 °C; IR (ATR) ν_{max} 3503, 1450, 1377, 1349, 1276, 1225, 1152, 1117, 1072, 1051 cm^{-1}; ^{1}H NMR (CDCl$_3$, 400 MHz) δ 5.35 (1H, m, H-6), 5.07 (1H, m, H-16), 3.53 (1H, m, H-3α), 2.99 (1H, dt, J = 11.1 Hz, J = 4.6 Hz, H-26α), 1.41 (3H, d, J = 7.2 Hz, H-21), 1.13 (3H, t, J = 7.5 Hz, C$\underline{H_3}$CH$_2$-C26), 1.03 (3H, s, H-19), 1.02 (3H, d, J = 6.8 Hz, H-27), 0.80 (3H, s, H-18); ^{13}C NMR (CDCl$_3$, 100 MHz) δ 140.8 (C), 121.2 (CH), 103.7 (C), 85.4 (CH), 71.7 (CH), 63.7 (CH), 62.7 (CH), 56.3 (CH), 49.9 (CH), 43.4 (CH), 42.2 (CH$_2$), 40.7 (C), 39.3 (CH$_2$), 37.1 (CH$_2$), 36.6 (C), 34.8 (CH$_2$), 33.8 (CH), 32.6 (CH$_2$), 31.9 (CH$_2$), 31.6 (CH$_2$), 31.4 (CH), 29.9 (CH$_2$), 20.8 (CH$_2$), 19.4 (CH$_3$), 19.0 (CH$_3$), 16.6 (CH$_2$), 16.2 (CH$_3$), 16.0 (CH$_3$), 11.8 (CH$_3$); HRMS m/z 491.3197 (calcd for C$_{29}$H$_{47}$O$_4$S$^+$, 491.3190).

3.2. Biology

3.2.1. Determination of Antimicrobial Activity

The antibacterial activity of tested compounds was assessed by monitoring the cell growth of *Staphylococcus aureus* 8325-4 and *Escherichia coli* ATCC 35218 using the broth microdilution method, conducted according to the National Committee for Clinical Laboratory Standards. First, compounds were dissolved in methanol and the solution was added to Mueller Hinton broth (MHB) for bacteria to give a final concentration of 2000 μg/mL. The samples were then serially two-fold diluted in bullion to obtain concentration ranging from 1000 to 0.12 μg/mL in 96-well microtiter plate with the final volumes of 100 μL. Next, 100 μL of bacteria suspension was injected into each well. The final bacteria cell concentration was 1×10^6 colony-forming units per ml (CFU/mL). The plates were incubated at 37 °C for 24 h. The MIC value was determined as the lowest concentration of an antibacterial agent that inhibited bacterial growth, as indicated by the absence of turbidity. The MBC value was determined as the lowest concentration of antibacterial agents for which no bacterial growth on the plates was observed [15].

3.2.2. Fluorescence Analysis of Diosgenin and Their Derivatives Interaction with Bacterial Cell Membranes

Bacteria strains: *S. aureus* and *E. coli* grown overnight at 37 °C in Mueller Hinton (MH) broth with shaking at 200 rpm. Next, bacteria suspensions were centrifuged (2300× g, 15 min) and resuspended in PBS after removing the supernatant. For experiments, bacteria suspensions (OD$_{600}$ = 0.1 in C = 10 mM PBS buffer, pH = 7.4) have been used in the presence (concentration range 0.5–2.0 μM) and without the studied compounds. The measurements were performed in quartz cuvette (1 cm × 1 cm). Fluorescence was monitored at the $\lambda_{exc.}$ = 295 nm and $\lambda_{em.}$ = 350 nm from tryptophan (Trp^{214}) residues in bacterial membrane proteins. The studies were carried out using Perkin-Elmer LS-55B (Perkin-Elmer, Pontyclun, UK) spectrofluorometer [25].

3.2.3. Cytotoxicity Tests

The studied compounds were evaluated for cytotoxicity in human cancer cell lines (cervical carcinoma HeLa, chronic myelogenous leukemia K562 and breast adenocarcinoma MCF7) and normal human retina cells RPE-1 (ca. 5.0×10^4 cells·mL^{-1}) after 72 h of treatment using resazurin (Merck/MilliporeSigma, St. Louis, MO, USA) as described earlier for Calcein AM dye [26]. The IC$_{50}$ values (μM) obtained from at least three independent experiments in triplicates are shown in Table 3.

4. Conclusions

26-Thiodiosgenin (**3**) is readily available from diosgenin acetate by treatment with hydrogen sulfide/BF$_3$·Et$_2$O followed by hydrolysis [10]. The chemical reactivity of this compound has been investigated. We found that the MCPBA oxidation of 26-thiodiosgenin (**3**) can be performed chemo- and stereoselectively at -78 °C. Under these conditions, the C5-C6 double bond is not affected and the axial (*S*)-sulfoxide **6** is formed in a large excess: diastereomeric excess (*de*) around 75–80%. With 2.2 equiv. of MCPBA at -40 °C the major reaction product is sulfone **8**. The pure epimeric sulfoxides **6** and **7** rapidly isomerize in solution to afford an equilibrium mixture in the ratio of 3:2. The sulfoxide **6** and the sulfone **8** undergo deprotonation with n-BuLi and then can be alkylated with methyl or ethyl iodides. 26-Thiodiosgenin (**3**) has a strong antimicrobial activity against Gram-positive and Gram-negative bacteria. However, the *S*-oxidation and further α-methylation of **3** decreases its antimicrobial potential. The sulfoxides **7** and **10**, as well as alkylated sulfones **11** and **12**, exhibit a weak cytotoxicity against HeLa cell lines.

Supplementary Materials: The following supporting information can be downloaded at: https://www.mdpi.com/article/10.3390/molecules28010189/s1.

Author Contributions: Conceptualization, A.M.T. and J.W.M.; methodology, S.S. and J.W.M.; validation, D.C.-S., S.S. and J.W.M.; formal analysis, A.M.T., D.C.-S. and S.S.; investigation, A.M.T., E.O.-G. and L.R.; resources, J.W.M. and S.S.; data curation, A.M.T., E.O.-G. and L.R.; writing—original draft preparation, J.W.M. and S.S.; writing— review and editing, J.W.M. and D.C.-S.; visualization, A.M.T. and E.O.-G.; supervision, J.W.M.; project administration, A.M.T.; funding acquisition, J.W.M. All authors have read and agreed to the published version of the manuscript.

Funding: This research received no external funding.

Institutional Review Board Statement: Not applicable.

Informed Consent Statement: Not applicable.

Data Availability Statement: Not applicable.

Acknowledgments: We are very grateful to the University of Bialystok for the continued support of our research program within grant no. BST-124. The authors are grateful to Leszek Siergiejczyk for recording NMR spectra and Jadwiga Maj for a skillful technical assistance.

Conflicts of Interest: The authors declare no conflict of interest.

References

1. Hostettmann, K.; Marston, A. *Saponins*; University Press: Cambridge, UK, 1995.
2. Patel, K.; Singh, R.B.; Patel, D.K.J. Medicinal significance, pharmacological activities, and analytical aspects of solasodine: A concise report of current scientific literature. *J. Acute Dis.* **2013**, *2*, 92–98. [CrossRef]
3. Hanson, J.R. *Natural Products: The Secondary Metabolites*; Royal Society of Chemistry: Cambridge, UK, 2003.
4. Cordell, G.A.; Quinn-Beattie, M.L.; Farnsworth, N.R. The potential of alkaloids in drug discovery. *Phytother. Res.* **2001**, *15*, 183–205. [CrossRef] [PubMed]
5. Hale, K.J. Steroidal alkaloids. In *Second Supplements to the 2nd Edition of Rodd's Chemistry of Carbon Compounds, Vol. IV: Heterocyclic Compounds*; Sainsbury, M., Ed.; Elsevier: Amsterdam, The Netherlands, 1998; Chapter 35; pp. 65–112.
6. Jiang, Q.-W.; Chen, M.-W.; Cheng, K.-J.; Yu, P.-Z.; Wei, X.; Shi, Z. Therapeutic Potential of Steroidal Alkaloids in Cancer and Other Diseases. *Med. Res. Rev.* **2016**, *36*, 119–143. [CrossRef] [PubMed]
7. Dey, P.; Kundu, A.; Chakraborty, H.J.; Kar, B.; Choi, W.S.; Lee, M.M.; Bhakta, T.; Atanasov, A.G.; Kim, H.S. Therapeutic value of steroidal alkaloids in cancer: Current trends and future perspectives. *Int. J. Cancer* **2018**, *145*, 1731–1744. [CrossRef] [PubMed]
8. Sucha, L.; Tomsik, P. The Steroidal Glycoalkaloids from Solanaceae: Toxic Effect, Antitumour Activity and Mechanism of Action. *Planta Med.* **2016**, *82*, 379–387. [CrossRef]
9. Uhle, F.C. Synthesis of a Diosgenin Ring F Thia Counterpart. *J. Org. Chem.* **1962**, *83*, 2797–2799. [CrossRef]
10. Wang, J.; Wu, J.; Tian, W. BF$_3$·Et$_2$O Promoted Sulfuration of Steroidal Sapogenins. *Chin. J. Chem.* **2015**, *33*, 632–636. [CrossRef]
11. Bhardwaj, N.; Tripathi, N.; Goel, B.; Jain, S.K. Anticancer Activity of Diosgenin and Its Semi-synthetic Derivatives: Role in Autophagy Mediated Cell Death and Induction of Apoptosis. *Mini Rev. Med. Chem.* **2021**, *21*, 1646–1665. [CrossRef]
12. Chen, P.; Wang, P.; Song, N.; Li, M. Convergent synthesis and cytotoxic activities of 26-thio- and selenodioscin. *Steroids* **2013**, *78*, 959–966. [CrossRef]

13. Quan, H.-J.; Koyanagi, J.; Ohmori, K.; Uesato, S.; Tsuchido, T.; Saito, S. Preparations of heterospirostanols and their pharmacological activities. *Eur. J. Med. Chem.* **2002**, *37*, 659–669. [CrossRef]
14. Ibrahim-Ouali, M.; Santelli, M. Recent advances in thiasteroids chemistry. *Steroids* **2006**, *71*, 1025–1044. [CrossRef] [PubMed]
15. Czajkowska-Szczykowska, D.; Olchowik-Grabarek, E.; Sękowski, S.; Żarkowski, J.; Morzycki, J.W. Concise synthesis of E/F ring spiroethers from tigogenin. Carbaanalogs of steroidal sapogenins and their biological activity. *J. Steroid Biochem. Mol. Biol.* **2022**, *224*, 106174. [CrossRef] [PubMed]
16. Kowalski, P.; Mitka, K.; Ossowska, K.; Kolarska, Z. Oxidation of sulfides to sulfoxides. Part 1: Oxidation using halogen derivatives. *Tetrahedron* **2005**, *61*, 1933–1953. [CrossRef]
17. Noyori, R.; Aoki, M.; Sato, K. Green oxidation with aqueous hydrogen peroxide. *Chem. Commun.* **2003**, 1977–1986. [CrossRef] [PubMed]
18. Xiong, Z.-G.; Zhang, J.; Hu, X.-M. Selective oxidation of spirolactone-related sulfides to corresponding sulfoxides and sulfones by hydrogen peroxide in the presence of N-hydroxysuccinimde. *Appl. Catal. A Gen.* **2008**, *334*, 44–50. [CrossRef]
19. Roy, K.-M. *Ullmann's Encyclopedia of Industrial Chemistry: Sulfones and Sulfoxides*; Wiley: Hoboken, NJ, USA, 2000; p. 711. [CrossRef]
20. Li, X.; Liu, S.; Qu, L.; Chen, Y.; Yuan, C.; Qin, A.; Liang, J.; Huang, Q.; Jiang, M.; Zou, W.J. Dioscin and diosgenin: Insights into their potential protective effects in cardiac diseases. *J. Ethnopharmacol.* **2021**, *274*, 114018. [CrossRef]
21. Cong, S.; Tong, Q.; Peng, Q.; Shen, T.; Zhu, X.; Xu, Y.; Qi, S. In vitro anti-bacterial activity of diosgenin on Porphyromonas gingivalis and Prevotella intermedia. *Mol. Med. Rep.* **2020**, *22*, 5392–5398. [CrossRef]
22. Brown, L.; Wolf, J.M.; Prados-Rosales, R.; Casadevall, A. Through the wall: Extracellular vesicles in Gram-positive bacteria, mycobacteria and fungi. *Nat. Rev. Microbiol.* **2015**, *13*, 620–630. [CrossRef]
23. Feng, M.; Tang, B.; Liang, S.H.; Jiang, X. Sulfur Containing Scaffolds in Drugs: Synthesis and Application in Medicinal Chemistry. *Curr. Top. Med. Chem.* **2016**, *16*, 1200–1216. [CrossRef]
24. Lakowicz, J.R. *Principles of Fluorescence Spectroscopy*, 2nd ed.; Kluwer Academic/Plenum Publishers: New York, NY, USA, 1999.
25. Olchowik-Grabarek, E.; Sękowski, S.; Kwiatek, A.; Płaczkiewicz, J.; Abdulladjanova, N.; Shlyonsky, V.; Swiecicka, I.; Zamaraeva, M. The Structural Changes in the Membranes of Staphylococcus aureus Caused by Hydrolysable Tannins Witness Their Antibacterial Activity. *Membranes* **2022**, *12*, 1124. [CrossRef]
26. Rárová, L.; Sedlák, D.; Oklestkova, J.; Steigerová, J.; Liebl, J.; Zahler, S.; Bartůněk, P.; Kolář, Z.; Kohout, L.; Kvasnica, M.; et al. The novel brassinosteroid analog BR4848 inhibits angiogenesis in human endothelial cells and induces apoptosis in human cancer cells in vitro. *J. Steroid Biochem. Mol. Biol.* **2016**, *159*, 154–169. [CrossRef] [PubMed]

Article

An Exemestane Derivative, Oxymestane-D1, as a New Multi-Target Steroidal Aromatase Inhibitor for Estrogen Receptor-Positive (ER⁺) Breast Cancer: Effects on Sensitive and Resistant Cell Lines

Cristina Amaral [1,2,*], Georgina Correia-da-Silva [1,2], Cristina Ferreira Almeida [1,2], Maria João Valente [3], Carla Varela [4,5], Elisiário Tavares-da-Silva [4], Anne Marie Vinggaard [3], Natércia Teixeira [1,2] and Fernanda M. F. Roleira [4,*]

[1] UCIBIO, REQUIMTE, Laboratory of Biochemistry, Department of Biological Sciences, Faculty of Pharmacy, University of Porto, Rua Jorge Viterbo Ferreira, n° 228, 4050-313 Porto, Portugal
[2] Associate Laboratory i4HB, Institute for Health and Bioeconomy, Faculty of Pharmacy, University of Porto, 4050-313 Porto, Portugal
[3] National Food Institute, Technical University of Denmark, 2800 Kongens Lyngby, Denmark
[4] Univ Coimbra, CIEPQPF, Faculty of Pharmacy, Laboratory of Pharmaceutical Chemistry, Azinhaga de Santa Comba, Pólo III, Pólo das Ciências da Saúde, 3000-548 Coimbra, Portugal
[5] CIEPQPF, Coimbra Institute for Clinical and Biomedical Research (iCBR), Clinic Academic Center of Coimbra (CACC), Faculty of Medicine, University of Coimbra, Azinhaga de Santa Comba, Pólo III Pólo das Ciências da Saúde, 3000-548 Coimbra, Portugal
[*] Correspondence: cristinamaralibd@gmail.com (C.A.); froleira@ff.uc.pt (F.M.F.R.); Tel.: +351-220428560 (C.A.); +351-239488400 (F.M.F.R.); Fax: +351-226093390 (C.A.); +351-239488503 (F.M.F.R.)

Citation: Amaral, C.; Correia-da-Silva, G.; Almeida, C.F.; Valente, M.J.; Varela, C.; Tavares-da-Silva, E.; Vinggaard, A.M.; Teixeira, N.; Roleira, F.M.F. An Exemestane Derivative, Oxymestane-D1, as a New Multi-Target Steroidal Aromatase Inhibitor for Estrogen Receptor-Positive (ER⁺) Breast Cancer: Effects on Sensitive and Resistant Cell Lines. *Molecules* **2023**, *28*, 789. https://doi.org/10.3390/molecules28020789

Academic Editors: Marina Savić, Erzsébet Mernyák, Jovana Ajdukovic and Suzana Jovanović-Šanta

Received: 21 December 2022
Revised: 6 January 2023
Accepted: 9 January 2023
Published: 12 January 2023

Abstract: Around 70–85% of all breast cancer (BC) cases are estrogen receptor-positive (ER⁺). The third generation of aromatase inhibitors (AIs) is the first-line treatment option for these tumors. Despite their therapeutic success, they induce several side effects and resistance, which limits their efficacy. Thus, it is crucial to search for novel, safe and more effective anti-cancer molecules. Currently, multi-target drugs are emerging, as they present higher efficacy and lower toxicity in comparison to standard options. Considering this, this work aimed to investigate the anti-cancer properties and the multi-target potential of the compound $1\alpha,2\alpha$-epoxy-6-methylenandrost-4-ene-3,17-dione (**Oxy**), also designated by Oxymestane-D1, a derivative of Exemestane, which we previously synthesized and demonstrated to be a potent AI. For this purpose, it was studied for its effects on the ER⁺ BC cell line that overexpresses aromatase, MCF-7aro cells, as well as on the AIs-resistant BC cell line, LTEDaro cells. **Oxy** reduces cell viability, impairs DNA synthesis and induces apoptosis in MCF-7aro cells. Moreover, its growth-inhibitory properties are inhibited in the presence of ERα, ERβ and AR antagonists, suggesting a mechanism of action dependent on these receptors. In fact, **Oxy** decreased ERα expression and activation and induced AR overexpression with a pro-death effect. Complementary transactivation assays demonstrated that **Oxy** presents ER antagonist and AR agonist activities. In addition, **Oxy** also decreased the viability and caused apoptosis of LTEDaro cells. Therefore, this work highlights the discovery of a new and promising multi-target drug that, besides acting as an AI, appears to also act as an ERα antagonist and AR agonist. Thus, the multi-target action of **Oxy** may be a therapeutic advantage over the three AIs applied in clinic. Furthermore, this new multi-target compound has the ability to sensitize the AI-resistant BC cells, which represents another advantage over the endocrine therapy used in the clinic, since resistance is a major drawback in the clinic.

Keywords: breast cancer; endocrine therapy; endocrine resistance; aromatase inhibitors; exemestane; oxymestane; anti-cancer properties; multi-target compounds; aromatase; estrogen receptor; androgen receptor

1. Introduction

In 2020, around 2.3 million new breast cancer cases were diagnosed [1], with about 70–85% of the cases being estrogen receptor-positive (ER$^+$). In this type of tumor, estrogens play a pivotal role in cell growth, with endocrine therapy being the main therapeutic approach used in clinic, by blocking estrogen synthesis through aromatase inhibitors (AIs), or preventing the activation of estrogen-signaling pathways, through the use of anti-estrogens [2,3]. In fact, the third-generation of AIs, Anastrozole (Ana), Letrozole (Let) and Exemestane (Exe) are the first-line treatment options in post-menopausal women, with early and metastatic stage, as well as in pre-menopausal women, after ovarian function suppression [4–6]. Nevertheless, despite their clinical benefit, they can induce several side effects, and their continued use may lead to the development of acquired resistance, which causes tumor re-growth and disease progression [2]. To improve treatment, some combined therapies with CDK4/6 or mTOR inhibitors and AIs have been applied in clinic. However, despite improving the overall therapeutic outcome, some of these combinations have worrisome cytotoxic effects or do not improve the expansion on overall survival, and as a result resistance continues to occur [2,7–9]. Furthermore, it has been reported for CDK4/6 inhibitors that 10% of patients develop de novo resistance, while others develop acquired resistance after 24–28 months when used as first-line therapy or after a shorter period when used as second-line therapy [10,11]. In addition to this concern, it is known that the endocrine therapy with AIs causes loss of bone mineral density, increasing the risk of bone fractures and osteoporosis [12,13]. Clinical application of bisphosphonates with AIs successfully prevents these adverse effects; however, it should be pointed out that their use is associated with some mild-to-severe side-effects [14,15]. Considering all this, the search for new, safer and more effective forms of treatment is crucial to improve therapy for ER$^+$ breast cancer.

In line with this, our group has designed and synthesized new potent steroidal AIs with promising anti-cancer properties in sensitive ER$^+$ breast cancer cells [16–21]. Some of these new AIs, in addition to acting on aromatase, have the ability to modulate ERα (selective ER modulator, SERM), as well as the androgen receptor (AR), acting as multi-target drugs in cancer cells [19,21]. Considering the key roles of aromatase and ERs on ER$^+$ breast tumors, the anti-cancer drugs with dual AI and SERM properties are pointed as a future therapeutic strategy for ER$^+$ breast cancer treatment [22]. In fact, the binding pockets of aromatase and ERs contain equivalent residues, suggesting that they can accommodate the same ligands [3]. In the past, this type of dual therapeutic strategy was studied using the combination of Ana with Tamoxifen, or Let with Fulvestrant, though AIs alone were more effective than when combined with a SERM or a selective ER down-regulator (SERD) [23,24]. Moreover, the uptake of different drugs enhances the risk of drug interactions and consequently leads to more pronounced side effects, which highlights the importance of a single compound with multi-target action [3]. Multi-target drugs are emerging because, in addition to being able to improve the overall tolerance to anti-cancer agents, they are pointed out to be more effective, more potent and less toxic in comparison to the standard options [3,25]. Several groups, including ours, have been working on the search for multi-target drugs for breast cancer treatment [19,21,22,26–29], with norendoxifen being the first multi-target compound identified [29]. As far as we know, only our group has demonstrated the in vitro anti-cancer properties of some of these molecules in ER$^+$ breast cancer cells [19,21,26]. Thus, following this line of research, in this work we investigated the anti-cancer properties and the multi-target potential of the compound 1α,2α-epoxy-6-methylenandrost-4-ene-3,17-dione (Figure 1), also known as Oxymestane-D1 (**Oxy**), which was previously designed and synthesized by our group [20]. **Oxy** was designed with the aim of preparing a molecule that simultaneously inhibits aromatase, but also possesses anti-proliferative activity through other mechanisms. For this, an epoxide function, which is a structural feature associated with anti-proliferative activity [30], was added at position 1,2 of Exe, a very potent AI. Actually, **Oxy** is a potent AI in human placental microsomes (IC$_{50}$ of 0.81 μM) and in ER$^+$ breast cancer cells (MCF-7aro) (IC$_{50}$ of 1.18 μM), reducing

MCF-7aro cells viability in an aromatase-dependent manner, being in this last case more potent than Exe [20]. In addition, our group demonstrated that **Oxy** can also reduce the viability of lung, liver, colon and prostate cancer cell lines, even being considered to be a more potent anti-cancer molecule than other conventional chemotherapeutic drugs [31].

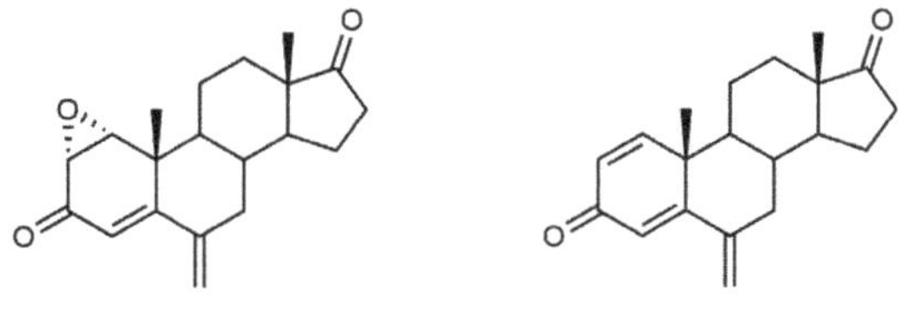

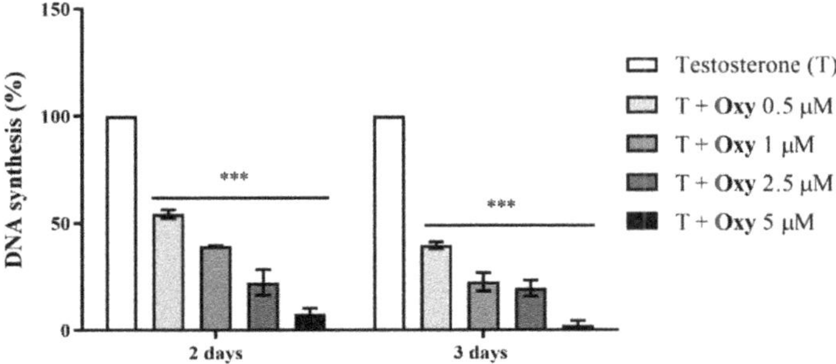

Figure 1. Chemical structures of 1α,2α-epoxy-6-methylenandrost-4-ene-3,17-dione, also known as Oxymestane-D1 (**Oxy**) (**A**) and its parent compound exemestane (Exe) (**B**).

2. Results

2.1. Effects of Oxy on Proliferation of ER⁺ Breast Cancer Cells

As previously reported by our group [20], **Oxy** significantly reduced the viability of MCF-7aro cells, in a dose- and time-dependent manner. To further understand the anti-cancer properties of the AI **Oxy** on ER⁺ breast cancer cells, its effects on MCF-7aro cell proliferation were explored by performing thymidine incorporation assays, as well as cell cycle analysis. MCF-7aro cells stimulated with T (1 nM) were treated with **Oxy** (0.1–5 μM) for 2 and 3 days. Results demonstrate that this AI dramatically reduced ($p < 0.001$) the rate of DNA synthesis in a dose- and time- dependent manner (Figure 2). Furthermore, as presented in Table 1, this compound impaired the progression of cell cycle by causing a significant ($p < 0.001$) cell cycle arrest on G_0/G_1 phase, which consequently significantly ($p < 0.001$) reduced the number of cells on the S phase.

Figure 2. Effects of **Oxy** on proliferation (rate of DNA synthesis) of sensitive breast cancer cells. MCF-7aro cells stimulated with T (1 nM) were treated with different concentrations of **Oxy** (0.5–5 μM), for 2 and 3 days. Cells only stimulated with T were used as control (100% of DNA synthesis). Results are expressed as a mean ± SEM of at least three independent experiments, each performed in triplicate. Significant differences between **Oxy**-treated cells and the control (T-stimulated cells) are shown by *** ($p < 0.001$).

Table 1. Effects of **Oxy** on MCF-7aro cell cycle progression.

	G_0/G_1	S	G_2/M
Testosterone	74.29 ± 0.68	11.34 ± 0.77	12.43 ± 0.74
T + **Oxy** (1 µM)	82.32 ± 0.72 ***	5.20 ± 0.20 ***	9.60 ± 0.89
T + **Oxy** (2.5 µM)	83.78 ± 0.58 ***	4.52 ± 0.37 ***	8.90 ± 0.65 *

Cells stimulated with T (1 nM) were treated with **Oxy** at 1 and 2.5 µM for 3 days. Cells were stained with PI (1 µg/mL) and analyzed by flow cytometry. Values are expressed as a percentage of single cell events in each stage of the cell cycle and are the mean ± SEM of at least three independent experiments, all performed in triplicate. Significant differences between the control and treated cells are shown by * ($p < 0.01$) and *** ($p < 0.001$).

2.2. Effects of *Oxy* on Cell Death of ER^+ Breast Cancer Cells

To understand if the effects of **Oxy** on viability of MCF-7aro cells [20] were a consequence of apoptosis, the translocation of PS to the outer surface of plasma membrane, the activities of caspases-7/-9/-8 and the mitochondrial transmembrane potential ($\Delta\Psi$m) were investigated in cells treated with **Oxy** at 1 µM for 3 days. This concentration was similar to the IC_{50} value of aromatase inhibition (IC_{50} of 1.18 µM) reported for **Oxy** in this cell model [20].

As presented in Figure 3, **Oxy** significantly increased ($p < 0.001$) the activities of caspase-7 (1.29 fold) (Figure 3A), caspase-9 (2.48 fold) (Figure 3B) and caspase-8 (1.96 fold) (Figure 3C) in comparison to control. The activation of all these caspases was significantly ($p < 0.001$) prevented by the incubation with Z-VAD-FMK, a pan-caspase inhibitor used as a negative control. In addition, by analyzing the $\Delta\Psi$m, it was observed that **Oxy** induced a 5.66 times ($p < 0.001$) greater $\Delta\Psi$m loss, when compared to control (Table 2). As mitochondria dysfunction can be linked with the formation of ROS, its generation was also evaluated. We verified (Figure 3D) that **Oxy** leads to a significant ($p < 0.05$) increase in production of ROS, when compared to control.

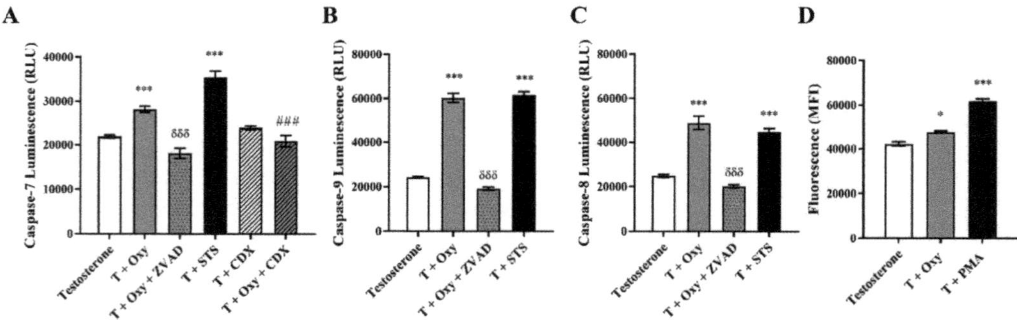

Figure 3. Involvement of apoptosis in the effects induced by **Oxy** on sensitive breast cancer cells. MCF-7aro cells stimulated with T (1 nM) were treated with **Oxy** (1 µM) with or without CDX (1 µM), for 3 days. After treatment, the activities of (**A**) caspase-7, (**B**) caspase-9 and (**C**) caspase-8 as well as (**D**) the intracellular production of ROS were analyzed. Cells treated only with T were considered as a control, while cells treated with STS (10 µM) and PMA (25 ng/mL) were used as positive controls for caspase activation assays and ROS, respectively. Z-VAD-FMK was used as a pan-caspase inhibitor in caspases activation assays. Results are shown as the mean ± SEM of at least three independent experiments, each performed in triplicate. Results are expressed as relative luminescence units (RLU) for caspases activation assays and as mean fluorescence intensity (MFI) for ROS. Significant differences between the control and **Oxy**-treated cells are shown by * ($p < 0.05$) and *** ($p < 0.001$), while differences between **Oxy**-treated cells with or without Z-VADFMK are indicated by $\delta\delta\delta$ ($p < 0.001$) and differences between **Oxy**-treated cells with or without CDX are indicated by ### ($p < 0.001$).

Table 2. Effects of **Oxy** on mitochondrial transmembrane potential ($\Delta\psi$m) in MCF-7aro cells.

	Viable Cells	**Cells with $\Delta\psi$m Loss**
Testosterone	94.66 ± 0.79	5.35 ± 0.78
T + **Oxy** (1 μM)	69.74 ± 0.74 ***	30.27 ± 0.73 ***
T + CCCP (10 μM)	41.43 ± 1.24 ***	58.56 ± 1.23 ***

Cells were stimulated with T (1 nM) and incubated with **Oxy** (1 μM) for 3 days. Treated cells were harvested and labelled with $DiOC_6(3)$ and PI followed by flow cytometry analysis. Data are presented as % of viable cells and cells with $\Delta\psi$m loss. Cells only cultured with T were considered as control and cells treated with T plus CCCP (10 μM) were considered as positive control. The data represent means $\pm$ SEM of three independent experiments conducted in triplicate. The ratio treatment/control is presented in bold within brackets. Significant differences between the control versus treated cells are indicated by *** ($p < 0.001$).

Moreover, by labelling MCF-7aro cells with Annexin V, it was detected that **Oxy** caused a significant ($p < 0.001$) increase (3.21 times) in the binding to Annexin V when compared to control (Table 3). This effect was also accompanied by an increase in 7-AAD$^+$ cells ($p < 0.001$).

Table 3. Effects of **Oxy** on Annexin V-FITC labelling in MCF-7aro cells.

	AnnexinV$^-$/7-AAD$^-$	**AnnexinV$^+$/7-AAD$^-$**	**AnnexinV$^+$/7-AAD$^+$**
Testosterone	85.45 ± 0.22	4.96 ± 0.73	8.70 ± 0.78
T + **Oxy** (1 μM)	61.74 ± 1.15 ***	15.93 ± 1.02 (3.21) ***	24.13 ± 0.87 ***
T + STS (10 μM)	59.20 ± 5.50 ***	21.99 ± 0.76 (4.43) ***	19.84 ± 2.76 *

Cells stimulated with T (1 nM) and incubated with **Oxy** (1 μM) for 3 days were labeled with Annexin V-FITC and 7-AAD followed by flow cytometry analysis. Data are presented as % of viable cells (Annexin V$^-$/7-AAD$^-$), % of early apoptotic (Annexin V$^+$/7-AAD$^-$) and late apoptotic or necrotic cells (Annexin V$^+$/7-AAD$^+$). Cells only treated with T were considered as control, while cells treated with STS (10 μM) were considered as positive control for apoptosis. The results are expressed as mean $\pm$ SEM of three independent experiments, performed in triplicate. The ratio treatment/control is presented in bold within brackets. Significant differences among the control and treated cells are denoted by * ($p < 0.05$), *** ($p < 0.001$).

2.3. Possible Mechanism of Action of *Oxy*: The Involvement of Aromatase, Estrogen Receptor and Androgen Receptor

Our group had previously described that **Oxy** presents an aromatase inhibition of 88.6% and an IC_{50} value of 1.18 μM in MCF-7aro cells and that it induces an aromatase-dependent reduction in cell viability [20]. Taking this into account and the fact that **Oxy** impaired cell growth and induced apoptosis, we further investigated its targets and mechanism of action by studying the involvement of aromatase, ERs and AR. MCF-7aro cells stimulated with T (1 nM) or E_2 (1 nM) were treated with **Oxy** (1 μM) with or without the ERα antagonist ICI 182 780 (fulvestrant) (100 nM), ERβ antagonist PHTPP (1 μM) or AR antagonist CDX (1 μM) for 3 days.

In relation to aromatase, and as previously reported [20], significant differences ($p < 0.05$, $p < 0.01$, and $p < 0.001$ for 0.5, 1 and 5 μM **Oxy**, respectively) were observed between T- versus E_2-treated cells, after 3 days of treatment (Figure 4A), a behaviour similar to Exe. However, contrary to Exe, **Oxy** caused no significant alteration to the expression levels of aromatase (Figure 4B).

Regarding ERα, the results demonstrated that, in the presence of the ERα antagonist ICI, **Oxy** was unable to reduce cell viability, and a significant increase ($p < 0.001$) in cell viability was even observed when compared with cells only treated with **Oxy** (Figure 5A). Moreover, this compound significantly reduced the protein expression levels of ERα ($p < 0.01$; Figure 5B), as well as the transcript levels of *ESR1* gene ($p < 0.001$; Figure 5C), the gene that encodes ERα, with this latter effect not observed when cells are treated with ICI. In addition, similarly to ICI, the new AI significantly decreased the transcript levels of the ERα-regulated genes (Figure 5C), *TFF1* ($p < 0.01$), *EGR3* ($p < 0.01$) and *PDZK1* ($p < 0.001$). Data on ER transactivation assay showed that **Oxy** acts as a potent ER antagonist in the presence of T (Figure 6A), with over 40% inhibition of T agonistic effect for all concentrations of **Oxy** tested ($p < 0.001$). Some inhibition was observed for the two lowest concentrations in the absence of the ER agonist ($p < 0.05$), meaning that **Oxy** does not appear to detain any ER agonistic potential. Furthermore, no effect on cell viability was observed under any of the tested conditions (Figure 6A).

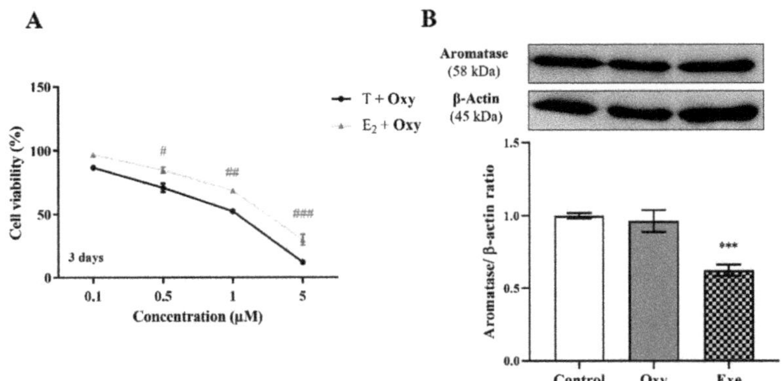

Figure 4. Involvement of aromatase in the effects induced by **Oxy** on breast cancer cells. (**A**) Effects of **Oxy** (0.1–5 μM) on viability of MCF-7aro cells stimulated with T (1 nM) or E_2 (1 nM), after 3 days of exposure. (**B**) Effects of **Oxy** (1 μM) on aromatase protein expression levels of MCF-7aro cells, after 8 h. Cells without **Oxy** treatment were considered as control. Exe at 10 μM was used as a reference AI. For Western Blot, β-actin was used as a loading control, with the densitometry results presented as aromatase/β-actin ratio. The protein expression obtained for treated cells was normalized in relation to protein expression of control. Results are presented as the mean ± SEM of at least three independent experiments, each performed in triplicate. Significant differences between T-treated and E_2-treated cells are denoted by # ($p < 0.05$), ## ($p < 0.01$) and ### ($p < 0.001$), while significant differences between the control and Exe-treated cells are shown by *** ($p < 0.001$).

On the other hand, in relation to the involvement of ERβ, our results showed that in the presence of the ERβ antagonist PTHPP, **Oxy** did not affect cell viability when compared to control (Figure 7A). Thus, significant differences ($p < 0.001$) in cell viability between **Oxy**-treated cells with and without PHTPP were noticed. Nevertheless, **Oxy** did not affect the expression levels of ERβ protein (Figure 7B) when compared to control.

In relation to the involvement of AR on **Oxy** action, results demonstrated that when AR is blocked by CDX, a significant ($p < 0.001$) increase in cell viability is detected when compared with cells treated only with **Oxy** (Figure 8A). In addition, by Western blot, it was demonstrated that this new molecule has the ability to significantly ($p < 0.001$) increase the expression levels of AR when compared to control (Figure 8B). To understand the role of this AR overexpression, the activity of caspase-7 was evaluated when AR was blocked and, as presented in Figure 3A, CDX significantly ($p < 0.001$) prevented the activation of caspase-7 induced by **Oxy**. AR activity of **Oxy** was confirmed through the AR-EcoScreen™ transactivation assay. As presented in Figure 6B, results show that **Oxy** acts as an AR agonist, with an induction of 1.8 and 2.9 over control for 0.5 and 1 μM **Oxy**, respectively ($p < 0.001$). Moreover, an induction of AR was also observed when cells were co-exposed to **Oxy** and the AR agonist R1881 (around 1.2-fold increase over control for the two highest concentrations tested; $p < 0.001$), which means that **Oxy** appears not to detain antagonistic potential. In addition, it should be pointed out that the concentrations studied were not cytotoxic in this cell model (Figure 6B).

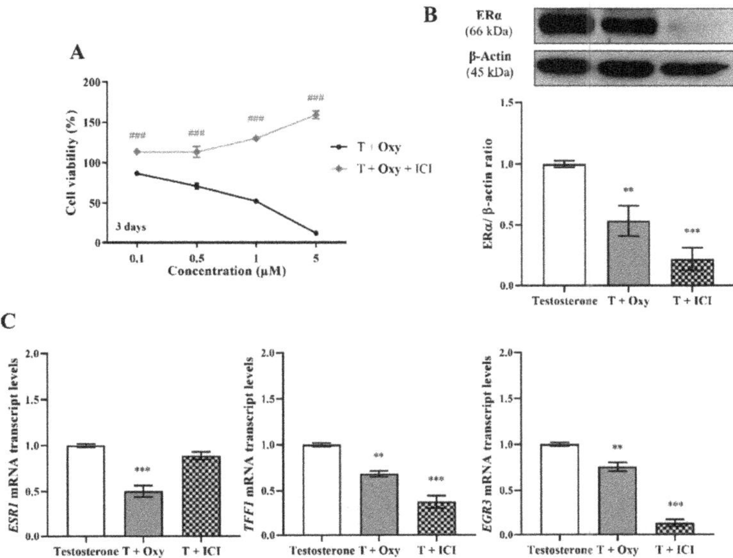

Figure 5. Involvement of ERα in the effects induced by **Oxy** on breast cancer cells. (**A**) Effects of **Oxy** (0.1–5 μM) on viability of MCF-7aro cells stimulated with T (1 nM) and treated with or without ICI (100 nM), after 3 days. Effects of **Oxy** (1 μM) on the expression levels of (**B**) ERα protein or on (**C**) mRNA transcript levels of *ESR1, TFF1, EGR3* and *PDZK1* genes, after 3 days. Cells without **Oxy** treatment were considered as control, while cells treated with ICI (100 nM) were designated as negative control. β-actin was used as a loading control, with the densitometry results presented as ERα/β-actin ratio. The protein expression obtained for treated cells was normalized in relation to protein expression of control (1 nM T). To quantify the mRNA transcript levels of *ESR1, TFF1, EGR3* and *PDZK1* genes, the housekeeping gene *TUBA1A* was used. The mRNA transcript levels of treated cells were normalized in relation to mRNA transcript levels of control (Testosterone). Results are presented as the mean ± SEM of at least three independent experiments, each performed in triplicate. Significant differences between cells treated with **Oxy** with or without ICI are denoted by ### ($p < 0.001$), while between the control and **Oxy**- or ICI-treated cells are shown by ** ($p < 0.01$) and *** ($p < 0.001$).

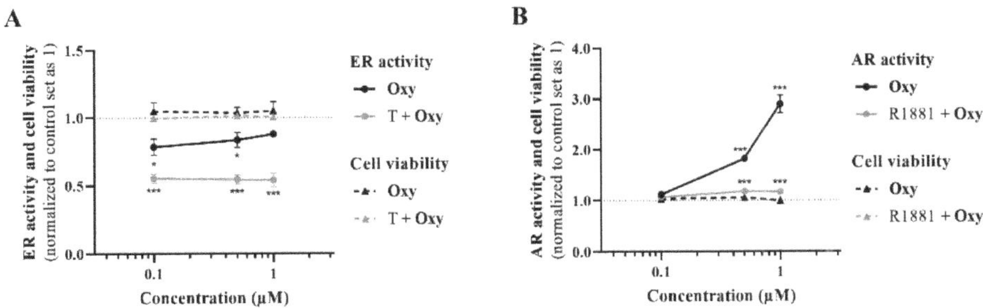

Figure 6. ER and AR transactivation assays. (**A**) Effects of **Oxy** (0.1–1 μM) on ER activation, in the presence or absence of 1 nM T, after 24 h of incubation. (**B**) Effects of **Oxy** (0.1–1 μM) on AR activation, in the presence or absence of 0.1 nM R1881, after 24 h of incubation. Data were normalized to control (cells not treated with **Oxy**), which was set as 1. Results are presented as mean ± SEM of four independent experiments, each performed in triplicate. Significant differences between the control and cells treated with **Oxy** are denoted by * ($p < 0.05$) and *** ($p < 0.001$).

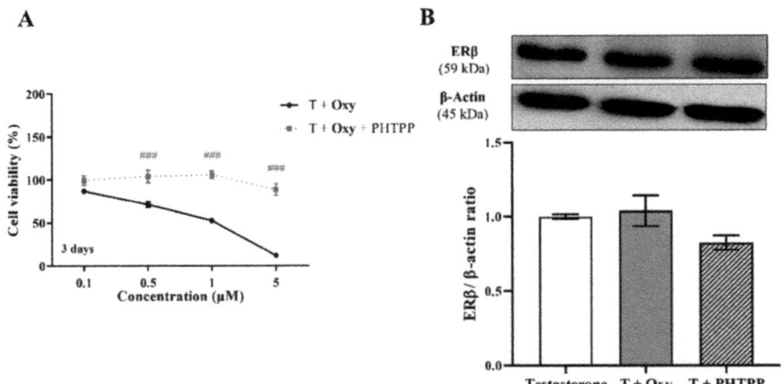

Figure 7. Involvement of ERβ in the effects induced by **Oxy** on breast cancer cells. (**A**) Effects of **Oxy** (0.1–5 μM) on viability of MCF-7aro cells stimulated with T (1 nM) and treated with or without PHTPP (1 μM) after 3 days. (**B**) Effects of **Oxy** (1 μM) on the expression levels of ERβ protein after 3 days. Cells without **Oxy** treatment were considered as control, while cells treated with PHTPP (1 μM) were designated as negative control. β-actin was used as a loading control, with the densitometry results presented as ERβ/β-actin ratio. The protein expression obtained for treated cells was normalized in relation to protein expression of control (1 nM T). Results are the mean ± SEM of at least three independent experiments, each performed in triplicate. Significant differences between cells treated with **Oxy** with or without PHTPP are denoted by ### ($p < 0.001$).

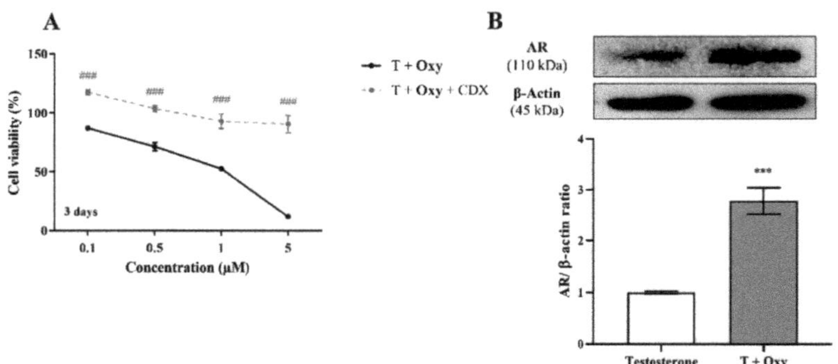

Figure 8. Involvement of AR in the effects induced by **Oxy** on breast cancer cells. (**A**) Effects of **Oxy** (0.1–5 μM) on viability of MCF-7aro cells stimulated with T (1 nM) and treated with or without CDX (1 μM) after 3 days. (**B**) Effects of **Oxy** (1 μM) on the expression levels of AR protein after 3 days. Cells without **Oxy** treatment were considered as control. β-actin was used as a loading control, with the densitometry results presented as AR/β-actin ratio. The protein expression obtained for treated cells was normalized in relation to protein expression of control (1 nM T). Results are the mean ± SEM of at least three independent experiments, each performed in triplicate. Significant differences between cells treated with **Oxy** with or without CDX are denoted by ### ($p < 0.001$), while between the control and **Oxy**-treated cells are shown by *** ($p < 0.001$).

2.4. Effects of *Oxy* on Resistant ER⁺ Breast Cancer Cells

In order to deepen the anti-cancer potential of **Oxy**, we explored its effects on viability of resistant breast cancer cells by performing MTT assay, as well as on the involvement of apoptosis by analyzing caspase-7 activity. LTEDaro cells were treated with **Oxy** (0.1–2.5 μM) for 3 and 6 days. As presented in Figure 9, **Oxy** induced a significant ($p < 0.001$) reduction on the viability of LTEDaro cells, with this effect being dose- and time-dependent. A significant ($p < 0.001$) increase in the activity of caspase-7 in the presence of

Oxy was also noted (Figure 9B). As expected, ZVAD-FMK significantly ($p < 0.001$) reverted the activation of caspase-7 induced by **Oxy**.

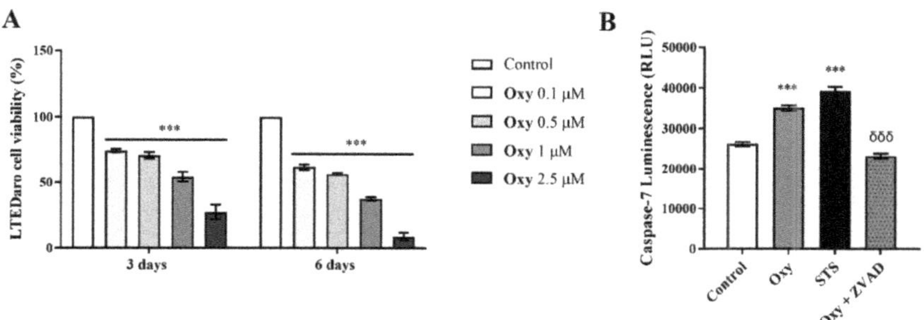

Figure 9. Effects of **Oxy** on viability and cell death of apoptosis of resistant breast cancer cells. (**A**) Effects of **Oxy** (0.1–2.5 μM) on viability of LTEDaro cells after 3 and 6 days. (**B**) Effects of **Oxy** (1 μM) on the activity of caspase-7 after 3 days of treatment. Cells without **Oxy** treatment were considered as control, while cells treated with STS (10 μM) were used as positive control. Z-VAD-FMK was used as a pan-caspase inhibitor. Cell viability effects of **Oxy** were normalized in relation to control (100% of cell viability). Results are the mean ± SEM of at least three independent experiments, each performed in triplicate. For caspase activation assays, results are expressed as relative luminescence units (RLU). Significant differences between **Oxy**-treated cells and the control are shown by *** ($p < 0.001$), while differences between **Oxy**-treated cells with or without Z-VADFMK are indicated by $\delta\delta\delta$ ($p < 0.001$).

3. Discussion

Due to the clinical limitations of ER$^+$ breast cancer therapy, in the last years many efforts have been made to find new therapeutic strategies or to discover more potent drugs with fewer side effects that may improve treatment. Recently, the interest in finding multi-target compounds for cancer has been rising, as they are more effective, more potent, less toxic and associated with reduced risks of drug interactions [3,25]. In fact, our group has been working on this and we have already discovered potent steroidal AIs that, in addition to aromatase inhibition and/or modulation of aromatase levels, also exhibit ERα- and AR-dependent effects [19] and modulate their expression [21] to induce breast cancer cell death. Moreover, we recently reported a non-steroidal molecule (tamoxifen bisphenol) that also acts as a multi-target compound in ER$^+$ breast cancer cells, since it reduces aromatase protein levels and acts as an ERα antagonist and induces ERβ up-regulation to inhibit growth and cause cancer cell death [26]. The ability of an anti-cancer molecule to simultaneously target aromatase, ERs or AR is, from a clinical point of view, very relevant, as these are the main therapeutic targets for these types of tumors. In fact, it should be pointed out that aromatase is the enzyme responsible for estrogen biosynthesis. ERα is responsible for growth, survival and proliferation of breast cancer cells, whereas ERβ display anti-proliferative properties by inhibiting the transcriptional activity of ERα, impairing cell cycle regulation and promoting apoptosis, acting, in that way, as tumor suppressors [3,32–35]. Depending on the hormonal status, breast cancer setting and cell type, AR may exhibit different roles, such as oncogenic or tumor suppressor [36–39]. Considering all this, and as an attempt to discover a new steroidal compound with these key properties, the mechanism of action of **Oxy**—a potent steroidal AI, derivative of Exe, designed and synthesized by our group—was investigated [20]. For this AI, it was reported that it affects the viability of ER$^+$ breast cancer cells [20] and presents promising anti-cancer properties in lung, liver, colon and prostate cancer cell lines by inducing cell cycle arrest, apoptosis and necrosis as well as DNA damage, and also by inhibiting the DNA damage response [31]. In addition, we previously verified that this molecule is more potent than Exe, the reference steroidal AI used in clinic, with regard to the decrease in cell viability [20,31] and other conventional chemotherapeutic drugs [31]. In this study, our results demonstrate that in sensitive MCF-7aro cells, **Oxy** dramatically affected the rate of DNA synthesis and impaired the progression of cell cycle by causing an arrest on G_0/G_1 phase. A similar effect on MCF-7aro cell cycle progression was also reported by our group for Exe [36,40], Ana and Let [41]. However, the G_0/G_1 cell cycle arrest caused by **Oxy** was not as pronounced as for the AIs used in the clinic, but was similar to the one induced by the main oxidized Exe metabolite 6-(hydroxymethyl)androsta-1,4,6-triene-3,17-dione (6-HME) [16]. On the other hand, by evaluating different biomarkers of apoptotic cell death, it was observed that **Oxy** induced $\Delta\Psi$m loss, translocation of PS to the outer surface of plasma membrane, caused activation of caspases-7, -9 and -8 and increased ROS generation. Thus, **Oxy** promoted apoptosis not only by activating the mitochondrial pathway, but also by activating caspase-8 through an unknown mechanism, with the mitochondria dysfunction being ROS-dependent. A similar behavior has also been reported for Exe in this cell line [36,40], although **Oxy** has an advantage, as in addition to activating the mitochondrial apoptotic pathway, it also activates caspase-8, which may promote a more efficient breast cancer cell death. A cross-talk between the intrinsic pathway and caspase-8 was already detected for other steroidal AIs studied by our group, namely Exe [36,40,42] and its metabolites [16], in sensitive and in resistant ER^+ breast cancer cells. For the non-steroidal AI Ana, it was also reported that the induction of apoptosis in breast cancer cells occurred via the mitochondrial pathway [41,43,44], but also by the up-regulation of caspase-8 by an unknown mechanism [44]. Several different mechanisms have been described for the interaction between mitochondria and caspase-8 activation [45–52], although this relationship is not totally elucidated. The ability of **Oxy** to impair cell cycle progression and activate apoptosis of cancer cells was also reported by our group, on lung, liver, colon and prostate cancer cell lines [31].

In order to understand the mechanism of action behind these biological effects and also to highlight the multi-target potential of **Oxy**, we investigated the involvement of aromatase, ERα, ERβ and AR. We previously reported that this AI presented an IC_{50} value for anti-aromatase activity of 1.18 μM in MCF-7aro cells and that it affected cell viability in an aromatase-dependent manner, a behavior similar to Exe [20]. Moreover, the IC_{50} values of **Oxy** and Exe on MCF-7aro cells are similar, as the IC_{50} value reported for Exe is 0.9 μM [20]. Nevertheless, contrary to Exe, which acts as an aromatase destabilizer [53], this study demonstrates that its derivative **Oxy** did not affect aromatase expression levels. Despite that, as in an estrogen-enriched environment, **Oxy** continues to affect MCF-7aro cell viability, the involvements of ERα and ERβ on these effects were explored. Our results demonstrated that in the presence of the ERα antagonist ICI or the ERβ antagonist PHTPP, the growth-inhibitory action induced by **Oxy** on MCF-7aro cells was impaired. Thus, these results suggest that **Oxy** acts on cells in an ERα- and ERβ-dependent manner. In fact, it was verified that **Oxy** decreased the gene and protein expression levels of ERα, as well as the levels of ERα-mediated transcription target genes, *TFF1*, *EGR3* and *PDZK1*. In relation to the effects of **Oxy** on Erβ, our data demonstrated that it did not affect the protein levels of ERβ. This comes in line with the absence of ER agonistic activity in the VM7Luc4E2 assay, as well as the apparent antagonism observed under basal conditions, which is most likely a result of the decrease in ERα expression. Importantly, the ER transactivation assays showed that **Oxy** presents a great ER antagonistic potential. Therefore, all these data indicate that this molecule blocks estrogen signaling, acting as a modulator and antagonist of ERα, a mechanism of action typical of SERMs [54]. This molecule does not seem to act as a down-regulator of ERα, since the reduction in ERα protein levels is a consequence of the decrease in the mRNA transcription of *ESR1* gene and not of ERα degradation, as the SERD ICI [55]. Either way, acting as a ERα antagonist is a therapeutic advantage, since by modulating ERα levels and activation, **Oxy** hampers the oncogenic properties of this receptor, impairing growth and proliferation of breast cancer cells [3,34,35]. This behavior induced by **Oxy** is an advantage in relation to Exe, since it is known that Exe presents weak estrogen-like effects in breast cancer cells [41,56]. In fact, we recently reported that besides reducing the expression levels of ERα on MCF-7aro cells, Exe does not affect the transcription levels of *ESR1* and *EGR3* genes, with this latter effect related to the weak-estrogen like effect induced by Exe [41]. Moreover, the ability of **Oxy** to decrease *EGR3* transcription levels is a clinical benefit, as this gene is correlated with poor response to therapy and decreased disease-free survival and overall survival [57]. In addition, the effects induced by **Oxy** on ERα are also an advantage over Ana and Let, since these non-steroidal AIs increase the expression of ERα at gene and protein levels on breast cancer cells [41]. Recently, clinical data proposed a relationship between leptin levels and the hormonal effects induced by the AIs Exe and Let [58]. Moreover, in MCF-7 cells, it was suggested that leptin may induce a functional activation of ERα through the $ERK_{1/2}$ pathway [59]. As Exe and **Oxy** present a similar steroidal chemical structure, **Oxy** may modulate leptin signaling, as reported for Exe [59], and thus may indirectly affect ERα expression and activation. In addition, considering the lack of cell death in the co-exposure with the ERβ antagonist PHTPP, our results also suggest that **Oxy** can act through ERβ to induce breast cancer cell death. Acting as an agonist of ERβ, it is also a therapeutic advantage, since ERβ is considered a breast tumor suppressor [3,32,33,35]. Nevertheless, it should be pointed that PHTPP exhibits 36-fold more selectivity for ERβ than ERα and that it abrogates estrogen action by acting on ERβ, presenting minimal effects through ERα [60].

It is known that AR is expressed in 85–95% of ER^+ breast cancer cases and in 77% of invasive breast cancers [61]. Furthermore, AR may have different functions after AI therapy. In fact, for the non-steroidal AI letrozole [41,62] and C7- [19] and C6-substituted steroidal AIs, [21] AR presents a pro-death role, while for the steroidal AI Exe it exhibits pro-survival and oncogenic functions [36]. Thus, considering the importance of AR in this type of breast tumor and the dual role of AR function after AI therapy, it is important to understand the involvement and the role of AR on **Oxy** action. Our results indicate that **Oxy** acts on cells in an AR-dependent manner, since the **Oxy** growth–inhibitory effect is compromised when AR is blocked by CDX, being observed an inhibition of apoptosis. Moreover, **Oxy** has the ability to increase the expression levels of AR and, using the AR-EcoScreen™ assay, we further showed that **Oxy** detains agonistic activity towards AR. Therefore, all these results indicate that **Oxy** modulates AR, acting as an AR agonist, which leads to ER^+ breast cancer cell death. This is a therapeutic advantage, as AR acts as a breast tumor suppressor, a mechanism of action similar to the observed for the non-steroidal AI letrozole [41,62] and to other C7- [19] and C6-substituted steroidal AIs [21] synthesized by our group. Interestingly, these AR-associated pro-death effects are not observed with Exe, as in this case, the AR has an oncogenic role [36].

Additionally, **Oxy** was able to re-sensitize the resistant ER^+ breast cancer cells, LTEDaro cells, since it was able to strongly reduce cell viability and increase the activity of the effector caspase-7. This behavior is an advantage of **Oxy** in relation to the AIs used in the clinic, since this cell model is characterized by mimicking long-term resistance to the AIs used in the clinic [16,17,36,42,63–65]. On the other hand, the effects induced by **Oxy** on this resistant cell line are much more pronounced and appealing than those induced by the Exe metabolites 6-HME and 17β-hydroxy-6-methylenandrosta-1,4-dien-3-one (17-βHE) [16].

In conclusion, to the best of our knowledge, this is the first work that describes an anti-cancer molecule, **Oxy**, that in addition to acting as an AI, also modulates both ERs and AR acting as an ERα antagonist and AR agonist, rendering **Oxy** a more effective anti-tumor profile. In fact, from a clinical point of view, these multi-target properties are very relevant for this type of tumor since they correspond to therapeutic targets with key roles in cancer growth, survival or promotion of cancer cell death. Thus, by acting as an AI that also modulates ERα, ERβ and AR, **Oxy** impaired cancer cell growth by disrupting cell cycle and DNA synthesis, and induced apoptosis of sensitive ER^+ breast cancer cells. This multi-target action of **Oxy** on ERs and AR is an advantage over the three AIs used in clinic, since none of these AIs act on these receptors in a therapeutically beneficial way. Additionally, it should be pointed that all these anti-cancer properties were observed at a dose ten times lower than the doses used in similar pre-clinical studies for the three AIs used in clinic [16,36,40,41]. Moreover, we had previously demonstrated that **Oxy** is more potent than Exe with regard to the decrease in cell viability [20,31] and even more potent than other conventional chemotherapeutic drugs [31]. In addition, this multi-target compound also has the ability to re-sensitize resistant breast cancer cells by activating apoptosis. All these findings emphasize the therapeutic potential of **Oxy**. Thus, this work, besides highlighting the importance of multi-target compounds for ER$^+$ breast cancer subtype, also allowed the discovery of a promising multi-target anti-cancer molecule for ER$^+$ breast cancer treatment, **Oxy**.

4. Material and Methods

4.1. Compound under Study

In this work, we studied the steroidal aromatase inhibitor (AI) 1α,2α-epoxy-6-methylenandrost-4-ene-3,17-dione, which was previously synthesized by our group [20] and further designated as Oxymestane-D1 (**Oxy**).

4.2. Cell Culture

In this study, two different ER$^+$ breast cancer cell lines were used to investigate the in vitro effects of AI **Oxy**, the sensitive MCF-7aro cells and the resistant LTEDaro cells. MCF-7aro cells are an ER$^+$ aromatase-overexpressing human breast cancer cell line, obtained from the parental human epithelial ER$^+$ breast cancer cell line (MCF-7 cells), after stable transfection with the human placental aromatase gene using Geneticin (G418) selection, as previously described [66,67]. These cells mimic the tumor microenvironment, being thus considered a suitable in vitro cell model to study ER$^+$ breast cancer and AIs [44]. Cells were maintained with Eagle's minimum essential medium (MEM) (Gibco Invitrogen Co., Paisley, Scotland, UK) supplemented with 1 mmol/L sodium pyruvate, 1% penicillin-streptomycin-amphotericin B, 100 μg/mL G418 and 10% heat-inactivated fetal bovine serum (FBS) (Gibco Invitrogen Co., Paisley, Scotland, UK). Three days before the experiments, cells were cultured in an estrogen-free MEM without phenol-red (Gibco Invitrogen Co., Paisley, Scotland, UK), containing 5% pre-treated charcoal heat-inactivated fetal bovine serum (CFBS), 1 mmol/L of sodium pyruvate, 1% of penicillin-streptomycin-amphotericin B and 2 mmol/L of L-glutamine (Gibco Invitrogen Co., Paisley, Scotland, UK), to avoid the interference of the hormones present in FBS and of the estrogen-like effects of phenol-red, as previously reported [36,40]. After this period, cells were stimulated with testosterone (T) or with estradiol (E$_2$) (Sigma-Aldrich Co., Saint Louis, MI, USA), the aromatase substrate or the aromatase product, respectively, at 1 nM, which were used as proliferation inducing agents [40,68] and treated with **Oxy**. All the experiments were performed under these conditions.

The long-term estrogen-deprived human ER$^+$ breast cancer cell line, LTEDaro cells, mimics the late-stage of acquired resistance to the AIs used in clinic, since they originated through long-term estrogen deprivation of the parental MCF-7aro cells, being thus considered a suitable in vitro cell model to study resistance [36,63,64]. These cells were cultured in Eagle's MEM without phenol-red and supplemented with Earle's salts and with 1 mmol/L sodium pyruvate, 1% penicillin–streptomycin-amphotericin B, 1% L-Glutamine, 100 μg/mL G418, and 10% of CFBS, as previously reported [36,65]. For the assays, cells were cultured in these conditions and treated with **Oxy**.

Both cell lines were kindly provided by Professor Shiuan Chen (Beckman Research Institute, City of Hope, Duarte, CA, USA) and were maintained at 37 °C and 5% CO$_2$ atmosphere.

Stock solutions of T and E$_2$ were prepared in absolute ethanol (Sigma-Aldrich Co., Saint Louis, MI, USA). On the other hand, **Oxy**, Exemestane (Exe) (Sigma-Aldrich Co., Saint Louis, MI, USA), ICI 182,780 (Sigma-Aldrich Co., Saint Louis, MI, USA), Casodex (CDX) (Sigma-Aldrich Co., Saint Louis, MI, USA) and 4-[2-phenyl-5,7,bis(trifluoromethyl)pyrazol [1,5-a]pyrimidin-3-yl]phenol (PHTPP) (Sigma-Aldrich Co., Saint Louis, MI, USA) were prepared in 100% DMSO (Sigma-Aldrich Co., Saint Louis, MI, USA). The AI **Oxy**, as well as, T, E_2, Exe, ICI, PHTPP and CDX were stored at $-20\ ^\circ$C, and fresh dilutions were prepared in medium before each experiment. Final concentrations of ethanol and DMSO in culture medium were lower than 0.05% and 0.01%, respectively. All the controls contained these vehicles in these culture conditions.

4.3. Cell Viability and Cell Proliferation

The sensitive MCF-7aro and the resistant LTEDaro cells were cultured in 96-well plates, with a cellular density of 2.5×10^4 cells/mL (2 and 3 days) and 1×10^4 cells/mL (6 days), and incubated with different concentrations of **Oxy** (0.1–5 µM) during 2, 3 and 6 days. Depending on the type of analysis, MCF-7aro cells were incubated with T (1 nM) or E_2 (1 nM), as well as, with ICI 182,780 (100 nM), PHTPP (1 µM) or CDX (1 µM). Cells without **Oxy** treatment were considered as control.

To explore the effects of **Oxy** on the viability of LTEDaro cells, the tetrazolium salt, 3-(4,5-dimethylthiazol-2-yl)-2,5-difenyltetrazolium (MTT) assay was performed. After each incubation time, MTT (0.5 mg/mL) (Sigma-Aldrich Co., Saint Louis, MI, USA) was added and quantified spectrophotometrically in a Biotek Synergy HTX Multi-Mode Microplate Reader (Biotek Instruments, Winowski, VT, USA).

To study the effects of **Oxy** on DNA synthesis of MCF-7aro cells, the ^{3}H-thymidine incorporation assay was performed. At each exposure time and for the final 8 h, ^{3}H-thymidine (0.5 µCi) (Amersham International, Amersham, UK) was added to each well. Cells were further harvested using a semi-automated cell harvester (Skatron Instruments, Lier, Norway), scintillation cocktail was added, and ^{3}H-thymidine incorporation was determined in a scintillation counter (LS 6500, Beckman Instruments, Brea, CA, USA).

All the results are expressed as relative percentage of the untreated control cells (100% of cell viability and cell proliferation).

4.4. Cell Cycle Analysis

To investigate the effects of **Oxy** on MCF-7aro cell cycle progression, the DNA content was assessed by flow cytometry. MCF-7aro cells (7×10^5 cells/mL) stimulated with T (1 nM) were incubated with **Oxy** (1 and 2.5 µM) for 3 days. Cells only treated with T were considered as control. After the incubation period, cells were fixed with 70% cold ethanol and stained with a DNA staining solution (5 µg/mL Propidium Iodide (PI), 0.1% Triton X-100 and 200 µg/mL DNase-free RNase A (Sigma-Aldrich Co., Saint Louis, MI, USA)), as previously described [19]. DNA content was analyzed by flow cytometry based on the acquisition of 40 000 events in a BD Accuri™ C6 cytometer (San Jose, CA, USA), equipped with a BD Accuri™ C6 analysis software. Detectors for the three fluorescence channels (FL-1, FL-2 and FL-3) and for forward (FSC) and side (SSC) light scatter were set on a linear scale. Debris, cell doublets and aggregates were gated out using a two-parameter plot of FL-2-Area to FL-2-Width of PI fluorescence. Data were analyzed using the BD Accuri™ C6 analysis software. The anti-proliferative effects were indicated by the percentage of cells in G_0/G_1, S and G_2/M phases of the cell cycle.

4.5. Analysis of Apoptosis

To understand the involvement of apoptosis on sensitive and resistant treated cells, the mitochondrial transmembrane potential ($\Delta\Psi m$), the activation of caspases-9, -8 and -7, the production of intracellular reactive oxygen species (ROS), as well as the translocation of phosphatidylserine (PS) were studied.

MCF-7aro and LTEDaro cells (7×10^5 cells/mL) were cultured in 6-well plates and treated with **Oxy** (1 µM) with or without CDX (1 µM) for 3 days. MCF-7aro cells were stimulated with T (1 nM). Cells without **Oxy** treatment were designated as control, while cells treated with Staurosporine (STS) (10 µM), carbonyl cyanide m-chlorophenylhydrazone (10 µM), or with phorbol 12-myristate 13-acetate (PMA) (25 ng/mL) (Sigma-Aldrich Co., Saint Louis, MI, USA) were used as positive controls.

The mitochondrial transmembrane potential ($\Delta\Psi$m) was evaluated by flow cytometry using 3,3'-dihexyloxacarbocyanine iodide ($DiOC_6(3)$) (Gibco Invitrogen Co., Paisley, Scotland, UK) at 10 nM, as previously described [19]. PI at 5 µg/mL was added prior to flow cytometry to discriminate between live cells that stain only with $DiOC_6(3)$ ($DiOC_6(3)^+/PI^-$), early apoptotic cells that lost the ability to accumulate DiOC6(3) ($DiOC_6(3)^-/PI^-$), and late apoptotic/necrotic cells that stained only with PI ($DiOC_6(3)^-/PI^+$). Flow cytometric analysis based on the acquisition of 40,000 events was carried out in a BD Accuri™ C6 cytometer (San Jose, CA, USA), equipped with a BD Accuri™ C6 analysis software. Detectors for FSC and SSC light scatter were set on a linear scale and all three fluorescence channels (FL-1, FL-2 and FL-3) were set on a logarithmic scale. FL-1 was used to measure $DiOC_6(3)$ at green fluorescence, while FL-2 and FL-3 were used to measure PI red fluorescence. Data were analyzed using BD Accuri™ C6 analysis software.

To study translocation of PS, cells were labelled with Annexin V-FITC Apoptosis Detection Kit (BioLegend Way, San Diego, CA, USA), according to the manufacturer's instructions, and analyzed based on the acquisition of 40,000 events in the BD Accuri™ C6 cytometer (San Jose, CA, USA), equipped with BD Accuri™ C6 analysis software, as previously described [36]. Detectors for all three fluorescence channels (FL-1, FL-2 and FL-3) were set on a logarithmic scale. Bivariant analysis of Annexin-FITC fluorescence (FL-1) and 7-amino-acitomycin (7-AAD) fluorescence (FL-3) distinguished different cell populations: Annexin $V^-/7-AAD^-$ were considered as viable cells, Annexin $V^+/7-AAD^-$ corresponded to apoptotic cells and Annexin $V^+/7-AAD^+$ were designated as late apoptotic and necrotic cells. Data were analyzed using BD Accuri™ C6 analysis software.

To evaluate caspase activities, luminescent assays with Caspase-Glo® 9, Caspase-Glo® 8 and Caspase-Glo® 3/7 (Promega Corporation, Madison, WI, USA), were performed according to the manufacturer's instructions and as previously described [18]. As a negative control, the pan-caspase inhibitor Z-VAD-FMK (50 µM) (Sigma-Aldrich Co., Saint Louis, MI, USA) was used. The resultant luminescence was measured in a Biotek Synergy HTX Multi-Mode Microplate Reader (Biotek Instruments, Winowski, VT, USA) and results were presented as relative luminescence units (RLU).

To detect the levels of intracellular ROS, the 2',7'-dichlorodihydrofluorescein diacetate ($DCFH_2$-DA) method was used by labelling cells with $DCFH_2$-DA (50 µM) (Sigma-Aldrich Co., Saint Louis, MI, USA), as previously described [18]. Fluorescence was measured using an excitation wavelength of 480 nm and an emission filter of 530 nm in the Biotek Synergy HTX Multi-Mode Microplate Reader (Biotek Instruments, Winowski, VT, USA) and data were presented as mean fluorescence intensity (MFI).

4.6. Western Blot Analysis

The expression levels of aromatase, estrogen receptor α (ERα), estrogen receptor β (ERβ) and androgen receptor (AR) were evaluated by Western Blot. For this purpose, MCF-7aro cells (7×10^5 cells/mL) were cultured in 6-well plates and incubated with **Oxy** (1 µM), during 8 h to study aromatase expression [19,21], and for 3 days to study the expression of ERα, ERβ and AR [26,36]. Cells without **Oxy** treatment were designated as control, while cells treated with Exe (10 µM), ICI 182 780 (100 nM), and PHTPP (1 µM) were used as positive controls. After treatment, cells were collected as previously reported [40] and 50 µg of protein sample was subjected to 10% of SDS-PAGE and then transferred onto nitrocellulose membranes. For the immunodetection, the mouse monoclonal CYP19A1 (1:200, sc-374176), mouse monoclonal ERα (1:200, sc-8002), mouse monoclonal AR (1:200, sc-7305) (Santa Cruz Biotechnology, Santa Cruz, CA, USA) and mouse monoclonal ERβ (1:200, PPZ0506) (Thermo Fisher, Waltham, MA, USA) were used as primary antibodies, whereas the peroxidase-conjugated goat anti-mouse (1:2000, G21040) (Thermo Fisher, Waltham, MA, USA) was used as a secondary antibody. A mouse monoclonal anti-β-tubulin antibody (1:500, sc-5274) (Santa Cruz Biotechnology, Santa Cruz, CA, USA) was used to control loading variations. Immunoreactive bands were visualized using a chemiluminescent substrate Super Signal West Pico (Pierce, Rockford, IL, USA) in a ChemiDoc™ Touch Imaging System (Bio-Rad, Laboratories Melville, NY, USA).

4.7. RNA Extraction and qPCR

Quantitative polymerase chain reaction (qPCR) analysis was performed to investigate the effects of **Oxy** on MCF-7aro cells in the transcription levels of *ESR1, EGR3, PDZK1* and *TFF1* genes, as previously reported [41]. MCF-7aro cells (7×10^5 cells/mL) were cultured in 6-well plates, stimulated with T (1 nM) and incubated with **Oxy** (1 µM) for 3 days. Cells without **Oxy** treatment were designated as control, while cells treated with ICI 182 780 (100 nM) were used as positive control. The RNA was extracted using the TripleXtractor reagent (GRiSP Research Solutions, Porto, Portugal), according to the manufacturer's protocol. RNA quality was measured with the Experion RNA StdSens Kit (Bio-Rad Laboratories), in the Experion analytical software (Bio-Rad Laboratories), and quantified using NanoDrop ND-1000 Spectrophotometer (NanoDrop Technologies, Inc, Wilmington, DE, USA). RNA was further converted into cDNA using the GRiSP Xpert cDNA Synthesis Mastermix (GRiSP Research Solutions, Porto, Portugal), containing reverse transcriptase, according to manufacturer's protocol. cDNA was amplified using GRiSP Xpert Fast SYBR (GRiSP Research Solutions, Porto, Portugal) in MiniOpticon Real-Time PCR Detection System (Bio-Rad Laboratories, Hercules, CA, USA), according to the manufacturer's protocol. Primer sequences (5'-3') are presented in Table 4. The fold change in gene expression was calculated using the $2^{-\Delta\Delta Ct}$ method, using as housekeeping genes, *TUBA1A* and *ACTB*.

Table 4. Primer sequences and qPCR conditions for housekeeping and target genes.

Target Gene	Primer Sequences (5'-3')		Ta/°C
	Sense	Anti-Sense	
ESR1	CCTGATCATGGAGGGTCAAA	TGGGCTTACTGACCAACCTG	55
EGR3	GACTCCCCTTCCAACTGGTG	GGATACATGGCCTCCACGTC	56
TFF1	GTGGTTTTCCTGGTGTCACG	AGGATAGAAGCACCAGGGGA	55
PDZK1	ACTCTGCAGGCTGGCTAAAG	ACCGCCCTTCTGTACCTCTT	56
ACTB	TACAGCTTCACCACCACAGC	AAGGAAGGCTGGAAGAGAGC	55
TUBA1A	CTGGAGCACTCTGATTGT	ATAAGGCGGTTAAGGTTAGT	55

4.8. AR and ER Transactivation Assays

In order to assess potential ER activity of **Oxy**, the VM7Luc ER transactivation assay, from Test No. 455 of the OECD Guidelines for the Testing of Chemicals [69], was performed. This assay uses VM7Luc4E2 cells, derived from the MCF-7 cell line, which endogenously express both human ERα and ERβ forms. Cells were maintained in Roswell Park Memorial Institute 1640 medium, supplemented with 1% penicillin/streptomycin, and 8% heat-inactivated FBS. Three days before the experiments, cells were split into an estrogen-free Dulbecco's Modified Eagle Medium (DMEM) without phenol red, supplemented with 1% penicillin/streptomycin, 4.5% CFBS, 2% L-glutamine, and 110 mg/mL sodium pyruvate (Gibco Invitrogen Co., Paisley, Scotland, UK). This medium was further used for seeding and exposure of cells. Briefly, cells (4×10^5 cells/mL) were exposed to 0.1–1 µM **Oxy** in 96-well white plates for 24 h. This exposure was conducted in the absence of any ER agonist to assess potential ER agonistic activity of **Oxy**, or in the presence of 1 nM T to assess antagonistic activity. Luminescence was read in EnSpire® multimode plate reader (Perkin Elmer, Inc., Waltham, MA, USA) using the Steady-Glo® Luciferase Assay System (Promega Corporation, Madison, WI, USA). ATP levels were assessed as an indirect measure of cell viability, using the CellTiter-Glo® Luminescent Cell Viability Assay (Promega Corporation, Madison, WI, USA). T-treated cells (781.2 pM–25.6 µM) were used as a positive control of ER agonism, with a maximum effect of 12.7-fold increase over control. Raloxifene

(12.0 pM–24.5 nM; Biosynth Ltd., Berkshire, United Kingdom) was used as a positive control of ER antagonism, with a maximum inhibitory effect of 42.4% in the presence of 1 nM T.

To confirm the potential activity of **Oxy** towards AR, the AR-EcoScreen™ assay, from Test No. 458 of the OECD Guidelines for the Testing of Chemicals [70], was conducted. The AR-EcoScreen™ cell line is derived from the Chinese hamster ovary CHO-K1 cell line, and it expresses three stably inserted constructs: the human AR; a firefly luciferase reporter construct bearing an androgen responsive element gene; and a renilla luciferase reporter construct for simultaneous assessment of cell viability. Cells were maintained in DMEM/Nutrient Mixture F-12 (DMEM/F12) medium without phenol red, supplemented with 1% penicillin/streptomycin, 10% heat-inactivated FBS, 200 µg/mL zeocin, and 100 µg/mL hygromycin. Cells were seeded (9×10^4 cells/mL) onto 96-well white plates and exposed to 0.1–1 µM **Oxy** for 24 h. Seeding and exposure of cells was performed in DMEM/F12 without phenol red, containing 1% penicillin/streptomycin and 5% CFBS. The exposure was conducted in the absence or presence of the potent AR agonist 0.1 nM methyltrienolone (R1881; AbMole BioScience, Houston, TX, USA) to assess potential AR agonistic and antagonistic activity of **Oxy**, respectively. Luminescence for both AR activity and cell viability was read in EnSpire® multimode plate reader (Perkin Elmer, Inc., Waltham, MA, USA) using the Dual-Glo® Luciferase Assay System (Promega Corporation, Madison, WI, USA). R1881 (7.8 pM–1 nM) was used as positive a control of AR agonism, showing a maximum effect of 5.5-fold increase over control. Hydroxyflutamide (OHF; 4.1 nM–9 µM; Sigma-Aldrich Co., Saint Louis, MI, USA) was used as positive control of AR antagonism, with a maximum inhibition of 69.8% in the presence of 0.1 nM R1881.

Data from four independent experiments were presented as fold change compared to control, which was set as 1. Stock solutions of T, raloxifene, R1881 and OHF were prepared in 100% DMSO and stored at −20 °C. Dilutions were prepared freshly in medium before each experiment. The final concentration of DMSO in exposure medium for all conditions was 0.06%.

4.9. Statistical Analysis

All the assays were performed in triplicate in at least three independent experiments, and data were expressed as the mean ± SEM. Statistical analysis was performed by using GraphPad Prism 7® software (GraphPad Software, Inc., San Diego, CA, USA) and by applying the analysis of variance (ANOVA), followed by multiple comparisons using Tukey and Bonferroni post hoc tests. Values that were $p < 0.05$ were designated as statistically significant.

Author Contributions: Conceptualization, C.A., G.C.-d.-S. and N.T.; investigation, C.A., C.F.A., M.J.V., C.V., E.T.-d.-S., A.M.V. and F.M.F.R.; methodology, C.A.; writing—original draft, C.A.; writing—review and editing, G.C.-d.-S., E.T.-d.-S., A.M.V., N.T. and F.M.F.R. All authors have read and agreed to the published version of the manuscript.

Funding: This research was funded by national funds from FCT—Fundação para a Ciência e a Tecnologia, I.P., in the scope of the project UIDP/04378/2020 and UIDB/04378/2020 of the Research Unit on Applied Molecular Biosciences—UCIBIO and the project LA/P/0140/2020 of the Associate Laboratory Institute for Health and Bioeconomy—i4HB.

Institutional Review Board Statement: Not applicable.

Informed Consent Statement: Not applicable.

Data Availability Statement: Not applicable.

Acknowledgments: Authors give thanks to Fundação para a Ciência e Tecnologia (FCT) for the Cristina Amaral contract under the funding program (DL 57/2016—Norma Transitória) and through the Post-doc grant (SFRH/BPD/98304/2013). This work is also financed by national funds from FCT—Fundação para a Ciência e a Tecnologia, I.P., in the scope of the project UIDP/04378/2020 and UIDB/04378/2020 of the Research Unit on Applied Molecular Biosciences—UCIBIO and the project LA/P/0140/2020 of the Associate Laboratory Institute for Health and Bioeconomy—i4HB. We also thank Shiuan Chen (Department of Cancer Biology, Beckman Research Institute of the City of Hope, Duarte, CA, USA) for kindly supplying MCF-7aro and LTEDaro cells.

Conflicts of Interest: The authors have no conflict of interest to declare.

Sample Availability: Samples of the compound Oxymestane-D1 (**Oxy**) are available from the authors.

References

1. Sung, H.; Ferlay, J.; Siegel, R.L.; Laversanne, M.; Soerjomataram, I.; Jemal, A.; Bray, F. Global cancer statistics 2020: GLOBOCAN estimates of incidence and mortality worldwide for 36 cancers in 185 countries. *CA Cancer J. Clin.* **2021**, *71*, 209–249. [CrossRef]
2. Augusto, T.; Correia-da-Silva, G.; Rodrigues, C.M.P.; Teixeira, N.; Amaral, C. Acquired-resistance to aromatase inhibitors: Where we stand! *Endocr. Relat. Cancer* **2018**, *25*, R283–R301. [CrossRef]
3. Almeida, C.F.; Oliveira, A.; Ramos, M.J.; Fernandes, P.A.; Teixeira, N.; Amaral, C. Estrogen receptor-positive (ER+) breast cancer treatment: Are multi-target compounds the next promising approach? *Biochem. Pharmacol.* **2020**, *177*, 113989. [CrossRef] [PubMed]
4. Cardoso, F.; Kyriakides, S.; Ohno, S.; Penault-Llorca, F.; Poortmans, P.; Rubio, I.T.; Zackrisson, S.; Senkus, E.; Committee, E.G. Early breast cancer: ESMO Clinical Practice Guidelines for diagnosis, treatment and follow-up. *Ann. Oncol.* **2019**, *30*, 1674. [CrossRef]
5. Cardoso, F.; Paluch-Shimon, S.; Senkus, E.; Curigliano, G.; Aapro, M.S.; Andre, F.; Barrios, C.H.; Bergh, J.; Bhattacharyya, G.S.; Biganzoli, L.; et al. 5th ESO-ESMO international consensus guidelines for advanced breast cancer (ABC 5). *Ann. Oncol.* **2020**, *31*, 1623–1649. [CrossRef]
6. Gennari, A.; Andre, F.; Barrios, C.H.; Cortes, J.; de Azambuja, E.; DeMichele, A.; Dent, R.; Fenlon, D.; Gligorov, J.; Hurvitz, S.A.; et al. ESMO Clinical Practice Guideline for the diagnosis, staging and treatment of patients with metastatic breast cancer. *Ann. Oncol.* **2021**, *32*, 1475–1495. [CrossRef]
7. Brufsky, A.M.; Dickler, M.N. Estrogen Receptor-Positive Breast Cancer: Exploiting Signaling Pathways Implicated in Endocrine Resistance. *Oncologist* **2018**, *23*, 528–539. [CrossRef]
8. Rasha, F.; Sharma, M.; Pruitt, K. Mechanisms of endocrine therapy resistance in breast cancer. *Mol. Cell Endocrinol.* **2021**, *532*, 111322. [CrossRef]
9. Saatci, O.; Huynh-Dam, K.T.; Sahin, O. Endocrine resistance in breast cancer: From molecular mechanisms to therapeutic strategies. *J. Mol. Med.* **2021**, *99*, 1691–1710. [CrossRef]
10. Portman, N.; Alexandrou, S.; Carson, E.; Wang, S.; Lim, E.; Caldon, C.E. Overcoming CDK4/6 inhibitor resistance in ER-positive breast cancer. *Endocr. Relat. Cancer* **2019**, *26*, R15–R30. [CrossRef]
11. Papadimitriou, M.C.; Pazaiti, A.; Iliakopoulos, K.; Markouli, M.; Michalaki, V.; Papadimitriou, C.A. Resistance to CDK4/6 inhibition: Mechanisms and strategies to overcome a therapeutic problem in the treatment of hormone receptor-positive metastatic breast cancer. *Biochim. Biophys. Acta Mol. Cell Res.* **2022**, *1869*, 119346. [CrossRef]
12. Tseng, O.L.; Spinelli, J.J.; Gotay, C.C.; Ho, W.Y.; McBride, M.L.; Dawes, M.G. Aromatase inhibitors are associated with a higher fracture risk than tamoxifen: A systematic review and meta-analysis. *Ther. Adv. Musculoskelet. Dis.* **2018**, *10*, 71–90. [CrossRef] [PubMed]
13. Awan, A.; Esfahani, K. Endocrine therapy for breast cancer in the primary care setting. *Curr. Oncol.* **2018**, *25*, 285–291. [CrossRef]
14. Goldvaser, H.; Amir, E. Role of Bisphosphonates in Breast Cancer Therapy. *Curr. Treat. Options Oncol.* **2019**, *20*, 26. [CrossRef]
15. Tanvetyanon, T.; Stiff, P.J. Management of the adverse effects associated with intravenous bisphosphonates. *Ann. Oncol.* **2006**, *17*, 897–907. [CrossRef]
16. Amaral, C.; Lopes, A.; Varela, C.L.; da Silva, E.T.; Roleira, F.M.; Correia-da-Silva, G.; Teixeira, N. Exemestane metabolites suppress growth of estrogen receptor-positive breast cancer cells by inducing apoptosis and autophagy: A comparative study with Exemestane. *Int. J. Biochem. Cell Biol.* **2015**, *69*, 183–195. [CrossRef]
17. Amaral, C.; Varela, C.; Azevedo, M.; da Silva, E.T.; Roleira, F.M.; Chen, S.; Correia-da-Silva, G.; Teixeira, N. Effects of steroidal aromatase inhibitors on sensitive and resistant breast cancer cells: Aromatase inhibition and autophagy. *J. Steroid. Biochem. Mol. Biol.* **2013**, *135*, 51–59. [CrossRef]
18. Amaral, C.; Varela, C.; Borges, M.; Tavares da Silva, E.; Roleira, F.M.; Correia-da-Silva, G.; Teixeira, N. Steroidal aromatase inhibitors inhibit growth of hormone-dependent breast cancer cells by inducing cell cycle arrest and apoptosis. *Apoptosis* **2013**, *18*, 1426–1436. [CrossRef]
19. Amaral, C.; Varela, C.L.; Mauricio, J.; Sobral, A.F.; Costa, S.C.; Roleira, F.M.F.; Tavares-da-Silva, E.J.; Correia-da-Silva, G.; Teixeira, N. Anti-tumor efficacy of new 7alpha-substituted androstanes as aromatase inhibitors in hormone-sensitive and resistant breast cancer cells. *J. Steroid Biochem. Mol. Biol.* **2017**, *171*, 218–228. [CrossRef]

20. Varela, C.L.; Amaral, C.; Tavares da Silva, E.; Lopes, A.; Correia-da-Silva, G.; Carvalho, R.A.; Costa, S.C.; Roleira, F.M.; Teixeira, N. Exemestane metabolites: Synthesis, stereochemical elucidation, biochemical activity and anti-proliferative effects in a hormone-dependent breast cancer cell line. *Eur. J. Med. Chem.* **2014**, *87*, 336–345. [CrossRef]

21. Augusto, T.V.; Amaral, C.; Varela, C.L.; Bernardo, F.; da Silva, E.T.; Roleira, F.F.M.; Costa, S.; Teixeira, N.; Correia-da-Silva, G. Effects of new C6-substituted steroidal aromatase inhibitors in hormone-sensitive breast cancer cells: Cell death mechanisms and modulation of estrogen and androgen receptors. *J. Steroid Biochem. Mol. Biol.* **2019**, *195*, 105486. [CrossRef]

22. Zhao, L.M.; Jin, H.S.; Liu, J.; Skaar, T.C.; Ipe, J.; Lv, W.; Flockhart, D.A.; Cushman, M. A new Suzuki synthesis of triphenylethylenes that inhibit aromatase and bind to estrogen receptors alpha and beta. *Bioorg. Med. Chem.* **2016**, *24*, 5400–5409. [CrossRef]

23. Baum, M.; Budzar, A.U.; Cuzick, J.; Forbes, J.; Houghton, J.H.; Klijn, J.G.; Sahmoud, T.; Group, A.T. Anastrozole alone or in combination with tamoxifen versus tamoxifen alone for adjuvant treatment of postmenopausal women with early breast cancer: First results of the ATAC randomised trial. *Lancet* **2002**, *359*, 2131–2139.

24. Jelovac, D.; Macedo, L.; Goloubeva, O.G.; Handratta, V.; Brodie, A.M. Additive antitumor effect of aromatase inhibitor letrozole and antiestrogen fulvestrant in a postmenopausal breast cancer model. *Cancer Res.* **2005**, *65*, 5439–5444. [CrossRef] [PubMed]

25. Petrelli, A.; Giordano, S. From single- to multi-target drugs in cancer therapy: When aspecificity becomes an advantage. *Curr. Med. Chem.* **2008**, *15*, 422–432. [PubMed]

26. Almeida, C.F.; Teixeira, N.; Oliveira, A.; Augusto, T.V.; Correia-da-Silva, G.; Ramos, M.J.; Fernandes, P.A.; Amaral, C. Discovery of a multi-target compound for estrogen receptor-positive (ER(+)) breast cancer: Involvement of aromatase and ERs. *Biochimie* **2020**, *181*, 65–76. [CrossRef]

27. Lv, W.; Liu, J.; Skaar, T.C.; Flockhart, D.A.; Cushman, M. Design and synthesis of norendoxifen analogues with dual aromatase inhibitory and estrogen receptor modulatory activities. *J. Med. Chem.* **2015**, *58*, 2623–2648. [CrossRef]

28. Lv, W.; Liu, J.; Skaar, T.C.; O'Neill, E.; Yu, G.; Flockhart, D.A.; Cushman, M. Synthesis of Triphenylethylene Bisphenols as Aromatase Inhibitors That Also Modulate Estrogen Receptors. *J. Med. Chem.* **2016**, *59*, 157–170. [CrossRef]

29. Lv, W.; Liu, J.; Lu, D.; Flockhart, D.A.; Cushman, M. Synthesis of mixed (E,Z)-, (E)-, and (Z)-norendoxifen with dual aromatase inhibitory and estrogen receptor modulatory activities. *J. Med. Chem.* **2013**, *56*, 4611–4618. [CrossRef]

30. Gomes, A.R.; Varela, C.L.; Tavares-da-Silva, E.J.; Roleira, F.M.F. Epoxide containing molecules: A good or a bad drug design approach. *Eur. J. Med. Chem.* **2020**, *201*, 112327. [CrossRef]

31. Pires, A.S.; Varela, C.L.; Marques, I.A.; Abrantes, A.M.; Goncalves, C.; Rodrigues, T.; Matafome, P.; Botelho, M.F.; Roleira, F.M.F.; Tavares-da-Silva, E. Oxymestane, a cytostatic steroid derivative of exemestane with greater antitumor activity in non-estrogen-dependent cell lines. *J. Steroid Biochem. Mol. Biol.* **2021**, *212*, 105950. [CrossRef] [PubMed]

32. Nakopoulou, L.; Lazaris, A.C.; Panayotopoulou, E.G.; Giannopoulou, I.; Givalos, N.; Markaki, S.; Keramopoulos, A. The favourable prognostic value of oestrogen receptor beta immunohistochemical expression in breast cancer. *J. Clin. Pathol.* **2004**, *57*, 523–528. [CrossRef]

33. Ramasamy, K.; Samayoa, C.; Krishnegowda, N.; Tekmal, R.R. Therapeutic Use of Estrogen Receptor beta Agonists in Prevention and Treatment of Endocrine Therapy Resistant Breast Cancers: Observations From Preclinical Models. *Prog. Mol. Biol. Transl. Sci.* **2017**, *151*, 177–194. [PubMed]

34. Wu, V.S.; Kanaya, N.; Lo, C.; Mortimer, J.; Chen, S. From bench to bedside: What do we know about hormone receptor-positive and human epidermal growth factor receptor 2-positive breast cancer? *J. Steroid Biochem. Mol. Biol.* **2015**, *153*, 45–53. [CrossRef] [PubMed]

35. Paterni, I.; Granchi, C.; Katzenellenbogen, J.A.; Minutolo, F. Estrogen receptors alpha (ERalpha) and beta (ERbeta): Subtype-selective ligands and clinical potential. *Steroids* **2014**, *90*, 13–29. [CrossRef]

36. Amaral, C.; Augusto, T.V.; Almada, M.; Cunha, S.C.; Correia-da-Silva, G.; Teixeira, N. The potential clinical benefit of targeting androgen receptor (AR) in estrogen-receptor positive breast cancer cells treated with Exemestane. *Biochim. Biophys. Acta Mol. Basis. Dis.* **2020**, *1866*, 165661. [CrossRef]

37. Basile, D.; Cinausero, M.; Iacono, D.; Pelizzari, G.; Bonotto, M.; Vitale, M.G.; Gerratana, L.; Puglisi, F. Androgen receptor in estrogen receptor positive breast cancer: Beyond expression. *Cancer Treat. Rev.* **2017**, *61*, 15–22. [CrossRef]

38. Feng, J.; Li, L.; Zhang, N.; Liu, J.; Zhang, L.; Gao, H.; Wang, G.; Li, Y.; Zhang, Y.; Li, X.; et al. Androgen and AR contribute to breast cancer development and metastasis: An insight of mechanisms. *Oncogene* **2017**, *36*, 2775–2790. [CrossRef]

39. Peters, A.A.; Buchanan, G.; Ricciardelli, C.; Bianco-Miotto, T.; Centenera, M.M.; Harris, J.M.; Jindal, S.; Segara, D.; Jia, L.; Moore, N.L.; et al. Androgen receptor inhibits estrogen receptor-alpha activity and is prognostic in breast cancer. *Cancer Res.* **2009**, *69*, 6131–6140. [CrossRef]

40. Amaral, C.; Borges, M.; Melo, S.; da Silva, E.T.; Correia-da-Silva, G.; Teixeira, N. Apoptosis and autophagy in breast cancer cells following exemestane treatment. *PLoS ONE* **2012**, *7*, e42398. [CrossRef]

41. Augusto, T.V.; Amaral, C.; Almeida, C.F.; Teixeira, N.; Correia-da-Silva, G. Differential biological effects of aromatase inhibitors: Apoptosis, autophagy, senescence and modulation of the hormonal status in breast cancer cells. *Mol. Cell. Endocrinol.* **2021**, *537*, 111426. [CrossRef] [PubMed]

42. Amaral, C.; Augusto, T.V.; Tavares-da-Silva, E.; Roleira, F.M.F.; Correia-da-Silva, G.; Teixeira, N. Hormone-dependent breast cancer: Targeting autophagy and PI3K overcomes Exemestane-acquired resistance. *J. Steroid Biochem. Mol. Biol.* **2018**, *183*, 51–61. [CrossRef] [PubMed]

43. Thiantanawat, A.; Long, B.J.; Brodie, A.M. Signaling pathways of apoptosis activated by aromatase inhibitors and antiestrogens. *Cancer Res.* **2003**, *63*, 8037–8050. [PubMed]
44. Itoh, T.; Karlsberg, K.; Kijima, I.; Yuan, Y.C.; Smith, D.; Ye, J.; Chen, S. Letrozole-, anastrozole-, and tamoxifen-responsive genes in MCF-7aro cells: A microarray approach. *Mol. Cancer Res.* **2005**, *3*, 203–218. [CrossRef]
45. De Vries, J.F.; Wammes, L.J.; Jedema, I.; van Dreunen, L.; Nijmeijer, B.A.; Heemskerk, M.H.; Willemze, R.; Falkenburg, J.H.; Barge, R.M. Involvement of caspase-8 in chemotherapy-induced apoptosis of patient derived leukemia cell lines independent of the death receptor pathway and downstream from mitochondria. *Apoptosis* **2007**, *12*, 181–193. [CrossRef]
46. Mutlu, A.; Gyulkhandanyan, A.V.; Freedman, J.; Leytin, V. Activation of caspases-9, -3 and -8 in human platelets triggered by BH3-only mimetic ABT-737 and calcium ionophore A23187: Caspase-8 is activated via bypass of the death receptors. *Br. J. Haematol.* **2012**, *159*, 565–571. [CrossRef]
47. Slee, E.A.; Harte, M.T.; Kluck, R.M.; Wolf, B.B.; Casiano, C.A.; Newmeyer, D.D.; Wang, H.G.; Reed, J.C.; Nicholson, D.W.; Alnemri, E.S.; et al. Ordering the cytochrome c-initiated caspase cascade: Hierarchical activation of caspases-2, -3, -6, -7, -8, and -10 in a caspase-9-dependent manner. *J. Cell Biol.* **1999**, *144*, 281–292. [CrossRef]
48. Cowling, V.; Downward, J. Caspase-6 is the direct activator of caspase-8 in the cytochrome c-induced apoptosis pathway: Absolute requirement for removal of caspase-6 prodomain. *Cell Death Differ.* **2002**, *9*, 1046–1056. [CrossRef]
49. Ferreira, C.G.; Span, S.W.; Peters, G.J.; Kruyt, F.A.; Giaccone, G. Chemotherapy triggers apoptosis in a caspase-8-dependent and mitochondria-controlled manner in the non-small cell lung cancer cell line NCI-H460. *Cancer Res.* **2000**, *60*, 7133–7141.
50. Sohn, D.; Schulze-Osthoff, K.; Janicke, R.U. Caspase-8 can be activated by interchain proteolysis without receptor-triggered dimerization during drug-induced apoptosis. *J. Biol. Chem.* **2005**, *280*, 5267–5273. [CrossRef]
51. Petak, I.; Houghton, J.A. Shared pathways: Death receptors and cytotoxic drugs in cancer therapy. *Pathol. Oncol. Res.* **2001**, *7*, 95–106. [CrossRef]
52. Cullen, S.P.; Martin, S.J. Caspase activation pathways: Some recent progress. *Cell Death Differ.* **2009**, *16*, 935–938. [CrossRef]
53. Wang, X.; Chen, S. Aromatase destabilizer: Novel action of exemestane, a food and drug administration-approved aromatase inhibitor. *Cancer Res.* **2006**, *66*, 10281–10286. [CrossRef]
54. Dutertre, M.; Smith, C.L. Molecular mechanisms of selective estrogen receptor modulator (SERM) action. *J. Pharm. Exp. Ther.* **2000**, *295*, 431–437.
55. Wakeling, A.E.; Dukes, M.; Bowler, J. A potent specific pure antiestrogen with clinical potential. *Cancer Res.* **1991**, *51*, 3867–3873.
56. Wang, X.; Masri, S.; Phung, S.; Chen, S. The role of amphiregulin in exemestane-resistant breast cancer cells: Evidence of an autocrine loop. *Cancer Res.* **2008**, *68*, 2259–2265. [CrossRef]
57. Vareslija, D.; McBryan, J.; Fagan, A.; Redmond, A.M.; Hao, Y.; Sims, A.H.; Turnbull, A.; Dixon, J.M.; Ó Gaora, P.; Hudson, L.; et al. Adaptation to AI Therapy in Breast Cancer Can Induce Dynamic Alterations in ER Activity Resulting in Estrogen-Independent Metastatic Tumors. *Clin. Cancer Res.* **2016**, *22*, 2765–2777. [CrossRef]
58. Bahrami, N.; Jabeen, S.; Tahiri, A.; Sauer, T.; Odegard, H.P.; Geisler, S.B.; Gravdehaug, B.; Reitsma, L.C.; Selsas, K.; Kristensen, V.; et al. Lack of cross-resistance between non-steroidal and steroidal aromatase inhibitors in breast cancer patients: The potential role of the adipokine leptin. *Breast Cancer Res. Treat.* **2021**, *190*, 435–449. [CrossRef]
59. Catalano, S.; Mauro, L.; Marsico, S.; Giordano, C.; Rizza, P.; Rago, V.; Montanaro, D.; Maggiolini, M.; Panno, M.L.; Ando, S. Leptin induces, via ERK1/ERK2 signal, functional activation of estrogen receptor alpha in MCF-7 cells. *J. Biol. Chem.* **2004**, *279*, 19908–19915. [CrossRef] [PubMed]
60. Compton, D.R.; Sheng, S.; Carlson, K.E.; Rebacz, N.A.; Lee, I.Y.; Katzenellenbogen, B.S.; Katzenellenbogen, J.A. Pyrazolo[1,5-a]pyrimidines: Estrogen receptor ligands possessing estrogen receptor beta antagonist activity. *J. Med. Chem.* **2004**, *47*, 5872–5893. [CrossRef] [PubMed]
61. Hu, R.; Dawood, S.; Holmes, M.D.; Collins, L.C.; Schnitt, S.J.; Cole, K.; Marotti, J.D.; Hankinson, S.E.; Colditz, G.A.; Tamimi, R.M. Androgen receptor expression and breast cancer survival in postmenopausal women. *Clin. Cancer Res.* **2011**, *17*, 1867–1874. [CrossRef] [PubMed]
62. Macedo, L.F.; Guo, Z.; Tilghman, S.L.; Sabnis, G.J.; Qiu, Y.; Brodie, A. Role of androgens on MCF-7 breast cancer cell growth and on the inhibitory effect of letrozole. *Cancer Res.* **2006**, *66*, 7775–7782. [CrossRef] [PubMed]
63. Masri, S.; Phung, S.; Wang, X.; Chen, S. Molecular characterization of aromatase inhibitor-resistant, tamoxifen-resistant and LTEDaro cell lines. *J. Steroid Biochem. Mol. Biol.* **2010**, *118*, 277–282. [CrossRef]
64. Masri, S.; Phung, S.; Wang, X.; Wu, X.; Yuan, Y.C.; Wagman, L.; Chen, S. Genome-wide analysis of aromatase inhibitor-resistant, tamoxifen-resistant, and long-term estrogen-deprived cells reveals a role for estrogen receptor. *Cancer Res.* **2008**, *68*, 4910–4918. [CrossRef]
65. Augusto, T.V.; Amaral, C.; Wang, Y.; Chen, S.; Almeida, C.F.; Teixeira, N.; Correia-da-Silva, G. Effects of PI3K inhibition in AI-resistant breast cancer cell lines: Autophagy, apoptosis, and cell cycle progression. *Breast Cancer Res. Treat.* **2021**, *190*, 227–240. [CrossRef]
66. Zhou, D.J.; Pompon, D.; Chen, S.A. Stable expression of human aromatase complementary DNA in mammalian cells: A useful system for aromatase inhibitor screening. *Cancer Res.* **1990**, *50*, 6949–6954.
67. Sun, X.Z.; Zhou, D.; Chen, S. Autocrine and paracrine actions of breast tumor aromatase. A three-dimensional cell culture study involving aromatase transfected MCF-7 and T-47D cells. *J. Steroid Biochem. Mol. Biol.* **1997**, *63*, 29–36. [CrossRef]

68. Masri, S.; Lui, K.; Phung, S.; Ye, J.; Zhou, D.; Wang, X.; Chen, S. Characterization of the weak estrogen receptor alpha agonistic activity of exemestane. *Breast Cancer Res. Treat.* **2009**, *116*, 461–470. [CrossRef]
69. OECD. *Test No. 455: Performance-Based Test Guideline for Stably Transfected Transactivation In Vitro Assays to Detect Estrogen Receptor Agonists and Antagonists*; OECD Publishing: Paris, France, 2021. [CrossRef]
70. OECD. *Test No. 458: Stably Transfected Human Androgen Receptor Transcriptional Activation Assay for Detection of Androgenic Agonist and Antagonist Activity of Chemicals*; OECD Publishing: Paris, France, 2020. [CrossRef]

molecules **MDPI**

Article

Synthesis and Antiproliferative Activity of Steroidal Diaryl Ethers

Édua Kovács [1], Hazhmat Ali [2], Renáta Minorics [2], Péter Traj [1], Vivien Resch [3], Gábor Paragi [3,4,5], Bella Bruszel [3], István Zupkó [2] and Erzsébet Mernyák [1,*]

1. Department of Organic Chemistry, University of Szeged, H-6720 Szeged, Hungary
2. Institute of Pharmacodynamics and Biopharmacy, University of Szeged, H-6720 Szeged, Hungary
3. Department of Medicinal Chemistry, University of Szeged, H-6720 Szeged, Hungary
4. Institute of Physics, University of Pécs, H-7625 Pécs, Hungary
5. Department of Theoretical Physics, University of Szeged, H-6720 Szeged, Hungary
* Correspondence: bobe@chem.u-szeged.hu; Tel.: +36-62-544277; Fax: +36-62-544200

Abstract: Novel 13α-estrone derivatives have been synthesized via direct arylation of the phenolic hydroxy function. Chan–Lam couplings of arylboronic acids with 13α-estrone as a nucleophilic partner were carried out under copper catalysis. The antiproliferative activities of the newly synthesized diaryl ethers against a panel of human cancer cell lines (A2780, MCF-7, MDA-MB 231, HeLa, SiHa) were investigated by means of MTT assays. The quinoline derivative displayed substantial antiproliferative activity against MCF-7 and HeLa cell lines with low micromolar IC_{50} values. Disturbance of tubulin polymerization has been confirmed by microplate-based photometric assay. Computational calculations reveal significant interactions of the quinoline derivative with the taxoid binding site of tubulin.

Keywords: Chan–Lam reaction; diaryl ether; 13α-estrone; antiproliferative effect; tubulin polymerization; molecular dynamics

check for **updates**

Citation: Kovács, É.; Ali, H.; Minorics, R.; Traj, P.; Resch, V.; Paragi, G.; Bruszel, B.; Zupkó, I.; Mernyák, E. Synthesis and Antiproliferative Activity of Steroidal Diaryl Ethers. *Molecules* **2023**, *28*, 1196. https://doi.org/10.3390/molecules28031196

Academic Editor: Carlotta Granchi

Received: 23 December 2022
Revised: 18 January 2023
Accepted: 18 January 2023
Published: 25 January 2023

1. Introduction

The development of highly efficient, environmentally friendly catalytic synthetic methods is one of the major goals of modern organic chemistry [1]. Carbon–heteroatom (C–X) bond formation is often a key challenge for organic chemists [2–8]. It is still desirable to develop mild but effective methods, which allow the construction of the C–X bond with high functional group tolerance. The Ullmann reaction is a traditional method using aryl halides and nucleophiles under transition metal catalysis [9,10]. Nevertheless, owing to its harsh reaction conditions, it is not applicable in certain cases. At the turn of the Millennium, Chan, Evans and Lam published the coupling reactions of arylboronic acids with nucleophiles under the copper salt catalysis, which were later called Chan–Lam coupling reactions [11–13]. These transformations are characterized by mild reaction conditions, low toxicity and appropriate stability. The extensions of Chan–Lam couplings allow the establishment of various C–X bonds. The proposed mechanism of the C–O coupling is depicted in Figure 1. After the coordination and transmetalation step (I.), disproportionation (II.) between a CuY_2 and a $Cu^{II}(Ar)Y$ occurs. A reductive elimination (III.) step provides the C–O coupled product. The terminal oxidant is responsible for the oxidation of Cu(I) salts. Application of such methodologies in selective transformations of biologically active compounds might allow feasible syntheses of drug candidates.

The literature describes certain copper-catalyzed methodologies for the synthesis of diphenyl ethers (DEs) [14]. The copper(II)-promoted coupling might even efficiently be carried out at room temperature using an organic or inorganic base in varied solvents. DE represents a compound group bearing two aromatic rings connected via a flexible oxygen bridge. The latter is an essential pharmacophore owing to its substantial hydrophobicity, good lipid solubility, cell membrane penetration and metabolic stability [15,16]. Both synthetic pharmaceuticals [17–20] and several biologically active natural products [21–24] contain the DE subunit. DEs, among others, possess anticancer [18–21], antiviral [25,26], anti-inflammatory [23], antibacterial [27], antiparasitic [28], fungicidal [29], herbicidal [30] or insecticidal [29] activities. Furthermore, the literature reports new efficient DE-based compounds for the therapeutic applications of devastating diseases affecting the central nervous or the cardiovascular system worldwide [14]. Figure 2 shows certain pharmacologically important DE-containing drugs, pesticides and natural products (**1–4**). Ibrutinib [17] and sorafenib [18] as small-molecule inhibitors of certain kinase enzymes belong to anticancer agents; however, nimesulide [20] is a nonsteroidal anti-inflammatory drug. Isoliensinine [21] displays a variety of biological activities, including anti-cancer, antioxidant and anti-HIV effects.

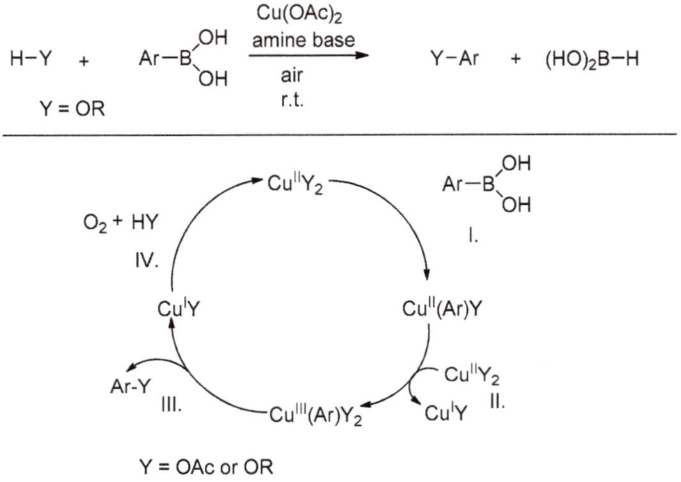

Figure 1. Chan–Lam coupling: C–O bond formation and its proposed mechanism [13].

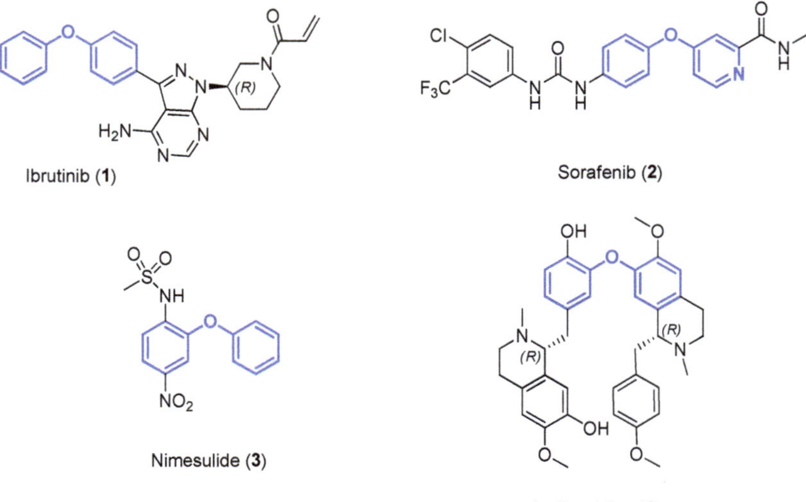

Ibrutinib (**1**)

Sorafenib (**2**)

Nimesulide (**3**)

Isoliensinine (**4**)

Figure 2. Structures of synthetic or natural DE-based drugs **1–4** (the diphenyl ether moiety is highlighted in blue).

Cancer still remains the leading cause of death around the world. The development of new, highly effective anticancer agents with high selectivity is still one of the major goals of medicinal chemistry. Antitubulin compounds are considered to be the most effective tools for cancer chemotherapy nowadays [31]. Certain antitubulin agents have already been approved by the Food and Drug Administration (FDA) [32], but their efficacy might be limited by the development of multidrug-resistant (MDR) cancer cells. Due to their crucial role in cell division, α- and β-tubulin are essential targets for the development of novel anticancer agents. Suppression of cell growth might be achieved by drugs that stabilize or destabilize microtubules (MTs) [33,34]. MT destabilizing agents (MDAs) prevent the polymerization of tubulin and they promote depolymerization, whereas MT stabilizing agents (MSAs) promote the polymerization of tubulin and they stabilize the polymer, preventing depolymerization. Six binding sites have been identified on tubulin [35,36]; however, MSAs generally bind reversibly to the taxoid binding site. Antimitotic drugs approved by FDA might be classified based on their binding site or their further modification (encapsulation or conjugation strategies) [32]. Vincristine sulfate (Oncovin), vinblastine sulfate (Velban) and vinorelbine (Navelbine) belong to the vinca alkaloid site binding drugs; however, colchicine binds to the site with the same name. Marquibo, as a vincristine sulfate liposome injection was developed to improve the pharmacodynamic properties of vincristine. T-DM1 is a maytansine derivative conjugated to trastuzumab applied in a second line breast cancer therapy. It can be stated that antitubulin agents bear a characteristic structure generally. It has two aryl rings and an ethylene, triazole or oxygen bridge, which determine the relative orientation of the rings [36,37]. The latter structural element, the diaryl ether scaffold, is present in certain potent antitubulin agents [15] including substituted or condensed variations (compounds **5–8**, Figure 3). Nevertheless, the development of antitubulin compounds possessing improved potency and selectivity is still a leading challenge.

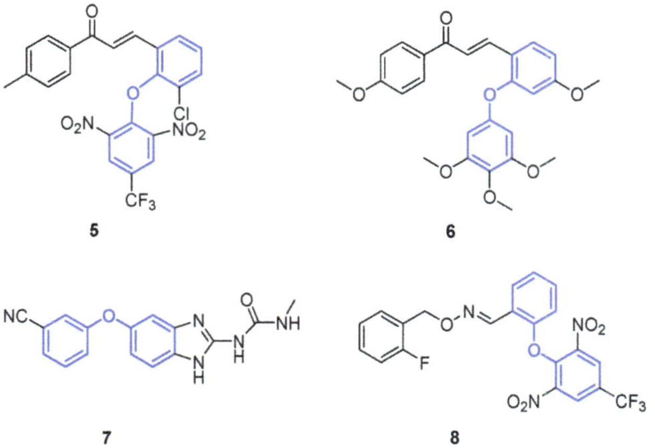

Figure 3. Structures of selected antitubulin agents (**5–8**) bearing diphenyl ether moieties highlighted in blue.

We recently published the syntheses and biochemical investigation of certain biologically active derivatives of the core-modified synthetic 13α-estrone **9**. Inversion of the C-13 in natural estrone derivatives substantially reduces their estrogenic activity [38–41]. The group of 13α-estrone derivatives proved to be promising concerning their enzyme inhibitory and/or antiproliferative properties. However, their biological activity greatly depends on their structure [33,42–48]. We have observed that the substitution pattern of ring A influences bioactivity markedly. The introduction of an apolar benzyl group onto the phenolic hydroxy function usually improves the cell growth-inhibitory action of 17-keto or 17-hydroxy 13α-estrone derivatives [42]. Accordingly, the presence of the apolar ether moiety at C-3 seemed to be advantageous. The low micromolar antiproliferative action of benzyl ethers could further be enhanced by inserting a polar triazole ring between the 3-OH and the benzylic moiety [42,48]. Compound **11a** displayed submicromolar IC_{50} values on HeLa, A2780, A431 and MCF-7 cancer cell lines. Introduction of a bromo substituent to the C-4 *ortho* position led to a compound (**11b**) with improved tumor selectivity, being most potent on A2780 cell line (Figure 4) [48]. Mechanistic investigations suggest that compound **11b** exerts a direct effect on microtubule formation. Molecular dynamics (MD, MMGBSA methods) were performed in order to calculate binding energy at an advanced level. Computational calculations revealed strong interactions of compound **11b** with both colchicine (CBS) and taxoid binding sites (TBS) of tubulin with a stronger interaction at the TBS. Consequently, triazole derivative **11b** might be considered as an MSA agent with remarkable tumor selectivity concerning the cell lines investigated. More recently we have shown that 3-deoxy-3-phenyl-13α-estra-1,3,5(10)-triene (**12**) displays weak antiproliferative action [47]. This observation suggests that direct phenylation at C-3 by the simultaneous removal of the hydroxy group is detrimental regarding the antiproliferative action. It follows that the oxygen-containing moiety at C-3 should be retained and its etherification with apolar benzyl or more polar triazolylbenzyl moieties might be highly beneficial. Our results indicate that the hormonally inactive 13α-estrane core with certain ring A modifications might be a suitable scaffold in the design of potent MT targeting agents.

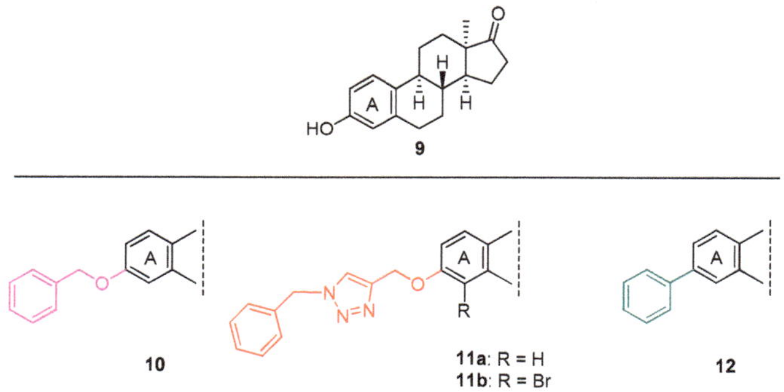

Figure 4. 13α-Estrone derivatives **9–12** synthesized recently (the modifications are indicated by different colours).

With these considerations in mind, here we aimed to perform the direct arylation of 13α-estrone at C-3-O via Chan–Lam coupling using arylboronic acids as reagents. The set of boronic acid coupling partners included not only substituted phenyl but heteroaryl derivatives too. The evaluation of antiproliferative action of the newly synthesized compounds against five human cancer cell lines was also planned. Mechanistic investigations concerning the direct effect of the most potent compound on microtubule formation were also intended. To gain insight into the interaction of the selected potent compound with the taxoid binding site of tubulin, computational calculations were performed.

2. Results and Discussion

2.1. Chemistry

First, the etherification of 13α-estrone **9** with phenylboronic acid **13a** was carried out. Arylboronic acids are outstanding coupling partners in several metal-catalyzed reactions, owing to their broad availability, low toxicity, high stability and extensive functional group tolerance. Based on literature data [12], a copper-promoted $C(sp^2)$–O coupling was performed, using 1 equiv. of $Cu(OAc)_2$ as the catalyst in dichloromethane solvent. K_2CO_3, trimethylamine (NEt_3) and diisopropylethylamine (DIPEA) were tested as bases but NEt_3 proved to be the best option based on yields. The protocol was extended to couplings of different arylboronic acids (**13a–l**) with 13α-estrone (**9**) (Scheme 1). In order to get important structure–activity information, substituted phenyl (**13a–h**) or condensed carbo or heterocyclic derivatives (**13i–l**) were chosen as reagents. C(sp^2)–O couplings proceeded with high isolated yields (Scheme 1).

The structures of the newly synthesized diaryl ethers (**14a–l**) were deduced from ^{1}H and ^{13}C NMR spectra.

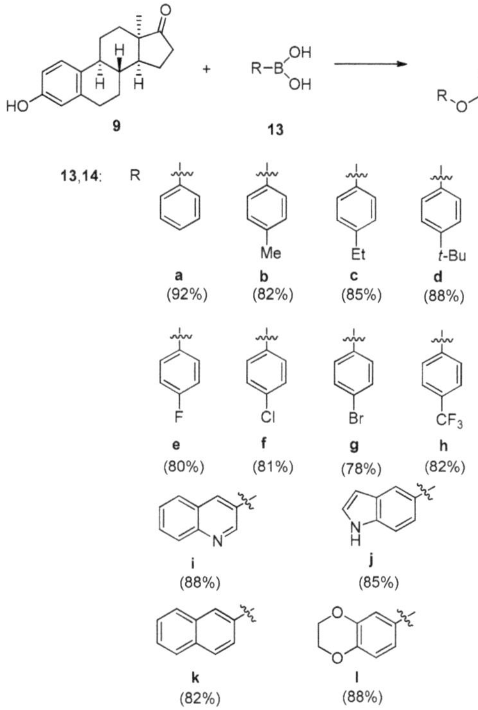

Scheme 1. Chan–Lam couplings of 13α-estrone (**9**) with arylboronic acids (**13a–l**).

2.2. Pharmacology

Here we investigated the in vitro cell growth-inhibitory properties of the newly synthesized diaryl ethers (**14a–l**) by the MTT assay [49] on a panel of human adherent cancer cell lines. The panel included different cervical (HeLa and SiHa), breast (MCF-7 and MDA-MB-231) and ovarian (A2780) cancer cell lines [50–56]. The cervical and breast cancer cell line pairs were selected on the basis of their different HPV or receptorial status. NIH/3T3 mouse fibroblast cell line was used for the determination of cancer selectivity.

We have recently described the antiproliferative properties of 13α-estrone-3-benzyl ether (**10**, Figure 4) and its triazolyl derivatives (**11a,b**) against HeLa, MCF-7 and A2780 cell lines [42,48]. The etherified compounds (**10; 11a,b**) inhibited the growth of cells more effectively than their 3-OH counterpart (**9**). In both compound types (**10; 11a,b**), the carbo or the heteroaromatic ring was connected to C-3-*O* via a methylene linker. Here we were interested in the investigation of the antiproliferative properties of 13α-estrone derivatives arylated directly at the C-3-*O* function. Although compound structures **14a** and **10** (IC$_{50}$ > 30 μM, HeLa [42]) differ only in a single methylene group, their antiproliferative action varies substantially (Table 1). Compound **14a** is more active, especially on the HPV-18 positive HeLa cervical cell line, displaying a low micromolar IC$_{50}$ value (5.53 μM, Table 1). In addition, the estrogen receptor positive breast cancer cell line MCF-7 proved to be sensitive to **14a** but the IC$_{50}$ value was twice as high as that mentioned previously. Interestingly, **14a** behaved differently on the pairs of breast and cervical cell lines.

Table 1. Antiproliferative activities of newly synthesized diaryl ethers (**14a–l**).

<table>
<tr><td colspan="8" align="center">Antiproliferative Activities of Newly Synthesized Diaryl Ethers (14a–l)</td></tr>
<tr><td rowspan="2">Compd.</td><td rowspan="2">Conc. (μM)</td><td colspan="6" align="center">Inhibition (%) ± SEM (Calculated IC₅₀ Value; μM)</td></tr>
<tr><td>MCF-7</td><td>MDA-MB-231</td><td>HeLa</td><td>SiHa</td><td>A2780</td><td>NIH/3T3</td></tr>
<tr><td rowspan="2">14a</td><td>10</td><td>37.52 ± 2.46</td><td>– *</td><td>60.24 ± 0.58</td><td>–</td><td>23.00 ± 1.32</td><td>-</td></tr>
<tr><td>30</td><td>81.42 ± 1.50
(10.12)</td><td>25.30 ± 1.82</td><td>92.02 ± 1.38
(5.53)</td><td>–</td><td>58.48 ± 1.03
(23.81)</td><td>38.75 ± 1.21</td></tr>
<tr><td rowspan="2">14b</td><td>10</td><td>–</td><td>–</td><td>56.69 ± 1.29</td><td>–</td><td>39.64 ± 1.84</td><td>–</td></tr>
<tr><td>30</td><td>29.28 ± 1.26</td><td>33.34 ± 1.75</td><td>83.29 ± 1.34
(7.99)</td><td>–</td><td>62.15 ± 1.15
(16.67)</td><td>43.34 ± 2.17</td></tr>
<tr><td rowspan="2">14c</td><td>10</td><td>43.99 ± 1.77</td><td>–</td><td>62.38 ± 1.19</td><td>27.24 ± 2.60</td><td>–</td><td>–</td></tr>
<tr><td>30</td><td>58.64 ± 1.08
(16.44)</td><td>33.16 ± 0.95</td><td>88.14 ± 0.79
(5.13)</td><td>43.55 ± 0.86</td><td>45.08 ± 1.75</td><td>21.99 ± 2.03</td></tr>
<tr><td rowspan="2">14d</td><td>10</td><td>24.75 ± 0.74</td><td>-</td><td>65.28 ± 1.02</td><td>-</td><td>30.84 ± 2.24</td><td>–</td></tr>
<tr><td>30</td><td>30.77 ± 1.67</td><td>32.93 ± 1.88</td><td>77.06 ± 0.93
(7.11)</td><td>41.14 ± 2.38</td><td>56.98 ± 1.07
(23.65)</td><td>24.07 ± 1.67</td></tr>
<tr><td rowspan="2">14e</td><td>10</td><td>49.02 ± 0.85</td><td>–</td><td>30.14 ± 3.05</td><td>–</td><td>–</td><td>–</td></tr>
<tr><td>30</td><td>60.22 ± 1.68
(13.28)</td><td>–</td><td>37.78 ± 3.86</td><td>–</td><td>46.50 ± 2.43</td><td>–</td></tr>
<tr><td rowspan="2">14f</td><td>10</td><td>-</td><td>-</td><td>60.72 ± 1.24</td><td>–</td><td>34.93 ± 2.85</td><td>–</td></tr>
<tr><td>30</td><td>37.51 ± 1.44</td><td>30.28 ± 2.24</td><td>75.40 ± 1.42
(5.78)</td><td>38.91 ± 1.69</td><td>51.48 ± 1.93
(26.17)</td><td>–</td></tr>
<tr><td rowspan="2">14g</td><td>10</td><td>48.82 ± 2.18</td><td>–</td><td>65.97 ± 0.61</td><td>20.31 ± 3.10</td><td>–</td><td>–</td></tr>
<tr><td>30</td><td>59.91 ± 1.73
(11.98)</td><td>28.28 ± 1.39</td><td>80.70 ± 0.76
(5.21)</td><td>38.42 ± 1.30</td><td>39.58 ± 2.60</td><td>30.92 ± 2.15</td></tr>
<tr><td rowspan="2">14h</td><td>10</td><td>35.08 ± 0.69</td><td>27.06 ± 2.93</td><td>67.46 ± 0.46</td><td>31.80 ± 2.77</td><td>36.13 ± 1.39</td><td>–</td></tr>
<tr><td>30</td><td>45.42 ± 3.30</td><td>39.67 ± 2.39</td><td>80.74 ± 0.58
(6.90)</td><td>45.55 ± 2.63</td><td>52.60 ± 1.00
(23.47)</td><td>45.47 ± 0.28</td></tr>
</table>

Table 1. *Cont.*

		Antiproliferative Activities of Newly Synthesized Diaryl Ethers (14a–l)					
		Inhibition (%) $\pm$ SEM (Calculated IC$_{50}$ Value; μM)					
Compd.	Conc. (μM)	MCF-7	MDA-MB-231	HeLa	SiHa	A2780	NIH/3T3
14i	10	49.68 $\pm$ 2.46	24.07 $\pm$ 2.47	76.18 $\pm$ 1.49	22.46 $\pm$ 2.36	48.66 $\pm$ 1.90	35.48 $\pm$ 1.33
	30	74.67 $\pm$ 0.51 (8.52)	58.28 $\pm$ 1.39 (22.95)	90.11 $\pm$ 0.80 (3.98)	48.33 $\pm$ 1.27	71.21 $\pm$ 1.35 (11.54)	49.87 $\pm$ 0.89
14j	10	–	–	50.99 $\pm$ 2.21	–	–	–
	30	42.25 $\pm$ 1.33	–	71.63 $\pm$ 1.45 (9.60)	–	28.02 $\pm$ 3.57	–
14k	10	–	29.61 $\pm$ 3.01	73.86 $\pm$ 1.48	28.80 $\pm$ 2.10	42.23 $\pm$ 0.71	–
	30	43.11 $\pm$ 2.035	41.87 $\pm$ 0.85	77.08 $\pm$ 1.34 (5.52)	54.33 $\pm$ 0.83 (23.90)	51.25 $\pm$ 1.43 (25.25)	24.15 $\pm$ 0.60
14l	10	29.08 $\pm$ 1.88	20.72 $\pm$ 1.33	61.30 $\pm$ 0.98	20.79 $\pm$ 3.34	48.44 $\pm$ 0.49	–
	30	63.02 $\pm$ 1.09 (18.91)	24.10 $\pm$ 3.15	74.01 $\pm$ 0.80 (9.16)	38.58 $\pm$ 2.61	69.17 $\pm$ 0.50 (11.47)	46.16 $\pm$ 2.11 –
cisplatin	10	53.03 $\pm$ 2.29	20.75 $\pm$ 0.81	42.61 $\pm$ 2.33	60.98 $\pm$ 0.92	83.57 $\pm$ 1.21	76.74 $\pm$ 1.26
	30	86.90 $\pm$ 1.24 (5.78)	74.47 $\pm$ 1.20 (19.13)	99.93 $\pm$ 0.26 (12.43)	88.95 $\pm$ 0.53 (4.29)	95.02 $\pm$ 0.28 (1.30)	96.90 $\pm$ 0.25 (4.73)

*: The inhibition value is less than 20% and not given numerically.

All test compounds seemed to be active against the HeLa cell line with IC_{50} values ranging from 5 to 10 μM with the exception of the 4-fluorophenyl derivative **14e** (Table 1). Substitution of a hydrogen with its fluorine bioisostere usually leads to unique biological activity of the fluorinated derivative. The highest electronegativity and the small van der Waals radius make fluorine derivatives generally more active than their unsubstituted counterparts. Nevertheless, in the case of **14e** the presence of fluorine seems to be disadvantageous. If we analyze the results obtained for *para*-tolyl (**14b**) and *para*-trifluoromethyl (**14h**) compounds, only slight differences appear. Accordingly, the effect of fluorine is more pronounced if it is connected directly to the phenyl group.

Introduction of a condensed bicyclic moiety onto the C-3-O site resulted in unique structure–activity results. The naphthyl derivative (**14k**) displayed growth-inhibitory actions similar to its phenyl counterpart (**14a**), except on cell lines MCF-7 and SiHa with SiHa being more sensitive to compound **14k**. The presence of a nitrogen heteroatom in the introduced moiety (compound **14i**) resulted in a significant improvement in the biological action. This compound should be highlighted as the most potent derivative on all cell lines investigated (except for SiHa). In contrast, the benzo-1,4-dioxane derivative (**14l**) seemed to be less potent than **14i** on four cancer cell lines. The ovarian cell line A2780 could not distinguish between the two heterocyclic compounds **14i** and **14l**. The other nitrogen-containing heterocyclic derivative (**14j**) was less effective than **14i**.

It should be emphasized that certain newly synthesized derivatives displayed more pronounced cell growth-inhibitory action than the reference compound cisplatin, especially on the HeLa cancer cell line.

In order to get information on the tumor selectivity of the antiproliferative action, the more promising derivatives were investigated on the mouse fibroblast cell line NIH/3T3. None of the compounds exerted inhibition above 50% even at a higher, 30 μM test concentration. In this regard, the reference agent cisplatin proved to be less selective.

It is particularly promising that slight structural modifications generated significant differences in inhibitory activities. This might be indicative of a cell type-dependent action. The broad toxic character of the test compounds might be excluded. On this basis, the compound group investigated might substantially contribute to lead-finding projects based on estrone derivatives. Nevertheless, the mechanistic investigation of antiproliferative action is necessary. The determination of the mechanism of action was inspired by our recent result [48]. We found earlier that certain 13α-estrone derivatives etherified on their phenolic hydroxy function might affect the microtubule formation. On this basis, here we performed an in vitro tubulin polymerization assay with the most potent compound **14i** in two different concentrations (125 and 250 μM). Paclitaxel, a clinically applied anticancer agent was used as a reference compound.

The obtained absorbance values indicate that **14i** disturbs the polymerization of tubulin by increasing the maximum rate of the procedure. This property was concentration-dependent and statistically significant even at the lower concentration range. The effect of **14i** at 250 μM was lower than that of the reference agent. Based on these in vitro results it can be concluded that the antiproliferative action of **14i** is elicited through the disturbance of tubulin polymerization (Figure 5).

2.3. Computational Investigations

Having a picture at the atomic level concerning the possible binding character of the **14i** ligand, MD calculations have been performed. To sample the conformational space of the ligand binding to the taxoid binding site, five independent simulations were run providing a 2.5 μs-long trajectory in total. The resulting protein–ligand interaction diagrams of all five trajectories are presented in Supplementary Materials (Figures S1–S5). According to the simulations, a significant hydrogen-bonding interaction was formed between the ligand and two amino acids (Gln-282 and Thr-276). In one of the trajectories, the Thr-276 amino acid shows extra stabilization with the 17-keto function of the sterane skeleton when the modified steroid finds a deeper position in the taxoid binding site. Both amino acids are located in a flexible loop region near to the taxoid binding pocket and the motion of the loop can affect the binding of the modified steroid. Interestingly, despite that two Gln amino acids located next to each other in position 281 and 282, the ligand had frequent interaction mainly with the second one in the simulations.

Finally, demonstrating the relative position of those amino acids, which had significant interaction with the protein or the guanosine diphosphate (GDP) co-factor, we present a representative ligand–protein complex in Figure 6.

It is worth to mention that the above-mentioned special hydrogen bonds with Gln-282 and Thr-276 always formed between one of the oxygen atoms of the ligand and never with the nitrogen of the quinoline skeleton. Concerning other secondary interactions of the ligand, the analysis showed two further connection types, namely water bridges and hydrophobic interactions. Interestingly, the nitrogen atom in the quinoline ring does not show any specific binding to the tubulin protein.

Finally, we would like to point out that the ligand remained in the binder pocket all along the trajectories, while the GDP left its binding pocket in a number of cases.

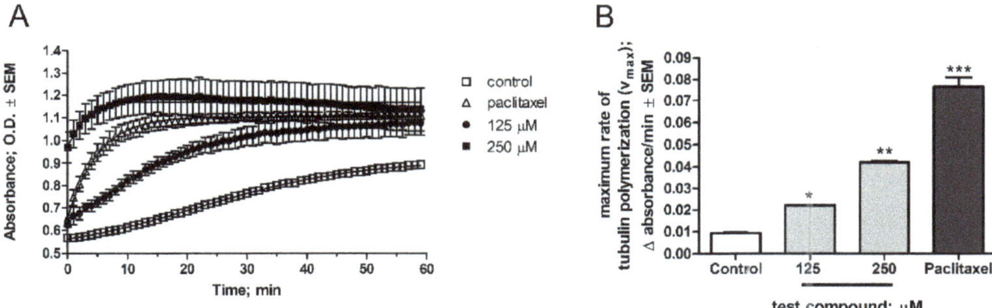

Figure 5. Effects of 14i and 10 µM paclitaxel on the calculated maximum reaction rate (Vmax) of in vitro microtubule formation. (**A**) Representative kinetic curves; (**B**) Calculated results. Control: untreated samples. The experiment was performed in two parallels and the measurements were repeated twice. Each bar denotes the mean ± SEM; *, ** and *** indicate $p < 0.05$, $p < 0.01$ and $p < 0.001$, respectively, compared with the control values.

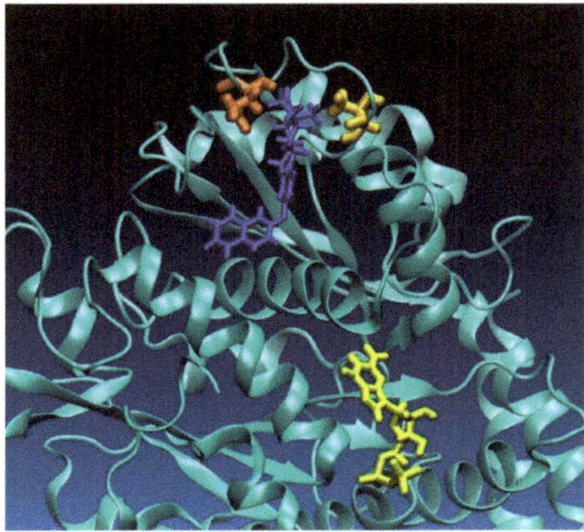

Figure 6. Relative position of compound **14i** (blue), Gln-282 (dark orange), Thr-276 (light orange) and GDP (yellow) nearby the taxoid binding site in the tubulin protein.

3. Materials and Methods

3.1. Chemistry

Melting points (Mp) were determined with a Kofler hot-stage apparatus. Perkin-Elmer CHN analyzer 2400 was used for the elemental analyses. Thin-layer chromatography was performed on silica gel 60 F_{254} (layer thickness 0.2 mm, Merck); eluent (ss): 30% ethyl acetate/70% hexanes. The spots were detected with I_2 or UV (365 nm) after spraying with 5% phosphomolybdic acid in 50% aqueous phosphoric acid and heating at 100–120 °C for 10 min. Flash chromatography was performed on silica gel 60, 40–63 μm (Merck). ^{1}H NMR spectra were recorded in DMSO-d_6 or CDCl$_3$ solution with a Bruker DRX-500 instrument at 500 MHz. ^{13}C NMR spectra were recorded with the same instrument at 125 MHz under the same conditions. Full scan mass spectra of the newly synthesized compounds were acquired in the range of 50 to 1000 m/z with a Finnigan TSQ-7000 triple quadrupole mass spectrometer (Finnigan-MAT, San Jose, CA, USA) equipped with a Finnigan electrospray ionization source. Analyses were achieved in positive ion mode applying flow injection mass spectrometry with a mobile phase of 50% aqueous acetonitrile containing 0.1 v/v% formic acid (flow rate: 0.3 mL/min). Five μL aliquot of the samples were loaded into the flow. The ESI capillary was adjusted to 4.5 kV and N_2 was used as a nebulizer gas.

3.2. General Procedure for the Synthesis of 3-aryloxy-13α-estra-1,3,5(10)-triene-17-ones

13α-Estrone **9** (50 mg, 0.185 mmol), Cu(OAc)$_2$ (33 mg, 1 equiv.), arylboronic acid (1 eq.) were dissolved in dichloromethane (5 mL) then triethylamine (125 μL, 5 equiv. mmol) was added. The reaction mixture was stirred at rt for 12–24 h, quenched with water (15 mL) and extracted with dichloromethane (3 × 15 mL). The organic phase was dried over anhydrous Na$_2$SO$_4$, concentrated in vacuum and the resulting residue was purified by column chromatography. Hexanes were used for crystallization if needed.

3-Phenoxy-13α-estra-1,3,5(10)-triene-17-one (**14a**). Reaction time: 16 h. The residue was purified by flash chromatography with hexanes/EtOAc = 6:1 (v/v) as eluent. Compound **14a** was isolated as a white solid (92%). Mp.: 118.5–119.5 °C; R$_f$: 0.38; Mr: 346.2; Anal. Calcd. for C$_{24}$H$_{26}$O$_2$: C, 83.20; H, 7.56. Found: C, 83.29; H, 7.52. ^{1}H NMR (500 MHz, DMSO-d_6) δ ppm: 0.98 (s, 3H, 13-CH$_3$); 2.75 (m, 2H, 6-H$_2$); 6.69 (d, 1H, J = 2.5 Hz, 4-H); 6.75 (dd, 1H, J = 8.5 Hz, J = 2.6 Hz, 2-H); 6.95 (d, 2H, J = 7.7 Hz, 2′- and 6′-H); 7.10 (t, 1H, J = 7.7 Hz, 4′-H); 7.28 (d, 1H, J = 8.6 Hz, 1-H); 7.36 (d, 2H, J = 7.7 Hz, J = 2.0 Hz, 3′- and 5′-H). ^{13}C NMR (DMSO-d_6) δ ppm: 20.4 (CH$_2$); 24.4 (C-18); 27.5 (CH$_2$); 27.8(CH$_2$); 29.4 (CH$_2$); 31.5 (CH$_2$); 32.8 (CH$_2$); 40.5 (CH); 40.8 (CH); 48.4 (CH); 49.3 (C-13); 116.1 (CH); 118.2 (2C, 2×CH); 118.3 (CH); 122.9 (CH); 127.3 (CH); 129.8 (2C, 2×CH); 134.8 (C-10); 138.6 (C-5); 154.1 (C); 156.9 (C); 220.5 (C=O). MS m/z (%) 347 (100, [M + H]$^+$).

3-(4-Tolyloxy)-13α-estra-1,3,5(10)-triene-17-one (**14b**). Reaction time: 20 h. The residue was purified by flash chromatography with hexanes/EtOAc = 6:1 (v/v) as eluent. Compound **14b** was isolated as a white solid (82%). Mp.: 147.9–148.7 °C; R$_f$: 0.40; M$_r$: 360.2; Anal. Calcd. for C$_{25}$H$_{28}$O$_2$: C, 83.29; H, 7.83. Found: C, 83.38; H, 7.79. ^{1}H NMR (500 MHz, DMSO-d_6) δ ppm: 0.97 (s, 3H, 13-CH$_3$); 2.27 (s, 3H, 4′-CH$_3$); 2.73 (m, 2H, 6-H$_2$); 6.63 (d, 1H, J = 2.5 Hz, 4-H); 6.70 (dd, 1H, J = 8.5 Hz, J = 2.5 Hz, 2-H); 6.85 and 7.15 (2×d, 2×2H, J = 8.5 Hz, 2′-, 3′-, 5′- and 6′-H); 7.24 (d, 1H, J = 8.6 Hz, 1-H). ^{13}C NMR (DMSO-d_6) δ ppm: 20.1 (CH$_3$); 20.4 (CH$_2$); 24.4 (C-18); 27.5 (CH$_2$); 27.8 (CH$_2$); 29.4 (CH$_2$); 31.5 (CH$_2$); 32.8 (CH$_2$); 40.5 (CH); 40.7 (CH); 48.4 (CH); 49.3 (C-13); 115.6 (CH); 117.7 (CH); 118.5 (2C, 2×CH); 127.2 (CH); 130.2 (2C, 2×CH); 132.1 (C); 134.3 (C); 138.4 (C); 154.4 (C); 154.7 (C); 220.5 (C=O). MS m/z (%) 361 (100, [M + H]$^+$).

3-(4-Ethylphenoxy)-13α-estra-1,3,5(10)-triene-17-one (**14c**). Reaction time: 20 h. The residue was purified by flash chromatography with hexanes/EtOAc = 6:1 (v/v) as eluent. Compound **14c** was isolated as a white solid (85%). Mp.: 131.7–132.7 °C; R$_f$: 0.43; M$_r$: 374.2; Anal. Calcd. for C$_{26}$H$_{30}$O$_2$: C, 83.38; H, 8.07. Found: C, 83.45; H, 8.01. ^{1}H NMR (500 MHz, DMSO-d_6) δ ppm: 0.99 (s, 3H, 13-CH$_3$); 1.18 (t, 3H, J = 7.6 Hz, -CH$_2$-CH$_3$); 2.58 (q, 2H, J = 7.6 Hz, -CH$_2$-CH$_3$); 2.75 (m, 2H, 6-H$_2$); 6.66 (d, 1H, J = 2.5 Hz, 4-H); 6.72 (dd, 1H, J = 8.6 Hz, J = 2.5 Hz, 2-H); 6.88 and 7.19 (2×d, 2×2H, J = 8.4 Hz, 2′-, 3′-, 5′- and 6′-H); 7.25

(d, 1H, J = 8.6 Hz, 1-H). ^{13}C NMR (DMSO-d_6) δ ppm: 15.4 (-CH$_2$-$\underline{C}$H$_3$); 20.3 (CH$_2$); 24.4 (C-18); 27.2 (CH$_2$); 27.4 (CH$_2$); 27.7 (CH$_2$); 29.3 (CH$_2$); 31.4 (CH$_2$); 32.7 (CH$_2$); 40.4 (CH); 40.7 (CH); 48.4 (CH); 49.2 (C-13); 115.6 (CH); 117.8 (CH); 118.3 (2C, 2×CH); 127.0 (CH); 128.8 (2C, 2×CH); 134.3 (C); 138.4 (2C, 2×C); 154.5 (C); 154.6 (C); 220.3 (C=O). MS m/z (%) 375 (100, [M + H]$^+$).

3-(4-tert-Butylphenoxy)-13α-estra-1,3,5(10)-triene-17-one (**14d**). Reaction time: 20 h. The residue was purified by flash chromatography with hexanes/EtOAc = 6:1 (*v/v*) as eluent. Compound **14d** was isolated as a white solid (88%). Mp.: 129.8–130.8 °C; R$_f$: 0.43; M$_r$: 402.3; Anal. Calcd. for C$_{28}$H$_{34}$O$_2$: C, 83.54; H, 8.51. Found: C, 83.63; H, 8.45. ^{1}H NMR (500 MHz, DMSO-d_6) δ ppm: 0.98 (s, 3H, 13-CH$_3$); 1.27 (s, 9H, *t*-Bu-CH$_3$); 2.75 (m, 2H, 6-H$_2$); 6.68 (d, 1H, J = 2.5 Hz, 4-H); 6.71 (dd, 1H, J = 8.5 Hz, J = 2.5 Hz, 2-H); 6.88 and 7.36 (2×d, 2×2H, J = 8.4 Hz, 2′-, 3′-, 5′- and 6′-H); 7.25 (d, 2H, J = 8.5 Hz, 1-H). ^{13}C NMR (DMSO-d_6) δ ppm: 20.3 (CH$_2$); 24.4 (C-18); 27.4 (CH$_2$); 27.7 (CH$_2$); 29.3 (CH$_2$); 31.1 (C($\underline{C}$H$_3$)$_3$); 31.4 (CH$_2$); 32.7 (CH$_2$); 33.8 ($\underline{C}$(CH$_3$)$_3$); 40.4 (CH); 40.7 (CH); 48.4 (CH); 49.2 (C-13); 115.8 (CH); 117.7 (2C, 2×CH); 118.0 (CH); 126.3 (2C, 2×CH); 127.0 (CH); 134.4 (C); 138.4 (C); 145.2 (C); 154.3 (C); 154.4 (C); 220.3 (C=O). MS m/z (%) 403 (100, [M + H]$^+$).

3-(4-Fluorophenoxy)-13α-estra-1,3,5(10)-triene-17-one (**14e**). Reaction time: 12 h. The residue was purified by flash chromatography with hexanes/EtOAc = 6:1 (*v/v*) as eluent. Compound **14e** was isolated as a white solid (80%). Mp.: 115.3–116.3 °C; R$_f$: 0.41; M$_r$: 364.2; Anal. Calcd. for C$_{24}$H$_{25}$FO$_2$: C, 79.09; H, 6.91. Found: C, 79.16; H, 6.86. ^{1}H NMR (500 MHz, DMSO-d_6) δ ppm: 0.98 (s, 3H, 13-CH$_3$); 2.75 (m, 2H, 6-H$_2$); 6.67 (d, 1H, J = 2.5 Hz, 4-H); 6.72 (dd, 1H, J = 8.5 Hz, J = 2.5 Hz, 2-H); 7.00 (overlapping doublets, 2H, J = 8.7 Hz, 2′- and 6′-H); 7.19 (t, 2H, J = 8.7 Hz, 3′- and 5′-H); 7.27 (d, 1H, J = 8.6 H, 1-Hz). ^{13}C NMR (DMSO-d_6) δ ppm: 20.4 (CH$_2$); 24.4 (CH$_3$); 27.5 (CH$_2$); 27.8 (CH$_2$); 29.4 (CH$_2$); 31.5 (CH$_2$); 32.8 (CH$_2$); 40.5 (CH); 40.7 (CH); 48.4 (CH); 49.3 (C); 115.6 (CH); 116.2 (CH); 116.4 (CH); 117.8 (CH); 120.1 (CH); 120.2 (CH); 127.3 (CH); 134.7 (C); 138.6 (C); 152.9 (C); 154.6 (C); 158.8 (C); 220.5 (C=O). MS m/z (%) 365 (100, [M + H]$^+$).

3-(4-Chlorophenoxy)-13α-estra-1,3,5(10)-triene-17-one (**14f**). Reaction time: 12 h. The residue was purified by flash chromatography with hexanes/EtOAc = 6:1 (*v/v*) as eluent. Compound **14f** was isolated as a white solid (81%). Mp.: 120.7–121.7 °C; R$_f$: 0.38; M$_r$: 380.2; Anal. Calcd. for C$_{24}$H$_{25}$ClO$_2$: C, 75.68; H, 6.62. Found: C, 75.77; H, 6.57. ^{1}H NMR (500 MHz, DMSO-d_6) δ ppm: 0.98 (s, 3H, 13-CH$_3$); 2.76 (m, 2H, 6-H$_2$); 6.72 (d, 1H, J = 2.5 Hz, 4-H); 6.78 (dd, 1H, J = 8.5 Hz, J = 2.5 Hz, 2-H); 6.97 (d, 2H, J = 8.9 Hz) and 7.39 (d, 2H, J = 8.9 Hz): 2′-, 3′-, 5′- and 6′-H; 7.29 (d, 1H, J = 8.6 Hz, 1-H). ^{13}C NMR (CDCl$_3$) δ ppm: 20.4 (CH$_2$); 24.4 (C-18); 27.4 (CH$_2$); 27.8 (CH$_2$); 29.4 (CH$_2$); 31.5 (CH$_2$); 32.8 (CH$_2$); 40.4 (CH); 40.8 (CH); 48.4 (CH); 49.3 (C-13); 116.3 (CH); 118.6 (CH); 119.7 (2C, 2×CH); 126.6 (C-4′); 127.4 (C-1); 129.6 (2C, 2xCH); 135.3 (C-10); 138.8 (C-5); 153.7 (C); 156.0 (C); 220.5 (C=O). MS m/z (%) 381 (100, [M + H]$^+$).

3-(4-Bromophenoxy)-13α-estra-1,3,5(10)-triene-17-one (**14g**). Reaction time: 13 h. The residue was purified by flash chromatography with hexanes/EtOAc = 6:1 (*v/v*) as eluent. Compound **14g** was isolated as a white solid (78%). Mp.: 121.3–122.3 °C; R$_f$: 0.39; M$_r$: 424.1; Anal. Calcd. for C$_{24}$H$_{25}$BrO$_2$: C, 67.77; H, 5.92. Found: C, 67.85; H, 5.88. ^{1}H NMR (500 MHz, DMSO-d_6) δ ppm: 0.98 (s, 3H, 13-CH$_3$); 2.76 (m, 2H, 6-H$_2$); 6.73 (d, 1H, J = 2.5 Hz, 4-H); 6.78 (dd, 1H, J = 8.6 Hz, J = 2.5 Hz, 2-H); 6.91 (d, 2H, J = 8.9 Hz) and 7.51 (d, 2H, J = 8.9 Hz): 2′-, 3′-, 5′- and 6′-H; 7.29 (d, 1H, J = 8.6 Hz, 1-H). ^{13}C NMR (DMSO-d_6) δ ppm: 20.3 (CH$_2$); 24.3 (C-18); 27.4 (CH$_2$); 27.7 (CH$_2$); 29.3 (CH$_2$); 31.4 (CH$_2$); 32.7 (CH$_2$); 40.4 (CH); 40.7 (CH); 48.4 (CH); 49.2 (C-13); 114.3 (C-4′); 116.2 (CH); 118.5 (CH); 120.0 (2C, 2×CH); 127.3 (C-1); 132.4 (2C, 2xCH); 135.3 (C-10); 138.7 (C-5); 153.5 (C); 156.4 (C); 220.3 (C=O). MS m/z (%) 425 (100, [M + H]$^+$).

3-(4-Trifluoromethylphenoxy)-13α-estra-1,3,5(10)-triene-17-one (**14h**). Reaction time: 14 h. The residue was purified by flash chromatography with hexanes/EtOAc = 6:1 (*v/v*) as eluent. Compound **14h** was isolated as a white solid (82%). Mp.: 123.7–124.7 °C; R$_f$: 0.37; M$_r$: 414.2; Anal. Calcd. for C$_{25}$H$_{25}$F$_3$O$_2$: C, 72.45; H, 6.08. Found: C, 72.52; H, 6.01. ^{1}H NMR (500 MHz, DMSO-d_6) δ ppm: 0.99 (s, 3H, 13-CH$_3$); 2.80 (m, 2H, 6-H$_2$); 6.82 (d, 1H, J = 2.5 Hz,

4-H); 6.86 (dd, 1H, *J* = 8.5 Hz, *J* = 2.5 Hz, 2-H); 7.09 (d, 2H, *J* = 8.7 Hz) and 7.69 (d, 2H, *J* = 8.7 Hz): 2′-, 3′-, 5′- and 6′-H; 7.34 (d, 1H, *J* = 8.5 Hz, 1-H). ^{13}C NMR (DMSO-d_6) δ ppm: 20.3 (CH$_2$); 24.4 (C-18); 27.4 (CH$_2$); 27.7 (CH$_2$); 29.3 (CH$_2$); 31.4 (CH$_2$); 32.7 (CH$_2$); 40.4 (CH); 40.7 (CH); 48.4 (CH); 49.2 (C-13); 117.1 (CH); 117.4 (2C, 2xCH); 119.4 (CH); 122.8 (q, *J* = 32.4 Hz, C); 124.1 (q, *J* = 271.4 Hz, C); 127.1 (q, 2C, *J* = 3.7 Hz, 2×CH); 127.4 (CH); 136.1 (C); 139.0 (C); 152.4 (C); 160.5 (C); 220.3 (C=O). MS *m/z* (%) 415 (100, [M + H]$^+$).

3-(Quinolinyl-3-oxy)-13α-estra-1,3,5(10)-triene-17-one (**14i**). Reaction time: 24 h. The residue was purified by flash chromatography with hexanes/EtOAc = 7:1 (*v/v*) as eluent. Compound **14i** was isolated as a white solid (88%). Mp.: 153.2–154.2 °C; R$_f$: 0.12; M$_r$: 397.2; Anal. Calcd. for C$_{27}$H$_{27}$NO$_2$: C, 81.58; H, 6.85. Found: C, 81.67; H, 6.79. ^{1}H NMR (500 MHz, DMSO-d_6) δ ppm: 1.00 (s, 3H, 13-CH$_3$); 2.80 (m, 2H, 6-H$_2$); 6.86 (d, 1H, *J* = 2.6 Hz, 4-H); 6.91 (dd, 1H, *J* = 8.6 Hz, *J* = 2.6 Hz, 2-H); 7.35 (d, 1H, *J* = 8.6 Hz, 1-H); 7.57 (t, 1H, *J* = 8.0 Hz); 7.67 (dt, 1H, *J* = 8.0 Hz, *J* = 1.2 Hz); 7.74 (d, 1H, *J* = 2.7 Hz); 7.90 (d, 1H, *J* = 8.0 Hz); 8.01 (d, 1H, *J* = 8.0 Hz); 8.77 (d, 1H, *J* = 2.7 Hz). ^{13}C NMR (DMSO-d_6) δ ppm: 20.3 (CH$_2$); 24.3 (C-18); 27.4 (CH$_2$); 27.7 (CH$_2$); 29.3 (CH$_2$); 31.4 (CH$_2$); 32.7 (CH$_2$); 40.4 (CH); 40.7 (CH); 48.4 (CH); 49.2 (C-13); 116.2 (CH); 118.5 (CH); 119.6 (CH); 127.0 (CH); 127.2 (CH); 127.4 (CH); 127.6 (CH); 128.1 (C); 128.4 (CH); 135.6 (C); 138.9 (C); 143.9 (C); 144.6 (CH); 150.6 (C); 153.4 (C); 220.3 (C=O). MS *m/z* (%) 398 (100, [M + H]$^+$).

3-(1H-Indol-5-yloxy)-13α-estra-1,3,5(10)-triene-17-one (**14j**). Reaction time: 24 h. The residue was purified by flash chromatography with hexanes/EtOAc = 7:1 (*v/v*) as eluent. Compound **14j** was isolated as a white solid (85%). Mp.: 205.3–206.3 °C; R$_f$: 0.09; M$_r$: 385.2; Anal. Calcd. for C$_{26}$H$_{27}$NO$_2$: C, 81.01; H, 7.06. Found: C, 81.10; H, 7.01. ^{1}H NMR (500 MHz, DMSO-d_6) δ ppm: 0.97 (s, 3H, 13-CH$_3$); 2.71 (m, 2H, 6-H$_2$); 6.36 (s, 1H); 6.57 (d, 1H, *J* = 2.5 Hz); 6.66 (dd, 1H, *J* = 8.6, *J* = 2.6 Hz); 6.78 (dd, 1H, *J* = 8.6 Hz, *J* = 2.6 Hz); 7.13 (d, 1H, *J* = 2.5 Hz); 7.19 (d, 1H, *J* = 8.6 Hz); 7.35 (t, 1H, *J* = 5.4 Hz); 7.37 (d, 1H, *J* = 8.6 Hz); 11.05 (s, 1H, NH). ^{13}C NMR (CDCl$_3$) δ ppm: 20.3 (CH$_2$); 24.4 (C-18); 27.5 (CH$_2$); 27.8 (CH$_2$); 29.4 (CH$_2$); 31.4 (CH$_2$); 32.7 (CH$_2$); 40.5 (CH); 40.6 (CH); 48.4 (CH); 49.2 (C); 100.9 (CH); 109.8 (CH); 112.1 (CH); 114.1 (CH); 114.5 (CH); 116.5 (CH); 126.3 (CH); 126.8 (CH); 128.0 (C); 132.6 (C); 133.1 (C); 138.0 (C); 149.0 (C); 156.5 (C); 220.3 (C=O). MS *m/z* (%) 386 (100, [M + H]$^+$).

3-(Naphthyl-2-oxy)-13α-estra-1,3,5(10)-triene-17-one (**14k**). Reaction time: 24 h. The residue was purified by flash chromatography with hexanes/EtOAc = 6:1 (*v/v*) as eluent. Compound **14k** was isolated as a white solid (82%). Mp.: 167.8–168.8 °C; R$_f$: 0.35; M$_r$: 396.2; Anal. Calcd. for C$_{28}$H$_{28}$O$_2$: C, 84.81; H, 7.12. Found: C, 84.88; H, 7.08. ^{1}H NMR (500 MHz, DMSO-d_6) δ ppm: 0.98 (s, 3H, 13-CH$_3$); 2.76 (m, 2H, 6-H$_2$); 6.76 (d, 1H, *J* = 2.6 Hz, 4-H); 6.82 (dd, 1H, *J* = 8.6 Hz, *J* = 2.6 Hz, 2-H); 7.24 (dd, 1H, *J* = 8.6 Hz, *J* = 2.6 Hz); 7.31 (d, 1H, *J* = 8.6 Hz); 7.33 (d, 1H, *J* = 2.6 Hz); 7.42 (t, 1H, *J* = 7.1 Hz); 7.47 (t, 1H, *J* = 7.1 Hz); 7.79 (d, 1H, *J* = 8.1 Hz); 7.89 (d, 1H, *J* = 8.1 Hz); 7.93 (d, 1H, *J* = 8.9 Hz). ^{13}C NMR (DMSO-d_6) δ ppm: 20.4 (CH$_2$); 24.4 (C-18); 27.5 (CH$_2$); 27.8 (CH$_2$); 29.5 (CH$_2$); 31.5 (CH$_2$); 32.8 (CH$_2$); 40.5 (CH); 40.8 (CH); 48.4 (CH); 49.4 (C-13); 113.1 (CH); 116.3 (CH); 118.5 (CH); 119.5 (CH); 124.6 (CH); 126.5 (CH); 126.9 (CH); 127.3 (CH); 127.5 (CH); 129.5 (C); 129.9 (CH); 133.8 (C); 135.0 (C); 138.7 (C); 154.1 (C); 154.8 (C); 220.5 (C=O). MS *m/z* (%) 397 (100, [M + H]$^+$).

3-(2,3-Dihydro-benzo[1,4]dioxin-6-yloxy)-13α-estra-1,3,5(10)-triene-17-one (**14l**). Reaction time: 24 h. The residue was purified by flash chromatography with hexanes/EtOAc = 6:1 (*v/v*) as eluent. Compound **14l** was isolated as a white solid (88%). Mp.: 177.2–178.0 °C; R$_f$: 0.26; M$_r$: 404.2; Anal. Calcd. for C$_{26}$H$_{28}$O$_4$: C, 77.20; H, 6.98. Found: C, 77.28; H, 6.93. ^{1}H NMR (500 MHz, DMSO-d_6) δ ppm: 0.98 (s, 3H, 13-CH$_3$); 2.74 (m, 2H, 6-H$_2$); 4.22 (m, 4H, 2′- and 3′-H$_2$); 6.45 (dd, 1H, *J* = 8.7 Hz, *J* = 2.8 Hz); 6.48 (d, 1H, *J* = 2.7 Hz); 6.63 (d, 1H, *J* = 2.4 Hz); 6.68 (dd, 1H, *J* = 8.5 Hz, *J* = 2.5 Hz); 6.83 (d, 1H, *J* = 8.7 Hz); 7.23 (d, 1H, *J* = 8.6 Hz). ^{13}C NMR (DMSO-d_6) δ ppm: 20.3 (CH$_2$); 24.4 (C-18); 27.4 (CH$_2$); 27.7 (CH$_2$); 29.4 (CH$_2$); 31.4 (CH$_2$); 32.7 (CH$_2$); 40.4 (CH); 40.6 (CH); 48.4 (CH); 49.2 (C-13); 63.6 (CH$_2$); 64.0 (CH$_2$); 107.7 (CH); 111.6 (CH); 115.1 (CH); 117.3 (CH); 117.4 (CH); 126.9 (CH); 134.1 (C); 138.3 (C); 139.4 (C); 143.6 (C); 150.2 (C); 155.0 (C); 220.3 (C=O). MS *m/z* (%) 405 (100, [M + H]$^+$).

4. Determination of Antiproliferative Activities

The antiproliferative activities of the currently presented molecules **14a–l** were determined against human adherent cancer cell lines of gynecological origin. MCF-7 and MDA-MB-231 cells were isolated from breast cancers, while A2780 is an ovarian cancer cell line. In addition, Hela and SiHa cells were isolated from cervical cancers containing HPV-18 and HPV-16, respectively. The tumor selectivity of molecules was determined using nonmalignant mouse embryo fibroblast cells (NIH/3T3). All cell lines were obtained from the European Collection of Cell Cultures (ECCAC, Salisbury, UK) except for SiHa (American Tissue Culture Collection, Manassas, VA, USA). Cells were maintained in minimal essential medium (MEM) completed with 10% fetal calf serum, 1% nonessential amino acids, and an antibiotic–antimycotic mixture. All media and supplements were purchased from Lonza Group Ltd., Basel, Switzerland. Near-confluent tumor and fibroblast cells were plated onto a 96-well microplate at 5000 cells/well density.

After overnight preincubation, a 200 µL new medium containing the tested compounds (at 10 or 30 µM) was added. After incubation (72 h, 37 °C, humidified air, 5% CO_2), the viable cells were determined by the addition of 3-(4,5-dimethylthiazol-2-yl)-2,5-diphenyltetrazolium bromide (MTT) solution (5 mg/mL, 20 µL). The reagent was metabolized by active mitochondrial reductase and precipitated as purple formazan during a 4 h contact period. Then the medium was discarded, and the crystals were solubilized in 100 µL of DMSO during a 60 min period of shaking at 37 °C. Finally, the produced formazan was assayed at 545 nm, using a microplate reader (SPECTROStar Nano, BMG Labtech, Offenburg, Germany), utilizing untreated cells as control [49]. In the case of the compounds eliciting higher than 50% growth inhibition at 30 µM, the assays were repeated with a set of dilutions and IC_{50} values were calculated from the determined data (sigmoidal curve, GraphPad Prism 5.01, GraphPad Software, San Diego, CA, USA). All experiments were performed on two microplates with at least five parallel conditions. Stock solutions of 10 mM were prepared from the investigated items in DMSO and the highest concentration of the solvent in the medium (0.3%) did not elicit any considerable action on cell growth. Cisplatin (Ebewe Pharma GmbH, Unterach, Austria) was included as a reference agent.

5. Tubulin Polymerization Assay

The effect of **14i** on tubulin polymerization was determined by means of a commercially available assay kit as described previously [48]. Briefly, 10 µL of a 125 or 250 µM solution of the tested analog was placed on a prewarmed (37 °C), UV-transparent microplate. Paclitaxel and general tubulin buffer were used as positive and negative control, respectively. A total volume of 100 µL of 3.0 mg/mL tubulin in 80 mM PIPES with 2 mM $MgCl_2$, 0.5 mM EGTA with 1 mM GTP at pH 6.9, was added to the samples to initiate the polymerization. A 60 min kinetic measurement protocol was used to describe the absorbance of the reaction mixture per minute at 340 nm (SpectoStarNano, BMG Labtech, Ortenberg, Germany). The maximum reaction rate (Vmax: Δabsorbance/min) was determined by calculating the moving averages of absorbances at three consecutive time points. The highest difference between two succeeding moving averages was considered as the Vmax of the tested molecule. Each sample was prepared in two parallels and the measurements were repeated twice. For statistical evaluation, Vmax data were analysed by the one-way ANOVA test with the Newmann–Keuls post-test by using Prism 5.01 software (GraphPad Software, San Diego, CA, USA).

6. Computational Simulations

For molecular dynamics (MD) simulations, a taxol-stabilized microtubule complex (PDB entry: 5SYF) was selected [57]. The original PDB structure was prepared for further calculations using the Protein Preparation Wizard module of the Schrodinger Maestro program [58,59]. It consists of the addition of missing hydrogen atoms under physiological pH condition as well as the missing loops or side chains. The prepared pdb structure was relaxed by a short (5 ns) MD simulation and the last frame of the relaxation running was selected as the target protein for the docking calculations. Docking was performed with the Glide package of the Schrödinger suite [60], using the extra-precision docking protocol. To sample the conformational space, 5 independent MD simulations of 500 ns length were run in a cubic box with a 10 Å buffer size and 0.15 M salt concentration by the Desmond package [61]. The initial structure of the production MD run was selected as the outcome of the docking calculation with the best docking score. In all MD runs, the OPLS4 force field and a simple point charge (SPC) water model were applied [62].

7. Conclusions

Antitubulin compounds represent the most effective class of anticancer agents. The more we understand about the structure–activity relationship of these compounds, the better options we have in utilizing them in fight against cancer. The literature reveals characteristic structural elements responsible for antitubulin effect, including the diaryl ether scaffold. In order to develop novel potential antitubulin compounds, we synthesized 13α-estrone 3-*O*-aryl derivatives via Chan–Lam coupling reactions. The copper-catalyzed etherification of the steroidal phenolic hydroxy function was achieved using arylboronic acids as coupling partners. Carbo or heterocyclic rings, bearing different substituents were introduced. The antitumoral properties of the newly synthesized 13α-estrone derivatives were investigated in vitro on a panel of human adherent cancer cell lines (A2780, MCF-7, MDA-MB 231, HeLa and SiHa). Certain compounds exerted more pronounced antiproliferative action than the reference agent cisplatin. The HeLa cancer cell line seemed to be the most sensitive to test compounds. The quinoline derivative **14i** should be highlighted as the most potent steroid against MCF-7 and HeLa cell lines. The tumor selectivity of the test compounds proved to be higher than that of the reference agent cisplatin. Compound **14i** might be regarded as a MT stabilizing agent, since it exerted its antiproliferative effect through the disturbance of tubulin polymerization. Significant interactions of the **14i** derivative with the taxoid binding site of tubulin were identified by computational simulations. Our results might contribute to the development of more potent antitubulin agents with high selectivity.

Supplementary Materials: The following supporting information can be downloaded at: https://www.mdpi.com/article/10.3390/molecules28031196/s1, Table S1: Antiproliferative activities of compounds **14a–l**. Figure S1: Presence of secondary interactions between the protein amino acids and **14i** ligand along the 1st trajectory. Figure S2: Presence of secondary interactions between the protein amino acids and **14i** ligand along the 2nd trajectory. Figure S3: Presence of secondary interactions between the protein amino acids and **14i** ligand along the 3rd trajectory. Figure S4: Presence of secondary interactions between the protein amino acids and **14i** ligand along the 4th trajectory. Figure S5: Presence of secondary interactions between the protein amino acids and **14i** ligand along the 5th trajectory. ^{1}H and ^{13}C NMR spectra of the newly synthesized compounds.

Author Contributions: P.T., É.K. and H.A. performed the experiments; E.M. and I.Z. contributed reagents, materials and analysis tools; E.M., R.M. and I.Z. conceived and designed the experiments; V.R. and G.P. designed and performed the docking calculations; E.M., G.P., R.M., P.T. and I.Z. analyzed the data; B.B. performed the MS experiments; E.M., R.M. and I.Z. wrote the paper. All authors have read and agreed to the published version of the manuscript.

Funding: This work was supported by National Research, Development and Innovation Office-NKFIH through projects OTKA SNN 139323 and TKP2021-EGA-17.

Institutional Review Board Statement: Not applicable.

Informed Consent Statement: Not applicable.

Data Availability Statement: Not applicable.

Conflicts of Interest: The authors declare no conflict of interest.

Sample Availability: Samples of the compounds are not available from the authors.

References

1. Chen, J.Q.; Li, J.H.; Dong, Z.B. A Review on the Latest Progress of Chan-Lam Coupling Reaction. *Adv. Synth. Catal.* **2020**, *362*, 3311–3331. [CrossRef]
2. Pal, T.; Lahiri, G.K.; Maiti, D. Copper in Efficient Synthesis of Aromatic Heterocycles with Single heteroatom. *Eur. J. Org. Chem.* **2020**, *44*, 6859–6869. [CrossRef]
3. De Nino, A.; Maiuolo, L.; Costanzo, P.; Algieri, V.; Jiritano, A.; Olivito, F.; Tallarida, M.A. Recent Progress in Catalytic Synthesis of 1,2,3-Triazoles. *Catalysts* **2021**, *11*, 1120. [CrossRef]
4. Yadav, P.; Bhalla, A. Recent Advances in Green Synthesis of Functionalized Quinolines of Medicinal Impact (2018–Present). *ChemistrySelect* **2022**, *7*, e202201721. [CrossRef]
5. Liu, C.; Zhang, H.; Shi, W.; Lei, A.W. Bond Formations between Two Nucleophiles: Transition Metal Catalyzed Oxidative Cross-Coupling Reactions. *Chem. Rev.* **2011**, *111*, 1780–1824. [CrossRef]
6. Mousseau, J.J.; Charette, A.B. Direct Functionalization Processes: A Journey from Palladium to Copper to Iron to Nickel to Metal-Free Coupling Reactions. *Acc. Chem. Res.* **2013**, *46*, 412–424. [CrossRef]
7. Liang, Y.; Zhang, X.; Macmillan, D.W.C. Decarboxylative sp^3 C–N coupling via dual copper and photoredox catalysis. *Nature* **2018**, *559*, 83–88. [CrossRef]
8. Kong, D.; Moon, P.J.; Bsharat, O.; Lundgren, R.J. Direct Catalytic Decarboxylative Amination of Aryl Acetic Acids. *Angew. Chem. Int. Ed.* **2020**, *59*, 1313–1319. [CrossRef]
9. Zhao, H.; Yang, K.; Zhen, H.Y.; Ding, R.C.; Yin, F.J.; Wang, N.; Li, Y.; Cheng, B.; Wang, H.F.; Zhai, H.B. A One-Pot Synthesis of Dibenzofurans from 6-Diazo-2-cyclohexenones. *Org. Lett.* **2015**, *17*, 5744–5747. [CrossRef]
10. Zhang, C.; Shi, Y.L.; Zhang, L.Y.; Yuan, D.P.; Ban, M.T.; Zheng, J.Y.; Liu, D.H.; Guo, S.N.; Cui, D.M. NaOH-promoted reaction of 1;1-dihaloalkenes and 1*H*-azoles: Synthesis of dihetaryl substituted alkenes. *New J. Chem.* **2018**, *42*, 17732–17739. [CrossRef]
11. Chan, D.M.T.; Monaco, K.L.; Wang, R.P.; Winters, M.P. New *N*- and *O*-arylations with phenylboronic acids and cupric acetate. *Tetrahedron Lett.* **1998**, *39*, 2933–2936. [CrossRef]
12. Evans, D.A.; Katz, J.L.; West, T.R. Synthesis of diaryl ethers through the copper-promoted arylation of phenols with arylboronic acids. An expedient synthesis of thyroxine. *Tetrahedron Lett.* **1998**, *39*, 2937–2940. [CrossRef]
13. Doyle, M.G.; Lundgren, R.J. Oxidative cross-coupling processes inspired by the Chan-Lam reaction. *ChemCommun* **2021**, *57*, 2724. [CrossRef] [PubMed]
14. Chen, T.; Xiong, H.; Yang, J.F.; Zhu, X.L.; Qu, R.Y.; Yang, G.F. Diaryl Ether: A Privileged Scaffold for Drug and Agrochemical Discovery. *J. Agric. Food Chem.* **2020**, *68*, 9839–9877. [CrossRef]
15. Bedos-Belval, F.; Rouch, A.; Vanucci-Bacque, C.; Baltas, M. Diaryl ether derivatives as anticancer agents—A review. *MedChemComm* **2012**, *3*, 1356–1372. [CrossRef]
16. Pitsinos, E.N.; Vidali, V.P.; Couladouros, E.A. Diaryl ether formation in the synthesis of natural products. *Eur. J. Org. Chem.* **2011**, *2011*, 1207–1222. [CrossRef]
17. Pan, Z.Y.; Scheerens, H.; Li, S.J.; Schultz, B.E.; Sprengeler, P.A.; Burrill, L.C.; Mendonca, R.V.; Sweeney, M.D.; Scott, K.C.K.; Grothaus, P.G.; et al. Discovery of selective irreversible inhibitors for Bruton's tyrosine kinase. *ChemMedChem* **2007**, *2*, 58–61. [CrossRef]
18. Llovet, J.M.; Ricci, S.; Mazzaferro, V.; Hilgard, P.; Gane, E.; Blanc, J.F.; de Oliveira, A.C.; Santoro, A.; Raoul, J.L.; Forner, A.; et al. Sorafenib in advanced hepatocellular carcinoma. *N. Engl. J. Med.* **2008**, *359*, 378–390. [CrossRef]
19. Gupta, N.; Wish, J.B. Hypoxia-inducible factor prolyl hydroxylase inhibitors: A potential new treatment for anemia in patients with CKD. *Am. J. Kidney Dis.* **2017**, *69*, 815–826. [CrossRef]
20. Rainsford, K.D. Nimesulide—A multifactorial approach to inflammation and pain: Scientific and clinical consensus. *Curr. Med. Res. Opin.* **2006**, *22*, 1161–1170. [CrossRef]
21. Shu, G.W.; Yue, L.; Zhao, W.H.; Xu, C.; Yang, J.; Wang, S.B.; Yang, X.Z. Isoliensinine; a bioactive alkaloid derived from embryos of Nelumbo nucifera; induces hepatocellular carcinoma cell apoptosis through suppression of NF-κB signaling. *J. Agric. Food Chem.* **2015**, *63*, 8793–8803. [CrossRef] [PubMed]
22. Ikeda, R.; Che, X.F.; Yamaguchi, T.; Ushiyama, M.; Zheng, C.L.; Okumura, H.; Takeda, Y.; Shibayama, Y.; Nakamura, K.; Jeung, H.C.; et al. Cepharanthine potently enhances the sensitivity of anticancer agents in K562 cells. *Cancer Sci.* **2005**, *96*, 372–376. [CrossRef]
23. Meng, Z.P.; Li, T.; Ma, X.X.; Wang, X.Q.; Van Ness, C.; Gan, Y.C.; Zhou, H.; Tang, J.F.; Lou, G.Y.; Wang, Y.F.; et al. Berbamine inhibits the growth of liver cancer cells and cancer-initiating cells by targeting Ca^{2+}/calmodulin-dependent protein kinase II. *Mol. Cancer Ther.* **2013**, *12*, 2067–2077. [CrossRef] [PubMed]
24. da Silva, A.; Maciel, D.; Freitas, V.P.; Conserva, G.A.A.; Alexandre, T.R.; Purisco, S.U.; Tempone, A.G.; Melhem, M.S.C.; Kato, M.J.; Guimaraes, E.F.; et al. Bioactivity-guided isolation of laevicarpin; an antitrypanosomal and anticryptococcal lactam from Piper laevicarpu (Piperaceae). *Fitoterapia* **2016**, *111*, 24–28. [CrossRef]
25. Hucke, O.; Coulombe, R.; Bonneau, P.; Bertrand-Laperle, M.; Brochu, C.; Gillard, J.; Joly, M.A.; Landry, S.; Lepage, O.; Llinas-Brunet, M.; et al. Molecular dynamics simulations and structure-based rational design lead to allosteric HCV NS5B polymerase thumb pocket 2 inhibitor with picomolar cellular replicon potency. *J. Med. Chem.* **2014**, *57*, 1932–1943. [CrossRef] [PubMed]

26. Beaulieu, P.L.; Coulombe, R.; Duan, J.M.; Fazal, G.; Godbout, C.; Hucke, O.; Jakalian, A.; Joly, M.A.; Lepage, O.; Llinas-Brunet, M.; et al. Structure-based design of novel HCV NS5B thumb pocket 2 allosteric inhibitors with submicromolar gt1 replicon potency: Discovery of a quinazolinone chemotype. *Bioorg. Med. Chem. Lett.* **2013**, *23*, 4132–4140. [CrossRef] [PubMed]

27. Yang, Y.H.; Wang, Z.L.; Yang, J.Z.; Yang, T.; Pi, W.Y.; Ang, W.; Lin, Y.N.; Liu, Y.Y.; Li, Z.C.; Luo, Y.F.; et al. Design, synthesis and evaluation of novel molecules with a diphenyl ether nucleus as potential antitubercular agents. *Bioorg. Med. Chem. Lett.* **2012**, *22*, 954–957. [CrossRef]

28. Phainuphong, P.; Rukachaisirikul, V.; Phongpaichit, S.; Preedanon, S.; Sakayaroj, J. Diphenyl ethers and indanones from the soil-derived fungus Aspergillus unguis PSU-RSPG204. *Tetrahedron* **2017**, *73*, 5920–5925. [CrossRef]

29. Luemmen, P.; Kunz, K.; Greul, J.; Guth, O.; Hartmann, B.; Ilg, K.; Moradi, W.A.; Seitz, T.; Mansfield, D.; Vors, J.P.; et al. Pesticide Phenyloxy Substituted Phenylamidine Derivatives. U.S. Patent Application 8,183,296 B2, 22 May 2012.

30. HRAC (Herbicide Resistance Action Committee). Available online: http://www.hracglobal.com (accessed on 17 May 2020).

31. Dumontet, C.; Jordan, M.A. Microtubule-binding agents: A dynamic field of cancer therapeutics. *Nat. Rev. Drug Discov.* **2010**, *9*, 790–803. [CrossRef] [PubMed]

32. Bates, D.; Eastman, A. Microtubule destabilising agents: Far more than just antimitotic anticancer drugs. *Br. J. Clin. Pharmacol.* **2017**, *83*, 255–268. [CrossRef] [PubMed]

33. Naaz, F.; Haider, M.R.; Shafi, S.; Yar, M.S. Anti-tubulin agents of natural origin: Targeting taxol; vinca; and colchicine binding domains. *Eur. J. Med. Chem.* **2019**, *171*, 310–331. [CrossRef] [PubMed]

34. Cao, Y.N.; Zheng, L.L.; Wang, D.; Liang, X.X.; Gao, F.; Zhou, X.L. Recent advances in microtubule-stabilizing agents. *Eur. J. Med. Chem.* **2018**, *143*, 806. [CrossRef] [PubMed]

35. Field, J.J.; Díaz, J.F.; Miller, J.H. The binding sites of microtubule stabilizing agents. *Chem. Biol.* **2013**, *20*, 301–315. [CrossRef] [PubMed]

36. Li, W.; Sun, H.; Xu, S.; Zhu, Z.; Xu, J. Tubulin inhibitors targeting the colchicine binding site: A perspective of privileged structures. *Future Med. Chem.* **2017**, *9*, 1765–1794. [CrossRef]

37. Beale, T.M.; Allwood, D.M.; Bender, A.; Bond, P.J.; Brenton, J.D.; Charnock-Jones, D.S.; Ley, S.V.; Myers, R.M.; Shearman, J.W.; Temple, J.; et al. A-ring dihalogenation increases the cellular activity of combretastatin templated tetrazoles. *ACS Med. Chem. Lett.* **2012**, *3*, 177–181. [CrossRef]

38. Butenandt, A.; Wolff, A.; Karlson, P. Uber Lumi-oestron. *Chem. Ber.* **1941**, *74*, 1308–1312. [CrossRef]

39. Yaremenko, F.G.; Khvat, A.V. A new one-pot synthesis of 17-oxo-13α-steroids of the androstane series from their 13β-analogues. *Mendeleev Commun.* **1994**, *187*, 187–188. [CrossRef]

40. Ayan, D.; Roy, J.; Maltais, R.; Poirier, D. Impact of estradiol structural modifications (18-methyl and/or 17-hydroxy inversion of configuration) on the in vitro and in vivo estrogenic activity. *J. Steroid Biochem. Mol. Biol.* **2011**, *127*, 324–330. [CrossRef]

41. Schonecker, B.; Lange, C.; Kotteritzsch, M.; Gunther, W.; Weston, J.; Anders, E.; Gorls, H. Conformational design for 13α-steroids. *J. Org. Chem.* **2000**, *65*, 5487–5497. [CrossRef] [PubMed]

42. Szabo, J.; Pataki, Z.; Wolfling, J.; Schneider, G.; Bózsity, N.; Minorics, R.; Zupkó, I.; Mernyák, E. Synthesis and biological evaluation of 13a-estrone derivatives as potential antiproliferative agents. *Steroids* **2016**, *113*, 14–21. [CrossRef]

43. Bacsa, I.; Herman, B.E.; Jojart, R.; Herman, K.S.; Wölfling, J.; Schneider, G.; Varga, M.; Tömböly, C.; Rižner, T.L.; Szécsi, M.; et al. Synthesis and structure–activity relationships of 2- and/or 4-halogenated 13β- and 13α-estrone derivatives as enzyme inhibitors of estrogen biosynthesis. *J. Enzym. Inhib. Med. Chem.* **2018**, *33*, 1271–1282. [CrossRef] [PubMed]

44. Jojart, R.; Pecsy, S.; Keglevich, G.; Szécsi, M.; Rigó, R.; Özvegy-Laczka, C.; Kecskeméti, G.; Mernyák, E. Pd-catalyzed microwave-assisted synthesis of phosphonated 13α-estrones as potential OATP2B1; 17β-HSD1 and/or STS inhibitors. *Beilstein J. Org. Chem.* **2018**, *14*, 2838–2845. [CrossRef] [PubMed]

45. Sinreih, M.; Jójárt, R.; Kele, Z.; Büdefeld, T.; Paragi, G.; Mernyák, E.; Rižner, T.L. Synthesis and evaluation of AKR1C inhibitory properties of A-ring halogenated oestrone derivatives. *J. Enzym. Inhib. Med. Chem.* **2021**, *36*, 1500–1508. [CrossRef] [PubMed]

46. Jójárt, R.; Laczkó-Rigó, R.; Klement, M.; Kőhl, G.; Kecskeméti, G.; Özvegy-Laczka, C.; Mernyák, E. Design, synthesis and biological evaluation of novel estrone phosphonates as high affinity organic anion-transporting polypeptide 2B1 (OATP2B1) inhibitors. *Bioorg. Chem.* **2021**, *112*, 104914–104925. [CrossRef]

47. Traj, P.; Abdolkhaliq, A.H.; Németh, A.; Dajcs, S.T.; Tömösi, F.; Lanisnik-Rizner, T.; Zupkó, I.; Mernyák, E. Transition metal-catalysed A-ring C-H activations and C(sp²)-C(sp²) couplings in the 13α-oestrone series and in vitro evaluation of antiproliferative properties. *J. Enzym. Inhib. Med. Chem.* **2021**, *36*, 895–902. [CrossRef]

48. Jójárt, R.; Tahaei, S.A.S.; Trungel-Nagy, P.; Kele, Z.; Minorics, R.; Paragi, G.; Zupkó, I.; Mernyák, E. Synthesis and evaluation of anticancer activities of 2- or 4-substituted 3-(N-benzyltriazolylmethyl)-13α-oestrone derivatives. *J. Enzym. Inhib. Med. Chem.* **2021**, *36*, 58–67. [CrossRef]

49. Mosmann, T. Rapid colorimetric assay for cellular growth and survival: Application to proliferation and cytotoxicity assays. *J. Immunol. Methods* **1983**, *65*, 55–63. [CrossRef]

50. Lee, Y.K.; Lim, J.; Yoon, S.Y.; Joo, J.C.; Park, S.J.; Park, Y.J. Promotion of cell death in cisplatin-resistant ovarian cancer cells through KDM1BDCLRE1B modulation. *Int. J. Mol. Sci.* **2019**, *20*, 2443. [CrossRef]

51. Schelz, Z.; Ocsovszki, I.; Bozsity, N.; Hohmann, J.; Zupkó, I. Antiproliferative effects of various furanoacridones isolated from Ruta graveolens on human breast cancer cell lines. *Anticancer Res.* **2016**, *36*, 2751–2758.

52. Chavez, K.J.; Garimella, S.V.; Lipkowitz, S. Triple negative breast cancer cell lines: One tool in the search for better treatment of triple negative breast cancer. *Breast Dis.* **2010**, *32*, 35–48. [CrossRef]
53. Anders, C.K.; Zagar, T.M.; Carey, L.A. The management of earlystage and metastatic triple-negative breast cancer: A review. *Hematol. Oncol. Clin. N. Am.* **2013**, *27*, 737–749. [CrossRef] [PubMed]
54. Suba, Z. Triple-negative breast cancer risk in women is defined by the defect of estrogen signaling: Preventive and therapeutic implications. *Onco Targets Ther.* **2014**, *7*, 147–164. [CrossRef]
55. Goodman, A. HPV testing as a screen for cervical cancer. *BMJ* **2015**, *350*, h2372. [CrossRef] [PubMed]
56. Ghittoni, R.; Accardi, R.; Chiocca, S.; Tommasino, M. Role of human papillomaviruses in carcinogenesis. *Ecancermedicalscience* **2015**, *9*, 526. [CrossRef] [PubMed]
57. Kellogg, E.H.; Hejab, N.M.; Howes, S.; Northcote, P.; Miller, J.H.; Diaz, J.F.; Downing, K.H.; Nogales, E. Insights into the Distinct Mechanisms of Action of Taxane and Non-Taxane Microtubule Stabilizers from Cryo-EM Structures. *J. Mol. Biol.* **2017**, *429*, 633–646. [CrossRef] [PubMed]
58. Schrödinger Release 2021-3: Protein Preparation Wizard; Epik, Schrödinger, LLC, New York, NY, USA, 2021; Impact, Schrödinger, LLC, New York, NY; Prime, Schrödinger, LLC, New York, NY, USA, 2021.
59. Maestro Schrödinger Release 2021-3: Maestro, Schrödinger, LLC, New York, NY, USA, 2021.
60. Schrödinger Release 2021-3: Glide, Schrödinger, LLC, New York, NY, USA, 2021.
61. Schrödinger Release 2021-3: Desmond Molecular Dynamics System, D. E. Shaw Research, New York, NY, USA, 2021. Maestro-Desmond Interoperability Tools, Schrödinger, New York, NY, USA, 2021.
62. Lu, C.; Wu, C.; Ghoreishi, D.; Chen, W.; Wang, L.; Damm, W.; Ross, G.; Dahlgren, M.; Russell, E.; Von Bargen, C.; et al. OPLS4: Improving Force Field Accuracy on Challenging Regimes of Chemical Space. *J. Chem. Theory Comput.* **2021**, *17*, 4291–4300. [CrossRef]

Review

The Structural Diversity and Biological Activity of Steroid Oximes

Ana R. Gomes [1,2], Ana S. Pires [2,3,4], Fernanda M. F. Roleira [1,*] and Elisiário J. Tavares-da-Silva [1,*]

[1] Univ Coimbra, CIEPQPF, Faculty of Pharmacy, Laboratory of Pharmaceutical Chemistry, Azinhaga de Santa Comba, Pólo III - Pólo das Ciências da Saúde, 3000-548 Coimbra, Portugal
[2] Univ Coimbra, Coimbra Institute for Clinical and Biomedical Research (iCBR) area of Environment Genetics and Oncobiology (CIMAGO), Institute of Biophysics, Faculty of Medicine, Azinhaga de Santa Comba, Pólo III - Pólo das Ciências da Saúde, 3000-548 Coimbra, Portugal
[3] Clinical Academic Center of Coimbra (CACC), Praceta Professor Mota Pinto, 3004-561 Coimbra, Portugal
[4] Univ Coimbra, Center for Innovative Biomedicine and Biotechnology (CIBB), Rua Larga, 3004-504 Coimbra, Portugal
* Correspondence: froleira@ff.uc.pt (F.M.F.R.); etavares@ff.uc.pt (E.J.T.-d.-S.);
Tel.: +351-239-488-400 (F.M.F.R. & E.J.T.-d.-S.); Fax: +351-239-488-503 (F.M.F.R. & E.J.T.-d.-S.)

Abstract: Steroids and their derivatives have been the subject of extensive research among investigators due to their wide range of pharmacological properties, in which steroidal oximes are included. Oximes are a chemical group with the general formula $R_1R_2C=N-OH$ and they exist as colorless crystals and are poorly soluble in water. Oximes can be easily obtained through the condensation of aldehydes or ketones with various amine derivatives, making them a very interesting chemical group in medicinal chemistry for the design of drugs as potential treatments for several diseases. In this review, we will focus on the different biological activities displayed by steroidal oximes such as anticancer, anti-inflammatory, antibacterial, antifungal and antiviral, among others, as well as their respective mechanisms of action. An overview of the chemistry of oximes will also be reported, and several steroidal oximes that are in clinical trials or already used as drugs are described. An extensive literature search was performed on three main databases—PubMed, Web of Science, and Google Scholar.

Keywords: steroids; oximes; chemistry; antitumor; anti-inflammatory; antibacterial; antifungal; antiviral

Citation: Gomes, A.R.; Pires, A.S.; Roleira, F.M.F.; Tavares-da-Silva, E.J. The Structural Diversity and Biological Activity of Steroid Oximes. *Molecules* **2023**, *28*, 1690. https://doi.org/10.3390/molecules28041690

Academic Editors: Marina Savić, Erzsébet Mernyák, Jovana Ajdukovic, Suzana Jovanović-Šanta and Antal Csámpai

Received: 5 January 2023
Revised: 1 February 2023
Accepted: 6 February 2023
Published: 10 February 2023

1. Introduction

Steroids belong to a class of natural or synthetic organic compounds, whose basic molecular structure is typically composed of 17 carbon atoms, bonded in four "fused" rings: three six-member cyclohexane rings (rings A, B and C) and one five-member cyclopentane ring (the D ring). (Figure 1). They play a crucial role in the human body, being responsible for the regulation of several biological processes. This fact, together with their interesting biochemical properties, such as the ability to penetrate cell membranes and bind to the nuclear and membrane receptors, makes them extremely attractive in the design of new potential drugs for the treatment of several diseases [1,2]. In fact, since their discovery in 1935, steroids have been widely used in the treatment of several conditions in the most variable areas of medicine, for example, for the treatment of autoimmune and inflammatory diseases and for the treatment of cancer [3,4]. Given the privileged scaffold of steroids and their suitability for structural modifications, steroidal derivatives have been arousing interest among medicinal chemists in the hunt for novel drug candidates. Slight alterations in the basic ring structure of steroids can elicit an enormous change in biological activity, giving rise to steroidal derivatives with a wide range of therapeutic activities [5,6].

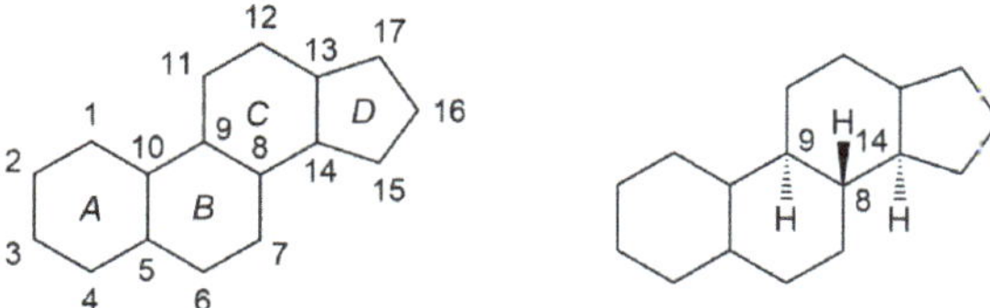

Figure 1. General structure of steroidal scaffold (left). Unless implied or stated to the contrary, the configuration of hydrogen atoms at the bridgehead positions 8, 9 and 14 are oriented as shown in the right formula (i.e., 8β, 9α, 14α).

Oximes are one of the most popular and extensively hailed nitrogen-containing biological compounds, presenting several biological and pharmacological applications [7]. They have achieved popularity due to their application as antidotes against nerve agents, which is attained by their capacity to reactivate acetylcholinesterase (AChE) [8]. Since that, these hydroxy-imine derivatives have also been associated with several other biological activities such as antibacterial, antifungal, anti-inflammatory, antioxidant, and finally, anticancer as described [7,9].

Employing the hydrophobic steroid core with a hydroxyimino group constitutes an advantage since this chemical group can increase the molecules' ability to interact with cell membranes, paving the way for enhanced biological activity [10]. For these reasons, in the last 20 years, a reasonable number of new steroidal oximes has been designed and synthesized and then evaluated for their biological activity. This review will focus mainly on the most active steroidal oximes developed as antitumor and antimicrobial agents. Additionally, a few examples of steroidal oximes with anti-inflammatory activity are also considered.

2. Chemistry of Steroidal Oximes

Oximes ($R_1R_2C=N-OH$) (Figure 2) are a nitrogen-rich group of compounds, which are produced in nature in the plant and animal kingdom. In plants, oximes and their derivatives play a fundamental role in the metabolism of plant growth and development and in a variety of biosynthetic pathways [9,11]. In animals, oximes are most commonly known for their participation in the olfactory communication between animals [12]. Oximes exist as colorless crystals, are poorly soluble in water, and are easily accessible in laboratories and in industry, which makes them very appealing [13,14]. Additionally, they are extensively used not only as protectors of carbonyl groups but also as intermediates in the Beckmann rearrangement to synthesize several lactam derivatives [9,15,16]. Furthermore, oximes have the particularity of being easily transformed into different chemical groups such as amines, nitro, and other heterocyclic compounds [16,17].

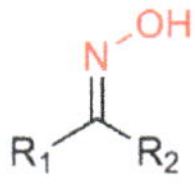

Figure 2. General structure of the hydroxyimino group of the oximes.

There are several ways to produce oxime derivatives and some reviews have been published regarding the chemistry of oximes [13,16,18]. The most classical method of oxime synthesis, which is the most used in the synthesis of steroidal oximes, involves the reaction of a carbonyl compound, a ketone or an aldehyde with hydroxylamine (NH_2OH) or a hydroxylammonium salt in the presence of a base (Figure 3). This type of reaction with aldehydes and non-symmetrical ketones can originate the two *E* or *Z* isomeric oxime forms, which can exist both as single compounds and or in mixture. Such chemical aspects can have a great impact on biological activity [7,12].

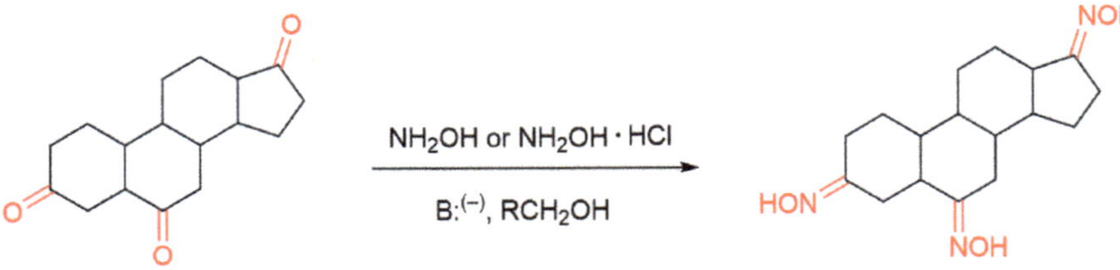

Figure 3. Classical synthesis of steroidal oximes which can take part in different positions in the steroidal scaffold.

Apart from the most commonly used synthetic strategy, there are other methods to prepare oximes involving non-carbonyl compounds, which consist of the reduction of nitroalkenes to create aldoximes and ketoximes. The reduction of α,β-unsaturated nitroalkenes gives rise to different oxime derivatives, depending if the nitro group is terminal or internal. More specifically, when the nitro group is terminal, aldoximes are produced, in mildly acidic conditions, in good yields. On the contrary, if the nitro group is internal, under basic conditions ketoximes are formed also in good yields [9,16]. Other variants of oxime synthesis include oxime ethers, esters, and amidoximes, all of which are of great biological and pharmacological importance. Oximes can act both as weak acids and weak bases. For this reason, the oxime anions can behave as ambident nucleophiles, which means that they can attack through two different sites, allowing them to be widely used for the synthesis of the above-mentioned class of compounds (ethers, nitrones, etc.) [13].

Another aspect of this chemical group is that it can behave both as hydrogen-bond donor (via OH group) and as hydrogen-bond acceptor (via nitrogen and oxygen atoms), which together with the high polarity of the oxime moiety can have a tremendous impact on the interaction with the receptor binding sites, enhancing biological activity, when compared to the carbonyl group [7,9]. This premise will be the focus of our review, in which we discuss several synthesized steroidal oximes with enhanced biological activity when compared with the parent carbonyl compounds.

3. Steroidal Oximes as Antitumor Agents

Steroidal compounds have been associated with antitumor activity for many years. Several reports focusing on their interesting properties as anticancer agents have been published [19–33]. The oxyimino group of oxime compounds is also a structural feature that confers very interesting biological properties among them antitumor activity [7,34]. For this reason, cytotoxic steroidal oximes have been extensively studied throughout the years [14,34]. In this review, we focused on the most active steroidal oximes, described in the literature, against several types of cancer. They are divided according to their steroidal motif: androstane, estrane, pregnane, cholestane, diosgenin, and bile acid derivatives.

3.1. Androstane Derivatives

5α-Reductase inhibitors have been widely studied for the treatment of diseases that are exacerbated by 5α-dihydrotestosterone (DHT). Finasteride and dutasteride are two 5α-reductase inhibitors approved by the FDA with several therapeutic applications such as for the treatment of benign prostatic hyperplasia and prostate cancer. Bearing this in mind, Dhingra et al. developed a series of 17-oxyimino -5-androsten-3β-yl esters and evaluated their cytotoxic activity against prostate cancer cells [35]. Compound **1a** (Figure 4) was the most active in DU145 prostate cancer cells with a percentage of growth inhibition of almost 91% at 5 μg/mL and an IC$_{50}$ value of 3.8 μM (Table 1), being more active than finasteride (78.51% of growth inhibition and IC$_{50}$ = 3.9 μM). Moreover, the authors also evaluated the compounds' acute toxicity using mouse macrophages, which indirectly allowed them to test for the compounds' selectivity towards cancer cells. The toxicity index value (LC$_{50}$) presented by compound **1a** was very high (LC$_{50}$ = 89.4 μM), proving that **1a** was non-toxic to mouse macrophages.

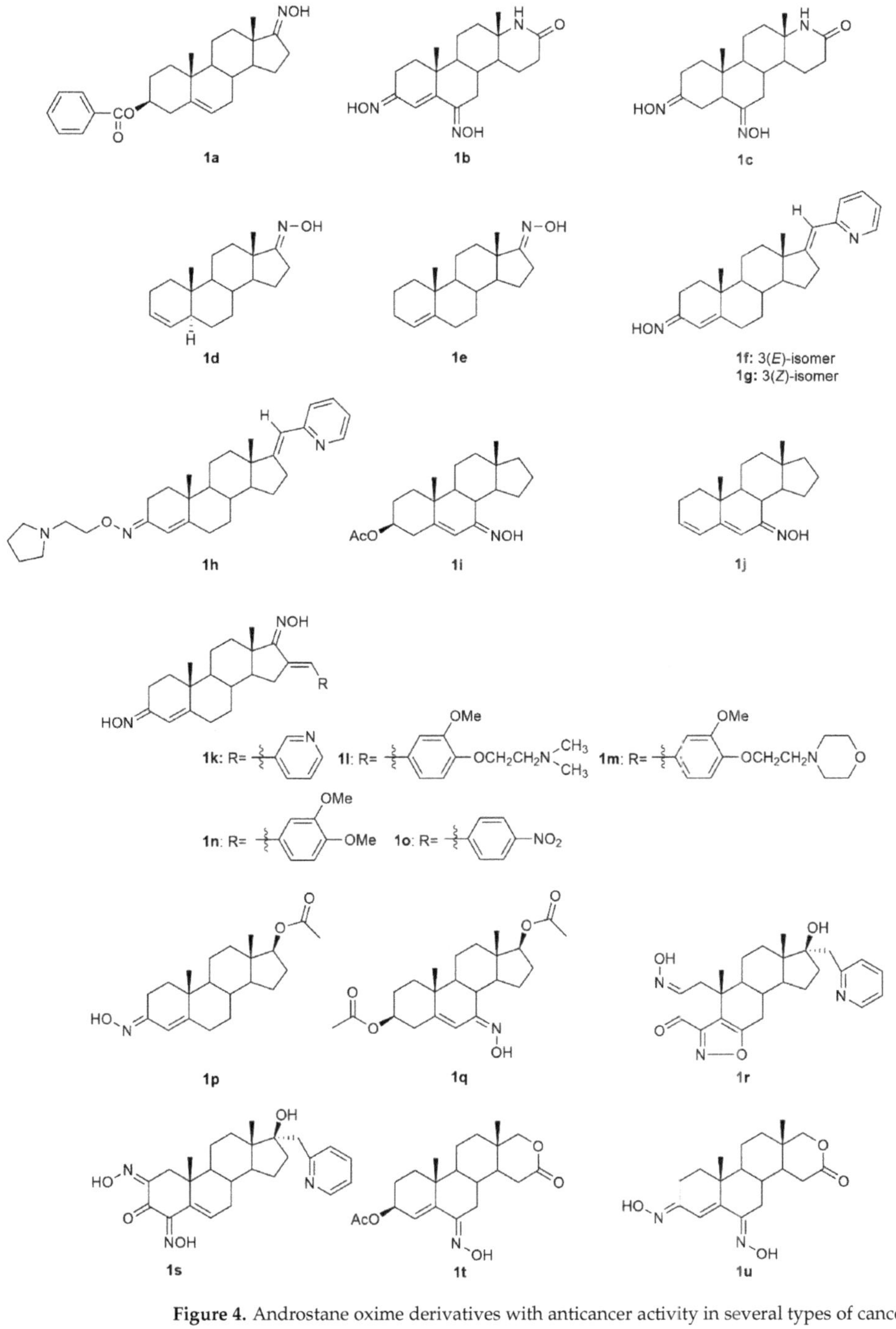

Figure 4. Androstane oxime derivatives with anticancer activity in several types of cancer cells.

The introduction of a heteroatom or the substitution of one or more carbon atoms in the steroid scaffold by a heteroatom can have a great impact on the compound's biological activity. Aza-homosteroids are a class of compounds with unusual structures, which have been associated with a wide range of biological activities such as antiparasitic, antifungal, and anticancer [36–38]. Following this line, Huang et al. synthesized a series of 17a-aza-D-homoandrostan-17-one derivatives, namely oximes **1b** and **1c** (Figure 4) [39]. Further cytotoxicity analysis by the 3-(4,5-dimethylthiazol-2-yl)-2,5-diphenyltetrazolium bromide (MTT) method revealed that both compounds were active against HeLa and SMMC7404 cells, being that compound **1b** stood out in HeLa cells with an IC_{50} of 15.1 μM (Table 1). Moreover, these oxime derivatives were in general more active than the compounds without the oxyimino group in their structure, which proves that this chemical group is important in conferring cytotoxicity.

Gomes et al. [34] designed and synthesized two novel steroidal oxime derivatives and evaluated them and two other previously synthesized oximes [40,41] in several cancer cell lines to assess their antiproliferative profile. Initial screening in WiDr, PC3, HepG2, and H1299 cancer cell lines revealed that oximes **1d** and **1e** (Figure 4) were able to decrease all cancer cells proliferation, being especially active against PC3 (IC_{50} = 13.8 μM for **1d** and 14.5 μM for **1e**) and WiDr (IC_{50} = 9.1 for **1d** and 16.1 μM for **1e**) cells (Table 1). Moreover, both oximes were even more potent than some of the chemotherapeutic drugs currently in clinical use for these types of cancer. Both oximes were able to induce cell cycle arrest at different phases, accompanied by cell death by apoptosis/necroptosis and oxidative stress (detected by an increase in ROS production) in both cell lines. Selectivity against cancer cells was also assessed by testing the compounds in normal human colon cells. Results demonstrated that both compounds are selective toward colon cancer cells [34].

Several novel oxime derivatives, such as (17E)-(pyridin-2-yl)methylidene 3-oximes **1f–1h** were designed and synthesized (Figure 4) [42]. After the synthesis, the antiproliferative activity of both compounds was assessed in a series of several human cancer cell lines (MCF7, MDA-MB-231, PC3, HeLa, HT29, A549) and a normal human cell line, MRC5. A549, HT29, and MDA-MB-231 cancer cells were the most sensitive cell lines to both oximes being that A549 was the one with the best IC_{50} values for all compounds, **1f** (IC_{50} = 1.5 μM) **1g** (IC_{50} = 1.8 μM) and **1h** (IC_{50} = 2.0 μM) (Table 1). Apoptosis induction analysis showed that these oximes induced apoptosis in A549 cells, while at the same time being non-toxic to normal lung fibroblasts MRC5. These results were very encouraging since lung cancer remains one of the most difficult cancers to treat.

Some steroidal compounds with a hydroxyimino group at position C-7 conjugated with an α,β-double bond in position C-5 were designed and synthesized [43]. After the synthesis, their antitumor activity was evaluated against several types of cancer such as cervical, gastric, epidermoid, and breast cancer. Results demonstrated that compounds **1i** and **1j** (Figure 4) were both able to decrease KB, HeLa, MKN-28, and MCF7 cancer cell proliferation (Table 1) and were more active than the corresponding parent ketones. Moreover, compound **1i** was especially active against MCF7 cells presenting an IC_{50} value of 10.2 μM.

Dubey et al. designed and synthesized novel dioximes of 16-benzylidene substituted derivatives [44]. The antitumor activity of these dioximes was then evaluated in terms of the percentage of growth inhibition of NCI-H460, MCF7, and SF268 cancer cells. Results demonstrated that compounds **1k–1o** (Figure 4) were considered active against these three cell lines, which encourages the need for further and more detailed studies.

A group of investigators focused their attention on androstene oximes and their O-alkylated derivatives [45]. These compounds and two previously synthesized oxime derivatives, compounds **1p** [46] and **1q** [47] (Figure 4) were evaluated in leukemia, colon, melanoma, and renal cancer cell lines. Only compounds **1p** and **1q** showed considerable cytotoxicity in all cell lines (percentages of growth of 10.07 to 75.01% at 10 μM), being this effect more pronounced in the leukemia cell lines, K562, HL60, and SR.

Aiming to evaluate the combined effect of the 17-heterocyclic ring and hydroxyimino function, a group of investigators designed and synthesized modified 17α-picolyl and 17(E)-picolinylidene androstane derivatives and their antiproliferative activity against breast, prostate, cervical, colon and lung adenocarcinoma, as well as normal fetal lung fibroblasts was assessed [48]. MTT assay results (Table 1) demonstrated that compound **1r** (Figure 4) was the most active compound in PC3 cells (IC_{50} = 6.6 µM), while compound **1s** (Figure 4) was more active, not only against PC3 cells (IC_{50} = 8.7 µM) but also against MCF7 cells (IC_{50} = 1.7 µM). Given these encouraging results, the authors went further ahead and studied deeply the potential mechanisms of action of compound **1s** in MCF7 cells. Results showed that **1s** induced apoptosis in breast cancer cells, assessed by alterations in the cells morphology, such as nuclear condensation, vacuolated cytoplasm, degradation of nuclei and cytoplasm, membrane blebbing, and apoptotic bodies formation [48].

Savić et al., designed and synthesized some novel D-homo lactone androstane derivatives and evaluated their antiproliferative activity against several cancer cell lines [49]. Among these compounds, the oxime derivative **1t** (Figure 4), demonstrated to have high activity (Table 1). Moreover, **1t** also revealed selectivity towards cancer cells since it presents a much higher IC_{50} in the normal human cell line, MRC5. A few years later, the same group of investigators designed, synthesized and evaluated the antitumor activity of some more new D-homo lactone androstane derivatives [50]. In vitro cytotoxicity assessment against cancer cells revealed that the steroidal oxime **1u** (Figure 4) was able to decrease the proliferation of PC3 (IC_{50} = 27.94 µM) and HeLa (IC_{50} = 13.86 µm) cells (Table 1).

Table 1. IC_{50} values (µM) of the synthesized androstane oxime derivatives.

Cell Line	Compounds													
	1a	1b	1c	1d	1e	1f	1g	1h	1i	1j	1r	1s	1t	1u
DU145	3.9	-	-	-	-	-	-	-	-	-	-	-	-	-
HeLa		15.1	75.7	-	-	>100	>100	22.6	12.8	22.2	>100	81.8	36.0	13.9
SMMC7404	-	>200	184	-	-	-	-	-	-	-	-	-	-	-
WiDr	-	-	-	9.1	16.1	-	-	-	-	-	-	-	-	-
PC3	-	-	-	13.8	14.5	>100	57.7	77.1	-	-	6.6	8.7	36.7	27.9
HepG2	-	-	-	23.9	18.2	-	-	-	-	-	-	-	-	-
H1299	-	-	-	18.6	19.2	-	-	-	-	-	-	-	-	-
MCF7	-	-	-	-	-	41.0	44.9	>100	10.2	19.8	50.4	1.7	81.3	>100
MDA-MB-231	-	-	-	-	-	47.3	5.2	4.7	-	-	25.3	40.1	11.9	>100
HT29	-	-	-	-	-	4.4	10.6	3.3	-	-	>100	10.3	4.0	>100
A549	-	-	-	-	-	1.5	1.8	2.0	-	-	>100	56.0	-	-
MRC5	-	-	-	-	-	>100	>100	>100	-	-	>100	>100	>100	>100
CEM	-	-	-	-	-	>50	>50	30.4	-	-	-	-	-	-
G361	-	-	-	-	-	45.3	46.6	8.9	-	-	-	-	-	-
BJ	-	-	-	-	-	>50	>50	25.3	-	-	-	-	-	-
KB	-	-	-	-	-	-	-	-	26	28.5	-	-	-	-
MKN-28	-	-	-	-	-	-	-	-	18.1	36.1	-	-	-	-
Ref.	[35]	[39]	[34]			[42]			[43]		[48]		[49]	[50]

These values were obtained through different techniques such as MTT and SRB assays.

3.2. Estrane Derivatives

Estrogen-3-*O*-sulfamates (EMATEs) are a class of steroidal compounds, which act as irreversible inhibitors of steroid sulfatase (STS), an enzyme involved in the development of estrogen-dependent breast cancer [51]. Given this, Leese et al. decided to design and synthesize D ring-modified 2-substituted EMATEs and evaluated their in vitro anticancer activity in MCF7 cancer cells [52]. Antiproliferative activity evaluation of the steroidal oximes **2a–2j** (Figure 5) revealed that in general, all oximes were very effective in inhibiting MCF7 cancer cell proliferation (Table 2). These results reinforce the importance of the modifications in the 17-position of the 2-substituted EMATEs, particularly the introduction of an oxyimino group, which increased the antiproliferative activity of this class of compounds [52].

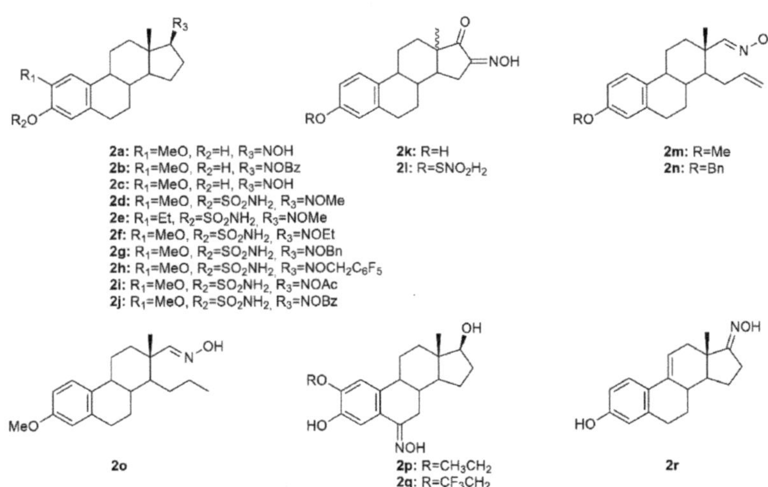

Figure 5. Steroidal estrane oxime derivatives with anticancer activity in several types of cancer cells.

A series of novel estrone-16-oxime ethers were designed and synthesized, and their antitumor activity was evaluated in HeLa, MCF7, and A431 cancer cells [53]. From all the synthesized compounds, oxime **2k** and **2l** (Figure 5) decreased cancer cell proliferation in a more pronounced way, being HeLa cells the most susceptible to both compounds (IC_{50} = 4.41 μM for **2k** and 4.04 μM for **2l**) (Table 2). Further evaluation to characterize the mechanisms of action of these compounds revealed both of them interfered with the cell cycle at G1 phase and induced apoptosis in HeLa cells, by the activation of caspase-3. Following this study, the same authors continued their research on this topic and designed and synthesized a series of novel D-secooxime derivatives in the 13β- and 13α-methylestrone series [54]. After MTT assays to assess antiproliferative activity in HeLa, MCF7, A2780, and A431 cancer cell lines, compounds **2m–2o** (Figure 5) stand out for displaying high cytotoxicity against all cell lines and being, generally, even more active than cisplatin as it can be seen in Table 2. Furthermore, compound **2n** was selected for additional analysis in A2780 cells, namely cell cycle analysis. Results showed that **2n** induced cell cycle arrest at the S phase, which in turn might be responsible for apoptosis induction [54].

Cushman et al. investigated several estradiol analogs to improve the anticancer activity of 2-methoxyestradiol, a naturally occurring tubulin polymerization inhibitor [55]. Starting from 2-ethoxyestradiol, an analog previously synthesized by the same group, from 2-methoxyestradiol [56], two novel steroidal oximes **2p** and **2q** (Figure 5) were designed and synthesized and their antitumor activity was assessed. Results revealed that both compounds were extremely toxic to all the cell lines studied (HOP62, HCT116, SF539, UACC62, OVCAR-3, SN12C and DU145) with GI_{50} (half growth inhibition) at values ranging from 0.010 to 0.066 μM (Table 2). Furthermore, **2p** and **2q** were able to inhibit tubulin polymerization. The oxime derivatives were the most active among all compounds synthesized, which points out the importance of the oxyimino functionality.

Aiming to develop new steroidal oximes with potential application in cancer treatment, a group of scientists designed and synthesized a series of estrone oxime derivatives and evaluated them in six cancer cell lines [57]. MTT results proved that compound **2r** (Figure 5) is the most active against all cell lines, being LNCaP cells the most sensitive to this molecule (IC_{50} = 3.59 μM) (Table 2). Given this, the cytotoxicity of this steroidal oxime was mediated by a cell cycle arrest at G2/M phases accompanied by cell death by apoptosis, with evidence of condensed and fragmented nuclei. Moreover, the authors speculated that this compound might interfere with β-tubulin [57].

Table 2. IC$_{50}$ values (µM) of the synthesized estrane oxime derivatives.

Cell Line	Compounds																	
	2a	2b	2c	2d	2e	2f	2g	2h	2i	2j	2k	2l	2m	2n	2o	2p	2q	2r
MCF7	5.87	6.47	0.24	0.17	0.23	0.23	1.33	5.14	0.17	>30	>30	2.6	2.6	>30	2.1	-	-	25.63
HeLa	-	-	-	-	-	-	-	-	-	-	-	-	1.2	7.1	1.7	-	-	-
A431	-	-	-	-	-	-	-	-	-	-	-	-	0.8	0.9	0.9	-	-	-
A2780	-	-	-	-	-	-	-	-	-	-	-	-	0.9	1.4	0.7	-	-	-
HOP62	-	-	-	-	-	-	-	-	-	-	-	-	-	-	-	0.017	1.2	-
HCT116	-	-	-	-	-	-	-	-	-	-	-	-	-	-	-	0.031	2.1	-
SF539																0.021	2.8	
UACC62	-	-	-	-	-	-	-	-	-	-	-	-	-	-	-	0.015	0.88	-
OVCAR3	-	-	-	-	-	-	-	-	-	-	-	-	-	-	-	0.016	5.4	-
SN12C	-	-	-	-	-	-	-	-	-	-	-	-	-	-	-	0.045	17	-
DU145	-	-	-	-	-	-	-	-	-	-	-	-	-	-	-	0.049	17	-
MDA-MB-231	-	-	-	-	-	-	-	-	-	-	-	-	-	-	-	0.010	6.5	-
MGM	-	-	-	-	-	-	-	-	-	-	-	-	-	-	-	0.066	4.2	-
T47D	-	-	-	-	-	-	-	-	-	-	-	-	-	-	-	-	-	43.45
LNCaP	-	-	-	-	-	-	-	-	-	-	-	-	-	-	-	-	-	3.59
HepaRG	-	-	-	-	-	-	-	-	-	-	-	-	-	-	-	-	-	18.35
Caco	-	-	-	-	-	-	-	-	-	-	-	-	-	-	-	-	-	24.33
NHDF	-	-	-	-	-	-	-	-	-	-	-	-	-	-	-	-	-	30.84
Ref.						[52]					[53]			[54]		[55]		[57]

These values were obtained through different techniques such as MTT assays and specific assay kits.

3.3. Pregnane Derivatives

Bearing in mind the importance of pregnenolone in biological systems, Choudhary et al. synthesized a series of novel pregnenolone derivatives, being that some of which were oximes [58]. After synthesis, the authors came up with compound **3a** (Figure 6) which was further evaluated in HepG2 and MDA-MB-231 cancer cells. Results demonstrated that this compound was quite active against both cell lines with IC$_{50}$ values of 4.50 and 6.76 µM in HepG2 and MDA-MB-231 (Table 3), respectively, which makes it very promising to be further analyzed.

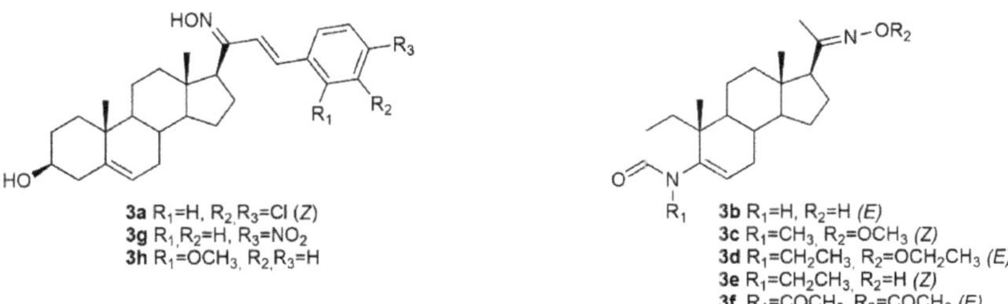

Figure 6. Steroidal pregnane oxime derivatives with antitumor activity in several types of cancer cells.

Table 3. IC$_{50}$ values (µM) of the synthesized pregnane oxime derivatives.

Cell Line	Compounds							
	3a	3b	3c	3d	3e	3f	3g	3h
HepG2	4.50	-	-	-	-	-	-	-
MDA-MB-231	6.76	-	-	-	-	-	-	-
T24	-	2.41	1.94	1.99	2.62	1.90	-	-
HT29	-	-	-	-	-	-	2.35	2.37
HCT15	-	-	-	-	-	-	0.31	0.65
SF295	-	-	-	-	-	-	1.67	5.44
HOP62	-	-	-	-	-	-	3.57	0.81
A549	-	-	-	-	-	-	3.56	7.17
MCF7	-	-	-	-	-	-	0.60	1.91
Ref.	[58]			[10]			[59]	

These values were obtained through different techniques such as MTT, MTS, and SRB assays.

Chen and collaborators did their research in 4-azasteroidal derivatives, namely 4-azasteroidal-20-oxime derivatives using progesterone as starting material [10]. The compounds were then evaluated by the MTS assay for their anticancer activity in human bladder carcinoma. After the synthesis, the authors came up with five very active oxime compounds (**3b–3f**, Figure 6), in T24 cells (Table 3). After structure-activity relationships (SAR) analysis, Chen et al. concluded that the methyl and ethyl oxime-ether derivatives were more active against the referred cell line when compared with the aryl oxime-ester derivatives.

Using pregnenolones as precursors, a group of scientists designed and synthesized a series of benzylidene pregnenolones and their oximes and further evaluated their potential antitumor activity in several cancer cell lines, namely HT29, HCT15, SF-295, HOP62, A549 and MCF-7 [59]. From all the synthesized oximes, compounds **3g** and **3h** (Figure 6) were the most active against HCT15 (IC$_{50}$ = 0.31 µM for **3g** and 0.65 µM for **3h**) and MCF7 (IC$_{50}$ = 0.60 µM for **3g** and 1.91 µM for **3h**) (Table 3) cancer cells, revealing a cell specificity. Moreover, the oxime derivatives were more potent than the corresponding precursors, which points out the importance of the oxyimino functionality in conferring cytotoxicity against cancer cells.

3.4. Cholestane Derivatives

Krstić et al. reported the design and synthesis of two novel steroidal oximes from cholesterol, compounds **4a** and **4b** (Figure 7) [60]. A few years later, the same group of scientists decided to evaluate the potential antitumor activity of these two oximes [61] against two human cancer cell lines (HeLa and K-562) and against non-stimulated and PHA-stimulated peripheral blood mononuclear cells (PBMC's) from healthy donors to assess selectivity. Both compounds showed a dose-dependent decrease in the proliferation of HeLa cancer cells (IC$_{50}$ = 35.24 ± 4.09 for **4a** and IC$_{50}$ = 20.68 ± 3.10 for **4b**) and K-562 cancer cells (IC$_{50}$ = 28.05 ± 10.18 for **4a** and IC$_{50}$ = 11.16 ± 1.24 for **4b**), while showing almost no effects in normal immunocompetent cells (Table 4). Further evaluation revealed that compound **4b** exerted its cytotoxicity by inducing apoptosis in HeLa cells. On the contrary, compound **4a** showed no evidence of apoptosis or necrosis, which can be explained by the difference in the *Z/E* stereochemistry in the hydroxyimino group of both compounds [61].

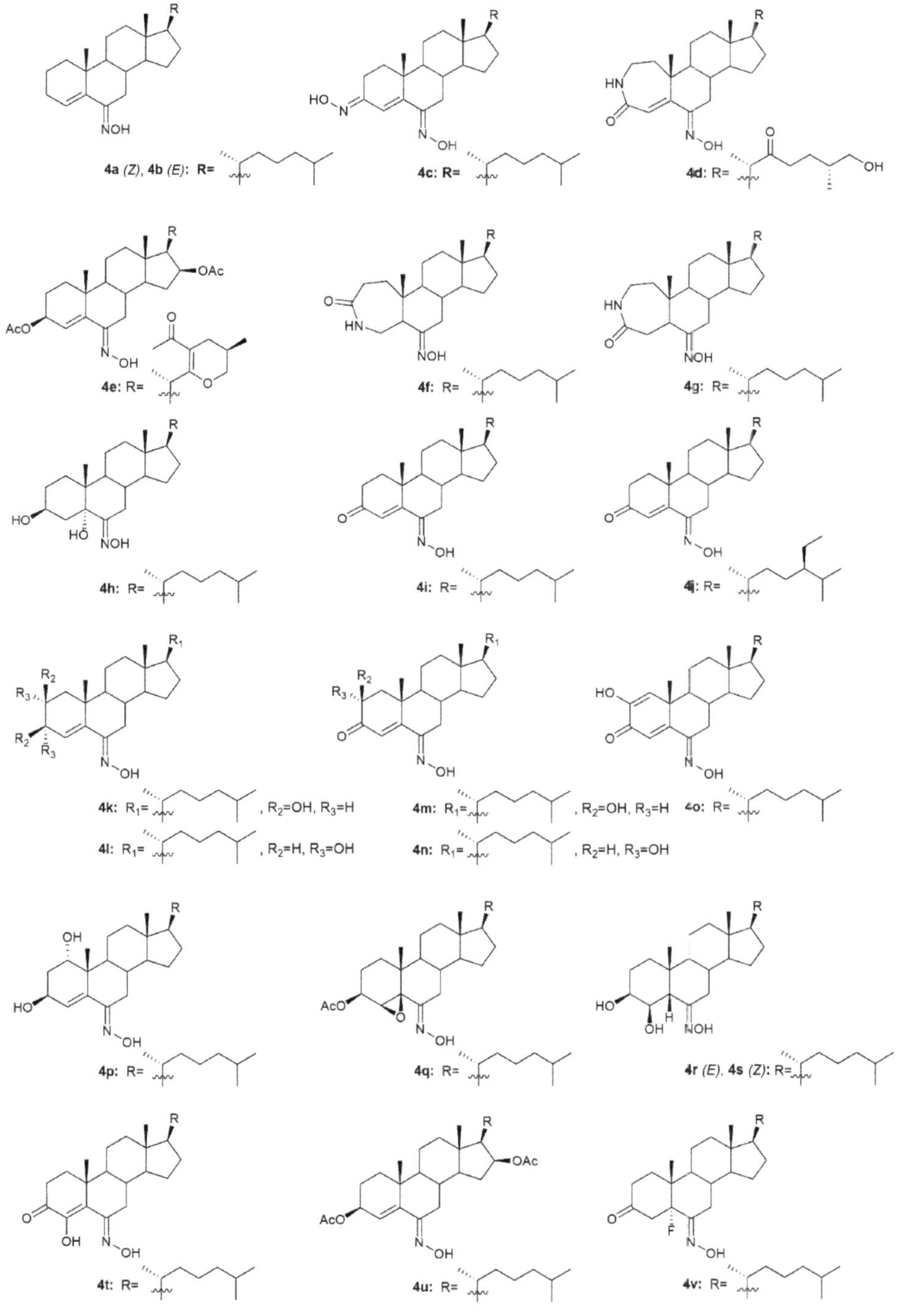

Figure 7. *Cont.*

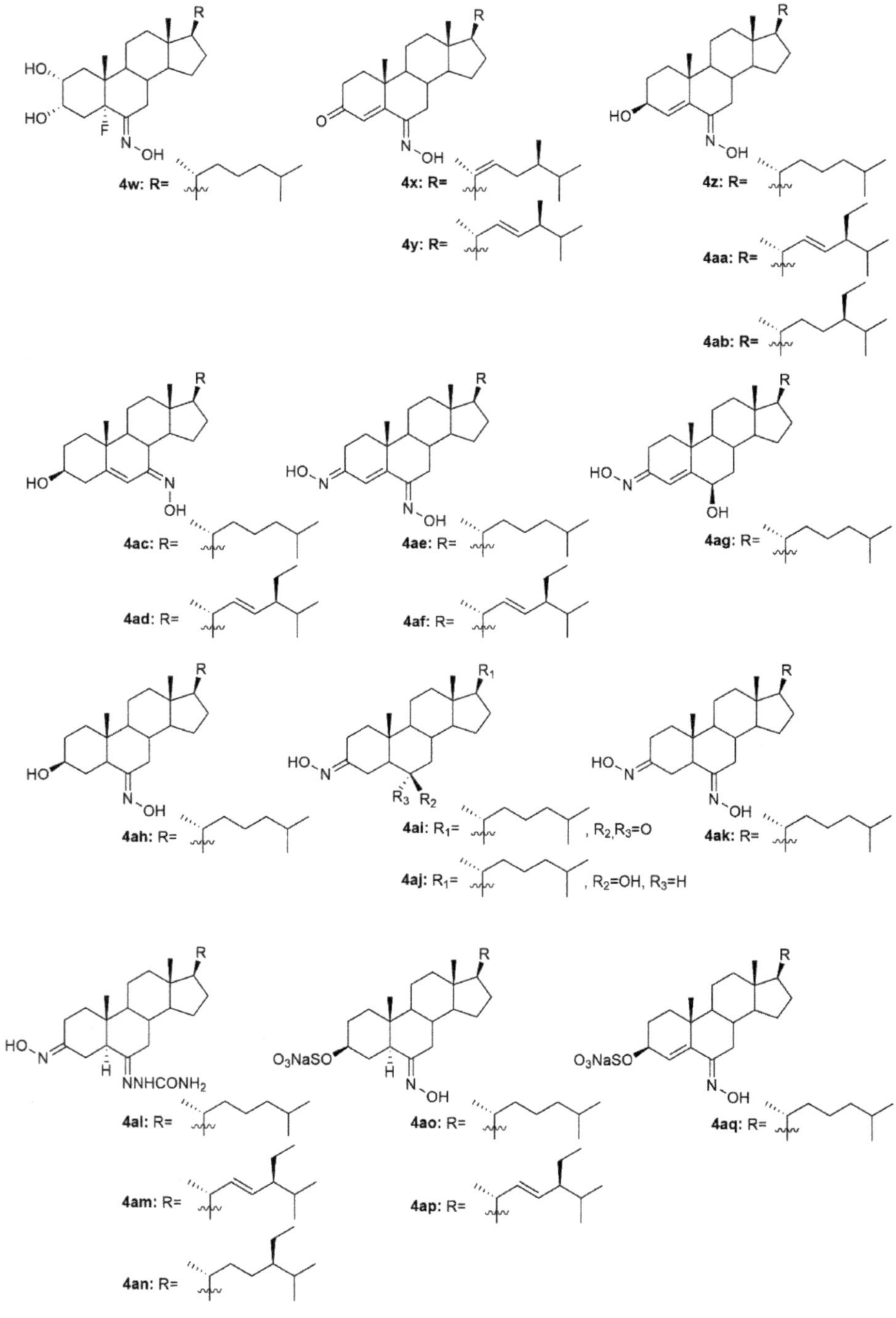

Figure 7. *Cont.*

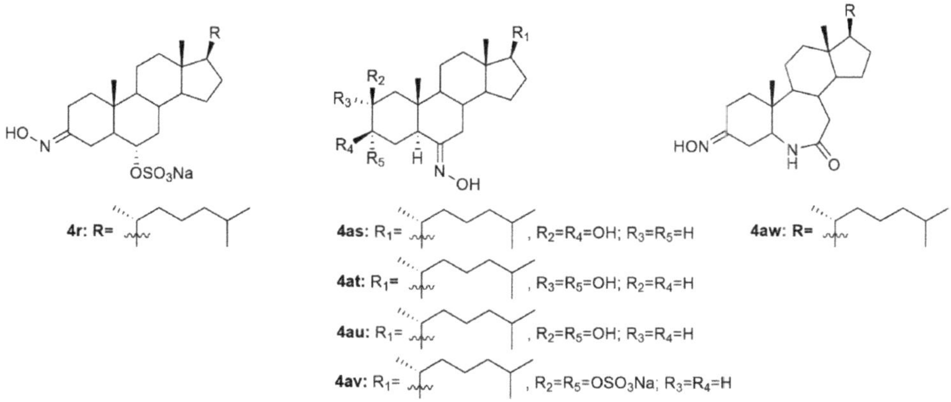

Figure 7. Steroidal cholestane oxime derivatives with antitumor activity in several types of cancer cells.

Table 4. IC$_{50}$ values (μM) of the synthesized cholestane oxime derivatives.

Cell Line	Compound															
	4a	4b	4c	4d	4e	4f	4g	4h	4p	4q	4r	4s	4t	4u	4v	4w
HeLa	35.2	20.7	-	-	-	22.8	5.6	-	-	-	-	-	-	-	-	-
K562	28.1	11.2	-	-	-	-	-	-	-	-	-	-	-	-	-	-
PBMC	110	34.6	>100	>100	>100	-	-	-	-	-	-	-	-	-	-	-
MCF7	-	-	8.2	9.5	7.9	-	-	-	-	-	-	-	-	-	-	-
MGC7901	-	-	-	-	-	12.8	16.3	-	-	-	-	-	-	-	-	-
SMMC7404	-	-	-	-	-	17.6	17.9	-	-	-	-	-	-	-	-	-
GNE2	-	-	-	-	-	12.1	-	-	-	-	-	-	-	-	-	-
SPC-A	-	-	-	-	-	74.5	-	-	-	-	-	-	-	-	-	-
Tu686	-	-	-	-	-	24.9	-	-	-	-	-	-	-	-	-	-
PC3	-	-	-	-	-	14.5	-	15.0	-	-	-	-	-	-	-	-
HT29	-	-	-	-	-	10.6	-	11.9	-	-	-	-	-	-	-	-
SH-SY5Y	-	-	-	-	-	-	-	16.8	-	-	-	-	-	-	-	-
HepG2	-	-	-	-	-	-	-	13.2	-	-	-	-	-	-	-	-
A549	-	-	-	-	-	-	-	15.0	11.6	1.06	11.5	11.5	11.7	19.4	11.5	2.22
ARPE19	-	-	-	-	-	-	-	23.7	-	-	-	-	-	-	-	-
BJ	-	-	-	-	-	-	-	17.1	-	-	-	-	-	-	-	-
HCT116	-	-	-	-	-	-	-	-	2.32	0.21	1.15	1.15	2.33	1.94	11.5	2.22
PSN1	-	-	-	-	-	-	-	-	2.32	1.06	11.5	23.1	>23.3	19.4	11.5	2.22
T98G	-	-	-	-	-	-	-	-	23.2	23.2	>23.1	23.1	23.3	19.4	11.5	2.22
Ref.		[46]		[47]		[48,49]		[51]				[54]				

A group of researchers investigated in aza-homosteroids derived from diosgenin and cholesterol-containing hydroxyimino and lactam groups in the A/B ring with four types of side chains: cholestane, spirostane, 22-oxocholestane and 22,26-epoxycholestene [62]. These compounds were further evaluated as potential antitumor agents in MCF7 cancer cells. Antitumor activity assessment revealed that from all the synthesized compounds, oximes **4c–4e** (Figure 7) were the most promising ones with an IC$_{50}$ of 8.2, 9.5, and 7.9 μM, respectively (Table 4). MCF7 cancer cells were more sensitive to the compounds containing a hydroxyimino group in comparison with the compounds without this chemical function. Moreover, Mora-Medina et al., evaluated these three oximes against PBMCs to test for the selectivity index of these compounds. Results showed that **4c-4e** were all remarkably selective for MCF-7 cells, which makes these compounds very promising [62].

Huang et al. synthesized a series of 6-hydroxyimino-substituted-3-aza- and 4-aza-A-homo-3-oxycholestanes using cholesterol as starting material [63]. Oximes **4f** and **4g** (Figure 7) were further evaluated in three different cancer cell lines, namely HeLa, SMMC7404, and MGC7901. Results showed that both compounds displayed cytotoxicity against these cell lines (Table 4), being more active than the referenced drug, cisplatin, in HeLa and SMMC7404 cells. Altogether, these results demonstrated that the introduction of the hydroxyimino group at position 6 was crucial for the compounds' cytotoxicity against cancer cells. Given the good outcomes obtained, these investigators decided to further analyze compound **4f** (Figure 7) and evaluated its antitumor activity in five more cancer cell lines (GNE2, SPC-A, Tu686, PC3, and HT29) [64]. Results indicate that oxime **4f** was also quite active in all cell lines with IC_{50} values ranging from 10.6 to 74.5 µM (Table 4). Furthermore, the molecular mechanisms by which **4f** decreased cancer cell proliferation were studied. Results unveiled that compound **4f** induced cancer cell apoptosis by activation of the intrinsic pathway, which was demonstrated by the annexin V labeling, activation of caspase-3, and release of cytochrome C. Moreover, this oxime was also evaluated in an in vivo model and proved to be able to inhibit tumor growth [64].

Oxysterols are a group of lipids derived from cholesterol, particularly interesting in the medicinal chemistry field due to their diverse biological effects [65]. Given this, Carvalho et al. designed and synthesized a series of oxysterols in which oxime **4h** (Figure 7), a 3β,5α,6β-trihydroxycholestanol derivative, was included [66]. After the synthesis of **4h**, the compound's cytotoxicity was studied in five human cancer cell lines (HT29, SH-SY5Y, HepG2, A549, PC3) and two human normal cell lines (ARPE-19 and BJ). Oxime **4h** was quite active in all cancer cell lines being that HT29 was the most sensitive (IC_{50} = 11.9 µM). Moreover, the compound revealed some selectivity towards HT29 cells (selectivity index of 1.99) since the IC_{50} displayed was higher in the normal cell lines (Table 4). This work contributed to deepening the understanding of oxysterols' cytotoxicity and shed some light on the SAR of these classes of compounds.

In 1997, two steroidal molecules with very interesting and unusual structures, (6*E*)-hydroxyiminocholest-4-en-3-one (**4i**, Figure 7) and its 24-ethyl analog (**4j**, Figure 7), were isolated from *Cinachyrella* marine sponges [67]. Analysis of their antitumor activity revealed that only compound **4i** was able to decrease the proliferation of P388 (IC_{50} = 1.25 µg/mL), A549 (IC_{50} = 1.25 µg/mL), HT29 (IC_{50} = 1.25 µg/mL), and MEL28 (IC_{50} = 2.5 µg/mL) cells (Table 5). Following this discovery, Deive et al. [68] further explored the SAR of this type of compound and prepared several derivatives of **4i** and **4j** with different structural features, namely with different side chains and degrees of unsaturation on ring A. From all the synthesized compounds, **4k–4o** (Figure 7) stood out with an IC_{50} ranging from 0.125 to 1.25 µg/mL (Table 5) in the above-mentioned cancer cell lines. These results allowed to shed some light regarding SAR analysis and demonstrated that the presence of a cholesterol-type side chain, a ketone group at C-3 and a high degree of oxidation in ring A might play a major role in the compounds' cytotoxicity [68]. A few years later, the same group continued their investigation in 6-hydroxyiminosteroids [69] and, bearing in mind the information about the SAR of these compounds, a series of new steroidal oximes were synthesized and evaluated as potential antitumor agents. After biological evaluation in four different cancer cell lines (A549, HCT116, PSN1, T98G), compounds **4p–4w** (Figure 7) were the most effective in decreasing cancer cell proliferation, demonstrating good IC_{50} values (Table 4). Once again, the oxygenation of ring A turned out to be very important in the increased cytotoxicity of the compounds against cancer cells [69].

Table 5. IC$_{50}$ values (μg/mL) of the synthesized cholestane oxime derivatives.

Cell line	Compound																			
	4i	4k	4l	4m	4n	4o	4x	4z	4aa	4ab	4ac	4ad	4ae	4af	4ag	4ag	4ah	4ai	4aj	4ak
P388	1.25	1.25	1.25	0.25	0.25	0.5	-	-	-	-	-	-	-	-	-	-	-	-	-	
A549	1.25	1.25	1.25	0.13	0.13	0.13	-	-	-	-	-	-	-	-	-	-	-	-	-	-
HT29	1.25	1.25	1.25	0.25	0.25	0.25	-	-	-	-	-	-	-	-	-	-	-	-	-	-
MEL28	2.5	1.25	1.25	0.13	0.13	0.13	-	-	-	-	-	-	-	-	-	-	-	-	-	-
Sk-Hep-1	-	-	-	-	-	-	43	43	20.1	37	45	25	76.8	24	57	34.5	39.5	33.7	35.4	48.8
H292	-	-	-	-	-	-	59.5	59.5	26.2	37	62.5	46	70	33	76	53	41.9	34.2	65.8	78.9
PC3	-	-	-	-	-	-	44	44	32.5	40.5	41.5	76	>90	36	66	52	49.8	103	60.1	62.7
Hey-1B	-	-	-	-	-	-	37	49	26.3	45	53	38	78	37	51	45	47.9	45.7	56.3	61.5
Ref.	[67]			[68]			[70]					[71]						[72]		

Another study by Cui et al. [70] also reported the synthesis and further biological evaluation of 6-hydroxyiminosteroidal cholestane derivatives. They not only developed a facile and efficient synthetic method for the synthesis of the natural compounds **4i** and **4j** (Figure 7) but also designed and synthesized a new steroidal oxime (**4x**, Figure 7). Further antitumor activity evaluation in four human cancer cell lines, namely Sk-Hep-1, H292, PC3, and Hey1B revealed that compound **4x** presented modest cytotoxicity against these cell lines with IC$_{50}$ values ranging from 37 to 59.5 μg/mL (Table 5). Once again and compared to the cytotoxicity displayed by compounds **4i** and **4j**, the structure of the side chains impacts the cytotoxicity of the compounds. The cholesterol-type side chain seems to be important for biological activity, which goes accordingly to the results obtained by other research groups [68]. The same group of investigators continued their research on this topic and synthesized more steroidal oximes with the hydroxyimino groups in different locations (A ring or B ring) and with different types of side chains at position 17 [71]. From all the derivatives, compounds **4y–4ag** (Figure 7) were the ones with the best antitumor activity against Sk-Hep-1, H292, PC3, and Hey1B (Table 5). This study reinforced the conclusions obtained in the previous ones [68–70], where it is stated that for enhanced cytotoxicity the compounds must have a cholesterol-type side chain, a hydroxyimino group on the B ring, and a hydroxy group on the A or B ring. The same authors went further ahead in the SAR and synthesized a series of derivatives similar to the ones synthesized in the previous study but without the 4,5-double bond [72]. After biological activity evaluation, compounds **4ah–4ak** (Figure 7) proved to be the most effective in decreasing Sk-Hep-1, H292, PC3, and Hey1B cancer cells proliferation with IC$_{50}$ values ranging from 35.4 to 103 μg/mL (Table 5). These compounds (without the 4,5-double bond) were more active than the equivalent ones with the 4,5-double bond [71], which indicates that the double bond in this particular position confers a negative effect in the biological activity displayed by these derivatives. All these studies were very important in unraveling the SAR of 6-hydroxyiminosteroids and might help shed some light on the design of novel chemotherapeutic drugs for the treatment of different types of cancer.

Gan et al. designed and synthesized some steroidal hydrazone derivatives with 3,6-disubstituted structure and different side chains at 17-position, being the hydroxyimino group one of these substitutions, namely in the 3-position [73]. Antiproliferative activity evaluation was carried out in vitro against gastric and liver cancer cells after 72 h of incubation. The synthetic routes gave rise to a series of novel molecules among them, compounds **4al–4an** (Figure 7), which were remarkably active against the two cancer cell lines used, Bel7404 and SGC7901 (Table 6). Compound **4am** was particularly active to Bel7404 cells (IC$_{50}$ = 7.4 μM) being three times more active than cisplatin (IC$_{50}$ = 22.3 μM), which makes **4am** a potential antitumor drug.

Table 6. - IC$_{50}$ values (μM) of the synthesized cholestane oxime derivatives.

Cell line	Compound												
	4al	4am	4an	4ao	4ap	4aq	4ar	4as	4at	4au	4av	4aw
HeLa	-	-	-	9.6	14.3	25.3	12.5	-	-	-	-	9.1
MGC7901	-	-	-	6.5	27.7	>100	33.3	-	-	-	-	-
SMMC7404	-	-	-	12.9	>100	62.2	10.2	-	-	-	-	-
GNE2	-	-	-	-	-	-	-	-	-	-	-	11.3
PC3	-	-	-	-	-	-	-	39.3	31.0	10.8	12.9	-
SGC-7901	13.2	32.3	26.2	-	-	-	-	-	-	-	-	-
Bel-7404	11.0	7.4	15.0	-	-	-	-	-	-	-	-	-
LNCaP	-	-	-	-	-	-	-	>100	26.2	44.8	13.9	-
Ref.	[73]			[74]				[75]				[60]

These values were obtained through different techniques such as MTT, KBR, alamar blue and SRB assays.

Huang et al. published a study describing the synthesis of new sulfated hydroxy-iminosterols as potential antitumor agents [74]. In vitro antiproliferative activity, assessed in HeLa, SMMC7404 and MGC7901 cancer cell lines revealed that compounds **4ao–4ar** (Figure 7) were all able to decrease all cancer cell lines proliferation (Table 6). Please note that compound **4ao** was even more cytotoxic than cisplatin against all the cancer cell lines studied.

A group of investigators from Argentina synthesized three new 6*E*-hydroxyiminosteroids and then assessed their antitumor activity against prostate cancer cells (PC3 and LNCaP) [75]. After antiproliferative activity evaluation, compounds **4as–4av** (Figure 7) proved to be quite active in both cancer cell lines with IC$_{50}$ values ranging from 10.8 to 44.8 μM (Table 6).

Using analogs of compounds **4i** and **4j** as precursors, Huang et al. designed and synthesized novel steroidal oximes and then evaluated their anticancer efficacy against a panel of six cancer cell lines [76]. Of all the synthesized compounds, oxime **4aw** (Figure 7) was the most powerful oxime, especially in HeLa and GNE2 cancer cell lines with IC$_{50}$ values of 9.1 and 11.3 μM, respectively (Table 6). Remarkably, this compound was even more active than cisplatin (IC$_{50}$ = 10.1 and 16.8 μM in HeLa and GNE2 cells, respectively).

3.5. Diosgenin Derivatives

Diosgenin is a natural steroidal sapogenin, which was first isolated from *Discorea tokoro* by Takeo Tsukamato in 1936 [77]. This molecule is widely used in the pharmaceutical industry as the main precursor in the synthesis of steroids. Given this, several investigators have been focusing their attention on diosgenin derivatives. Sánchez et al. designed and synthesized two novel oxime derivatives (**5a** and **5b**, Figure 8) using diosgenin as a precursor and then tested their antitumor activity against HeLa and CaSki cancer cell lines [78]. Results revealed that both compounds caused a dose-dependent decrease in HeLa and CaSki cell proliferation with IC$_{50}$ ranging from 10.9 to 48.18 μM (Table 7) and compound **5a** was even more active than diosgenin, which reinforces the importance of the hydroxyimino group, positioned in the side chain, in conferring cytotoxicity. Further analysis of both compounds demonstrated that they exerted their antitumor activity by interfering with the cell cycle, and causing cell death by apoptosis mediated by the activation of caspase-3, which, in turn, is responsible for DNA fragmentation [78].

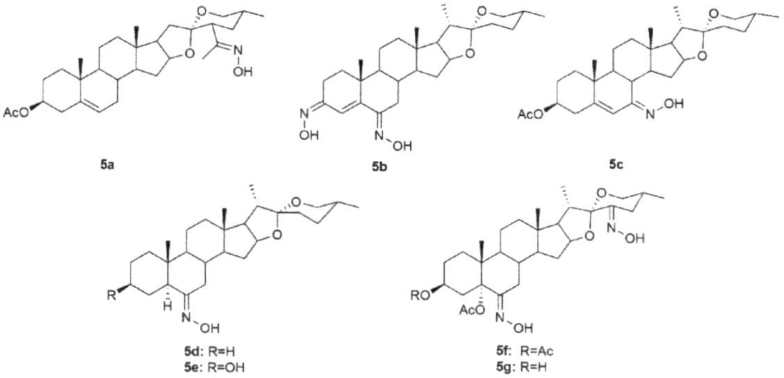

Figure 8. Steroidal diosgenin oxime derivatives with antitumor activity in several types of cancer cells.

Table 7. IC$_{50}$ values (µM) of the synthesized diosgenin oxime derivatives.

Cell Line	Compounds						
	5a	**5b**	**5c**	**5d**	**5e**	**5f**	**5g**
CaSki	10.51	48.18	-	-	-		
HeLa	10.9	41.61	37.5	-	-	11	19
MDA-MB-231	-	-	22.3	9.3	9.3		
HCT15	-	-	36.2	-	-		
MCF7	-	-	-	78.5	78		
A549						20	44
HBL-100						19	78
SW1573						22	>100
T47D						12	29
WiDr						14	17
Ref.		[78]		[79]		[80]	[81]

These values were obtained through different techniques such as crystal violet staining and XTT assays.

Another group of investigators synthesized a series of diosgenin derivatives by introducing new modifications in rings A and B [79]. Among all these compounds, oxime **5c** (Figure 8) was evaluated in HeLa, MDA-MB-231, and HCT-15 cancer cells to assess its antitumor activity. Results showed that **5c** was quite active in all cell lines with IC$_{50}$ of 18.23, 10.83, and 17.56 µg/mL in HeLa, MDA-MB-231, and HCT-15 (Table 7), respectively. Furthermore, this compound was not toxic to PBMC which can point to a selectivity towards cancer cells.

Carballo et al. designed and synthesized a series of novel hydroxyimino steroidal derivatives and evaluated their antiproliferative activity in MCF7 and MDA-MB-231 breast cancer cells [80]. Compounds **5d–5e** (Figure 8) were able to decrease cell proliferation with IC$_{50}$ values ranging from 9.3 to 11.8 µM (Table 7). Interestingly, compounds **5d** and **5e** (spirostan derivatives) were more active against the triple-negative cells, a subtype of breast cancer associated with a worse prognosis and fewer therapeutic options.

More recently, a group of scientists designed and synthesized a series of steroidal oximes from diosgenin and evaluated their antitumor activity against six cancer cell lines (A549, HBL100, HeLa, SW1573, T47D, and WiDr) [81]. Of all the compounds synthesized, **5f** and **5g** (Figure 8) were the most active against all cancer cell lines tested (Table 7). Moreover, compound **5f** was more active than cisplatin in T47D and WiDr cells and **5g** was more active only in WiDr cells. These results encourage more research to unravel the mechanisms of action behind the cytotoxicity displayed by these compounds against cancer cells.

3.6. Bile Acids Derivatives

Bile acids have shown good biological and chemical properties, which make them very useful in the designing of new pharmacological entities [82]. Given this, a group of investigators reported the synthesis of new 3-aza-A-homo-4-one bile acid and 7-deoxycholic acid derivatives, among them some with the hydroxyimino group in their structure [83]. Of all the compounds, **6a** and **6b** (Figure 9) were the most promising ones being very effective in decreasing the MGC7901, HeLa, and SMMC7404 cells proliferation. Please note that the most sensitive cell line to both **6a** and **6b** was the HeLa, ovarian cancer cell line, with an IC_{50} of 14.3 and 24.3 μM, respectively. Compound **6a** was even more cytotoxic against cancer cells than the reference drug, cisplatin (IC_{50} = 20.6 μM). These results enlighten, once again, the relationship between the hydroxyimino group and biological activity [83].

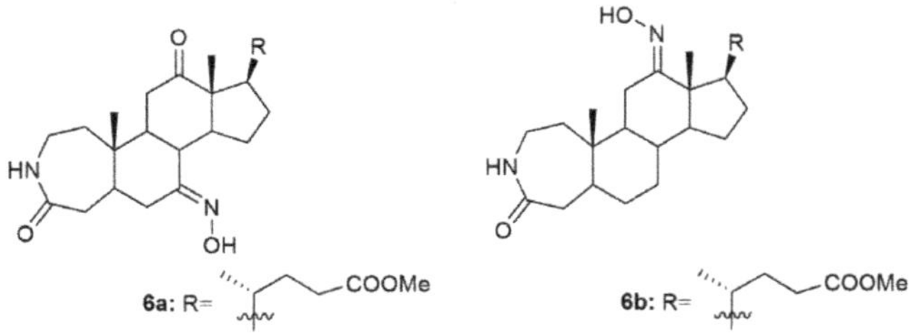

Figure 9. Steroidal bile acid oxime derivatives with antitumor activity in several types of cancer cells.

4. Steroidal Oximes as Antimicrobial Agents

Infectious diseases are a public health problem and are among the top ten leading causes of death worldwide according to WHO [84]. The need for novel therapeutic options to treat this type of disease is of great importance. Steroids and hydroxyimino group-bearing compounds have been shown to be effective against bacteria, fungi, and some viruses [12,85]. In this section, we will describe the most potent steroidal oximes with antibacterial, antifungal, and antiviral activity.

4.1. Androstane Derivatives

A series of androstane derivatives containing the hydroxyimino group (**7a–7e**, Figure 10) were designed, synthesized, and further analyzed as antibacterial and antifungal drug candidates [86] against different strains of bacteria and fungi. Results demonstrated that all oximes presented excellent antibacterial and antifungal activity, being in general more toxic to the pathogens than the referenced drugs, as seen by the minimal inhibitory (MIC) and minimal bactericidal/fungicidal (MBC/MFC) concentrations described in Tables 8 and 9.

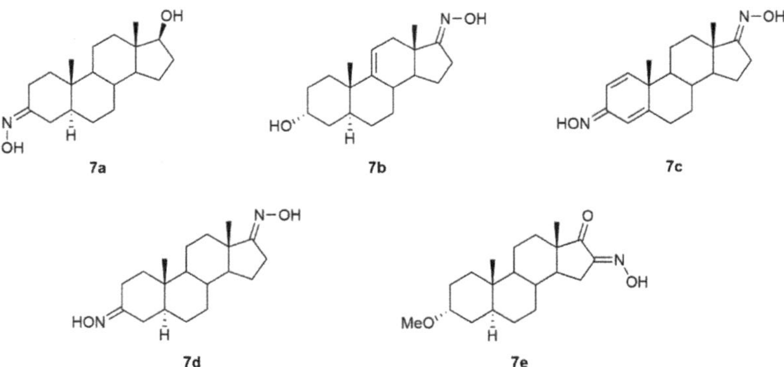

Figure 10. Steroidal androstane oxime derivatives with antibacterial and antifungal activity.

Table 8. MIC and MBC (mg/mL) values of the synthesized androstane oxime derivatives. Adapted from [86].

Compound		MIC and MBC (mg/mL)							
		S. a.	*MRSA*	*L. m.*	*P. a.*	*P.aO1*	*E. c.*	*E. c. res*	*S. t.*
7a	MIC	-	0.020	0.15	0.10	0.10	0.060	0.15	0.20
	MBC	-	0.040	0.30	0.15	0.15	0.080	0.30	0.30
7b	MIC	0.20	0.030	0.15	0.08	0.10	0.04	0.08	0.15
	MBC	0.30	0.040	0.30	0.15	0.15	0.08	0.15	0.30
7c	MIC	-	0.20	0.30	0.15	-	0.15	-	0.30
	MBC	-	0.30	0.60	0.30	-	0.30	-	0.60
7d	MIC	0.30	0.075	0.30	0.10	0.10	0.20	-	0.30
	MBC	0.60	0.15	0.60	0.15	0.15	0.30	-	0.60
7e	MIC	0.10	0.0037	0.10	0.075	0.050	0.037	0.015	0.15
	MBC	0.15	0.015	0.15	0.15	0.075	0.075	0.037	0.30
Streptomycin	MIC	0.10	0.10	0.15	0.10	0.05	0.10	0.10	0.10
	MBC	0.20	-	0.30	0.20	0.1	0.05	0.20	0.20
Ampicillin	MIC	0.10	-	0.15	0.30	0.2	0.15	0.20	0.10
	MBC	0.15	-	0.30	0.50	-	0.20	-	0.20

S. a.—S. aureus; MRSA—methicillin resistant S. aureus; L. m.—L. monocytogenes; P. a.—P. aeruginosa; P.aO1—P. aeruginosa resistant; E. c.—E.coli; E. c. res—E. coli resistant; S. t.—S. typhimurium.

Table 9. MIC and MFC (mg/mL) values of the synthesized androstane oxime derivatives. Adapted from [86].

Compound		MIC and MFC (mg/mL)							
		A. fum.	*A. v.*	*A. o.*	*A. n.*	*T. v.*	*P. f.*	*P. o.*	*P. v.c.*
7a	MIC	0.037	0.015	0.015	0.037	0.007	0.015	0.02	0.05
	MFC	0.075	0.037	0.037	0.075	0.015	0.075	0.037	0.037
7b	MIC	0.015	0.015	0.007	0.037	0.007	0.015	0.037	0.05
	MFC	0.037	0.075	0.015	0.075	0.015	0.037	0.075	0.075
7c	MIC	0.075	0.015	0.075	0.15	0.10	0.15	0.20	0.20
	MFC	0.15	0.037	0.15	0.30	0.15	0.30	0.30	0.30
7d	MIC	0.015	0.037	0.10	0.075	0.015	0.15	0.075	0.15
	MFC	0.037	0.15	0.15	0.15	0.037	0.30	0.15	0.30
7e	MIC	0.037	0.037	0.075	0.075	0.050	0.075	0.15	0.10
	MFC	0.075	0.075	0.15	0.15	0.075	0.15	0.30	0.15
Ketoconazole	MIC	0.20	0.20	0.15	0.20	1.00	0.20	1.50	0.30
	MFC	0.50	0.50	0.20	0.50	1.50	0.50	1.50	0.30
Bifonazole	MIC	0.15	0.10	0.15	0.15	0.15	0.20	0.10	0.10
	MFC	0.20	0.20	0.20	0.20	0.20	0.25	0.25	0.20

A. fum.—A. fumigatus; A. v.—A. versicolor; A. o.—A. ochraceus; A. n.—A niger; T. v.—T. viride; P. f.—P. funiculosum; P. o.—P. ochrochloron; P. v.c.—P. verucosum var. cyclopium.

4.2. Pregnane Derivatives

Using pregnenolone as starting material, Prabpayak et al. synthesized a novel oxime derivative (**8a**, Figure 11) and evaluated its antibacterial activity against a series of different strains of bacteria [87]. Results revealed that **8a** was able to decrease bacterial growth, as seen by the calculation of the zone of inhibition which was 12.5 mm for *S. mutans* ATCC 1275 and 25 mm for *C. diptheriae*.

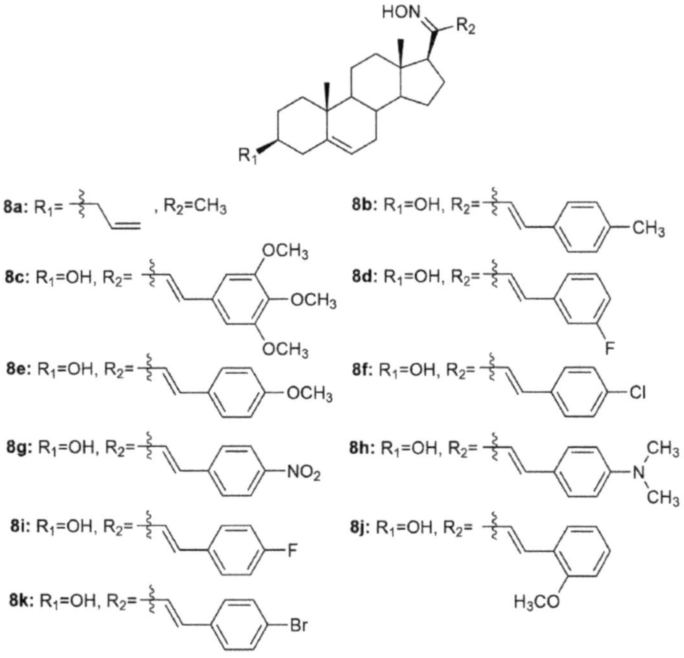

Figure 11. Pregnane oxime derivatives with antimicrobial activity.

Lone et al. designed and synthesized nine oximes of steroidal chalcones (**8b–8k**, Figure 11) and screened them for in vitro antimicrobial activity against different strains of bacteria and fungi [88]. All compounds displayed good antimicrobial activity against all strains studied (Table 10). Moreover, the oximes showed enhanced antimicrobial activity when compared with the corresponding chalcones with a carbonyl group instead of a hydroxyimino group. These results highlight the importance of this hydroxyimino group for the toxicity displayed against pathogens.

Table 10. MIC values of (μg/mL) of the synthesized pregnane oxime derivatives. Adapted with permission from Ref [88], 2023, Ana S. Pires.

Compound	MIC (μg/mL)					
	B. subtilis	*S. epidermidis*	*P. vulgaris*	*P. aeruginosa*	*A. niger*	*P. chrysogenum*
8b	≤64	≤128	>512	≤256	≤128	≤128
8c	≤128	≤64	>512	≤128	≤64	≤256
8d	≤64	≤64	>512	≤256	≤128	≤128
8e	≤128	≤128	>512	≤128	≤64	≤256
8f	≤64	≤128	>512	≤256	≤128	≤128
8g	≤128	≤64	>512	≤512	≤128	≤256
8h	≤64	≤128	>512	≤256	-	≤256
8i	≤128	≤256	>512	≤128	≤256	-
8j	≤128	≤64	>512	≤256	≤64	≤128
8k	≤512	≤128	>512	≤512	≤128	≤256

Bacteria: *Bacillus subtilis* (MTCC 619), *Staphylococcus epidermidis* (MTCC 435), *Proteus vulgaris* (MTCC 426), *Pseudomonas aeruginosa* (MTCC 424). Fungi: *A. niger* (MTCC 1344), *P. chrysogenum* (MTCC 947).

4.3. Cholestane Derivatives

A group of scientists designed and synthesized three novel steroidal oximes (**9a–9c**, Figure 12) from cholestane [89]. After synthesis, the antibacterial and antifungal activity in different strains of bacteria (*S. pyogenes*, *S. aureus*, *S. typhi*, *P. aeruginosa* and *E. coli*) and fungi (*P. marneffei*, *A. fumigatus*, *T. mentagrophytes*, *C. albicans* and *C. krusei*) was assessed through analysis of the diameter of zone of inhibition (mm). Anthelmintic activity was also analyzed against earthworms. Generally, all compounds presented good antibacterial and antifungal activity with compound **9c** being the most active against both bacterial (zone of inhibition ranging from 17.3 to 23.9 mm) and fungi strains (zone of inhibition ranging from 15.1 and 25.5 mm). This compound was also the most effective anthelmintic compound showing great early paralysis and lethal times. Further docking studies demonstrated that **9c** had not only better affinity to the receptor, but also presented the best docking score, making it a very promising antimicrobial [89].

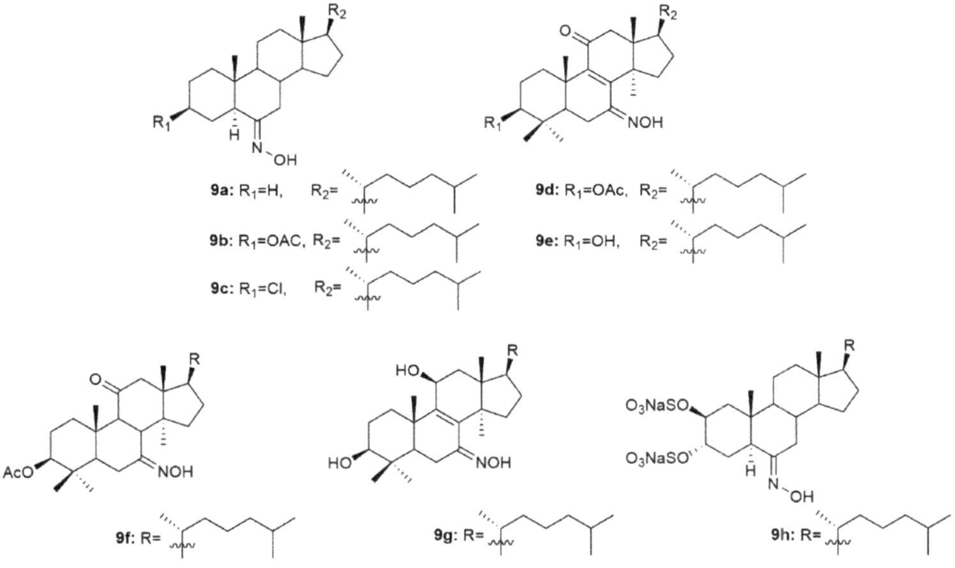

Figure 12. Cholestane oxime derivatives with antimicrobial activity.

Oxysterols apart from displaying antitumor activity have also been associated with antimicrobial activity [90]. Compounds **9d–9g** (Figure 12) were designed and synthesized and further evaluated against a series of fungal strains [91]. In general, all compounds displayed good antifungal activity against all the strains used in this study (MIC values ranging from 2 to 64 µg/mL, Table 11). *C. neoformans* (CN1) was particularly sensitive to all oximes synthesized, being the MIC values (2–4 µg/mL) presented even lower than the referenced drugs fluconazole (16 µg/mL) and amphotericin B (32 µg/mL).

Aiming to develop novel therapeutic options for the treatment of herpes simplex virus, Pujol et al. synthesized a series of sulfated steroids derived from 5α-cholestanes [92]. The oxime **9h** (Figure 12) was the most promising compound, showing the best inhibitory values of HSV-1, HSV-2, and pseudorabies virus (PrV) strains, including acyclovir-resistant strains in human and monkey cell lines (EC$_{50}$ values ranging from 16.7 to 25 µg/mL, Table 11). Moreover, the authors went further ahead and decided to investigate the mechanism of action of **9h** in HSV-1. After using the virucidal assay, authors concluded that the **9h** was not able to affect the initial steps of virus entry but rather inhibited an ensuing event in the infection process of this virus [92].

Table 11. MIC values (µg/mL) of cholestane oxime derivatives.

Microorganism	Compound				
	9d	**9e**	**9f**	**9g**	**9h**
C. albicans	64	8	32	32	
C. neoformans (CN1)	4	4	4	2	
B. poitrasii	32	8	16	16	
Y. lipolytica	8	64	16	16	-
F. oxysporum	>64	16	8	8	-
HSV-1 strain B2006	-	-	-	-	19.5
HSV-2 strain G	-	-	-	-	17.9
PrV strain RC79	-	-	-	-	17.2
HSV-1 strain F	-	-	-	-	16.7
Ref.			[91]		[92]

4.4. Bile Acids Derivatives

Hepatitis B is a major global health problem and is caused by the hepatitis B virus (HBV). This condition affects the liver, leading to cirrhosis and liver cancer, which can lead to patient death [93]. In this context, a group of scientists focused their attention on designing and synthesizing oxime derivatives of dehydrocholic acid as potential HBV drugs [94]. After anti-hepatitis B virus activity evaluation in HepG 2.2.15 cells, compounds **10a–10c** (Figure 13) exhibited more cytotoxicity against the virus than the referenced drug, entecavir, as seen by the most effective inhibition of HBeAg (hepatitis B -antigen). Compound **10b** was the most active compound showing significant anti-HBV activity on inhibition secretion of HBeAg (IC_{50} = 96.64 µM) when compared, once again, with entecavir (IC_{50} = 161.24 µM). Moreover, docking studies were also performed to evaluate the potential mechanisms behind the activity of these compounds. Results point out a possible interaction with protein residues of heparan sulfate proteoglycan (HSPG) in host hepatocytes and bile acid receptors [94].

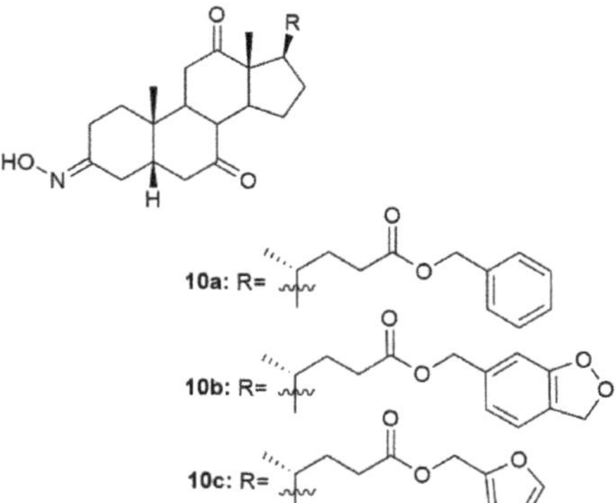

Figure 13. Bile acid oxime derivatives with antiviral activity.

5. Steroidal Oximes as Anti-Inflammatory Agents

5.1. Cholestane Derivatives

Since inflammation plays a critical role in tumor progression, Díaz et al. designed and synthesized a series of 22-oxocholestane oximes as potential anti-inflammatory agents in an acute inflammation mouse ear model [95]. Five oxime derivatives (**11a–11e**, Figure 14) stood out from the rest as being the most promising, being able to reduce ear inflammation and edema. Moreover, these compounds also repressed the expression of pro-inflammatory genes such as TNF-α, COX-2, and IL-6, making them very promising lead candidates for further assessment.

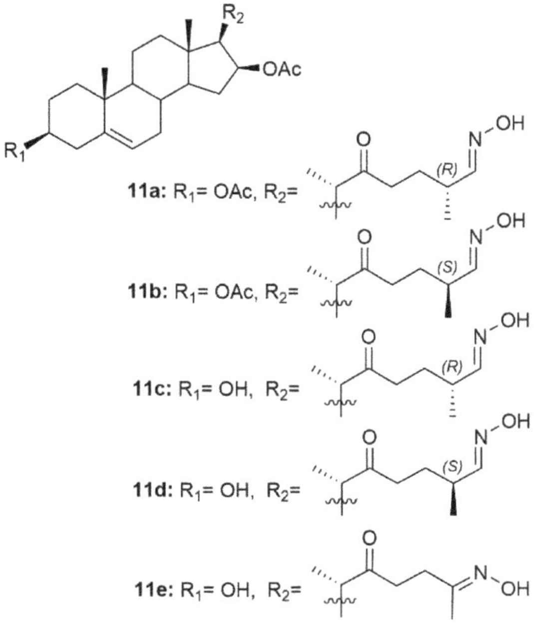

Figure 14. Cholestane oxime derivatives with anti-inflammatory activity.

5.2. Bile Acid Derivatives

The need for anti-inflammatory drugs such as glucocorticoids has been increasing, especially since the COVID-19 pandemic, because this type of compounds is the standard treatment for this disease. Bearing this in mind, Bjedov and collaborators designed, synthesized, and screened the binding activity of a series of bile acid derivatives for the ligand-binding domain of glucocorticoid receptor (GR-LBD), the main receptor in the anti-inflammatory process [96]. Of all the synthesized compounds, oxime **12a** (Figure 15) presented the best relative binding affinity for GR-LBD. Even though the authors were not able to perform molecular docking and predict the interactions between compound **12a** and the enzyme, they suggest that this binding affinity could be attributed not only to the C-24 carboxylic group but also to the hydroxyimino groups. Moreover, the C-12 hydroxyimino group could also be used as an alternative hydrogen donor to the 11β-OH group of the glucocorticoids and the hydroxyimino group at C-3 may also be beneficial for glucocorticoid receptor affinity through the establishment of a hydrogen bond. This study was very important and helped shed some light on the possible interactions of steroidal oximes with their targets.

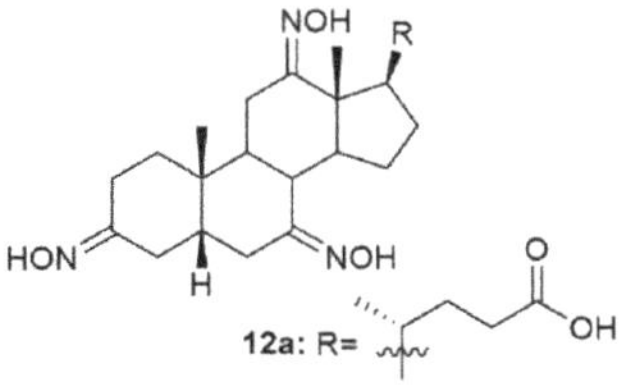

Figure 15. Bile acid derivative with anti-inflammatory activity.

6. Steroidal Oximes in Clinical Development

Istaroxime

Istaroxime (Figure 16) is an inhibitor of sodium/potassium adenosine triphosphatase (Na+/K+ ATPase), which is currently under phase 2 clinical trial development for the treatment of acute decompensated heart failure [97,98]. Apart from this, istaroxime has also been evaluated as a potential antitumor agent in several types of cancer [99,100].

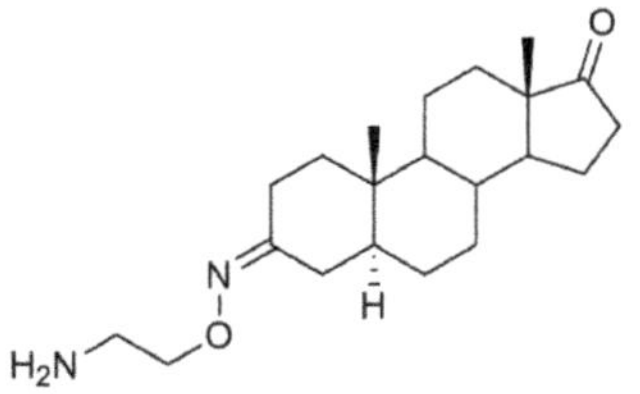

Figure 16. Istaroxime.

7. Steroidal Oximes in Clinical Use

7.1. Norgestimate

Norgestimate (Figure 17), a steroidal oxime, (brand names, Ortho Tri-Cyclen or, Previfem, among others) is a progestin, which is used in hormonal contraception and menopausal hormone therapy for the treatment of some menopausal symptoms [101]. Norgestimate is not available as a single therapy, being used together with ethinylestradiol in birth control pills and in combination with estradiol in menopausal hormone therapy [102]. It was first introduced in the USA in 1999 and is one of the most prescribed birth control pills worldwide. This compound is sold as a mixture of the two *E* and *Z* conformers.

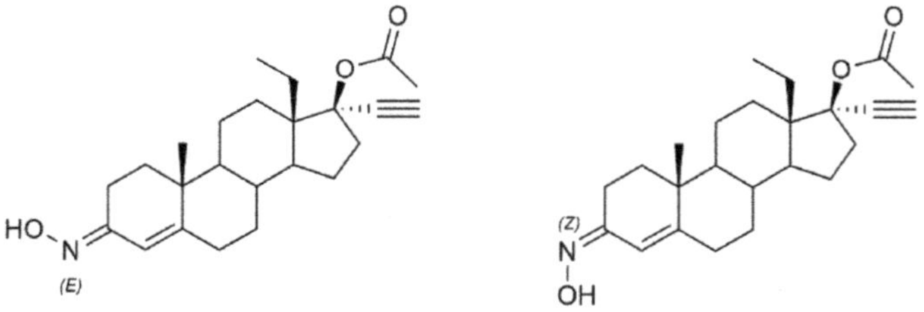

Figure 17. Norgestimate.

This oxime has been subject to further biological evaluation such as antibacterial assessment and proved to be a promising lead compound to treat biofilm-associated infections and to resensitize bacterial strains resistant to some antibiotics [103].

7.2. Norelgestromin

Norelgestromin (Figure 18), or norelgestromine (brand names, Evra or Ortho Evra, among others) is also a progestin used for birth control. Like norgestimate, norelgestromin is not available as a single drug but rather in combination with an estrogen, ethinyl estradiol in the form of contraceptive patches [104]. It was first introduced to the market in 2002. This compound is also sold as a mixture of two *E* and *Z* isomers.

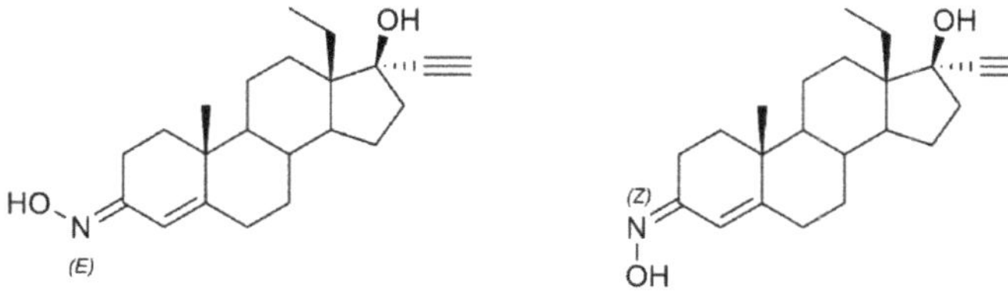

Figure 18. Norelgestromin.

8. Mechanisms of Action of Steroidal Oximes

Throughout this review, we came upon a wide variety of synthesized steroidal oximes, starting from androstane to estrane, pregnane, cholestane, diosgenin, and bile acids derivatives with very different mechanisms of action.

Most of the compounds summarized here are being evaluated as potential antitumor agents and exert their cytotoxicity against cancer cells mainly by inducing apoptosis; however, the pathways leading to it differ from compounds. For example, the androstane derivatives appear to induce apoptosis [34,42,48] by cell cycle arrest at different phases and increased ROS production [34]. Necroptosis might also be a mechanism involved in the compounds' cytotoxicity against cancer cells [34]. Oxime estrane derivatives induced cell death by apoptosis through cell cycle interference at G1 [53], S [54] and G2/M phases [57] and through activation of caspase-3 [53]. This class of compounds also interferes with microtubules by interfering with β-tubulin [55,57]. These are often related to cell cycle arrest and, consequently, a decrease in cell division and cell death. Concerning oxime pregnane, diosgenin, and some cholestane derivatives, they cause cell cycle alterations, and cell death by the activation of the apoptotic intrinsic pathway mediated by caspase-3 activation [64,78] and release of cytochrome C [61,64,78]. Additionally, and since inflammation plays an important role in carcinogenesis, inhibition of pro-inflammatory genes such as TNF-α, COX, and IL6 is also another important mechanism displayed by these steroidal oximes [95]. Additionally, for some of the cholestane derivatives, their mechanism of action is not well elucidated since most of the studies were conducted mainly to infer about SAR, rather than go deeper into their mode of action. However, the breakthroughs reached in this field are of great importance and encourage the need to further analyze their mechanisms of action.

Aside from antitumor activity, steroidal oximes also display antimicrobial activity. However, little is known about the mode of action of these compounds. Regarding antiviral activity, it seems that steroidal oximes exert their effect by inhibition of the infection process of the virus after it enters the cells [92] and by a possible interaction with protein residues of heparan sulfate proteoglycan (HSPG) in host hepatocyte and bile acid receptor [94].

Concerning the compounds in clinical trials and already in clinical use, their mechanisms are very different. Istaroxime is an inhibitor of the Na$^+$/K$^+$ ATPase pump [98], whereas norgestimate and norelgestromin, which are already in the market for birth control, act as progesterone agonists, inhibiting ovulation [101,102].

9. Structure-Activity Relationships of Steroidal Oximes

The biological activity of a compound is closely related to its chemical structure. Steroidal oximes exert several types of biological activities, which can vary depending on the position of the oxime functionality on the steroid scaffold. Given this, some SAR can be elucidated especially for the antitumor and antimicrobial activity to help in the design and synthesis of novel molecules with pharmaceutical potential (Figures 19–21). However, the conclusions that can be drawn from Figures 19–21 must be analyzed carefully, as they are just trends that were observed from the compounds studied in this review article, which were the most active compounds of each paper, and were evaluated with different methodologies, different times, and in different cancer cells.

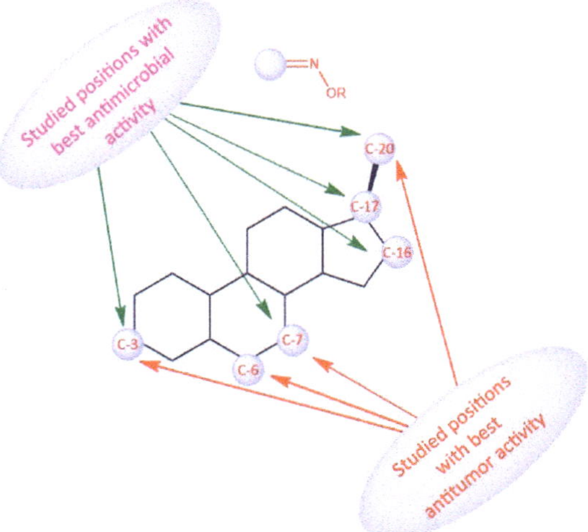

Figure 19. Overview of the most common SAR of steroidal oximes with antitumor and antimicrobial activity.

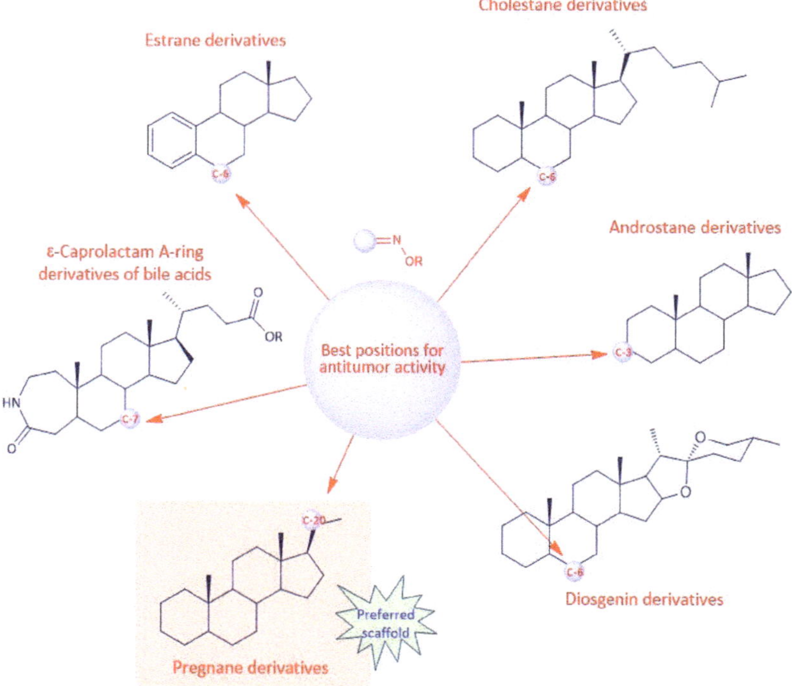

Figure 20. Most common SAR of steroidal oximes with antitumor activity.

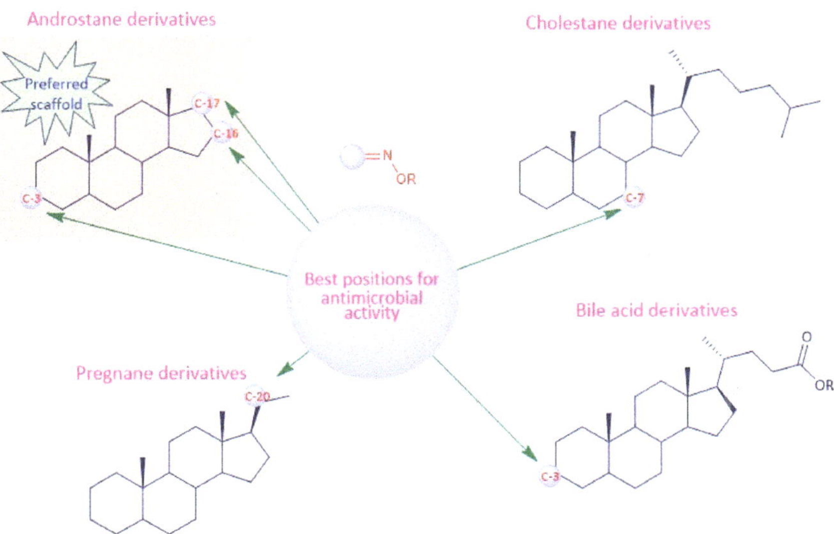

Figure 21. Most common SAR of steroidal oximes with antimicrobial activity.

Regarding the antitumor activity, when comparing the different IC_{50} values displayed by the androstane oxime derivatives, and despite the presence of other functional groups, it seems that position C-3 of the steroidal scaffold is preferable since the compounds with a hydroxyimino group in this position were, in general, the ones with the lower IC_{50} values (**1f**, **1g**, **1h**). Compound **1h**, which has a substituted oxyimino group at position C-3 was generally the compound with the best IC_{50} values in all cell lines reported in Table 1. Interestingly, the compounds that, in addition to having a hydroxyimino group at position C-3, also had another hydroxyimino group at position C-6 (**1b**, **1c** and **1u**) or at position C-17 (**1k–1o**) were slightly less active, suggesting that only a single hydroxyimino group at position C-3 is preferable and might be enough for the antitumor activity. Furthermore, in some types of cancer, namely in breast and prostate cancers, a hydroxyimino group at position C-2 was also important for the cytotoxicity displayed against cancer cells (compound **1r** and **1s**). Compound **1s** combines another hydroxyimino group at position C-4, which proved to be better than with just a single hydroxyimino group at position C-2. Positions C-17 and C-7 seem to be less favorable than the other positions mentioned above. Regarding the estrane series, position C-6 seems to be the most preferable concerning antitumor activity since compounds **2p** and **2q** were the ones with best IC_{50} values. In general, compounds with a hydroxyimino group at position C-17 were also very active. Concerning the pregnane series, all oximes (substituted or not) present high activity with very low IC_{50} values (IC_{50} values ranging from 0.31 to 7.17 μM. A hydroxyimino group at position C-6 appears to be the most advantageous position, regarding the cholestane series, since they present the lowers IC50s. With regard to the diosgenin series, the compounds with just a single hydroxyimino group were slightly more cytoxic against cancer cells than the ones containing two hydroxyimino groups (compounds **5b**, **5f** and **5g** present higher IC_{50} values than compounds **5a**, **5d** and **5e**). Finally, introduction of a hydroxyimino group in the B ring was clearly more beneficial than at C-ring, considering that compound **6a** presented a lower IC_{50} value than compound **6b**, for the case of bile acid derivatives.

Considering the antimicrobial activity, the androstane oxime derivatives with just a single hydroxyimino group were slightly more potent than the compounds with two hydroxyimino groups, suggesting that a single hydroxyimino group is better for antimicrobial activity rather than two hydroxyimino groups. Furthermore, a hydroxyimino group at position C-7 might be a better option when designing and synthesizing oximes in the cholestane series, since the compounds with a hydroxyimino group at this position were slightly more toxic against pathogens than compounds with a hydroxyimino group at position C-6. Pregnane oxime derivatives were more active than the parent pregnane compounds.

To sum up, for the antitumor activity, SAR analysis points out that the pregnane scaffold might be the best option when designing novel molecules with antitumor activity. This steroidal oxime series presented the best IC_{50} values against the cancer cell lines studied when compared with the other above-mentioned series. As for the antimicrobial activity, and since the number of studies is considerably less, it turns out to be more difficult to establish robust SAR, but it seems that the androstane scaffold may be the most favorable option. In Figure 19 we can see an overview of the most common SAR of steroidal oximes for the antitumor and antimicrobial biological activities.

As mentioned at the beginning of this chapter, SAR are extremely import for the design and synthesis of novel drugs. Given this, it is our hope that this review might help guide researchers to design more potent and selective chemical entities and to help them understand the pharmaceutical potential of steroidal oximes.

10. Conclusions

This review sums up the work that has been developed regarding the most promising steroidal oximes in recent years. The oxyimino group is a privileged chemical function being, for that reason, very popular among the medicinal chemistry community.

Most steroidal oximes in preclinical settings present antitumor activity. In vitro and in vivo studies have demonstrated the significant antiproliferative activity and elucidated some of the mechanisms by which steroidal oximes display cytotoxicity against cancer cells, which include induction of cell cycle arrest, apoptosis, necroptosis, increased ROS production and interference with microtubules. The described targets for steroidal oximes exerting antitumor activity are caspase-3 and β-tubulin. Moreover, steroidal oximes also exert antimicrobial activity, sometimes by interfering with HSPG, resulting in tremendous cytotoxicity against bacteria, fungi, and viruses. The anti-inflammatory activity is associated with TNF-α, COX-2, and IL-6 genes expression. Finally, there are currently two steroidal oximes in the market used as birth control drugs and one in clinical trials for acute heart failure, acting by progesterone agonists and Na+/K+ ATPase pump inhibitors, respectively.

The acquired knowledge about the diversified biological activities of steroidal oximes only emphasizes the need to deeper understand the mechanisms behind their activity and consequently, unravel which are their targets. In addition, molecular interactions with targets need to be elucidated, robust SAR must be constructed, and ADME properties have to be studied in order to design more potent and selective compounds. Steroidal oximes can be, in fact, very promising lead compounds for the development of drugs for the treatment of several diseases and deserve to be better explored.

Author Contributions: Conceptualization, F.M.F.R., E.J.T.-d.-S. and A.R.G., Writing—original draft, A.R.G., Writing—review and editing, F.M.F.R., E.J.T.-d.-S., A.R.G. and A.S.P. All authors have read and agreed to the published version of the manuscript.

Funding: This research was funded by Foundation for Science and Technology (Portugal) through the Strategic Projects UIDB/04539/2020 and UIDP/04539/2020 (CIBB) and through PhD grant of Ana R. Gomes (UI/BD/150865/2021).

Institutional Review Board Statement: Not applicable.

Informed Consent Statement: Not applicable.

Data Availability Statement: Not applicable.

Acknowledgments: The authors would like to thank the Foundation for Science and Technology (FCT), Portugal for financial support through the PhD grant of Ana R. Gomes (UI/BD/150865/2021).

Conflicts of Interest: The authors declare no conflict of interest.

Nomenclature

A2780	human ovarian cancer cell line
A431	human skin epidermoid cancer cell line
A549	human lung carcinoma cell line
Bel7404	human liver cancer cell line
Ca Ski	human cervical cancer cell line
DU145	human prostate cancer cell line
GNE2	human nasopharyngeal cancer cell line
H1299	human non-small lung cancer cell line
H292	human lung cancer cell line
HBL100	human normal breast cell line
HCT116	human colon cancer cell line
HCT15	human colorectal cancer cell line
HeLa	human cervical cancer cell line
HepG2	human liver cancer cell line
Hey1B	human ovarian cancer cell line
HL60	human leukemic cell line
HOP62	human lung cancer cell line
HT29	human colorectal cancer cell line
K562	human leukemic cell line
KB	human nasopharyngeal cancer cell line
LNCaP	human prostate cancer cell line
MCF7	human breast cancer cell line
MDA-MB-231	human breast triple-negative cancer cell line
MEL28	human melanoma cancer cell line
MGC7901	human gastric cancer cell line
MKN-28	human gastric cancer cell line
NCI-H460	human lung large cell
OVCAR-3	human ovarian cancer cell line
P388	human lymphoma cell line
PC3	human prostate cancer cell line
PSN1	human pancreatic cancer cell line
SF268	human astrocytoma cancer cell line
SF539	human gliosarcoma cancer cell line
SGC7901	human papillomavirus-related endocervical adenocarcinoma cell line
SH-SY5Y	human neuroblastoma cell line
SK-Hep-1	human liver cancer cell line
SMMC7404	human liver cancer cell line
SN12C	human renal cancer cell line
SPC-A	human papillomavirus-related endocervical adenocarcinoma cell line
SR	human lymphoma cell line
SW1573	human lung cancer cell line
T24	human urinary bladder cancer cell line
T47D	human breast ductal carcinoma cell line
T98G	human glioblastoma cancer cell line
Tu686	human laryngeal squamous cell carcinoma cell line
UACC62	human skin cancer cell line
WiDr	human colorectal cancer cell line

References

1. Bhatti, H.N.; Khera, R.A. Biological transformations of steroidal compounds: A review. *Steroids* **2012**, *77*, 1267–1290. [CrossRef] [PubMed]
2. Gupta, A.; Kumar, B.S.; Negi, A.S. Current status on development of steroids as anticancer agents. *J. Steroid Biochem. Mol. Biol.* **2013**, *137*, 242–270. [CrossRef] [PubMed]
3. Ericson-Neilsen, W.; Kaye, A.D. Steroids: Pharmacology, complications, and practice delivery issues. *Ochsner J.* **2014**, *14*, 203–207. [PubMed]
4. Rasheed, A.; Qasim, M. A Review of natural steroids and their applications. *Int. J. Pharm. Sci. Res.* **2013**, *4*, 520–531.

5. Bansal, R.; Suryan, A. A Comprehensive Review on Steroidal Bioconjugates as Promising Leads in Drug Discovery. *ACS Bio Med Chem Au* **2022**, *2*, 340–369. [CrossRef]

6. Bai, C.; Schmidt, A.; Freedman, L.P. Steroid Hormone Receptors and Drug Discovery: Therapeutic Opportunities and Assay Designs. *ASSAY Drug Dev. Technol.* **2003**, *1*, 843–852. [CrossRef]

7. Schepetkin, I.; Plotnikov, M.; Khlebnikov, A.; Plotnikova, T.; Quinn, M. Oximes: Novel Therapeutics with Anticancer and Anti-Inflammatory Potential. *Biomolecules* **2021**, *11*, 777. [CrossRef]

8. Grob, D.; Johns, R.J. Use of oximes in the treatment of intoxication by anticholinesterase compounds in normal subjects. *Am. J. Med.* **1958**, *24*, 497–511. [CrossRef]

9. Dhuguru, J.; Zviagin, E.; Skouta, R. FDA-Approved Oximes and Their Significance in Medicinal Chemistry. *Pharmaceuticals* **2022**, *15*, 66. [CrossRef]

10. Chen, S.-R.; Shen, F.-J.; Feng, G.-L.; Yuan, R.-X. Synthesis and Anticancer Activity of 4-azasteroidal-20-oxime Derivatives. *J. Chem. Res.* **2015**, *39*, 527–530. [CrossRef]

11. Sørensen, M.; Neilson, E.H.; Møller, B.L. Oximes: Unrecognized Chameleons in General and Specialized Plant Metabolism. *Mol. Plant* **2017**, *11*, 95–117. [CrossRef] [PubMed]

12. Surowiak, A.K.; Lochyński, S.; Strub, D.J. Unsubstituted oximes as potential therapeutic agents. *Symmetry* **2020**, *12*, 575. [CrossRef]

13. Aakeröy, C.B.; Sinha, A.S.; Epa, K.N.; Chopade, P.D.; Smith, M.M.; Desper, J. Structural Chemistry of Oximes. *Cryst. Growth Des.* **2013**, *13*, 2687–2695. [CrossRef]

14. Canario, C.; Silvestre, S.; Falcão, A.; Alves, G. Steroidal Oximes: Useful Compounds with Antitumor Activities. *Curr. Med. Chem.* **2018**, *25*, 660–686. [CrossRef]

15. Donaruma, L.G.; Heldt, W.Z. The Beckmann rearrangement. *Org. React.* **2011**, *15*, 1–156.

16. Ābele, E.; Lukevics, E. Recent advances in the chemistry of oximes. *Org. Prep. Proced. Int.* **2000**, *32*, 235–264. [CrossRef]

17. Bolotin, D.S.; Bokach, N.A.; Demakova, M.Y.; Kukushkin, V.Y. Metal-Involving Synthesis and Reactions of Oximes. *Chem. Rev.* **2017**, *117*, 13039–13122. [CrossRef]

18. Elliott, M.C. Imines and Their N-Substituted Derivatives: Hydrazones and Other N,N-Derivatives Including Diazo Compounds. *Cheminform* **2005**, *36*, 469–523. [CrossRef]

19. Savić, M.P.; Ajduković, J.J.; Plavša, J.J.; Bekić, S.S.; Ćelić, A.S.; Klisurić, O.R.; Jakimov, D.S.; Petri, E.T.; Djurendić, E.A. Evaluation of A-ring fused pyridine ᴅ-modified androstane derivatives for antiproliferative and aldo–keto reductase 1C3 inhibitory activity. *Medchemcomm* **2018**, *20*, 969–981. [CrossRef]

20. Savić, M.P.; Škorić, D.Đ.; Kuzminac, I.Z.; Jakimov, D.S.; Kojić, V.V.; Rárová, L.; Strnad, M.; Djurendić, E.A. New A-homo lactam ᴅ-homo lactone androstane derivative: Synthesis and evaluation of cytotoxic and anti-inflammatory activities in vitro. *Steroids* **2020**, *157*, 108596. [CrossRef]

21. Savić, M.P.; Kuzminac, I.Z.; Škorić, D.; Jakimov, D.S.; Rárová, L.; Sakač, M.N.; Djurendić, E.A. New oxygen-containing androstane derivatives: Synthesis and biological potential. *J. Chem. Sci.* **2020**, *132*, 1–10. [CrossRef]

22. Kulmány, Á.E.; Herman, B.E.; Zupkó, I.; Sinreih, M.; Rižner, T.L.; Savić, M.; Oklješa, A.; Nikolić, A.; Nagy, V.; Ocsovszki, I.; et al. Heterocyclic androstane and estrane d-ring modified steroids: Microwave-assisted synthesis, steroid-converting enzyme inhibition, apoptosis induction, and effects on genes encoding estrogen inactivating enzymes. *J. Steroid Biochem. Mol. Biol.* **2021**, *214*, 105997. [CrossRef] [PubMed]

23. Kovačević, S.Z.; Podunavac-Kuzmanović, S.O.; Jevrić, L.R.; Jovanov, P.T.; Djurendić, E.A.; Ajduković, J.J. Comprehensive QSRR modeling as a starting point in characterization and further development of anticancer drugs based on 17α-picolyl and 17(E)-picolinylidene androstane structures. *Eur. J. Pharm. Sci.* **2016**, *93*, 1–10. [CrossRef] [PubMed]

24. Jójárt, R.; Ali, H.; Horváth, G.; Kele, Z.; Zupkó, I.; Mernyák, E. Pd-catalyzed Suzuki–Miyaura couplings and evaluation of 13α-estrone derivatives as potential anticancer agents. *Steroids* **2020**, *164*, 108731. [CrossRef] [PubMed]

25. Pires, A.S.; Varela, C.L.; Marques, I.A.; Abrantes, A.M.; Gonçalves, C.; Rodrigues, T.; Matafome, P.; Botelho, M.F.; Roleira, F.M.; Tavares-Da-Silva, E. Oxymestane, a cytostatic steroid derivative of exemestane with greater antitumor activity in non-estrogen-dependent cell lines. *J. Steroid Biochem. Mol. Biol.* **2021**, *212*, 105950. [CrossRef]

26. Augusto, T.V.; Amaral, C.; Varela, C.L.; Bernardo, F.; da Silva, E.T.; Roleira, F.F.; Costa, S.; Teixeira, N.; Correia-Da-Silva, G. Effects of new C6-substituted steroidal aromatase inhibitors in hormone-sensitive breast cancer cells: Cell death mechanisms and modulation of estrogen and androgen receptors. *J. Steroid Biochem. Mol. Biol.* **2019**, *195*, 105486. [CrossRef] [PubMed]

27. Roleira, F.M.F.; Varela, C.; Amaral, C.; Costa, S.C.; Correia-da-Silva, G.; Moraca, F.; Costa, G.; Alcaro, S.; Teixeira, N.A.A.; da Silva, E.J.T. C-6α- vs C-7α-substituted steroidal aromatase inhibitors: Which is better? Synthesis, biochemical evaluation, docking studies, and structure–activity relationships. *J. Med. Chem.* **2019**, *62*, 3636–3657. [CrossRef]

28. Amaral, C.; Augusto, T.V.; Tavares-Da-Silva, E.; Roleira, F.M.; Correia-Da-Silva, G.; Teixeira, N. Hormone-dependent breast cancer: Targeting autophagy and PI3K overcomes Exemestane-acquired resistance. *J. Steroid Biochem. Mol. Biol.* **2018**, *183*, 51–61. [CrossRef]

29. Amaral, C.; Varela, C.L.; Maurício, J.; Sobral, A.F.; Costa, S.C.; Roleira, F.M.; da Silva, E.T.; Correia-Da-Silva, G.; Teixeira, N. Anti-tumor efficacy of new 7α-substituted androstanes as aromatase inhibitors in hormone-sensitive and resistant breast cancer cells. *J. Steroid Biochem. Mol. Biol.* **2017**, *171*, 218–228. [CrossRef]

30. Amaral, C.; Lopes, A.; Varela, C.L.; da Silva, E.T.; Roleira, F.M.; Correia-Da-Silva, G.; Teixeira, N. Exemestane metabolites suppress growth of estrogen receptor-positive breast cancer cells by inducing apoptosis and autophagy: A comparative study with Exemestane. *Int. J. Biochem. Cell Biol.* **2015**, *69*, 183–195. [CrossRef]

31. Varela, C.L.; Amaral, C.; Correia-Da-Silva, G.; Carvalho, R.A.; Teixeira, N.A.; Costa, S.C.; Roleira, F.M.; Tavares-Da-Silva, E.J. Design, synthesis and biochemical studies of new 7α-allylandrostanes as aromatase inhibitors. *Steroids* **2013**, *78*, 662–669. [CrossRef]

32. Amaral, C.; Varela, C.; Correia-Da-Silva, G.; da Silva, E.T.; Carvalho, R.A.; Costa, S.C.; Cunha, S.C.; Fernandes, J.O.; Teixeira, N.; Roleira, F.M. New steroidal 17β-carboxy derivatives present anti-5α-reductase activity and anti-proliferative effects in a human androgen-responsive prostate cancer cell line. *Biochimie* **2013**, *95*, 2097–2106. [CrossRef] [PubMed]

33. Amaral, C.; Varela, C.; Borges, M.; da Silva, E.T.; Roleira, F.M.F.; Correia-Da-Silva, G.; Teixeira, N. Steroidal aromatase inhibitors inhibit growth of hormone-dependent breast cancer cells by inducing cell cycle arrest and apoptosis. *Apoptosis* **2013**, *18*, 1426–1436. [CrossRef] [PubMed]

34. Gomes, A.R.; Pires, A.S.; Abrantes, A.M.; Gonçalves, A.C.; Costa, S.C.; Varela, C.L.; Silva, E.T.; Botelho, M.F.; Roleira, F.M. Design, synthesis, and antitumor activity evaluation of steroidal oximes. *Bioorg. Med. Chem.* **2021**, *46*, 116360. [CrossRef] [PubMed]

35. Dhingra, N.; Bhardwaj, T.R.; Mehta, N.; Mukhopadhyay, T.; Kumar, A.; Kumar, M. 17-Oximino-5-androsten-3β-yl esters: Synthesis, antiproliferative activity, acute toxicity, and effect on serum androgen level. *Med. Chem. Res.* **2011**, *20*, 817–825. [CrossRef]

36. Gros, L.; Lorente, S.O.; Jimenez, C.J.; Yardley, V.; Rattray, L.; Wharton, H.; Little, S.; Croft, S.L.; Ruiz-Perez, L.M.; Gonzalez-Pacanowska, D.; et al. Evaluation of Azasterols as Anti-Parasitics. *J. Med. Chem.* **2006**, *49*, 6094–6103. [CrossRef]

37. Burbiel, J.; Bracher, F. Azasteroids as antifungals. *Steroids* **2003**, *68*, 587–594. [CrossRef]

38. Birudukota, N.; Mudgal, M.M.; Shanbhag, V. Discovery and development of azasteroids as anticancer agents. *Steroids* **2019**, *152*, 108505. [CrossRef]

39. Huang, Y.; Cui, J.; Zhong, Z.; Gan, C.; Zhang, W.; Song, H. Synthesis and cytotoxicity of 17a-aza-d-homo-androster-17-one derivatives. *Bioorg. Med. Chem. Lett.* **2011**, *21*, 3641–3643. [CrossRef]

40. Cepa, M.; Tavares da Silva, E.J.; Correia-da-Silva, G.; Roleira, F.M.F.; Teixeira, N.A.A. Synthesis and biochemical studies of 17-substituted androst-3-enes and 3,4-epoxyandrostanes as aromatase inhibitors. *Steroids* **2008**, *73*, 1409–1415. [CrossRef]

41. Andrade, L.C.R.; de Almeida, M.J.M.; Roleira, F.M.F.; Varela, C.L.; da Silva, E.J.T. 5α-Androst-3-en-17-one oxime. *Acta Cryst. C* **2008**, *64*, o508–o510. [CrossRef] [PubMed]

42. Ajduković, J.J.; Jakimov, D.S.; Rárová, L.; Strnad, M.; Dzichenka, Y.U.; Usanov, S.; Škorić, D.; Jovanović-Šanta, S.S.; Sakač, M.N. Novel alkylaminoethyl derivatives of androstane 3-oximes as anticancer candidates: Synthesis and evaluation of cytotoxic effects. *RSC Adv.* **2021**, *11*, 37449–37461. [CrossRef] [PubMed]

43. Yao, J.; Ye, W.; Liu, J.; Liu, J.; Wang, C. Synthesis and cytotoxicity of (3β)-3-acetyloxy-5(6)-androsten-7-one oxime and 3,5(6)-androstadien-7-one oxime. *Med. Chem. Res.* **2014**, *23*, 1839–1843. [CrossRef]

44. Dubey, P.P.A.D.P.J.S.; Piplani, P.; Jindal, D. Synthesis and Evaluation of Some 16-Benzylidene Substituted 3,17- Dioximino Androstene Derivatives as Anticancer Agents. *Lett. Drug Des. Discov.* **2005**, *2*, 537–545. [CrossRef]

45. Acharya, P.C.; Bansal, R. Synthesis and Antiproliferative Activity of Some Androstene Oximes and Their O-Alkylated Derivatives. *Arch. Pharm.* **2013**, *347*, 193–199. [CrossRef]

46. Jindal, D.P.; Chattopadhaya, R.; Guleria, S.; Gupta, R. Synthesis and antineoplastic activity of 2-alkylaminoethyl derivatives of various steroidal oximes. *Eur. J. Med. Chem.* **2003**, *38*, 1025–1034. [CrossRef]

47. Bansal, R.; Guleria, S.; Ries, C.; Hartmann, R.W. Synthesis and Antineoplastic Activity of O-Alkylated Derivatives of 7-Hydroximinoandrost-5-ene Steroids. *Arch. Pharm.* **2010**, *343*, 377–383. [CrossRef] [PubMed]

48. Ajduković, J.J.; Penov Gaši, K.M.; Jakimov, D.S.; Klisurić, O.R.; Jovanović-Šanta, S.S.; Sakač, M.N.; Aleksić, L.D.; Djurendić, E.A. Synthesis, structural analysis and antitumor activity of novel 17α-picolyl and 17(E)-picolinylidene A-modified androstane derivatives. *Bioorg. Med. Chem.* **2015**, *23*, 1557–1568. [CrossRef]

49. Savić, M.P.; Djurendić, E.A.; Petri, E.T.; Ćelić, A.; Klisurić, O.R.; Sakač, M.N.; Jakimov, D.S.; Kojićd, V.V.; Gašia, K.M.P. Synthesis, structural analysis and antiproliferative activity of some novel D-homo lactone androstane derivatives. *RSC Adv.* **2013**, *3*, 10385–10395. [CrossRef]

50. Savić, M.P.; Klisurić, O.R.; Penov Gaši, K.M.; Jakimov, D.S.; Sakač, M.N.; Djurendić, E.A. Synthesis, structural analysis and cytotoxic activity of novel A- and B-modified D-homo lactone androstane derivative. *J. Chem. Crystallogr.* **2016**, *46*, 84–92. [CrossRef]

51. Purohit, A.; Woo, L.; Chander, S.; Newman, S.; Ireson, C.; Ho, Y.; Grasso, A.; Leese, M.; Potter, B.; Reed, M. Steroid sulphatase inhibitors for breast cancer therapy. *J. Steroid Biochem. Mol. Biol.* **2003**, *86*, 423–432. [CrossRef] [PubMed]

52. Leese, M.P.; Leblond, B.; Newman, S.P.; Purohit, A.; Reed, M.J.; Potter, B.V. Anti-cancer activities of novel D-ring modified 2-substituted estrogen-3-O-sulfamates. *J. Steroid Biochem. Mol. Biol.* **2005**, *94*, 239–251. [CrossRef] [PubMed]

53. Berényi, Á.; Minorics, R.; Iványi, Z.; Ocsovszki, I.; Ducza, E.; Thole, H.; Messinger, J.; Wölfling, J.; Mótyán, G.; Mernyák, E.; et al. Synthesis and investigation of the anticancer effects of estrone-16-oxime ethers in vitro. *Steroids* **2013**, *78*, 69–78. [CrossRef] [PubMed]

54. Mernyák, E.; Fiser, G.; Szabó, J.; Bodnár, B.; Schneider, G.; Kovács, I.; Ocsovszki, I.; Zupkó, I.; Wölfling, J. Synthesis and in vitro antiproliferative evaluation of d-secooxime derivatives of 13β- and 13α-estrone. *Steroids* **2014**, *89*, 47–55. [CrossRef] [PubMed]

55. Cushman, M.; He, H.-M.; Katzenellenbogen, J.A.; Varma, R.K.; Hamel, E.; Lin, C.M.; Ram, S.; Sachdeva, Y.P. Synthesis of Analogs of 2-Methoxyestradiol with Enhanced Inhibitory Effects on Tubulin Polymerization and Cancer Cell Growth. *J. Med. Chem.* **1997**, *40*, 2323–2334. [CrossRef]

56. Cushman, M.; He, H.M.; Katzenellenbogen, J.A.; Lin, C.M.; Hamel, E. Synthesis, antitubulin and antimitotic activity, and cytotoxicity of analogs of 2-methoxyestradiol, an endogenous mammalian metabolite of estradiol that inhibits tubulin polymerization by binding to the colchicine binding site. *J. Med. Chem.* **1995**, *38*, 2041–2049. [CrossRef]

57. Canário, C.; Matias, M.; Brito, V.; Santos, A.; Falcão, A.; Silvestre, S.; Alves, G. New Estrone Oxime Derivatives: Synthesis, Cytotoxic Evaluation and Docking Studies. *Molecules* **2021**, *26*, 2687. [CrossRef]

58. Choudhary, M.I.; Alam, M.S.; Yousuf, S.; Wu, Y.C.; Lin, A.S.; Shaheen, F. Pregnenolone derivatives as potential anticancer agents. *Steroids* **2011**, *76*, 1554–1559. [CrossRef]

59. Banday, A.H.; Akram, S.; Shameem, S.A. Benzylidine pregnenolones and their oximes as potential anticancer agents: Synthesis and biological evaluation. *Steroids* **2014**, *84*, 64–69. [CrossRef]

60. Krstić, N.M.; Bjelaković, M.S.; Dabović, M.M.; Lorenc, L.B.; Pavlović, V.D. Photochemical and Beckmann rearrangement of (Z)-cholest-4-en-6-one oxime. *J. Serb. Chem. Soc.* **2004**, *69*, 413–420. [CrossRef]

61. Krstić, N.M.; Bjelaković, M.S.; Žižak, Ž.; Pavlović, M.D.; Juranić, Z.D.; Pavlović, V.D. Synthesis of some steroidal oximes, lactams, thiolactams and their antitumor activities. *Steroids* **2007**, *72*, 406–414. [CrossRef] [PubMed]

62. Mora-Medina, T.L.; Martínez-Pascual, R.; Peña-Rico, M.; Viñas-Bravo, O.; Montiel-Smith, S.; Pérez-Picaso, L.; Moreno-Díaz, H. Preparation and cytotoxic evaluation of new steroidal oximes and aza-homosteroids from diosgenin and cholesterol. *Steroids* **2022**, *182*, 109012. [CrossRef] [PubMed]

63. Huang, Y.; Cui, J.; Chen, S.; Gan, C.; Zhou, A. Synthesis and antiproliferative activity of some steroidal lactams. *Steroids* **2011**, *76*, 1346–1350. [CrossRef] [PubMed]

64. Huang, Y.; Cui, J.; Zheng, Q.; Zeng, C.; Chen, Q.; Zhou, A. 6-Hydroximino-4-aza-A-homo-cholest-3-one and related analogue as a potent inducer of apoptosis in cancer cells. *Steroids* **2012**, *77*, 829–834. [CrossRef] [PubMed]

65. Vejux, A.; Lizard, G. Cytotoxic effects of oxysterols associated with human diseases: Induction of cell death (apoptosis and/or oncosis), oxidative and inflammatory activities, and phospholipidosis. *Mol. Asp. Med.* **2009**, *30*, 153–170. [CrossRef]

66. Carvalho, J.F.S.; Silva, M.M.C.; Moreira, J.N.; Simões, S.; e Melo, M.L.S. Selective Cytotoxicity of Oxysterols through Structural Modulation on Rings A and B. Synthesis, in Vitro Evaluation, and SAR. *J. Med. Chem.* **2011**, *54*, 6375–6393. [CrossRef]

67. Rodriguez, J.; Nufiez, L.; Peixinho, S.; Jimnez, C. Isolation and synthesis of the first natural 6-hydroximino 4-en-3-one-steroids from the sponges *Cinachyrella* spp. *Tetrahedron Lett.* **1997**, *38*, 1833–1836. [CrossRef]

68. Deive, N.; Rodríguez, J.; Jiménez, C. Synthesis of Cytotoxic 6E-Hydroximino-4-ene Steroids: Structure/Activity Studies. *J. Med. Chem.* **2001**, *44*, 2612–2618. [CrossRef]

69. Poza, J.; Rega, M.; Paz, V.; Alonso, B.; Rodríguez, J.; Salvador, N.; Fernández, A.; Jiménez, C. Synthesis and evaluation of new 6-hydroximinosteroid analogs as cytotoxic agents. *Bioorg. Med. Chem.* **2007**, *15*, 4722–4740. [CrossRef]

70. Cui, J.; Huang, L.; Fan, L.; Zhou, A. A facile and efficient synthesis of some (6E)-hydroximino-4-en-3-one steroids, steroidal oximes from *Cinachyrella* spp. sponges. *Steroids* **2008**, *73*, 252–256. [CrossRef]

71. Cui, J.G.; Fan, L.; Huang, L.L.; Liu, H.L.; Zhou, A.M. Synthesis and evaluation of some steroidal oximes as cytotoxic agents: Structure/activity studies (I). *Steroids* **2009**, *74*, 62–72. [CrossRef] [PubMed]

72. Cui, J.; Fan, L.; Huang, Y.; Xin, Y.; Zhou, A. Synthesis and evaluation of some steroidal oximes as cytotoxic agents: Structure/activity studies (II). *Steroids* **2009**, *74*, 989–995. [CrossRef] [PubMed]

73. Gan, C.; Cui, J.; Su, S.; Lin, Q.; Jia, L.; Fan, L.; Huang, Y. Synthesis and antiproliferative activity of some steroidal thiosemicarbazones, semicarbazones and hydrozones. *Steroids* **2014**, *87*, 99–107. [CrossRef]

74. Huang, Y.; Cui, J.; Li, Y.; Fan, L.; Jiao, Y.; Su, S. Syntheses and antiproliferative activity of some sulfated hydroximinosterols. *Med. Chem. Res.* **2012**, *22*, 409–414. [CrossRef]

75. Richmond, V.; Careaga, V.P.; Sacca, P.; Calvo, J.C.; Maier, M.S. Synthesis and cytotoxic evaluation of four new 6E-hydroximinosteroids. *Steroids* **2014**, *84*, 7–10. [CrossRef]

76. Huang, Y.; Cui, J.; Chen, S.; Lin, Q.; Song, H.; Gan, C.; Su, B.; Zhou, A. Synthesis and Evaluation of Some New Aza-B-homocholestane Derivatives as Anticancer Agents. *Mar. Drugs* **2014**, *12*, 1715–1731. [CrossRef] [PubMed]

77. Tsukomoto, T.; Ueno, Y.; Ohta, Z. Constitution of diosgenin I. Glucoside of Dioscorea tokoro Makino. *J. Pharm. Soc. Jpn.* **1936**, *57*, 985–991.

78. Sánchez-Sánchez, L.; Hernández-Linares, M.G.; Escobar, M.L.; López-Muñoz, H.; Zenteno, E.; Fernández-Herrera, M.A.; Guerrero-Luna, G.; Carrasco-Carballo, A.; Sandoval-Ramírez, J. Antiproliferative, Cytotoxic, and Apoptotic Activity of Steroidal Oximes in Cervicouterine Cell Lines. *Molecules* **2016**, *21*, 1533. [CrossRef]

79. Martínez-Gallegos, A.A.; Guerrero-Luna, G.; Ortiz-González, A.; Cárdenas-García, M.; Bernès, S.; Hernández-Linares, M.G. Azasteroids from diosgenin: Synthesis and evaluation of their antiproliferative activity. *Steroids* **2020**, *166*, 108777. [CrossRef]

80. Carrasco-Carballo, A.; Guadalupe Hernández-Linares, M.; Cárdenas-García, M.; Sandoval-Ramírez, J. Synthesis and biological in vitro evaluation of the effect of hydroxyimino steroidal derivatives on breast cancer cells. *Steroids* **2021**, *166*, 108787. [CrossRef]

81. Martínez-Pascual, R.; Meza-Reyes, S.; Vega-Baez, J.L.; Merino-Montiel, P.; Padrón, J.M.; Mendoza, Á.; Montiel-Smith, S. Novel synthesis of steroidal oximes and lactams and their biological evaluation as antiproliferative agents. *Steroids* **2017**, *122*, 24–33. [CrossRef] [PubMed]

82. Salunke, D.; Hazra, B.G.; Pore, V.S. Steroidal Conjugates and Their Pharmacological Applications. *Curr. Med. Chem.* **2006**, *13*, 813–847. [CrossRef]

83. Huang, Y.; Chen, S.; Cui, J.; Gan, C.; Liu, Z.; Wei, Y.; Song, H. Synthesis and cytotoxicity of A-homo-lactam derivatives of cholic acid and 7-deoxycholic acid. *Steroids* **2011**, *76*, 690–694. [CrossRef] [PubMed]

84. The Top 10 Causes of Death—Factsheet. Available online: https://www.who.int/news-room/fact-sheets/detail/the-top-10 -causes-of-death (accessed on 25 October 2022).

85. Hanson, J.R. Steroids: Partial synthesis in medicinal chemistry. *Nat. Prod. Rep.* **2010**, *27*, 887–899. [CrossRef] [PubMed]

86. Amiranashvili, L.; Nadaraia, N.; Merlani, M.; Kamoutsis, C.; Petrou, A.; Geronikaki, A.; Pogodin, P.; Druzhilovskiy, D.; Poroikov, V.; Ciric, A.; et al. Antimicrobial Activity of Nitrogen-Containing 5-α-Androstane Derivatives: In Silico and Experimental Studies. *Antibiotics* **2020**, *9*, 224. [CrossRef]

87. Prabpayak, W.; Charoenying, P.; Laosinwattana, C.; Aroonrerk, N. Antibacterial activity of pregnenolone derivatives. *Curr. Appl. Sci. Technol.* **2006**, *6*, 466–470.

88. Lone, I.H.; Khan, K.Z.; Fozdar, B.I.; Hussain, F. Synthesis antimicrobial and antioxidant studies of new oximes of steroidal chalcones. *Steroids* **2013**, *78*, 945–950. [CrossRef]

89. Alam, M.; Lee, D.-U. Green synthesis, biochemical and quantum chemical studies of steroidal oximes. *Korean J. Chem. Eng.* **2014**, *32*, 1142–1150. [CrossRef]

90. Brunel, J.; Loncle, C.; Vidal, N.; Dherbomez, M.; Letourneux, Y. Synthesis and antifungal activity of oxygenated cholesterol derivatives. *Steroids* **2005**, *70*, 907–912. [CrossRef]

91. Shingate, B.B.; Hazra, B.G.; Salunke, D.B.; Pore, V.S.; Shirazi, F.; Deshpande, M.V. Synthesis and antimicrobial activity of novel oxysterols from lanosterol. *Tetrahedron* **2013**, *69*, 11155–11163. [CrossRef]

92. Pujol, C.A.; Sepúlveda, C.S.; Richmond, V.; Maier, M.S.; Damonte, E.B. Polyhydroxylated sulfated steroids derived from 5α-cholestanes as antiviral agents against herpes simplex virus. *Arch. Virol.* **2016**, *161*, 1993–1999. [CrossRef] [PubMed]

93. You, C.R.; Lee, S.W.; Jang, J.W.; Yoon, S.K. Update on hepatitis B virus infection. *World J. Gastroenterol. WJG* **2014**, *20*, 13293. [CrossRef] [PubMed]

94. Wei, Z.; Tan, J.; Cui, X.; Zhou, M.; Huang, Y.; Zang, N.; Chen, Z.; Wei, W. Design, Synthesis and Bioactive Evaluation of Oxime Derivatives of Dehydrocholic Acid as Anti-Hepatitis B Virus Agents. *Molecules* **2020**, *25*, 3359. [CrossRef] [PubMed]

95. Zeferino-Díaz, R.; Olivera-Castillo, L.; Dávalos, A.; Grant, G.; Kantún-Moreno, N.; Rodriguez-Canul, R.; Bernès, S.; Sandoval-Ramírez, J.; Fernández-Herrera, M.A. 22-Oxocholestane oximes as potential anti-inflammatory drug candidates. *Eur. J. Med. Chem.* **2019**, *168*, 78–86. [CrossRef] [PubMed]

96. Bjedov, S.; Bekic, S.; Marinovic, M.; Skoric, D.; Pavlovic, K.; Celic, A.; Petri, E.; Sakac, M. Screening the binding affinity of bile acid derivatives for the glucocorticoid receptor ligand-binding domain. *J. Serb. Chem. Soc.* **2023**, *88*, 123–139. [CrossRef]

97. The Safety and Efficacy of Istaroxime for Pre-Cardiogenic Chock—Full Text View—ClinicalTrials.gov. Available online: https://clinicaltrials.gov/ct2/show/NCT04325035?term=istaroxime&draw=2&rank=2 (accessed on 26 October 2022).

98. Aditya, S.; Rattan, A. Istaroxime: A rising star in acute heart failure. *J. Pharmacol. Pharmacother.* **2012**, *3*, 353–355. [CrossRef]

99. Stagno, M.J.; Zacharopoulou, N.; Bochem, J.; Tsapara, A.; Pelzl, L.; Al-Maghout, T.; Kallergi, G.; Alkahtani, S.; Alevizopoulos, K.; Dimas, K.; et al. Istaroxime Inhibits Motility and Down-Regulates Orai1 Expression, SOCE and FAK Phosphorylation in Prostate Cancer Cells. *Cell. Physiol. Biochem.* **2017**, *42*, 1366–1376. [CrossRef]

100. Alevizopoulos, K.; Dimas, K.; Papadopoulou, N.; Schmidt, E.M.; Tsapara, A.; Alkahtani, S.; Honisch, S.; Prousis, K.C.; Alarifi, S.; Calogeropoulou, T.; et al. Functional characterization and anti-cancer action of the clinical phase II cardiac Na+/K+ ATPase inhibitor istaroxime: In vitro and in vivo properties and cross talk with the membrane androgen receptor. *Oncotarget* **2016**, *7*, 24415–24428. [CrossRef]

101. Bringer, J. Norgestimate: A clinical overview of a new progestin. *Am. J. Obstet. Gynecol.* **1992**, *166*, 1969–1979. [CrossRef]

102. Henzl, M.R. Norgestimate. From the laboratory to three clinical indications. *J. Reprod Med.* **2001**, *46*, 647–661.

103. Yoshii, Y.; Okuda, K.I.; Yamada, S.; Nagakura, M.; Sugimoto, S.; Nagano, T.; Okabe, T.; Kojima, H.; Iwamoto, T.; Kuwano, K.; et al. Norgestimate inhibits staphylococcal biofilm formation and resensitizes methicillin-resistant Staphylococcus aureus to β-lactam antibiotics. *NPJ Biofilms Microbiomes* **2017**, *3*, 18. [CrossRef] [PubMed]

104. Philibert, D.; Bouchoux, F.; Degryse, M.; Lecaque, D.; Petit, F.; Gaillard, M. The pharmacological profile of a novel norpregnane progestin (trimegestone). *Gynecol. Endocrinol.* **1999**, *13*, 316–326. [CrossRef] [PubMed]

Review

Biological Activity and Structural Diversity of Steroids Containing Aromatic Rings, Phosphate Groups, or Halogen Atoms

Valery M. Dembitsky [†] 

Centre for Applied Research, Innovation and Entrepreneurship, Lethbridge College, 3000 College Drive South, Lethbridge, AB T1K 1L6, Canada; valery.dembitsky@lethbridgecollege.ca or dvmioch@gmail.com
[†] Current address: Bio-Geo-Chem Laboratories, 23615 El Toro Rd X, POB 049, Lake Forest, CA 92630, USA.

Abstract: This review delves into the investigation of the biological activity and structural diversity of steroids and related isoprenoid lipids. The study encompasses various natural compounds, such as steroids with aromatic ring(s), steroid phosphate esters derived from marine invertebrates, and steroids incorporating halogen atoms (I, Br, or Cl). These compounds are either produced by fungi or fungal endophytes or found in extracts of plants, algae, or marine invertebrates. To assess the biological activity of these natural compounds, an extensive examination of referenced literature sources was conducted. The evaluation encompassed in vivo and in vitro studies, as well as the utilization of the QSAR method. Numerous compounds exhibited notable properties such as strong anti-inflammatory, anti-neoplastic, anti-proliferative, anti-hypercholesterolemic, anti-Parkinsonian, diuretic, anti-eczematic, anti-psoriatic, and various other activities. Throughout the review, 3D graphs illustrating the activity of individual steroids are presented alongside images of selected terrestrial or marine organisms. Additionally, the review provides explanations for specific types of biological activity associated with these compounds. The data presented in this review hold scientific interest for academic science as well as practical implications in the fields of pharmacology and practical medicine. The analysis of the biological activity and structural diversity of steroids and related isoprenoid lipids provides valuable insights that can contribute to advancements in both theoretical understanding and applied research.

Keywords: steroids; triterpenoids; isoprenoid lipids; anti-neoplastic; anti-inflammatory; anti-fungal; anti-bacterial; anti-viral; fungal endophytes; plants; marine invertebrates

Citation: Dembitsky, V.M. Biological Activity and Structural Diversity of Steroids Containing Aromatic Rings, Phosphate Groups, or Halogen Atoms. *Molecules* **2023**, *28*, 5549. https://doi.org/10.3390/molecules28145549

Academic Editor: Pierangela Ciuffreda

Received: 27 June 2023
Revised: 10 July 2023
Accepted: 13 July 2023
Published: 20 July 2023

1. Introduction

Natural steroids belong to the class of isoprenoid lipids [1,2]. These metabolites, which can originate from animals, fungi, and plants, exhibit high biological activity and contain a sterane skeleton composed of isoprenoid precursors [3–6]. Steroids are characterized by the presence of a fused tetracyclic system, such as androstane (**1A**) and related structures, estrane (**1B**), gonane (**1C**), cholestane (**2**), and protostane (**3**) (refer to Figure 1 for their structures) [7,8]. The androstane, cholestane, and/or protostane cores in steroids or triterpenoids can be saturated or partially unsaturated and may incorporate alkyl, hydroxyl, carbonyl, or carboxyl groups [7–9]. Isoprenoid lipids, on the other hand, are natural metabolites derived from isoprene molecules and serve various physiological functions while exhibiting a wide range of biological activities [1–6].

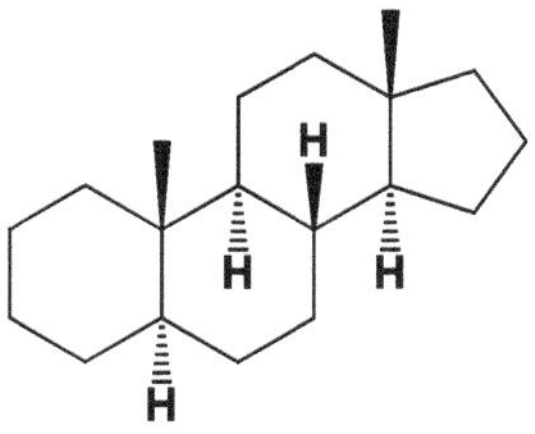

1A 5α-Androstane

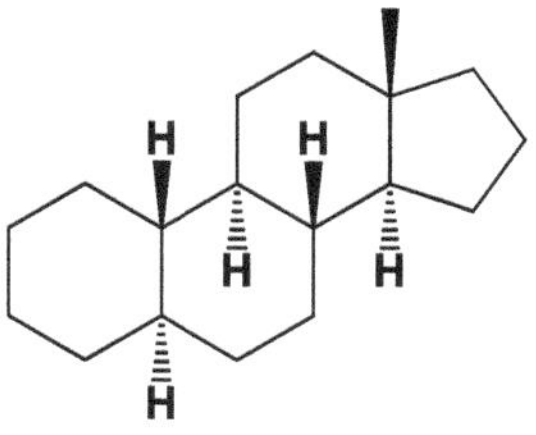

1B 5α-Estrane

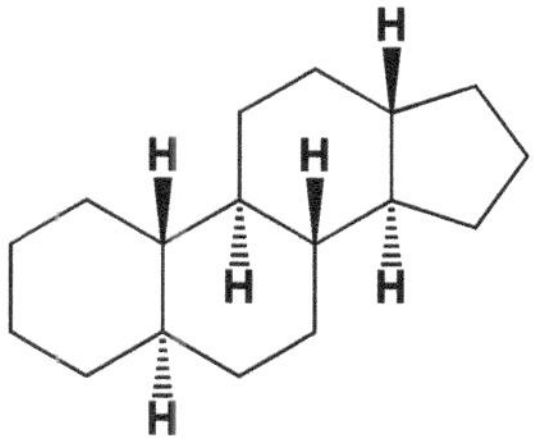

1C 5α-Gonane

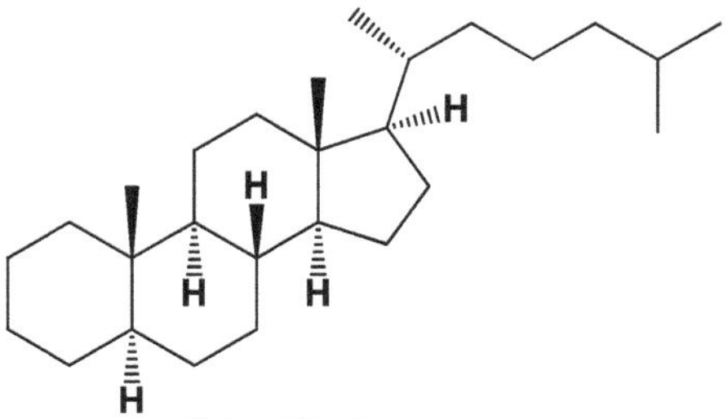

2 5α-Cholestane

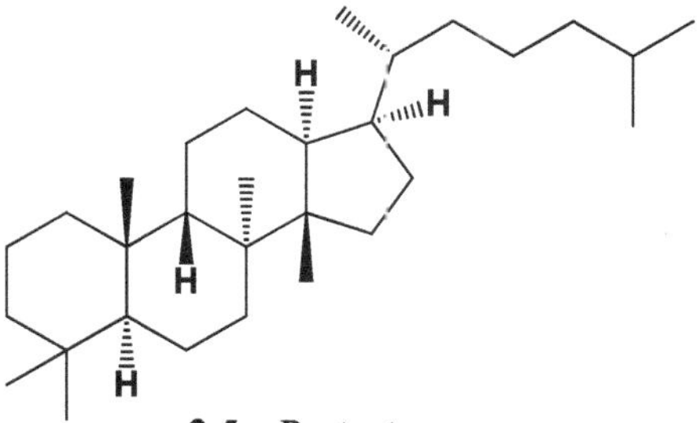

3 5α-Protostane

Figure 1. Androstane, cholestane, and protostane are steroid or triterpenoid core structures.

Protostane-type triterpenoids, predominantly found in plants of the genus *Alisma*, exhibit diverse carbon skeletons and intriguing biological activities [10]. Furthermore, marine- and plant-derived steroids can incorporate various halogens, including chlorine, bromine, or iodine [11–14]. Notably, seaweeds possess significant nutritional value and have been integral to the diets of many cultures throughout history (depicted in Figure 2). Seaweed extracts are notably abundant in natural growth hormones, known as phytosterols, as well as essential nutrients and trace elements. Algal-derived sterols contribute substantially as the principal lipid component of plant cell membranes and display a broad spectrum of biological activities [15–20].

This review provides an overview of the biological activities of steroids and isoprenoid lipids derived from diverse natural sources. Given the extensive number of natural steroids and isoprenoid lipids, we have focused on compounds with established biological activities through experimental studies and computational analyses. This selection aims to cater to pharmacologists, chemists, and researchers from various disciplines who utilize steroids for medicinal purposes.

Figure 2. The red and brown algae macrophytes are abundant sources of biologically active metabo-lites, including steroids. Here, we highlight some representative examples: (**a**) *Laurencia pacifica* (red alga, Rhodophyceae). This species is known for producing halogenated metabolites, such as sesquiterpenes, diterpenes, triterpenes, and C15 acetogenins. (**b**) *Laminaria digitata* (brown alga, Phaeophyceae). A commonly consumed brown algae, particularly in coastal regions, with kelp that is rich in terpenoids, essential amino acids, polyunsaturated fatty acids, carbohydrates, vitamins, and minerals such as iron and calcium. (**c**) *Sargassum* sp. (brown alga, Fucales). Various species of the *Sargassum* genus are utilized for human nutrition and serve as a valuable source of steroids, proteins, vitamins, carotenoids, and minerals, and Professor Dembitsky collected the biological material in Southern California, summer 2018. (**d**) *Ulva lactuca* (sea lettuce, green alga, Ulvaceae). Cultivated in China, Republic of Korea, and Japan, sea lettuce is consumed by manatees, sea slugs, and shellfish. Extracts of this edible green algae contain bioactive components, including steroids and triterpenoids. (**e**) *Enteromorpha intestinalis* (green bait, sea lettuce, green alga, Ulvaceae). This green alga, commonly known as green bait or sea lettuce, produces a wide range of terpenoids, including steroids. (**f**) *Gracilaria pacifica* (red spaghetti, red alga). Widely used in the cosmetic industry for shampoos, creams, soaps, and sunscreens, this red alga contains terpenoids and carotenoids and serves as a source of high-quality agar.

2. Steroids Bearing Aromatic Ring(s)

Steroids bearing aromatic rings are a distinct subgroup within the larger family of steroids, which are characterized by their fused ring structure [21–23]. The presence of one or more aromatic rings in these steroids imparts unique chemical and biological properties, making them of particular interest in various fields of research, including medicinal chemistry and drug discovery. Steroids bearing aromatic rings represent a fascinating subgroup of steroids that possess distinct chemical and biological characteristics [22–24]. Their unique structural features and diverse pharmacological profiles make them promising candidates for drug development and therapeutic applications. Continued research in this field will expand our knowledge of their biological activities and unlock their potential in various areas of medicine and biology. Natural steroids and triterpenoids that contain one or more aromatic rings in their structure are referred to as aromatic steroids. They are a diverse group of lipid molecules synthesized by bacteria, fungi, plants, invertebrates, and animals [21–26]. These aromatic steroids have been identified in various sources, including geological samples, marine sediments, and oil [27–31].

A comprehensive analysis of the literature reveals that the most prevalent subgroup among natural lipids is mono-aromatic steroids and triterpenoids, with an aromatic ring in either position A (approximately 200 metabolites) or position B (around 20 steroids) [32]. Additionally, a small number of di-aromatic steroids have been identified in living organisms, geological samples, marine sediments, and oil, while only a few tri-aromatic steroid hydrocarbons have been found in living organisms, marine sediments, and oil [28–30,33,34].

2.1. Steroids Bearing Aromatic Ring A in Plants

Steroids bearing an aromatic ring in position A (aromatic ring A) are commonly found in plants, and this contributes to their diverse biological activities. These aromatic steroids play important roles in plant growth, development, and defense mechanisms. Here, we explore the occurrence and functions of steroids with aromatic ring A in plants.

Estrone (**4**, or estra-1,3,5(10)-triene-3-ol-17-one), estradiol (**5**), estriol (**6**), equilin (**7**), hippulin (**8**), and their derivatives (**9**, **10**, **11**, and **12**) represent the well-known examples of mono-aromatic steroids (refer to Figure 3 for their structures). Table 1 provides an overview of their biological activities. Estrone, a female sex hormone, was initially discovered in the 1920s by independent groups of scientists from the USA and Germany [35–39].

Female sex hormonal steroids, specifically estrogens (**4–10**), were initially discovered in plants in 1926 by Dohrn and colleagues [40]. Subsequently, other researchers also identified these compounds [41–43]. It is noteworthy that hormones such as 17β-estradiol, androsterone, testosterone, and progesterone were found in approximately 80% of the plant species investigated [41]. Estrone (**4**) has been isolated from various plant sources, including the seeds and pollen of *Glossostemon bruguieri*, *Hyphaene thebaica*, *Malus pumila*, *Phoenix dactylifera*, *Punica granatum*, and *Salix caprea*. A sample plant (*Glossostemon bruguieri*) is depicted in Figure 4. Additionally, 17β-estradiol (**8**) was found in the seeds of *Phaseolus vulgaris*, along with estrone (**4**). The distribution of biological activity, exemplified by estrone, is shown in Figure 5. Furthermore, estriol (**6**) has been identified in *Glycyrrhiza glabra* and *Salix* sp. [41–43].

Various plant species, including *Brassica campestris*, *Ginkgo biloba*, *Lilium davidii*, and *Zea mays*, have been found to contain total estrogens (**4–7**) and 17β-estradiol (**8**) in their pollen and style [44]. Additionally, testosterone has been detected in the pollen of *Pinus bungeana*, *Ginkgo biloba*, and *P. tabulaeformis* [45]. Furthermore, holaromine (**13**), a steroidal alkaloid, has been isolated from the ornamental shrub *Holarrhena floribunda* [46]. Figure 6 illustrates a 3D graph showcasing the predicted and calculated activity of estrone (**4**) as an ovulation inhibitor.

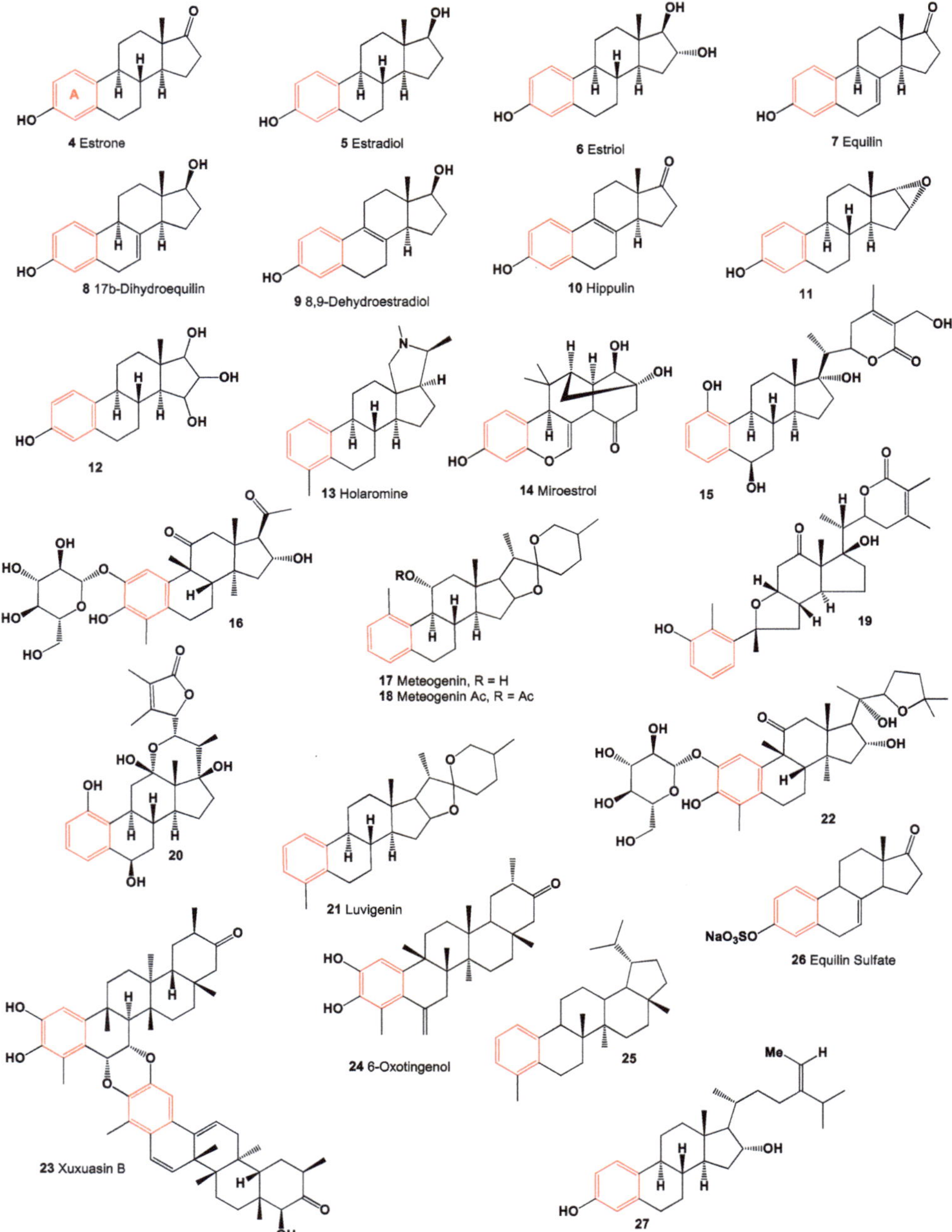

Figure 3. Steroids bearing ring A derived from plants.

Table 1. Biological activities of mono-aromatic steroids (**4–27**).

No.	Dominated Biological Activity (Pa) *	Additional Predicted Activities (Pa) *
4	Ovulation inhibitor (0.942) Cardiovascular analeptic (0.924) Apoptosis agonist (0.750)	Anti-hypercholesterolemic (0.871) Lipid metabolism regulator (0.788) Prostate disorders treatment (0.737)
5	Anti-hypercholesterolemic (0.894) Ovulation inhibitor (0.889) Anesthetic general (0.868)	Respiratory analeptic (0.851) Acute neurologic disorders treatment (0.793) Prostate disorders treatment (0.729)
6	Anesthetic general (0.845) Ovulation inhibitor (0.832)	Acute neurologic disorders treatment (0.822) Neuroprotector (0.815)
7	Anti-hypercholesterolemic (0.856) Ovulation inhibitor (0.847) Cardiovascular analeptic (0.842)	Lipid metabolism regulator (0.788) Apoptosis agonist (0.750) Prostate disorders treatment (0.725)
8	Anti-hypercholesterolemic (0.885) Apoptosis agonist (0.801)	Hepatic disorders treatment (0.739) Ovulation inhibitor (0.726)
9	Acute neurologic disorders treatment (0.871) Respiratory analeptic (0.843) Vasoprotector (0.811)	Neuroprotector (0.785) Anesthetic general (0.753) Ovulation inhibitor (0.740)
10	Cardiovascular analeptic (0.882) Ovulation inhibitor (0.860)	Respiratory analeptic (0.846) Acute neurologic disorders treatment (0.844)
11	Respiratory analeptic (0.879) Ovulation inhibitor (0.765)	Neuroprotector (0.762) Cardiovascular analeptic (0.692)
12	Acute neurologic disorders treatment (0.849) Vasoprotector (0.795)	Anti-inflammatory (0.788) Ovulation inhibitor (0.778)
13	Psychotropic (0.815) Ovulation inhibitor (0.586)	Attention deficit/hyperactivity disorder treatment (0.744)
14	Postmenopausal disorders treatment (0.945)	Anti-inflammatory (0.669)
15	Lipid metabolism regulator (0.913) Cytostatic (0.891) Anti-neoplastic (0.876)	Hepatoprotectant (0.845) Immunosuppressant (0.792) Apoptosis agonist (0.784)
16	Chemopreventive (0.919) Proliferative diseases treatment (0.914)	Anti-neoplastic (0.337) Vasoprotector (0.824)
17	Apoptosis agonist (0.893) Anti-neoplastic (0.827)	Anti-inflammatory (0.873) Hypolipemic (0.854)
18	Apoptosis agonist (0.883) Anti-neoplastic (0.826)	Hypolipemic (0.863) Anti-inflammatory (0.855)
19	Anti-neoplastic (0.879) Apoptosis agonist (0.775)	Immunosuppressant (0.744) Anti-inflammatory (0.715)
20	Anti-neoplastic (0.782)	Genital warts treatment (0.736)
21	Apoptosis agonist (0.896) Anti-neoplastic (0.843)	Hypolipemic (0.850) Anti-inflammatory (0.814)
22	Chemopreventive (0.887) Anti-neoplastic (0.794)	Anti-inflammatory (0.819) Proliferative diseases treatment (0.784)
23	Anti-neoplastic (0.909) Apoptosis agonist (0.790)	Anti-inflammatory (0.822) Immunosuppressant (0.727)
24	Anti-neoplastic (0.888) Apoptosis agonist (0.847)	Anti-inflammatory (0.830) Immunosuppressant (0.739)
25	Anti-neoplastic (0.802) Apoptosis agonist (0.789)	Anti-inflammatory (0.786) Prostate disorders treatment (0.685)
26	Acute neurologic disorders treatment (0.867) Anti-neoplastic (0.812)	Diuretic (0.813) Male reproductive dysfunction treatment (0.759)
27	Anti-hypercholesterolemic (0.959)	Anti-neoplastic (0.832)

* Only activities with Pa > 0.7 are shown. The main biological activity has a value where Pa is more than 0.7.

Figure 4. (**a**) *Glossostemon bruguieri*: *G. bruguieri* (also known as Moghat) is a shrub native to Iraq and Iran. In the past, it was cultivated in Egypt for its edible roots. The dried and peeled roots have been used in folk medicine to treat conditions such as gout and spasms and as a tonic and nourishment. Additionally, powdered Moghat has been traditionally consumed as a tonic and lactagogic remedy by women after childbirth. (**b**) *Hyphaene thebaica*: *H. thebaica* is a plant species commonly known as doum palm. It is native to regions of Africa and the Middle East. The seeds and pollen of *H. thebaica* are a source of estrone (**4**). (**c**) *Malus pumila*: *M. pumila*, commonly known as apple, is a fruit-bearing tree cultivated worldwide. Estrone (**4**) has been isolated from the seeds and pollen of *M. pumila*. (**d**) *Punica granatum*: *P. granatum*, or pomegranate, is a fruit-bearing shrub or small tree. It has been associated with various health benefits and may help prevent or treat conditions such as high blood pressure, high cholesterol, oxidative stress, hyperglycemia, and inflammatory activity. Estrone (**4**) has been found in *P. granatum*. Note: all photos used in this figure are obtained from sites where permission is granted for non-commercial use.

Deoxymiroestrol (**14**), a phytoestrogen, has been isolated from the Thai herb *Pueraria mirifica* [47]. Withanolides (**15**, **19**, and **20**), which are steroids, have been found in various parts of different plants [48]. Jaborosalactone-7 was extracted from *Jaborosa leucotricha*, while jaborosalactone-45 was identified in *Jaborosa laciniata* [49]. In the extract of *Fevillea trilobata* seeds, andirobicin B glucoside (**16**) was discovered [50]. Furthermore, 1-methyl-19-nor-25-D-spirosta-1,3,5(10)-trien-11α-ol (**17**) and its acetate (**18**) were found in the rhizome of *Metanarthecium luteoviride* [51]. The predicted biological activity for mono-aromatic steroids isolated from plants is presented in Table 1. Additionally, Figure 7 illustrates a 3D graph depicting the predicted and calculated anti-neoplastic activity of mono-aromatic ring A plant steroids (**16**, **17**, **21**, **23**, and **24**).

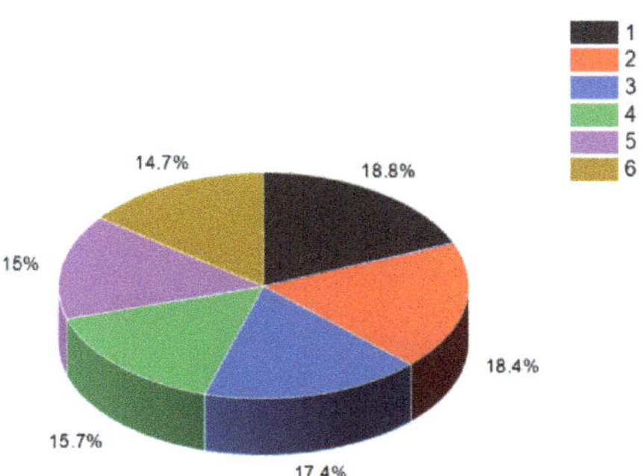

Figure 5. The percentage distribution of various biological activities associated with estrone (**4**), a compound known for its diverse pharmacological properties. The activities and their corresponding percentages are as follows: (1) ovulation inhibitor (18.8%); (2) cardiovascular analeptic (18.4%); (3) anti-hypercholesterolemic (17.4%); (4) apoptosis agonist (15.7%); (5) lipid metabolism regulator (15%); (6) prostate disorders treatment (14.7%). Estrone (**4**), which is a steroid bearing an aromatic ring A, is present in the pollen and seeds of numerous plants and plays a role in the reproductive development of these plants [35–43].

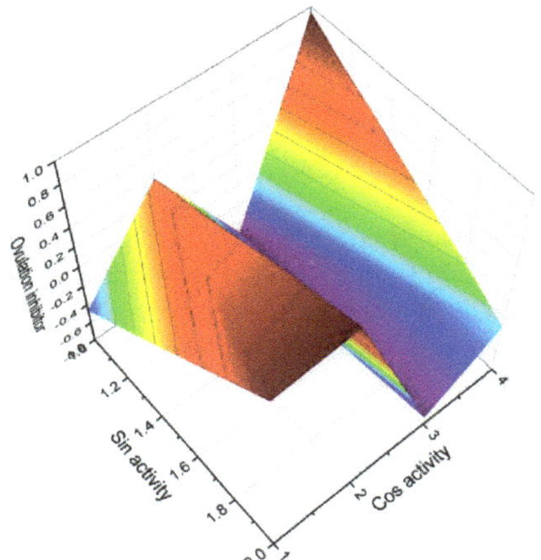

Figure 6. A 3D graph that illustrates the predicted and calculated activity of estrone (**4**, or estra-1,3,5(10)-triene-3-ol-17-one) as an ovulation inhibitor, with a confidence level exceeding 94%. This steroid has been isolated from the seeds and pollen of various plants, including *Glossostemon bruguieri*, *Hyphaene thebaica*, *Malus pumila*, *Phoenix dactylifera*, *Punica granatum*, and *Salix caprea*.

Luvigenin (**21**), a steroid, has been detected in the leaves of *Metanarthecium luteoviride* [52], *Yucca gloriosa* [53], and *Allium giganteum* [54]. Additionally, a cancer-fighting steroid called cayaponoside A4 (**22**) was isolated from the roots and bark of the *Tayuya* tree, which can be found in the Amazon rainforest across Bolivia, Brazil, and Peru [55–57].

An unusual triterpene dimer, xuxuasin B (**23**), was isolated from the Brazilian medicinal plant *Maytenus chuchuhuasca* [58]. The leaf extracts and root of *Maytenus ilicifolia* also demonstrated anti-cancer activity and contained a steroid called 6-oxotingenol (**24**) [59–61]. In an interesting discovery, an aromatic triterpenoid (**25**) was found in the cones of *Taxodium balticum* extract [62], and it has also been identified among terpenoids in Eocene and Miocene conifer fossils [63]. Furthermore, the bark extract of *Terminalia catappa* contained various compounds, including estrone (**4**), estriol (**6**), equilin (**7**), equilin sulfate (**26**), and a steroid (**27**) [64].

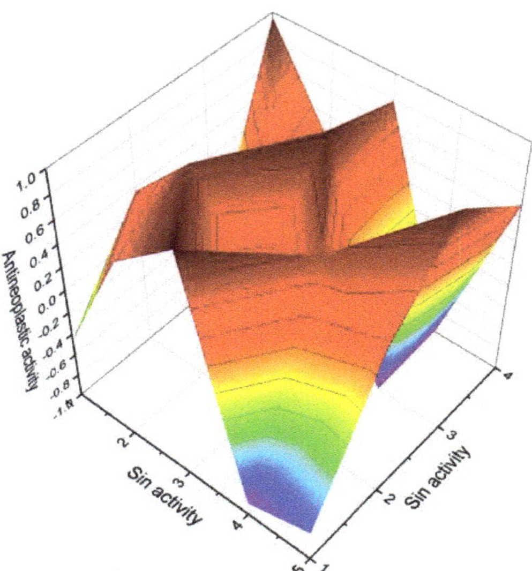

Figure 7. A 3D graph presenting the predicted and calculated anti-neoplastic activity of mono-aromatic ring A plant steroids (**16**, **17**, **21**, **23**, and **24**) with a confidence level exceeding 86%.

Steroids Bearing A, B, C, or D Aromatic Ring

Steroids can be categorized based on the presence of an aromatic ring in different positions, such as A, B, C, or D rings [1,7–9]. The following are some examples of steroids bearing aromatic rings in these positions. Aromatic A-ring steroids: estradiol: a natural estrogen hormone found in both males and females. Testosterone: the primary male sex hormone responsible for male sexual development and function. Aromatic B-ring steroids: progesterone: a female sex hormone involved in the menstrual cycle and pregnancy. Cortisol: a stress hormone involved in regulating metabolism and immune response. Aromatic C-ring steroids: aldosterone: a hormone that regulates electrolyte balance and blood pressure. Prednisone: a synthetic corticosteroid used as an anti-inflammatory and immunosuppressant. Aromatic D-ring steroids: vitamin D: a group of fat-soluble vitamins important for calcium and phosphate absorption. Calcitriol: the active form of vitamin D involved in calcium regulation and bone health. These are just a few examples of steroids with aromatic rings in different positions. Steroids play various roles in the body, including regulating physiological processes, acting as hormones, and serving as building blocks for other molecules [1–16].

The compound 3-Hydroxy-19-nor-1,3,5(10),22-cholatetraen-24-oic acid (**25**) is classified as a ring A aromatic bile acid and was discovered in an extract of the Australian sponge *Sollasella moretonensis* [65]. It was also found earlier in human intestinal flora, likely produced by bacteria [66]. Another steroid, a 4-hydroxy-6-oxopregnane-3-glycoside (**29**), which possesses an aromatic ring A, was isolated from a Pohnpei sponge called *Cribrochalina olemda*. Figure 8 depicts the 3D graph representing this compound [67]. Moreover, the extract of the marine sponge *Topsentia* sp. contains geodisterol-3-O-sulfite (**30**), which exhibits anti-fungal activity against *Candida albicans* [68]. In addition to these, a compound named 24,26-cyclo-19-norcholesta-1,3,5(10),22-tetraen-3-ol (**31**) was discovered in the Hainan soft coral *Dendronephthya studeri* [69]. Furthermore, an anti-tumor steroid thioester known as parathiosteroid C (**32**) was identified in the 2-propanol extract of another soft coral species, *Paragorgia* sp. [70].

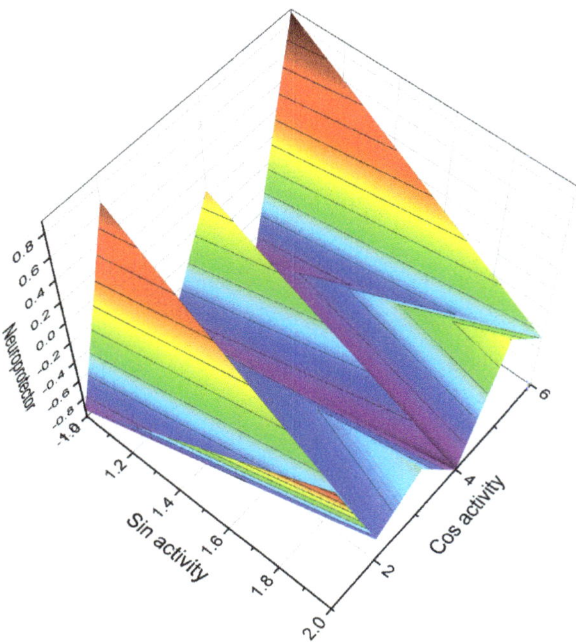

Figure 8. A 3D graph illustrating the predicted and calculated activity of the compound 4-hydroxy-6-oxopregnane-3-glycoside (**29**) as a neuroprotector. The graph demonstrates the relationship between the compound's structure and its predicted neuroprotective activity with a confidence level exceeding 97%. This steroid, containing an aromatic ring A, was isolated from a Pohnpei sponge known as *Cribrochalina olemda*. The graph provides insight into the relationship between the molecular structure of the compound and its predicted efficacy as a neuroprotector. By analyzing the graph, one can observe how variations in the structural features of the compound may impact its potential neuroprotective effects. The high confidence level of over 97% suggests a strong reliability in the predicted activity of this steroid as a neuroprotector. Understanding the neuroprotective activity of compounds is crucial for the development of potential treatments or interventions for neurodegenerative disorders, brain injuries, and other conditions that affect the health and function of the nervous system. Neuroprotector activity refers to the ability of a compound to protect and preserve the health and function of neurons in the brain and nervous system.

Mono-aromatic B-ring steroids are a rare group of steroids that can be synthesized by various types of fungi or fungal endophytes. They have also been found in marine sediments and oil deposits. One example is the 19-norergostane skeleton with an aromatic B-ring, known as phycomysterols A (**33**) and C (**34**), which are found in the filamentous fungus *Phycomyces blakesleeanus*. Phycomysterol A has shown anti-HIV activity, as demonstrated by activity analysis. Figure 9 illustrates the 3D graph representing phycomysterol A [71]. The lipid extract of the pathogenic fungus *Fusarium roseum*, also known as *Gibberella zeae*, contained (22*E*,24*R*)-1(10→6)-abeoergosta-5,7,9,22-tetraen-3α-ol (**35**) [72].

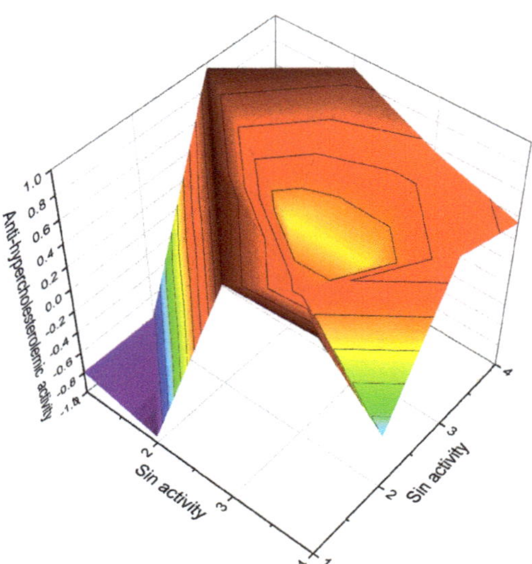

Figure 9. A 3D graph that illustrates the predicted and calculated anti-hypercholesterolemic activity of mono-aromatic ring A plant steroids (33, 34, 35, and 37) with a confidence level exceeding 91%. Anti-hypercholesterolemic activity refers to the ability of a compound to help reduce high cholesterol levels in the body. The graph showcases the relationship between the chemical structures of these mono-aromatic ring A plant steroids and their predicted efficacy in combating hypercholesterolemia. By analyzing the graph, it is possible to observe the correlation between the structural features of these compounds and their potential anti-hypercholesterolemic effects. The confidence level of over 91% indicates a high degree of reliability in the predicted activity of these steroids.

Asperfloketal B (**36**), featuring a trioxahexaheterocyclic ring system, was isolated from the sponge-associated fungus *Aspergillus flocculosus* 16D-1 [73]. Furthermore, an aromatic B-ring compound called topsentisterol E1 (**37**) was detected in the bioactive fraction of a marine sponge, *Topsentia* sp. (a sample of the sponge is shown in Figure 10) [74]. Another interesting aromatic B-ring steroid called phomarol (**38**) was produced by a cultured fungus, *Phoma* sp., derived from the giant jellyfish *Nemopilema nomurai* [75]. Additionally, an antibacterial lanostanoid, 19-nor-lanosta-5(10),6,8,24-tetraene-1α,3β,12β,22S-tetraol (**39**), was produced by an endophytic fungus called *Diaporthe* sp. LG23, which inhabits the leaves of the Chinese medicinal plant *Mahonia fortunei* [76].

Mono-aromatic C- and D-ring steroids form a rare group of compounds that have been discovered in various sources such as vegetable oils, marine sediments, and petroleum. In the Alberta oil sands, the C20 C-ring mono-aromatic hydroxy steroid acids (**40** and **41**) were found, and it was observed that these compounds can also be synthesized by soil fungi [77]. Steroidal hydrocarbons (**42** and **46**) have been detected in sediments and petroleum samples [78]. An unprecedented sesterterpenoid called phorone A (**43**), featuring an aromatic D ring, was identified in extracts of the Korean sea sponge *Phorbas* sp. [79]. Furthermore, the anti-cancer compound nakiterpiosinone (**44**), which is a C-nor-D homosteroid, was isolated from the sponge *Terpios hoshinota* [80]. Additionally, an intriguing compound called akaol A (**45**), classified as a sesquiterpene quinol, was associated with marine sponges of the genus Aka. The structure of akaol A is depicted in Figure 11 [81].

The extract of *Salpichroa origanifolia* plants, harvested in the provinces of Buenos Aires and Cordoba in Argentina, was found to contain two minor steroids with an aromatic E ring (**47** and **48**) [82]. From the marine sponge *Haliclona* sp., two compounds were identified: terpene-ketide haliclotriol A (**49**) and halicloic acid B (**50**) [83,84]. Steroidal hydrocarbons (**51** and **52**) were isolated from marine sediments and petroleum sources [85–87]. Table 2 displays the predicted biological activity for mono-aromatic steroids isolated from various sources, including plants, fungi, invertebrates, marine sediments, and oils. This table provides insights into the potential biological effects and activities associated with these mono-aromatic steroids.

Figure 10. Steroid (**35**) was found in the mycelium of pathogenic fungus *Fusarium roseum* (**a**); steroid (**37**) was isolated from a marine sponge *Topsentia* sp. (**b**); steroid (**38**) is a metabolite from the giant jellyfish *Nemopilema nomurai* (**c**); and steroid (**39**) is produced by an endophytic fungus, *Diaporthe* sp., which inhabits leaves of the Chinese medicinal plant *Mahonia fortunei* (**d**).

Table 2. Biological activities of mono-aromatic steroids (**28–52**).

No.	Dominated Biological Activity (Pa) *	Additional Predicted Activities (Pa) *
28	Anti-hypercholesterolemic (0.961) Proliferative diseases treatment (0.711)	Anti-neoplastic (0.840) Apoptosis agonist (0.787)
29	Neuroprotector (0.979) Respiratory analeptic (0.970) Anti-neoplastic (0.888)	Anti-hypercholesterolemic (0.953) Anti-infective (0.933) Anti-protozoal (*Leishmania*) (0.922)
30	Anti-hypercholesterolemic (0.860) Anti-inflammatory (0.754)	Anti-neoplastic (0.805) Chemopreventive (0.721)
31	Anti-hypercholesterolemic (0.907) Anti-inflammatory (0.765)	Anti-neoplastic (0.836) Apoptosis agonist (0.788)
32	Anti-hypercholesterolemic (0.764)	Anti-inflammatory (0.695)
33	Anti-hypercholesterolemic (0.929)	Respiratory analeptic (0.885)
34	Anti-hypercholesterolemic (0.935)	Apoptosis agonist (0.850)

Table 2. *Cont.*

35	Anti-hypercholesterolemic (0.950) Anti-Parkinsonian, rigidity relieving (0.875)	Apoptosis agonist (0.898) Anti-neoplastic (0.880)
36	Anti-hypercholesterolemic (0.806)	Anti-neoplastic (0.729)
37	Anti-hypercholesterolemic (0.914) Hypolipemic (0.858)	Apoptosis agonist (0.894) Anti-neoplastic (0.879)
38	Anti-neoplastic (0.922)	Immunosuppressant (0.774)
39	Anti-neoplastic (0.899) Apoptosis agonist (0.896)	Anti-inflammatory (0.795)
40	Neuroprotector (0.829)	Anti-allergic (0.731)
41	Anti-convulsant (0.877)	
42	Apoptosis agonist (0.828) Anti-neoplastic (0.798)	Anti-inflammatory (0.813)
43	Anti-neoplastic (0.782)	Anti-bacterial (0.736)
44	Acute neurologic disorders treatment (0.867)	Anti-neoplastic (0.797)
45	Anti-inflammatory (0.825)	Apoptosis agonist (0.793)
46	Anti-neoplastic (0.884)	Apoptosis agonist (0.848)
47	Anti-neoplastic (0.799)	Apoptosis agonist (0.716)
48	Anti-neoplastic (0.858)	Anti-hypercholesterolemic (0.839)
49	Anti-neoplastic (0.858)	Cell adhesion molecule inhibitor (0.795)
50	Anti-neoplastic (0.841)	Immunosuppressant (0.722)
51	Anti-neoplastic (0.844)	Apoptosis agonist (0.792)
52	Apoptosis agonist (0.706)	Acute neurologic disorders treatment (0.768)

* Only activities with Pa > 0.7 are shown.

2.2. Steroids Bearing Two or Three Aromatic Rings Derived from Natural Sources

Steroids bearing two or three aromatic rings derived from natural sources can be found in various organisms and have diverse biological activities. These are just a few examples of steroids bearing two or three rings that are derived from natural sources. Steroids with complex ring systems can be found in a wide range of organisms and play important roles in biological processes [1,9,78].

Di- and tri-aromatic steroids (**53–83**, structures see in Figure 12) represent a small group of natural lipids. These compounds have been isolated and identified in various sources such as marine sediments, oils, and sedimentary rocks [78,85,88]. It is worth noting that di-aromatic steroids, which contain a naphthalene ring, are primarily synthesized by fungi or fungal endophytes [89]. These unique steroids with di-aromatic or tri-aromatic structures contribute to the diversity of natural lipids and their distribution in different environments. Their presence in marine sediments, oils, and sedimentary rocks suggests their relevance in geological and ecological contexts.

In 1936, Canadian biochemist Desmond Beall isolated 6,8-Didehydroestrone (**53**) from the urine of pregnant mares [90]. Additionally, another steroidal hormone called equilenin, specifically estra-1,3,5(10),6,8-pentaen-3-ol-17-one, was also discovered in the urine of pregnant mares in the same year. Subsequently, in 1938, equilenin sulfate (**54**) was isolated from the urine of pregnant mares by Schachter and Marrian [91]. In 1939, it was further synthesized by Bachmann et al. [92]. Moreover, derivatives of equilenin, including 17α-Dihydroequilenin (**55**) and estra-1,3,5,7,9-pentaen-17-one (**56**), were found to be excreted in the urine of horses [93]. These compounds contribute to the understanding of hormonal compositions and metabolic pathways in horses.

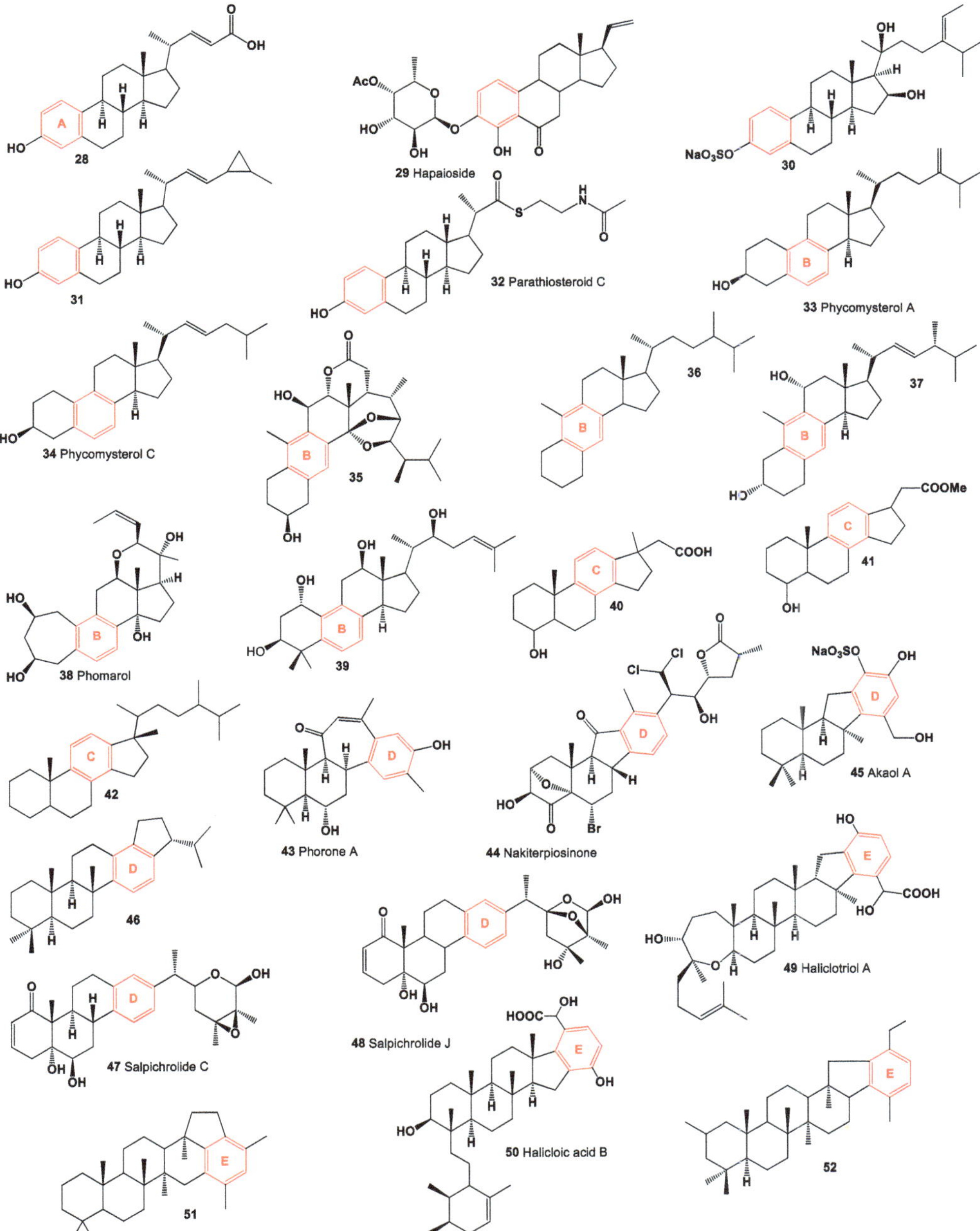

Figure 11. Steroids bearing ring A, B, C, D, and E in natural sources.

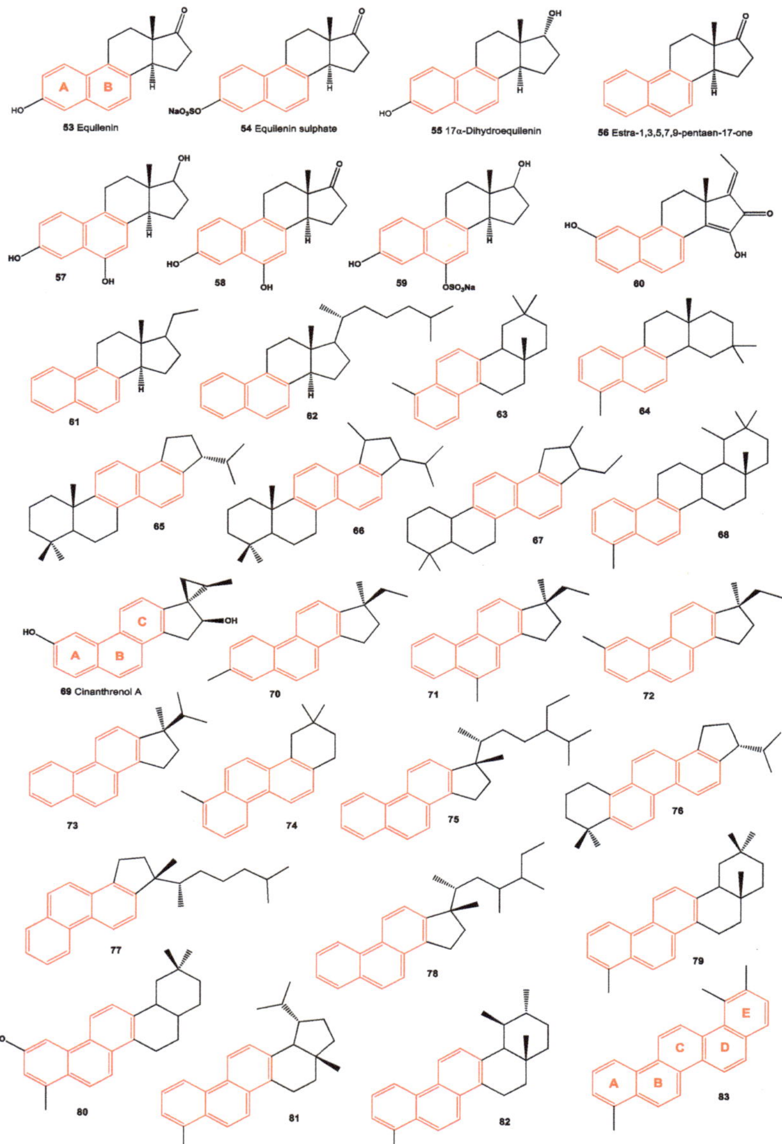

Figure 12. Di- and tri-aromatic steroids and terpenoids derived from fungi, invertebrates, sediments, and petroleum.

The distribution and biological activity of mono-, di-, and tri-aromatic steroids in nature are well-documented. These aromatic steroids are produced by various sources, including microorganisms, fungi, marine invertebrates, plants, animals, marine sediments, and karst deposits. These compounds have demonstrated significant biological activities, including anti-tumor, anti-inflammatory, and neuroprotective effects. The reliability of these activities ranges from 78% to 92%, indicating a high level of confidence in their observed effects. The wide occurrence of aromatic steroids across different natural sources highlights their importance and potential therapeutic applications. Further research and exploration of these compounds could lead to the discovery of novel drugs and therapeutic interventions.

Rare naphthalene-containing steroids (**56–59**) have been discovered in the bark of the *Terminalia catappa* tree. It is believed that these naphthalene steroids are synthesized by fungal endophytes that are associated with these plants [89]. Extensive studies of these plants have revealed a wide variety of fungal endophytes present, including species such as *Cercospora* spp., *Cercospora olivascens*, *Colletotrichum gloeosporioides*, *Diaporthe* sp., *Fusarium* sp., *Lasiodiplodia theobromae*, *Pestalotiopsis* spp., *Penicillium* sp., *Penicillium chermesinum*, *Xylaria* sp., *Phoma microchlamidospora*, and *Phomopsis* sp. [94,95]. In addition, a rare di-aromatic steroid (**60**) that contains an unusual naphthyl A/B ring system, resembling equilenin, was isolated from a Hawaiian sponge belonging to the genus *Strongylophora* [96]. Furthermore, a di-aromatic steroid known as (17β,20R,22E,24R)-19-norergosta-1,3,5,7,9,14,22-heptaene (**62**) is produced by the ascomycete fungus *Daldinia concentrica* [97]. These compounds contribute to the diversity of rare di-aromatic steroids and highlight their presence in unique natural sources.

A diverse range of naphthalene steroid hydrocarbons (**63–68**) have been discovered in various natural sources, including marine sediments, fossil plants and algae, ancient fossils, and petroleum [78,98–101]. These compounds contribute to the wide array of naphthalene-based steroids found in different geological and biological contexts. In contrast, tri-aromatic steroids, or phenanthrene-containing steroids (**69–73**) are relatively rare in nature and are found in only a limited number of specimens. One intriguing example is the phenanthrene-containing steroid called cinanthrenol A, which was identified in the marine sponge *Cinachyrella* sp. (a sample of the sponge is depicted in Figure 13). Cinanthrenol A has demonstrated cytotoxic activity against P-388 and HeLa cells and has also shown inhibitory effects on estrogen receptors [102]. These unique phenanthrene-containing steroids exemplify the fascinating diversity of naturally occurring compounds and their potential for various biological activities. Further exploration of these compounds could lead to the discovery of novel therapeutic agents or insights into biological processes.

Figure 13. Di-aromatic steroid (**60**) was found in the marine sponge *Strongylophora* sp. (**a**); another di-aromatic steroid (**62**) was produced by the ascomycete *Daldinia concentrica* (**b**); and tri-aromatic steroids or phenanthrene-containing steroids (**69–73**) were found in the marine sponge *Cinachyrella* sp. (**c**).

Acute neurological disorders refer to a group of sudden-onset conditions that affect the nervous system, including the brain, spinal cord, and peripheral nerves. These disorders can arise due to various factors such as infections, trauma, vascular events, metabolic imbalances, autoimmune reactions, or toxic exposures. They are characterized by rapid onset and can lead to severe neurological symptoms and impairments. Figure 14, a 3D graph, illustrates the predicted and calculated activity of an aromatic steroid (**81**) as a potential treatment for acute neurological disorders. The graph demonstrates the relationship between the activity of the compound and its efficacy in treating these disorders. The predicted and calculated activity values, shown on the axes of the graph, represent the potency or effectiveness of the compound in addressing the neurological symptoms associated with acute disorders. The graph also mentions a confidence level of over 92%. This indicates a high degree of certainty in the accuracy of the predicted and calculated activity values. Such confidence levels are typically derived from statistical analysis or predictive modeling techniques used in drug discovery and development. It is important to note that without additional context or information about the specific compound (aromatic steroid **81**), its mechanism of action, and the specific acute neurological disorders being targeted, it is difficult to provide a detailed interpretation of the graph. Further research, clinical trials, and scientific investigation would be necessary to validate the efficacy and safety of the compound as a potential treatment for acute neurological disorders.

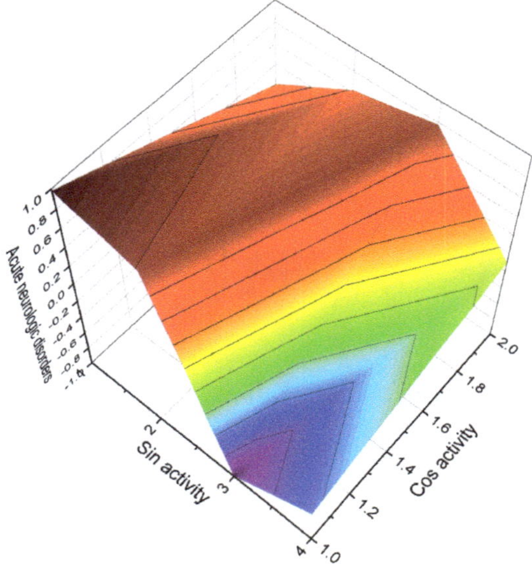

Figure 14. A 3D graph showing the predicted and calculated activity as a treatment for acute neurological disorders of aromatic steroid (**81**) with over 92% confidence.

Tri-aromatic and/or polyaromatic steroid hydrocarbons are a class of organic compounds that contain three or more aromatic rings fused together with a steroid structure [1,9,78]. These compounds have been identified in various natural sources, including lipid extracts of fossil plants and algae, marine sediments, and petroleum. The presence of tri-aromatic and polyaromatic steroid hydrocarbons in these sources suggests that they have a natural origin and may be formed through the diagenesis and maturation processes of organic matter over time. These compounds often exhibit complex and diverse chemical structures due to the multiple aromatic rings and steroid backbone. The identification and characterization of these compounds have been facilitated by analytical techniques such as gas chromatography-mass spectrometry (GC-MS) and nuclear magnetic resonance (NMR) spectroscopy. Their presence in various geological and biological samples suggests that they may have ecological, physiological, or pharmacological relevance. Investigating their biological activities and potential applications can contribute to our understanding of their functions in nature and may uncover new possibilities for their utilization in various fields.

Tri-aromatic and/or polyaromatic steroid hydrocarbons (**70–83**) have been detected in lipid extracts obtained from various sources such as fossil plants, algae, marine sediments, and petroleum [78,85,98,99,103,104]. Among these compounds, an oleanane-related triterpenoid (**80**) with a unique C-2 oxygenated functionality has been identified as the most abundant triterpenoid in a 4900-year-old oak wood sample that was buried in freshwater sediment [105,106]. In addition, other triterpenoids containing phenanthrene structures (**79, 81**, and **82**) have been found, along with stigmast-4-ene, stigmast-5-ene, stigmastanol, stigmastanol-3-one, 24-ethylcholesta-4,6,22-triene, and β-sitosterol, in fossil cones of *Taxodium balticum*. Stigmastanol-3-one has also been identified in *T. dubium* [107]. Table 3 presents the reported biological activities of mono-aromatic steroids that have been isolated from various sources including fungi, invertebrates, marine sediments, and petroleum. This table provides information on the observed biological effects or properties exhibited by these compounds. However, the specific details of the biological activities mentioned in Table 3 are not provided in the given text.

Table 3. Biological activities of aromatic steroids (**53–83**).

No.	Dominated Biological Activities (Pa) *	Additional Predicted Activities (Pa) *
53	Ovulation inhibitor (0.866)	Anti-neoplastic (0.824)
54	Acute neurologic disorders treatment (0.925) Anti-neoplastic (0.790)	Diuretic (0.824) Male reproductive dysfunction treatment (0.791)
55	Acute neurologic disorders treatment (0.826) Anti-neoplastic (0.818)	Respiratory analeptic (0.811) Neuroprotector (0.807)
56	Ovulation inhibitor (0.846); male reproductive dysfunction treatment (0.815)	Anti-neoplastic (0.821)
57	Neuroprotector (0.837) Anti-neoplastic (0.833)	Acute neurologic disorders treatment (0.828)
58	Ovulation inhibitor (0.843) Lipid metabolism regulator (0.723)	Anti-neoplastic (0.839) Neuroprotector (0.829)
59	Acute neurologic disorders treatment (0.932) Anti-neoplastic (0.810)	Laxative (0.833) Diuretic (0.751)
60	Apoptosis agonist (0.924) Anti-neoplastic (0.868)	Antioxidant (0.776) Neuroprotector (0.728)
61	Anti-osteoporotic (0.837)	Anti-neoplastic (0.735)
62	Anti-hypercholesterolemic (0.860)	Respiratory analeptic (0.847)
63	Anti-osteoporotic (0.776)	Anti-neoplastic (0.732)
64	Apoptosis agonist (0.758) Anti-neoplastic (0.733)	Anti-inflammatory (0.744)
65	Apoptosis agonist (0.758) Anti-neoplastic (0.733)	Anti-inflammatory (0.744)
66	Anti-inflammatory (0.807)	Apoptosis agonist (0.746); anti-neoplastic (0.726)
67	Anti-infertility, female (0.796)	Anti-inflammatory (0.794)
68	Anti-neoplastic (0.697)	Ovulation inhibitor (0.683)
69	Prostate disorders treatment (0.699)	Anti-inflammatory (0.661)

Table 3. *Cont.*

70	Anti-neoplastic (0.825) Alzheimer's disease treatment (0.824)	Neurodegenerative diseases treatment (0.809) Psychotropic (0.700)
71	Anti-eczematic (0.767)	Anti-dyskinetic (0.670)
72	Anti-eczematic (0.695)	Autoimmune disorders treatment (0.652)
73	Anti-eczematic (0.767)	Anti-dyskinetic (0.670)
74	Anti-eczematic (0.782) Anti-psoriatic (0.619)	Anti-neurotic (0.709)
75	Neuroprotector (0.685)	Acute neurologic disorders treatment (0.647)
76	Hypolipemic (0.724)	Anti-convulsant (0.649)
77	Anti-eczematic (0.885) Anti-psoriatic (0.757)	Anti-inflammatory (0.735)
78	Anti-eczematic (0.709) Anti-psoriatic (0.632)	Anti-convulsant (0.661)
79	Anti-eczematic (0.691) Anti-psoriatic (0.622)	Psychotropic (0.611) Anti-convulsant (0.570)
80	Apoptosis agonist (0.758) Anti-neoplastic (0.733)	Anti-inflammatory (0.744)
81	Acute neurologic disorders treatment (0.778)	Neuroprotector (0.733)
82	Anti-inflammatory (0.650)	Menopausal disorders treatment (0.628)
83	Anti-inflammatory (0.782)	Anti-eczematic (0.771)

* Only activities with Pa > 0.7 are shown.

Further research is needed to fully understand the roles and significance of tri-aromatic and polyaromatic steroid hydrocarbons in natural systems. Their presence in various geological and biological samples suggests that they may have ecological, physiological, or pharmacological relevance. Investigating their biological activities and potential applications can contribute to our understanding of their functions in nature and may uncover new possibilities for their utilization in various fields.

3. Steroids Bearing Phosphate Esters

Phosphorus, with an atomic number of 15, is a prevalent chemical element found in both the earth's crust and seawater [108–110]. Its discovery dates back approximately 350 years [111]. Due to its high reactivity, phosphorus is typically found in nature in the form of phosphates, which are salts of phosphoric acid [112]. Apatite, a mineral compound, is considered one of the most significant sources of phosphorus [113,114].

Steroids bearing phosphate esters are a class of organic compounds that combine the structure of steroids with phosphate groups attached to specific positions. These phosphate esters can be covalently linked to the steroid molecule, typically through ester bonds. The addition of phosphate esters to steroids introduces new chemical properties and functional groups, which can have significant effects on the compound's biological activity and physiological functions. Phosphate esters play important roles in cellular signaling, energy metabolism, and various biochemical processes. Phosphate esters in steroids can also serve as important intermediates in metabolic pathways. For instance, in the biosynthesis of steroid hormones, phosphate esters are involved in the conversion of cholesterol to various hormone precursors, such as pregnenolone. Furthermore, some steroid-based drugs utilize phosphate esters to enhance their pharmacological properties. By introducing phosphate groups, these compounds can exhibit improved solubility, bioavailability, and targeted delivery to specific tissues or cells. Overall, steroids bearing phosphate esters are biologically significant molecules that contribute to cellular processes, membrane structure, and the modulation of hormonal activities. Understanding their synthesis, functions, and interactions is crucial in unraveling the complexities of biological systems and developing therapeutic interventions [115–121].

Steroid Phosphate Esters in Marine Invertebrates

Steroid phosphates (**84–87**), as shown in Figure 15, were first discovered by Italian scientists from the University of Federico II approximately three decades ago. Their discovery came during the study of polar lipids extracted from the deep marine starfish *Tremaster novaecaledoniae* [122]. The isolated glycosides obtained from this research were named tremasterols A–C (**84**, activity is shown in Table 4), along with compounds **85** and **86**. Figure 16 illustrates the distribution of biological activity, specifically for tremasterol (**84**), represented as a percentage. This graph provides insights into the effectiveness or impact of tremasterol in various biological contexts. The identification and characterization of these steroid phosphates from the marine starfish *T. novaecaledoniae* represent significant contributions to the field of natural product research. Further investigations are likely needed to fully understand the biological activities and potential applications of these compounds, including their mechanisms of action and potential therapeutic benefits.

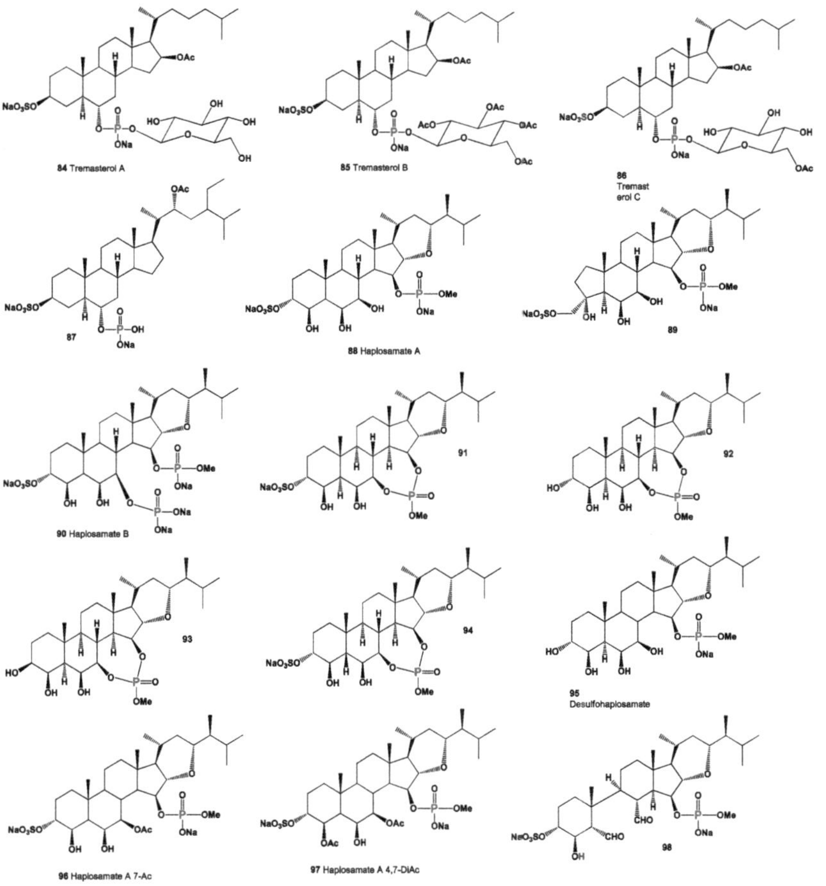

Figure 15. Natural bioactive steroid phosphate esters.

Table 4. Biological activities of steroid phosphate esters (**84–98**).

No.	Dominated Biological Activity (Pa) *	Additional Predicted Activities (Pa) *
84	Wound-healing agent (0.975) Hepatoprotectant (0.961) Analeptic (0.952) Laxative (0.933)	Anti-hypercholesterolemic (0.926) Anti-carcinogenic (0.912) Hemostatic (0.853) Anti-neoplastic (0.841)
85	Hepatoprotectant (0.874) Analeptic (0.874)	Anti-carcinogenic (0.861) Anti-neoplastic (0.848)
86	Wound-healing agent (0.947) Analeptic (0.941) Hepatoprotectant (0.932)	Anti-carcinogenic (0.915) Anti-hypercholesterolemic (0.912) Anti-neoplastic (0.843)
87	Anti-hypercholesterolemic (0.894) Hepatoprotectant (0.853) Wound-healing agent (0.844)	Anti-neoplastic (0.816) Anti-inflammatory (0.782) Cholesterol synthesis inhibitor (0.778)
88	Anti-hypercholesterolemic (0.894) Hepatoprotectant (0.853) Wound-healing agent (0.844)	Anti-neoplastic (0.816) Anti-inflammatory (0.782) Cholesterol synthesis inhibitor (0.778)
89	Anti-neoplastic (0.845) Anti-fungal (0.814)	Anti-inflammatory (0.693) Anti-bacterial (0.651)
90	Anti-fungal (0.837)	Anti-neoplastic (0.824)
91	Anti-neoplastic (0.827)	Anti-fungal (0.663)
92	Anti-neoplastic (0.852) Anti-neoplastic (liver cancer) (0.790)	Anti-eczematic (0.730) Anti-allergic (0.650)
93	Anti-neoplastic (0.852) Anti-neoplastic (liver cancer) (0.790)	Anti-eczematic (0.730) Anti-allergic (0.650)
94	Anti-neoplastic (0.827) Anti-neoplastic (liver cancer) (0.607)	Anti-fungal (0.663) Anti-bacterial (0.636)
95	Anti-neoplastic (0.841)	Anti-fungal (0.799)
96	Anti-fungal (0.850) Anti-bacterial (0.717)	Anti-neoplastic (0.832) Anti-carcinogenic (0.707)
97	Anti-fungal (0.850) Anti-bacterial (0.717)	Anti-neoplastic (0.832) Anti-carcinogenic (0.707)
98	Anti-fungal (0.858) Anti-bacterial (0.739)	Anti-neoplastic (0.842) Anti-carcinogenic (0.733)

* Only activities with Pa > 0.7 are shown.

Phosphorylated sterol sulfates, known as haplosamates A (**88**) and B (**90**) and minor secosteroid (**89**), were discovered in a marine sponge species called *Cribrochalina* sp. [123]. Haplosamate A is distinguished by its unique C28 sterol structure, featuring a sulfate group at C-3 and a methyl phosphate at position 15. Haplosamate B, on the other hand, contains two phosphate groups at positions 7 and 15 [123]. The 3D graph illustrating the activity of haplosamate A (**88**) is depicted in Figure 17. Further semi-synthetic analogues, including compounds **91–94**, have also been isolated and studied. Desulfohaplosamate (**95**), haplosamate A (**88**), and other steroid analogues (**96–99**) were evaluated for their interaction with CB1 and CB2 cannabinoid receptors through binding tests [124]. It is worth noting that both steroids containing a phosphate group, namely **88** and **90**, were discovered in the polar organic fraction of an Indonesian sponge species called *Dasychalina* sp. (shown in Figure 18) [124]. The identification and evaluation of these phosphorylated sterol sulfates and their analogues provide valuable insights into their potential biological activities and interactions. Further research is necessary to fully understand their mechanisms of action, therapeutic potential, and roles within marine ecosystems.

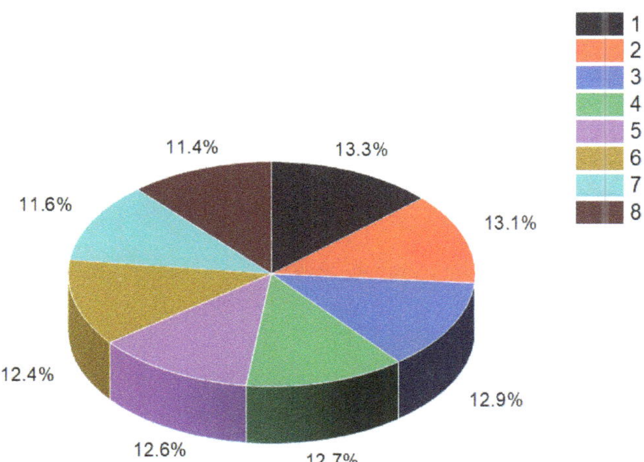

Figure 16. Illustration of the percentage distribution of biological activities on the example of tremasterol A (**84**), which is derived from the marine starfish *Tremaster novaecaledoniae*, and this steroid has a wide range of pharmacological properties. Activities are indicated under the numbers: 1, wound-healing agent (13.3%); 2, hepatoprotectant (13.1%); 3, analeptic (12.9%); 4, laxative (12.7%); 5, anti-hypercholesterolemic (12.6%); 6, anti-carcinogenic (12.4%); 7, hemostatic (11.6%), and 8, anti-neoplastic (11.4%).

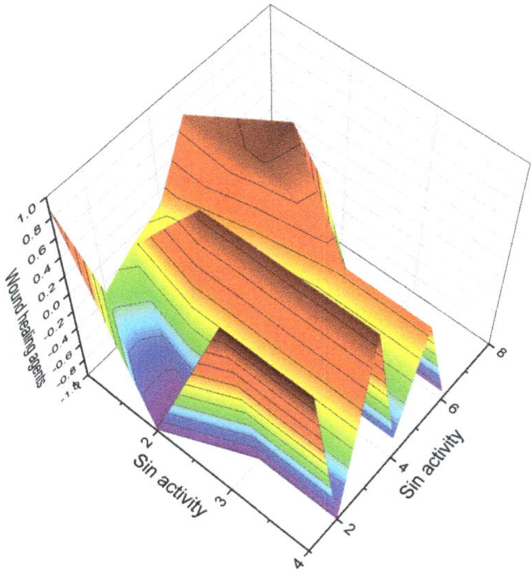

Figure 17. A 3D graph showing the predicted and calculated activity as wound-healing agents of steroid phosphate esters (**84**, **86**, **87**, and **88**) with over 89% confidence. Wound-healing agents are substances or treatments that promote the healing of wounds. These agents can be in the form of medications, dressings, or therapies that aid in the different stages of the wound-healing process. The wound-healing process involves a series of complex biological events that aim to restore the damaged tissue and close the wound. The stages of wound healing include hemostasis (stopping bleeding), inflammation, proliferation, and remodeling.

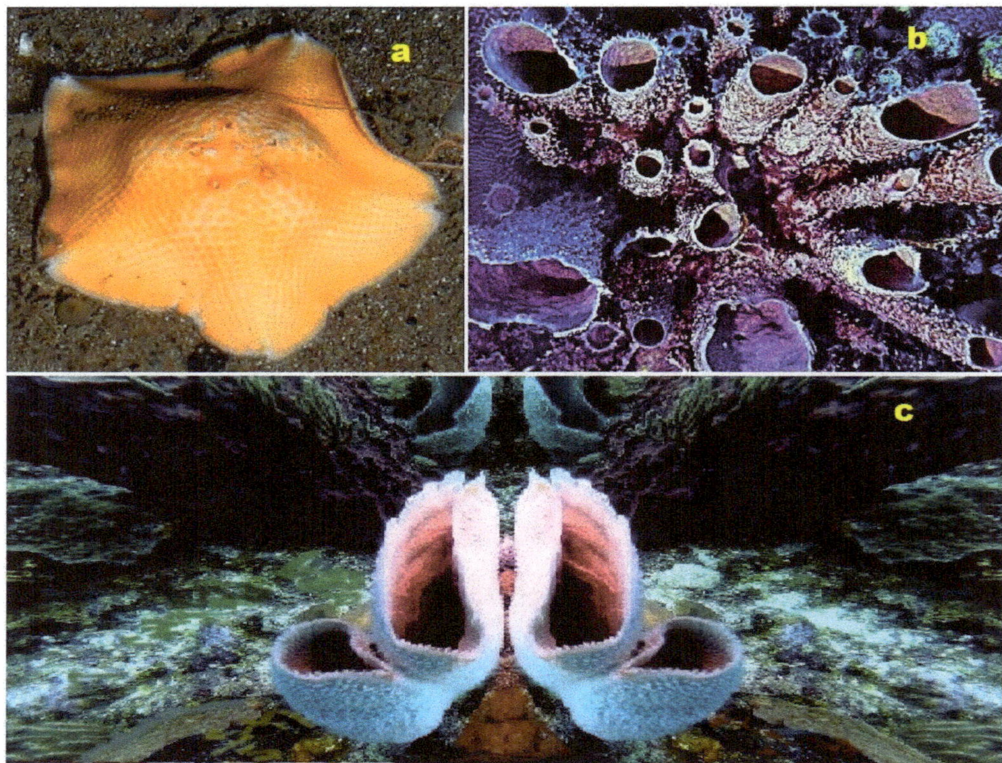

Figure 18. Steroid phosphate esters (**84–87**) were isolated from the deep marine starfish *Tremaster novaecaledoniae* (**a**); steroids (**88**), (**89**), and (**90**) are found in a marine sponge *Cribrochalina* sp. (**b**); and steroids (**88 and 90**) were also found in the Indonesian sponge *Dasychalina* sp. (**c**).

For comparing biological activity, several semi-synthetic steroids have been selected. These include prednisone phosphate (**99**), testosterone 17β-phosphate (**100**), cortisol 21-phosphate (**101**), and cholesterol 3β-phosphate (**102**). Prednisone phosphate (**99**) has been shown to possess anti-inflammatory activity [125]. This property makes it useful in the treatment of various inflammatory conditions. Testosterone 17β-phosphate (**100**) is an androgen and belongs to the class of anabolic steroids. It is commonly used for intramuscular injections and is known for its anabolic effects on muscle growth. Additionally, it serves as a substrate for phosphatases in the phosphatase pool of the prostate [126]. Cortisol 21-phosphate (**101**) is a glucocorticoid that plays a crucial role in regulating various physiological processes. It is involved in the regulation of metabolism, immune responses, and stress responses. The phosphate group attached to cortisol 21 enhances its solubility and may influence its activity. Cholesterol 3β-phosphate (**102**) is a modified form of cholesterol with a phosphate group attached to its 3β position. The addition of the phosphate group introduces new chemical properties to cholesterol, potentially influencing its functions and interactions within the body. These semi-synthetic steroids have been selected for comparison with steroids isolated from marine invertebrates in order to gain insights into their biological activities and potential applications. Further research is necessary to fully understand the specific mechanisms of action and therapeutic implications of these compounds.

Cortisol 21-phosphate (**101**), as depicted in Figure 19, belongs to the glucocorticoid class of hormones. It functions to increase blood sugar levels through gluconeogenesis and promotes the metabolism of fats, proteins, and carbohydrates. Additionally, cortisol

21-phosphate serves as a substrate for alkaline phosphatase and finds utility in enzyme immunoassays for human chorionic gonadotropin, human growth hormone, α-fetoprotein, and estradiol [127]. The activities and properties of cortisol 21-phosphate can be found in Table 5, and its 3D graph is illustrated in Figure 20.

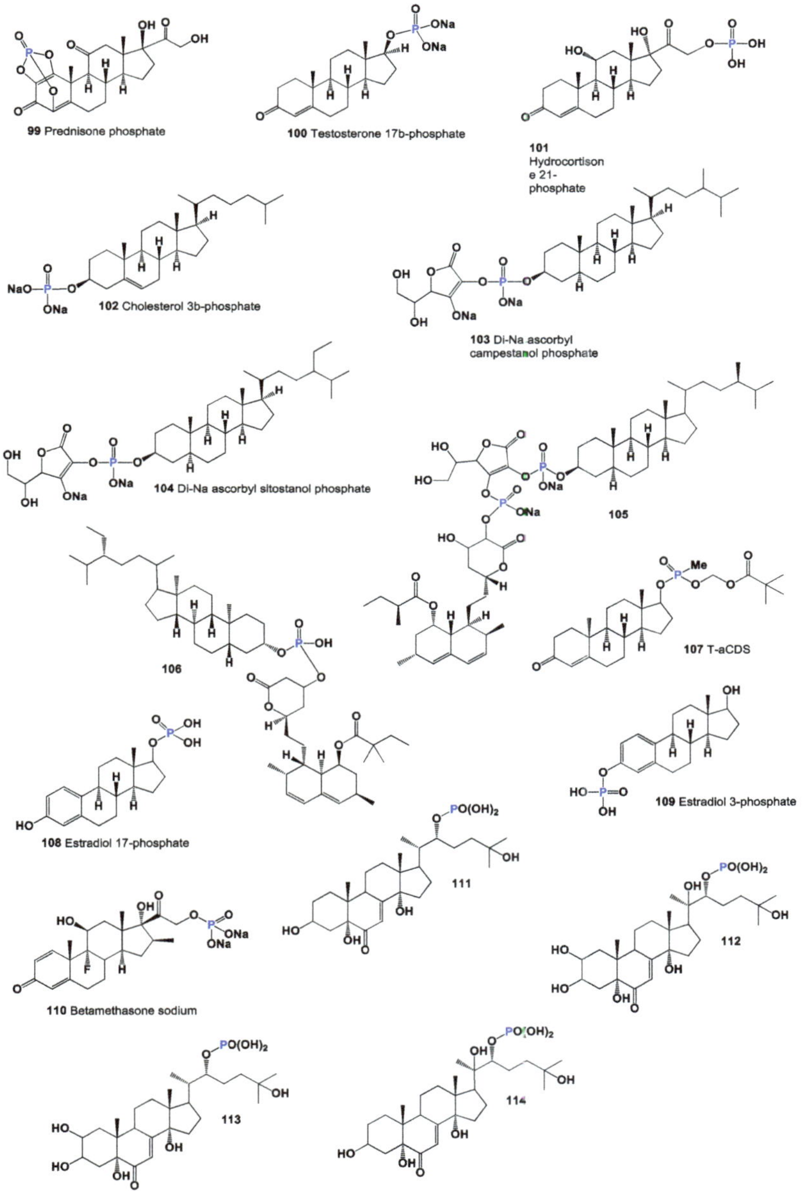

Figure 19. Natural and semi-synthetic bioactive steroid phosphate esters.

Table 5. Biological activities of steroid phosphate esters (**99–114**).

No.	Dominated Biological Activity (Pa) *	Additional Predicted Activities (Pa) *
99	Anti-inflammatory (0.910) Anesthetic general (0.908)	Respiratory analeptic (0.904) Anti-osteoporotic (0.878)
100	Neuroprotector (0.987) Anesthetic general (0.959)	Respiratory analeptic (0.944) Anti-hypercholesterolemic (0.909)
101	Anesthetic general (0.991) Neuroprotector (0.976) Anti-inflammatory (0.906)	Respiratory analeptic (0.990) Anti-hypercholesterolemic (0.894)
102	Respiratory analeptic (0.979) Anesthetic general (0.973) Neuroprotector (0.972)	Anti-hypercholesterolemic (0.971) Wound-healing agent (0.913) Anti-neoplastic (0.826)
103	Respiratory analeptic (0.995) Anesthetic general (0.948) Wound-healing agent (0.897)	Anti-hypercholesterolemic (0.945) Neuroprotector (0.932) Hemostatic (0.910)
104	Respiratory analeptic (0.995) Anti-hypercholesterolemic (0.967) Anesthetic general (0.954)	Hemostatic (0.928) Wound-healing agent (0.921) Neuroprotector (0.909)
105	Anti-hypercholesterolemic (0.996) Cholesterol absorption inhibitor (0.976) Cholesterol synthesis inhibitor (0.952) Lipid metabolism regulator (0.952)	Acute neurologic disorders treatment (0.948) Anti-hyperlipoproteinemic (0.920) Hypolipemic (0.919) Respiratory analeptic (0.908)
106	Anti-hypercholesterolemic (0.999) Anti-hyperlipoproteinemic (0.986) Hypolipemic (0.974)	Cholesterol absorption inhibitor (0.957) Lipid metabolism regulator (0.954) Cholesterol synthesis inhibitor (0.916)
107	Anti-neoplastic (0.822)	Anti-inflammatory (0.645)
108	Neuroprotector (0.982) Anesthetic general (0.931)	Anti-hypercholesterolemic (0.909)
109	Anesthetic general (0.970) Neuroprotector (0.965)	Respiratory analeptic (0.961) Acute neurologic disorders treatment (0.916)
110	Anti-inflammatory (0.979) Anti-allergic (0.959)	Anti-asthmatic (0.951) Anti-arthritic (0.944)
111	Respiratory analeptic (0.929) Anti-ischemic, cerebral (0.907)	Anesthetic general (0.897) Anti-neoplastic (0.847)
112	Anti-ischemic, cerebral (0.979) Respiratory analeptic (0.919)	Anti-osteoporotic (0.843) Anesthetic general (0.830)
113	Respiratory analeptic (0.937) Anti-ischemic, cerebral (0.922)	Anesthetic general (0.897)
114	Anti-ischemic, cerebral (0.978) Respiratory analeptic (0.911)	Anti-osteoporotic (0.852)

* Only activities with Pa > 0.7 are shown.

Cholesterol 3β-phosphate (**102**) is known for its role in promoting the normalization of blood pressure and its involvement in atherogenesis, the process of plaque formation in arteries [128,129]. Two cholesterol-lowering agents, sodium ascorbyl campestanol phosphate (**103**) and sodium ascorbyl sitostanol phosphate (**104**), have been derived from cholesterol and extensively studied [130]. Furthermore, two semi-synthetic steroidal phosphate esters (**105** and **106**, 3D graph is illustrated in Figure 21), are identified as inhibitors of cholesterol biosynthesis. These compounds show potential for the treatment or prevention of atherosclerosis, a major contributor to cardiovascular disease [131]. The investigation and understanding of these steroidal phosphate compounds contribute to advancements in the field of hormone research and lipid metabolism and the development of potential therapeutic interventions for various conditions, including atherosclerosis and related cardiovascular disorders.

Compound (**107**) is a steroid phosphate ester that incorporates pivalic acid. This compound, known as the anionic chemical delivery system (ACDS), was specifically developed to facilitate the delivery of testosterone to the brain. By enhancing its lipophilicity, systemically administered T-ACDS can passively traverse the blood–brain barrier. The effectiveness of this tested drug has been demonstrated [132]. Estradiol phosphates (**108** and **109**) are esters of estrogen that are combined with phosphoric acid. These compounds serve as prodrugs of estradiol within the human body. In medical practice, both drugs have been utilized for the treatment of prostate cancer [133]. Betamethasone sodium phosphate (**110**) has been synthesized and is employed in the treatment of various conditions such as asthma, allergies, arthritis, Crohn's disease, ulcerative colitis, and adrenal disease [134]. The development and utilization of these compounds highlight the ongoing advancements in drug development and therapeutic approaches. However, it is crucial to consult with healthcare professionals for proper guidance and administration of these medications, considering individual patient factors and specific medical conditions.

Several steroid phosphate esters, namely compounds **111** to **114** (3D graph is shiwn in Figure 22), have been identified in the eggs of the desert locust, *Schistocera gregaria*. It is intriguing to note the presence of these steroids in deferred eggs, although their specific origin remains unknown [135]. The detection of these compounds in locust eggs raises interesting questions about their potential roles and functions in the reproductive processes of the species. However, it is important to highlight that the biological activity of these compounds obtained from locust eggs has not been investigated or characterized.

Understanding the presence and activities of steroid phosphate esters in locust eggs may contribute to our knowledge of reproductive biology, insect development, and the hormonal regulation of insect populations. Further research is warranted to explore the biological properties and potential functions of these compounds in the context of locust biology.

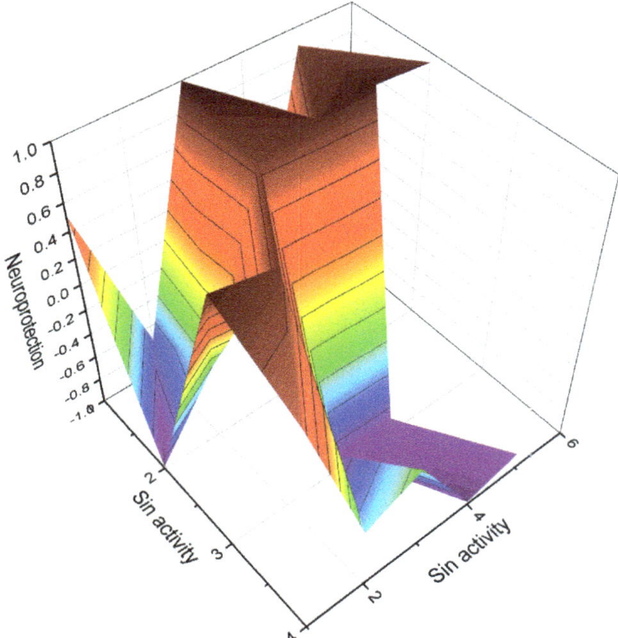

Figure 20. A 3D graph presenting the predicted and calculated activity for neuroprotection of steroid phosphate esters, specifically compounds **100**, **101**, **108**, and **109**. The graph demonstrates the relationship between the activity of these compounds and their efficacy in promoting neuroprotection.

The predicted and calculated activity values, shown on the axes of the graph, represent the potency or effectiveness of the steroid phosphate esters in terms of their neuroprotective properties. With a confidence level of over 96%, the graph indicates a high degree of certainty in the accuracy of the predicted and calculated activity values. Neuroprotection is a critical aspect of research and development in the field of neuroscience, aiming to identify compounds that can preserve and protect neurons from damage or degeneration. The evaluation of steroid phosphate esters for their neuroprotective activity provides valuable insights into their potential applications in treating neurological disorders or promoting overall brain health. The concept of neuroprotection has gained significant attention in the field of neuroscience and neurology, particularly in the context of neurodegenerative diseases, stroke, traumatic brain injury, and other conditions that involve neuronal damage. Ayurveda, a centuries-old Indian traditional medicine practice, incorporates the use of herbal extracts and plant-based remedies to address a range of neuropsychiatric disorders [136–138]. This ancient healing system recognizes the potential of natural compounds derived from herbs and plants in promoting neurological and mental well-being. In recent times, scientific research has provided evidence supporting the neuroprotective properties of steroid phosphate esters derived from invertebrates or their semi-synthetic analogues. These compounds have demonstrated significant efficacy in safeguarding neurons and mitigating neurodegenerative processes. The exploration of steroid phosphate esters derived from invertebrates, or their synthetic counterparts, as potential neuroprotective agents is an exciting area of research. These compounds hold promise in the development of novel therapeutic interventions for various neurological conditions and disorders.

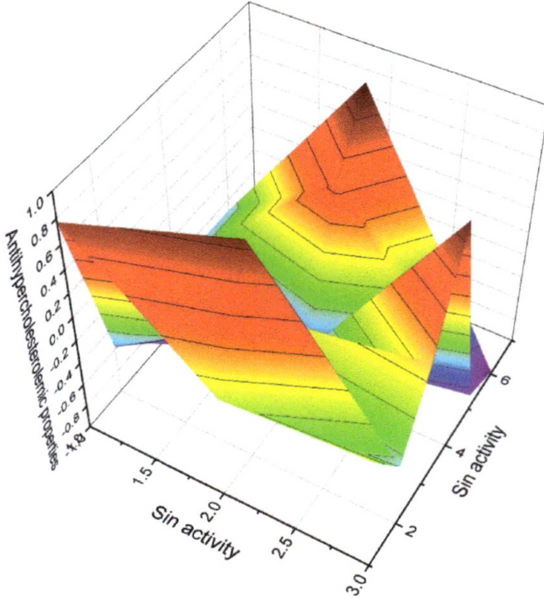

Figure 21. A 3D graph of the predicted and calculated anti-hypercholesterolemic activity of the steroid phosphate esters **104, 105, 106**. With a confidence level of over 96%, the graph reflects a high degree of certainty in the accuracy of the predicted and calculated activity values. It is noteworthy that various plants, such as *Hemidesmus indicus*, *Pulicaria gnaphalodes*, *Pandanus tectorius fruits*, *Buchholzia coriacea*, and *Swietenia mahagoni*, have been recognized for their anti-hypercholesterolemic properties, as demonstrated by their extracts [138–142]. Additionally, extracts from brown algae and the cyanobacterium *Arthrospira platensis* have also shown an anti-hypercholesterolemic effect. Interestingly, steroid phosphate esters derived from invertebrates exhibit strong anti-hypercholesterolemic properties. This highlights the potential of these compounds as promising candidates for the development of therapeutic interventions aimed at managing high cholesterol levels.

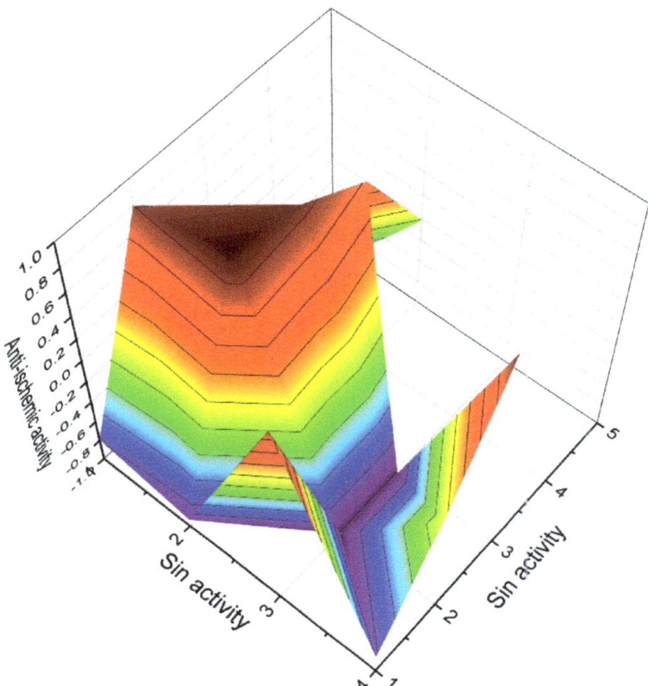

Figure 22. A 3D graph illustrating the predicted and calculated anti-ischemic activity of steroid phosphate esters, specifically compounds **111**, **112**, **113**, and **114**. The graph depicts the relationship between the activity of these compounds and their effectiveness in reducing myocardial oxygen consumption during ischemic conditions. The predicted and calculated activity values, depicted on the axes of the graph, represent the potency or efficacy of the steroid phosphate esters in terms of their anti-ischemic properties. The confidence level of over 93% indicates a high degree of certainty in the accuracy of the predicted and calculated activity values. Anti-ischemic activity refers to the ability of a compound to mitigate the detrimental effects of reduced blood flow and inadequate oxygen supply to the heart muscle. By reducing myocardial oxygen consumption, these steroid phosphate esters hold promise in preventing or alleviating ischemic episodes and related cardiac complications. It is important to note that further research, including experimental validation and clinical trials, is necessary to fully understand the mechanisms of action and optimal applications of these steroid phosphate esters as anti-ischemic agents. Their potential therapeutic implications in the context of ischemic heart disease warrant exploration to develop effective treatments for this condition.

4. Steroids Bearing a Halogen Atom (Cl, Br, or I)

Natural halogenated steroids are a class of organic compounds that contain halogen atoms (such as fluorine, chlorine, bromine, or iodine) attached to the steroid structure. These compounds can be found in various natural sources, including marine organisms, plants, and microorganisms [143–147].

Halogenated steroids often exhibit unique chemical and biological properties due to the presence of halogen atoms. The incorporation of halogens into the steroid structure can affect the compound's stability, lipophilicity, and interactions with biological systems. Marine organisms, particularly marine sponges, are known to produce a wide range of halogenated steroids. These compounds are believed to play important roles in the defense mechanisms of these organisms, protecting them against predators and pathogens. Halogenated steroids from marine sources have been the subject of extensive research due to their diverse biological activities and potential therapeutic applications. Some of these compounds have demonstrated anti-microbial, anti-viral, anti-inflammatory, and anti-cancer properties [143–156].

4.1. Chlorinated Plant Steroids

Chlorinated plant steroids are a specific subset of plant steroids that contain chlorine atoms attached to their chemical structure. These compounds are derived from plants and exhibit unique properties and potential biological activities due to the presence of chlorine atoms [143,146,148,157]. These chlorinated plant steroids can be found in various plant species, particularly those that have adapted to environments with high chlorine levels, such as salt marshes or coastal areas. These compounds are believed to play a role in the plants' adaptation to such environments, helping them cope with salinity stress or other ecological factors.

Chlorinated plant steroids have also been investigated for their potential as bioactive compounds with pharmacological applications. However, further research is needed to fully understand their mechanisms of action, physiological functions, and potential therapeutic uses. It is important to note that the presence and biological activities of chlorinated plant steroids can vary among different plant species. Studying these compounds can provide valuable insights into plant adaptations to challenging environments and may contribute to the discovery of novel bioactive compounds with pharmaceutical or agricultural significance. Research on chlorinated plant steroids is still relatively limited compared to other classes of plant steroids. However, some studies have identified and characterized specific chlorinated plant steroids and explored their potential biological activities [145–156].

The discovery of chlorine-containing steroids began with the isolation of jaborosalactone C (**115**) and jaborosalactone E (**116**) from the leaves of the *Jaborosa integrifolia* plant, which belongs to the Solanaceae family (a representative plant is shown in Figure 23) [158]. These compounds represent the first identified chlorine-containing steroids. In addition, the *Acnistus breviflorus* plant has been found to produce steroids such as compound **116** and compound **117**, which possess cytostatic activity. Similarly, cytotoxic withanolide (**117**, structure seen in Figure 24) has been isolated from *Withania frutescens*, another plant from the Solanaceae family [159].

Physalolactone C (**118**), displayed in a 3D graph in Figure 25, was identified in the fruits of *Physalis peruviana* (Cape gooseberry) [160]. This compound is structurally similar to the aforementioned steroids and exhibits cytotoxic properties. Additionally, from the same plant, physalolactone (**119**) was obtained from the roots, and a minor steroid of the leaves, 4-deoxyphysalolactone (**120**), was extracted [161].

Physaguline B (**121**, activity shown in Table 6) was discovered in *Physalis angulata* [162]. This compound represents a chlorinated sterol found in the plant, expanding our knowledge of the chemical diversity within *Physalis* species. Withanolide D chlorohydrin (**122**), presented in a 3D graph in Figure 26, was identified in *Withania somnifera*, commonly known as Ashwagandha, while (**119**) and (**123**) were discovered in *Acnistus breviflorus* [163,164]. Further research on *W. somnifera* revealed the presence of withanolide C (**123**), (**119**), and (**124**). These compounds were also found in *Dunalia tubulosa*, which belongs to the Solanaceae family, closely related to the plants [165].

Jaborochlorodiol (**125**) and jaborochlorotriol (**126**), representing a new structural type of chlorinated steroid, were identified in extracts from *Jaborosa magellanica*, a flowering plant of the Solanaceae family found in Punta Arenas, Chile [166]. Furthermore, the aerial parts of *Tolpis proustii* and *T. lagopoda*, native to La Gomera, Canary Islands, led to the isolation of chlorinated sterols: 30-chloro-3β-acetoxy-22α-hydroxyl-20(21)-taraxastene (**127**) and its acetylated analogue (**128**). In vitro antioxidant activities of the extracts were evaluated using the DPPH and ABTS scavenging methods. The cytotoxicity of isolated compounds demonstrated activity against the human myeloid leukemia K-562 and K-562/ADR cell lines [167].

Withanolide Z (**129**) was isolated from *Withania somnifera* as an inhibitor of topoisomerase I from the parasite *Leishmania donovani*, suggesting its potential in anti-parasitic applications [168]. Cytotoxic phyperunolides C (**130**) were found in the leaves of *Physalis peruviana* [169,170], highlighting their potential cytotoxic properties. Hsieh et al. [171] isolated cytotoxic tubocapsenolide G (**131**) from *Tubocapsicum anomalum*.

Physagulin I (**132**, the 3D graph is shown in Figure 27), a 14β-hydroxywithanolide, has been isolated from *Physalis* species and possesses an α-oxygenated functionality at position 15 [172]. Additionally, jaborosalactol 23 (**133**), another 14β-hydroxywithanolide, has been identified in *Jaborosa bergii*, a flowering plant in the Solanaceae family [173]. Nicotra et al. [174] reported the isomeric chlorohydrin, jaborosalactone 37 (**134**, structure seen in Figure 28, and activity see in Table 7), from *Jaborosa rotacea*, and jaborosalactone T (**135**) was isolated from *Jaborosa sativa* (synonym *Trechonaetes sativa*) collected in Argentina [175]. Anomanolide D (**136**), identified as the 16α-hydroxy substituent, was discovered in the fruits of *Tubocapsicum anomalum* collected in Japan [176]. Additionally, tubonolide A (**137**, the 3D graph is shown in Figure 29), a 16,17-dihydroxylated withajardin, was found in the same plant [177].

Figure 23. Various plant species wherein sterols containing a chlorine atom have been discovered. Chlorinated steroids (**115** and **116**) were isolated from the leaves of the *Jaborosa integrifolia* (**a**); steroids (**116**) and (**117**) were found in *Acnistus breviflorus* (**b**); withanolide Z (**129**) was isolated from *Withania somnifera* (**c**); and steroid (**140**) was found in *Tubocapsicum anomalum* (**d**).

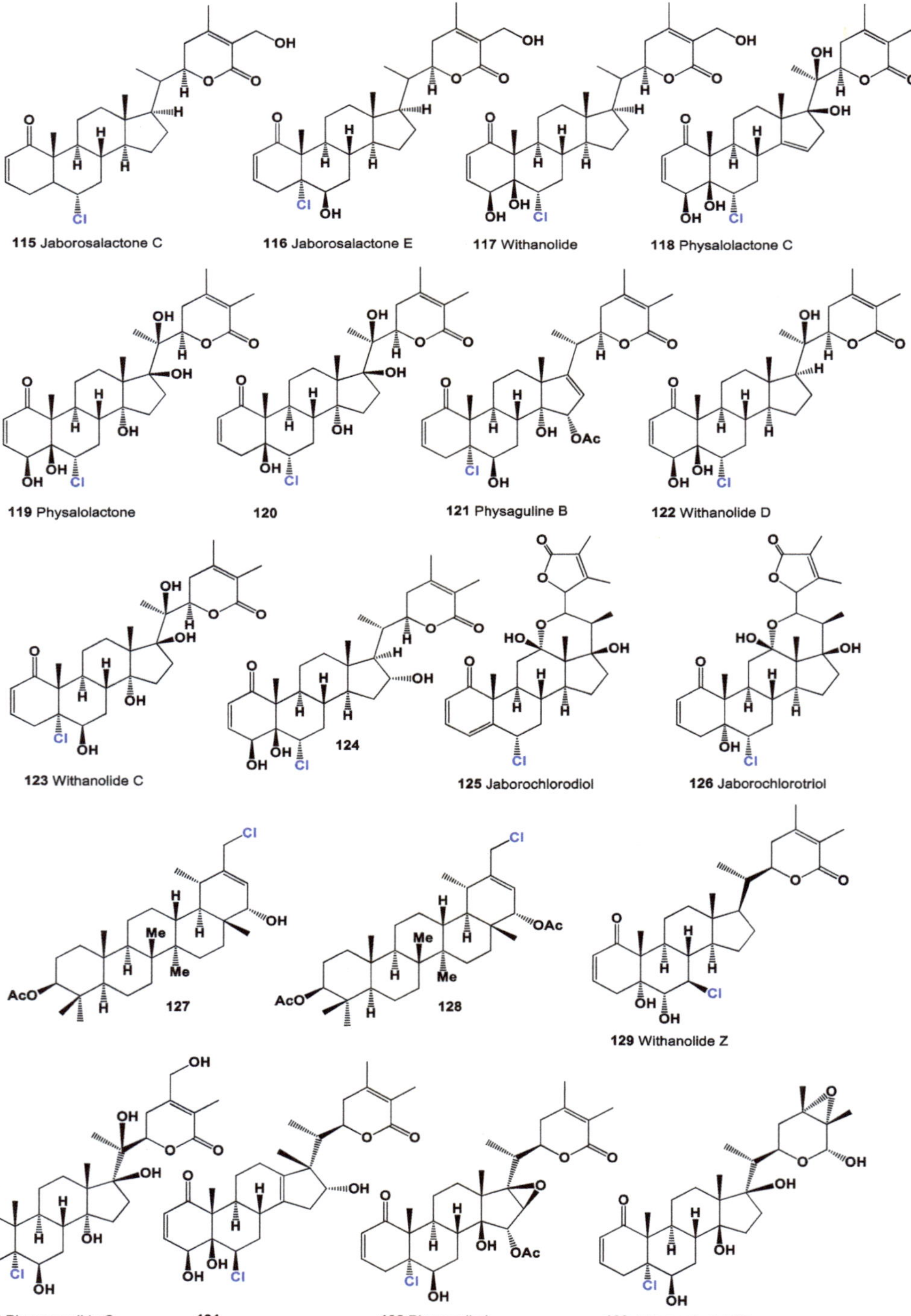

Figure 24. Chlorinated steroids and triterpenoids derived from plants.

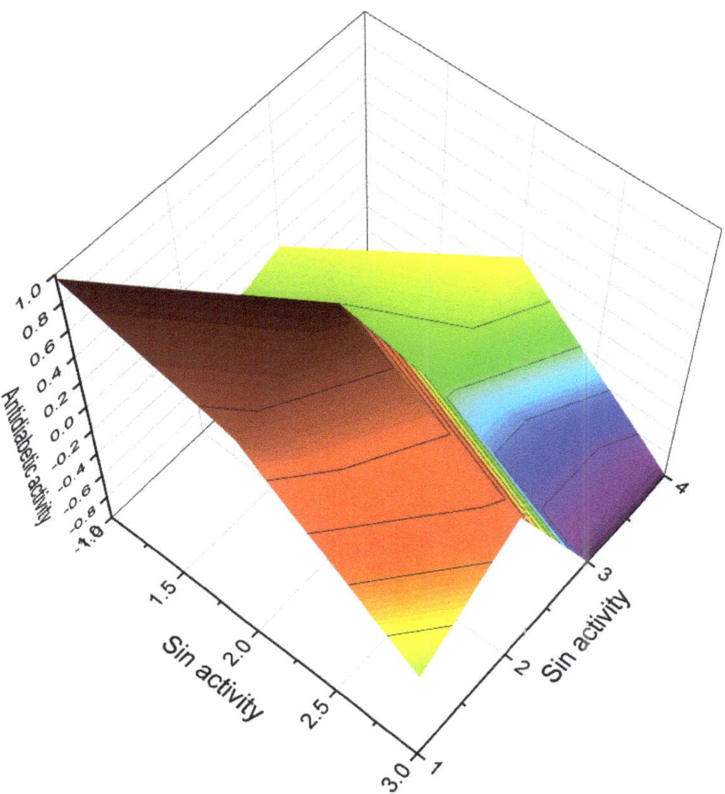

Figure 25. A 3D graph displaying the predicted and calculated anti-diabetic activity of chlorinated steroids, specifically compounds **118**, **119**, and **120**. The graph provides insights into the relationship between the activity of these compounds and their potential efficacy in managing diabetes. Anti-diabetic activity refers to the ability of a compound to help manage or control diabetes, a metabolic disorder characterized by high blood sugar levels. Compounds with anti-diabetic activity can exert various effects on glucose metabolism, insulin sensitivity, and other related pathways. The predicted and calculated activity values depicted on the graph represent the potency or effectiveness of the chlorinated steroids in terms of their anti-diabetic properties. With a confidence level of over 94%, the graph indicates a high degree of certainty in the accuracy of the predicted and calculated activity values. The exploration of chlorinated steroids for their anti-diabetic activity is of great interest in the field of diabetes research. Identifying compounds with potential anti-diabetic properties can contribute to the development of new treatment approaches and therapies for individuals living with diabetes. It is important to note that further research, including in vitro and in vivo studies, is necessary to fully understand the mechanisms of action, optimal dosage, and potential applications of these chlorinated steroids in managing diabetes. Additionally, clinical trials would be required to assess their safety and efficacy in human subjects. The study of chlorinated steroids and their anti-diabetic activity holds promise in advancing our understanding of natural compounds that may help in the management of diabetes and related metabolic disorders.

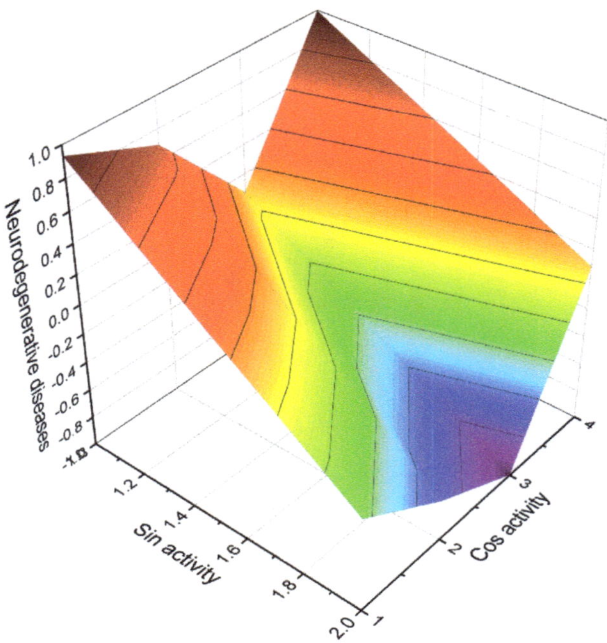

Figure 26. A 3D graph depicting the predicted and calculated activity of a specific chlorinated steroid (**122**) as a potential treatment for neurodegenerative diseases, specifically Alzheimer's and Parkinson's diseases. The graph provides insights into the relationship between the activity of the compound and its potential efficacy in treating these conditions. Neurodegenerative diseases, such as Alzheimer's disease and Parkinson's disease, are characterized by the progressive loss of structure and function of neurons in the central nervous system. These diseases often lead to cognitive decline, motor impairments, and various neurological symptoms. The predicted and calculated activity values shown on the graph represent the potency or effectiveness of the chlorinated steroid in terms of its activity against neurodegenerative diseases. With a confidence level of over 92%, the graph indicates a high degree of certainty in the accuracy of the predicted and calculated activity values. The exploration of chlorinated steroids for their potential therapeutic effects in neurodegenerative diseases is an important area of research. These compounds may interact with various molecular targets and pathways involved in the pathogenesis of these diseases, potentially slowing down or preventing neuronal degeneration, reducing inflammation, or promoting neuroprotective mechanisms. It is crucial to note that further research, including preclinical and clinical studies, is necessary to fully understand the mechanisms of action, therapeutic potential, and safety profile of the specific chlorinated steroid (**122**) and other compounds in the treatment of neurodegenerative diseases. Developing effective treatments for Alzheimer's and Parkinson's diseases remains a significant challenge, and ongoing research is vital in advancing our understanding and finding novel therapeutic strategies. The study of chlorinated steroids and their potential role in neurodegenerative diseases provides hope for the development of new therapeutic interventions that can improve the quality of life for individuals affected by these devastating conditions.

Table 6. Biological activities of chlorinated plant steroids (**115–133**).

No.	Dominated Biological Activity (Pa) *	Additional Predicted Activities (Pa) *
115	Hepatic disorders treatment (0.940) Anti-eczematic (0.924)	Macular degeneration treatment (0.921) Cytostatic (0.904)
116	Hepatic disorders treatment (0.933) Anti-eczematic (0.932)	Macular degeneration treatment (0.926) Cytostatic (0.875)
117	Anti-eczematic (0.919) Hepatic disorders treatment (0.908)	Cytostatic (0.921) Macular degeneration treatment (0.912)
118	Anti-diabetic (0.938) Myocardial infarction treatment (0.823)	Anti-eczematic (0.902) Alzheimer's disease treatment (0.664)
119	Anti-diabetic (0.981) Lipoprotein disorders treatment (0.938)	Anti-eczematic (0.902) Alzheimer's disease treatment (0.666)
120	Anti-diabetic (0.980) Lipoprotein disorders treatment (0.939)	Anti-eczematic (0.897) Alzheimer's disease treatment (0.696)
121	Apoptosis agonist (0.888) Anti-neoplastic (0.860)	Anti-eczematic (0.910) Cytostatic (0.643)
122	Neurodegenerative diseases treatment (0.913) Alzheimer's disease treatment (0.889)	Anti-eczematic (0.926) Anti-Parkinsonian (0.856)
123	Lipoprotein disorders treatment (0.968) Anti-diabetic (0.953)	Anti-eczematic (0.912) Alzheimer's disease treatment (0.670)
124	Anti-eczematic (0.930) Myocardial infarction treatment (0.872)	Anti-neoplastic (0.866) Cytostatic (0.819)
125	Anti-eczematic (0.823) Allergic conjunctivitis treatment (0.629)	Anti-neoplastic (0.785) Anti-inflammatory (0.731)
126	Myocardial infarction treatment (0.825) Anti-neoplastic (0.707)	Anti-eczematic (0.815) Allergic conjunctivitis treatment (0.618)
127	Anti-neoplastic (0.918) Apoptosis agonist (0.793) Anti-neoplastic (myeloid leukemia) (0.520)	Respiratory analeptic (0.757) Anti-secretoric (0.755) Lipid metabolism regulator (0.677)
128	Anti-neoplastic (0.892) Apoptosis agonist (0.796) Anti-metastatic (0.551)	Hepatoprotectant (0.739) Hepatic disorders treatment (0.701) Dermatologic (0.614)
129	Cytostatic (0.863) Anti-neoplastic (0.826) Apoptosis agonist (0.797)	Anti-eczematic (0.929) Macular degeneration treatment (0.856) Alzheimer's disease treatment (0.729)
130	Lipoprotein disorders treatment (0.952) Anti-diabetic (0.943) Anti-asthmatic (0.593)	Anti-eczematic (0.904) Anti-neoplastic (0.765) Anti-leukemic (0.651)
131	Insulin promoter (0.986) Myocardial infarction treatment (0.868) Anti-neoplastic (0.833) Apoptosis agonist (0.768)	Anti-eczematic (0.910) Anti-fungal (0.670) Anti-psoriatic (0.582) Anti-bacterial (0.535)
132	Anti-eczematic (0.914) Anti-fungal (0.795) Anti-parasitic (0.756)	Anti-neoplastic (0.854) Apoptosis agonist (0.786) Cytostatic (0.722)
133	Anti-neoplastic (0.914) Apoptosis agonist (0.823)	Anti-asthmatic (0.834) Anti-allergic (0.828)

* Only activities with Pa > 0.7 are shown.

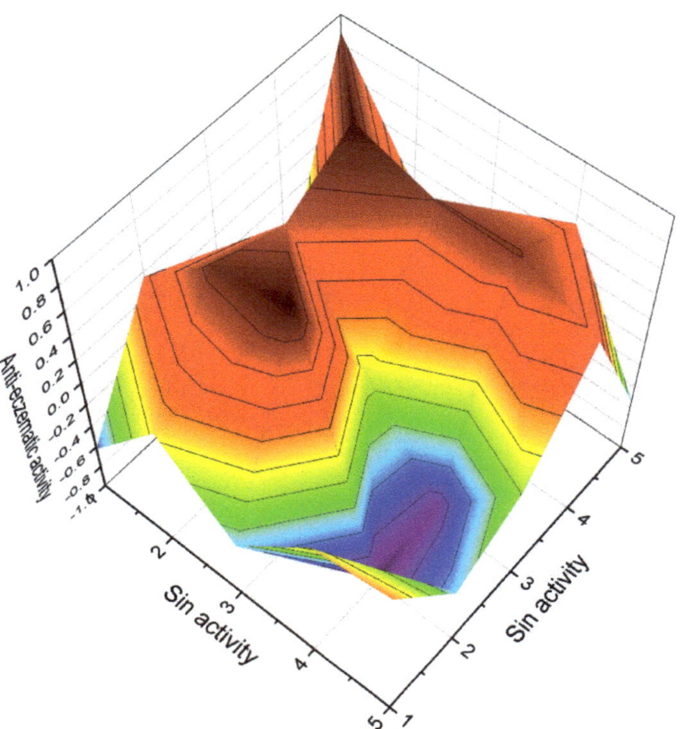

Figure 27. A 3D graph illustrating the predicted and calculated anti-eczematic activity of chlorinated steroids, specifically compounds **115**, **116**, **117**, **124**, and **132**. The graph provides insights into the relationship between the activity of these compounds and their potential efficacy in treating eczema. Anti-eczematic activity refers to the ability of a compound to alleviate or manage symptoms associated with eczema, a chronic inflammatory skin condition characterized by itching, redness, and rash. Compounds with anti-eczematic activity can help reduce inflammation, relieve itching, and promote skin healing. The predicted and calculated activity values depicted on the graph represent the potency or effectiveness of the chlorinated steroids in terms of their anti-eczematic properties. With a confidence level of over 91%, the graph indicates a high degree of certainty in the accuracy of the predicted and calculated activity values. The exploration of chlorinated steroids for their anti-eczematic activity holds promise in the field of dermatology and skin health. Identifying compounds that can effectively reduce inflammation, alleviate itching, and promote skin repair can significantly improve the management of eczema. It is important to note that further research, including in vitro and clinical studies, is necessary to fully understand the mechanisms of action, optimal dosage, and potential applications of these chlorinated steroids in treating eczema. Additionally, comprehensive safety evaluations would be required to assess their suitability for use in human subjects.

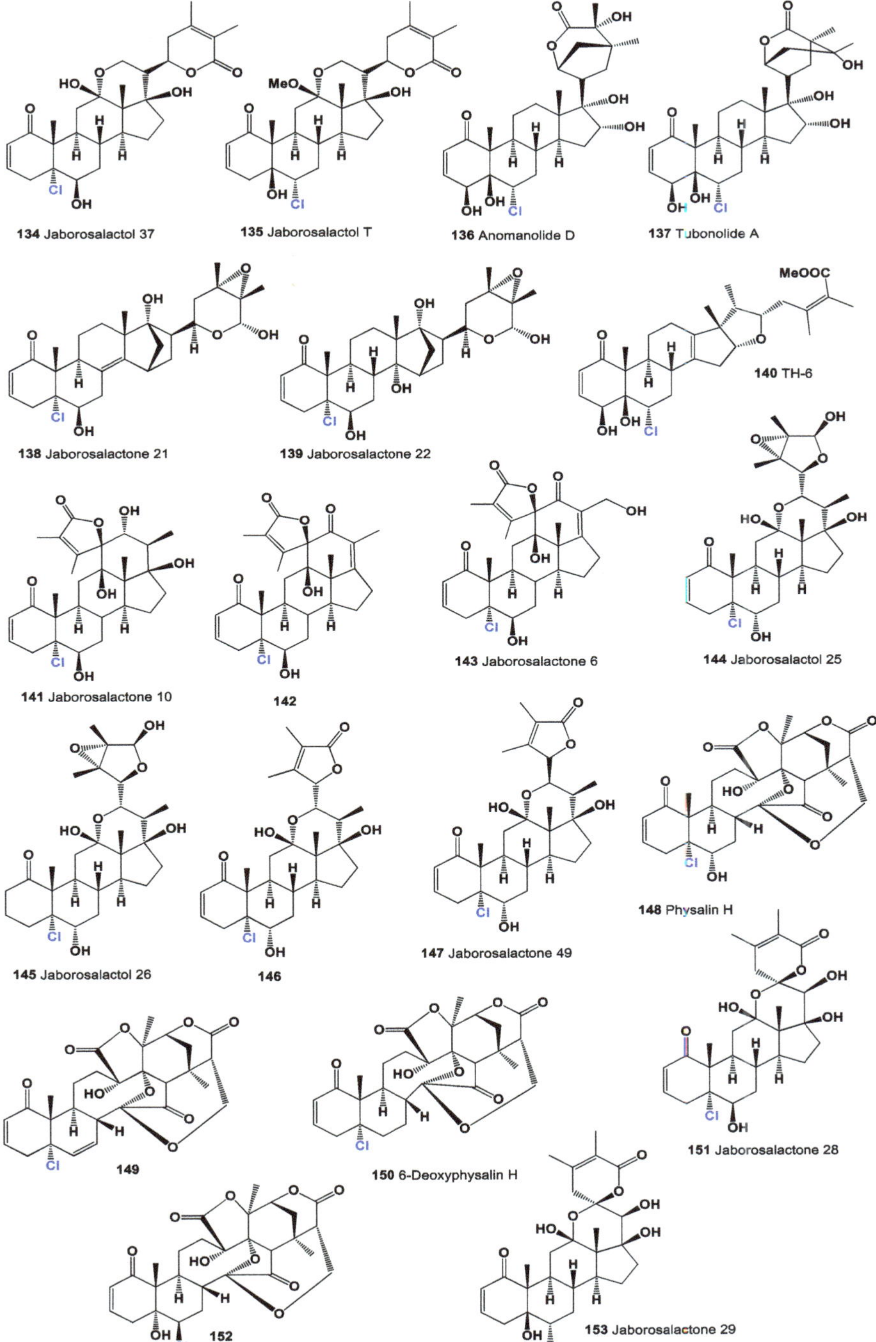

Figure 28. Chlorinated highly oxygenated steroids derived from plants.

Table 7. Biological activities of chlorinated plant steroids (**134–153**).

No.	Dominated Biological Activity (Pa) *	Additional Predicted Activities (Pa) *
134	Apoptosis agonist (0.806) Anti-neoplastic (0.803)	Genital warts treatment (0.724) Anti-eczematic (0.718)
135	Insulin promoter (0.981) Myocardial infarction treatment (0.819)	Anti-neoplastic (0.797) Apoptosis agonist (0.695)
136	Insulin promoter (0.986) Myocardial infarction treatment (0.899)	Anti-neoplastic (0.866) Apoptosis agonist (0.772)
137	Insulin promoter (0.986) Myocardial infarction treatment (0.899)	Anti-neoplastic (0.839) Apoptosis agonist (0.696)
138	Anti-neoplastic (0.875) Apoptosis agonist (0.795)	Anti-asthmatic (0.816) Anti-allergic (0.533)
139	Anti-neoplastic (0.885) Apoptosis agonist (0.824)	Anti-psoriatic (0.595) Anti-allergic (0.539)
140	Anti-neoplastic (0.806) Apoptosis agonist (0.634)	Myocardial infarction treatment (0.781) Hypolipemic (0.599)
141	Hepatic disorders treatment (0.934) Immunosuppressant (0.691)	Anti-allergic (0.618) Allergic conjunctivitis treatment (0.543)
142	Hepatic disorders treatment (0.942) Anti-neoplastic (0.782)	Anti-allergic (0.758) Anti-asthmatic (0.728)
143	Hepatic disorders treatment (0.930) Anti-neoplastic (0.753)	Anti-allergic (0.711) Allergic conjunctivitis treatment (0.597)
144	Anti-neoplastic (0.888) Apoptosis agonist (0.761)	Anti-inflammatory (0.815) Anti-fungal (0.629)
145	Anti-neoplastic (0.907) Apoptosis agonist (0.673)	Anti-inflammatory (0.824) Anti-fungal (0.597)
146	Anti-eczematic (0.850) Anti-neoplastic (0.765)	Allergic conjunctivitis treatment (0.649) Anti-allergic (0.641)
147	Anti-eczematic (0.850) Anti-pruritic (0.787)	Allergic conjunctivitis treatment (0.649) Anti-allergic (0.641)
148	Anti-protozoal (0.956) Genital warts treatment (0.824)	Anti-neoplastic (0.761) Anti-metastatic (0.530)
149	Anti-protozoal (0.954) Genital warts treatment (0.805)	Anti-neoplastic (0.759) Apoptosis agonist (0.540)
150	Anti-protozoal (0.958) Anti-protozoal (*Plasmodium*) (0.953)	Genital warts treatment (0.798) Anti-neoplastic (0.766)
151	Insulin promoter (0.984) Cytostatic (0.907)	Anti-eczematic (0.907) Anti-fungal (0.752)
152	Insulin promoter (0.982) Cytostatic (0.921)	Anti-eczematic (0.919) Macular degeneration treatment (0.912)
153	Anti-eczematic (0.922) Macular degeneration treatment (0.913)	Anti-neoplastic (0.868) Cytostatic (0.866)

* Only activities with Pa > 0.7 are shown.

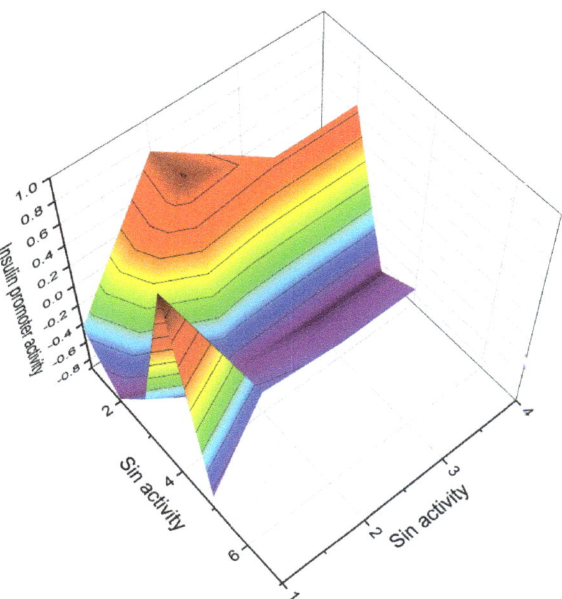

Figure 29. A 3D graph illustrating the predicted and calculated insulin promoter activity of chlorinated steroids, specifically compounds **131**, **135**, **136**, and **137**. The graph provides insights into the relationship between the activity of these compounds and their potential efficacy in promoting insulin production. Insulin promoter activity refers to the ability of a compound to enhance the production or secretion of insulin, a hormone that plays a crucial role in regulating blood sugar levels. Compounds with insulin promoter activity can help improve glucose metabolism and enhance insulin signaling, which is beneficial for individuals with conditions such as diabetes. The predicted and calculated activity values displayed on the graph represent the potency or effectiveness of the chlorinated steroids in terms of their insulin promoter properties. With a confidence level of over 98%, the graph indicates a high degree of certainty in the accuracy of the predicted and calculated activity values. The exploration of chlorinated steroids for their insulin promoter activity holds significant promise in the field of diabetes research. Identifying compounds that can enhance insulin production or secretion can contribute to the development of new strategies for managing diabetes and improving glycemic control.

Unusual 15,21-cyclowithanolides of the norbornane type, jaborosalactols 21 (**138**) and 22 (**139**), were isolated from *Jaborosa bergii* [178]. Furthermore, the acid hydrolysate of a methanolic extract of *Tubocapsicum anomalum* contained TH-6 (**140**) [179]. These discoveries highlight the occurrence of chlorine-containing steroids in plants, particularly in the Solanaceae family. The identification and characterization of these compounds contribute to our understanding of the chemical diversity of natural products and their potential biological activities. Further research is needed to explore the mechanisms of action and therapeutic applications of these chlorine-containing steroids in various fields, including medicine and agriculture.

A group of spiranoid withanolides with a 17(20)-ene-22-keto system, namely jaborosalactones 3 (**142**) and 6 (**143**), were isolated from *Jaborosa runcinata* collected in Argentina [180]. These compounds represent chlorinated steroids with unique structural features. Additionally, jaborosalactone 10 (**141**), presented in a 3D graph in Figure 30, was found in both *J. runcinata* and *J. odonelliana* [181]. This compound further expands our understanding of the chemical diversity within the *Jaborosa* genus.

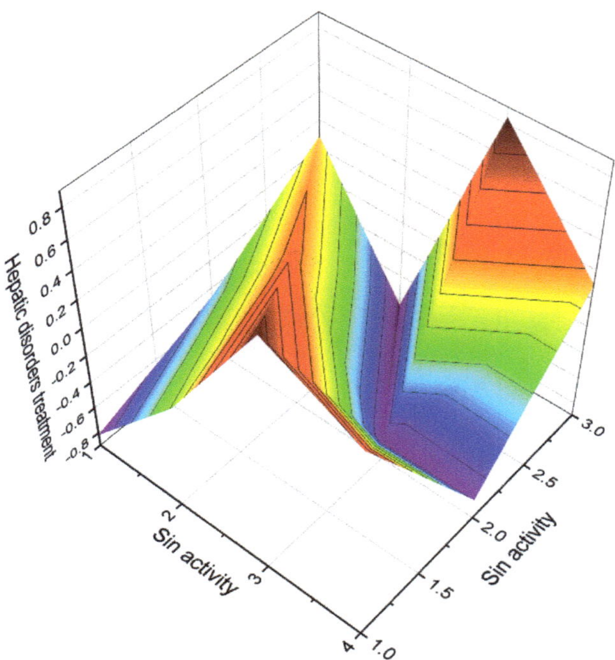

Figure 30. A 3D graph depicting the predicted and calculated activity of chlorinated steroids, specifically compounds **141**, **142**, and **143**, as potential treatments for liver disease. The graph provides insights into the relationship between the activity of these compounds and their potential efficacy in treating liver diseases. Liver disease refers to a wide range of conditions that affect the liver, impairing its normal functioning. These conditions can include liver inflammation (hepatitis), fatty liver disease, cirrhosis, liver cancer, and others. Treatment options for liver disease are diverse, including medications that can help manage symptoms, slow down disease progression, or promote liver regeneration. The predicted and calculated activity values displayed on the graph represent the potency or effectiveness of the chlorinated steroids in terms of their activity against liver disease. With a confidence level of over 93%, the graph indicates a high degree of certainty in the accuracy of the predicted and calculated activity values. The exploration of chlorinated steroids for their potential therapeutic effects in liver disease is an area of active research. These compounds may interact with various molecular targets and pathways involved in liver function, inflammation, and regeneration, potentially offering benefits in the management of liver diseases.

Two chlorinated 24,25-epoxy-γ-lactols (**144** and **145**) were isolated from plants of *Jaborosa parviflora* [182]. These compounds possess a chlorine atom and an epoxy group within their structures, contributing to their distinctive properties. Furthermore, the chlorohydrins jaborosalactone 42 (**146**) and jaborosalactone 49 (**147**) were detected in *Jaborosa caulescens* var. *bipinnatifida* [183] and *Jaborosa laciniata* [184]. These compounds exhibit a chlorohydrin moiety, further enhancing the chemical diversity within the *Jaborosa* species.

A group of constituents called physalins, which belong to the 13,14-seco-16,24-cycloergostane class of compounds, have been identified in extracts of *Brachistus stramoniifolius*, *Margaranthus solanaceous* (sub nom. *Physalis solanaceous*), and *Schraderanthus viscosus* (sub nom. *Saracha viscosa*) [185–187]. These compounds, including physalins **148**, **149**, **150**, and **151** (the 3D graph is shown in Figure 31), exhibit unique structural characteristics within the 13,14-seco-16,24-cycloergostane framework.

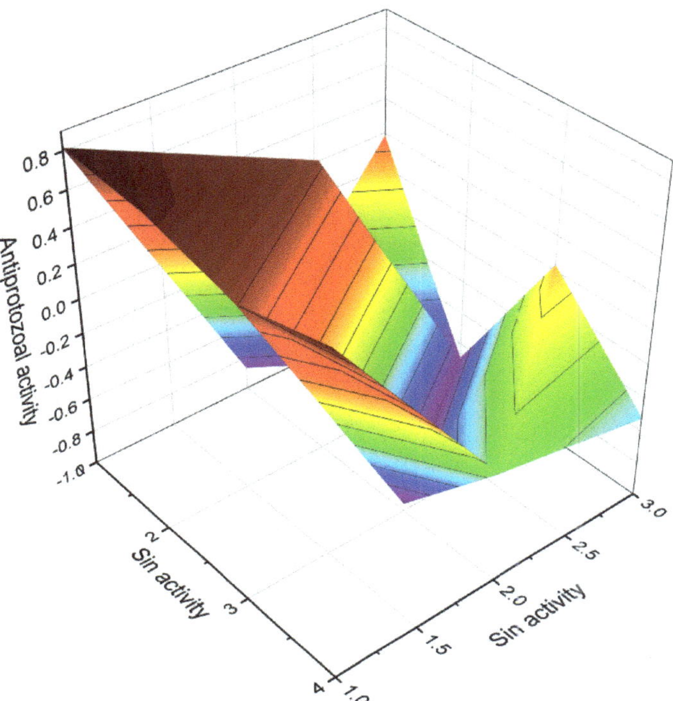

Figure 31. A 3D graph illustrating the predicted and calculated anti-protozoal activity of chlorinated steroids, specifically compounds **148**, **149**, and **150**. The graph provides insights into the relationship between the activity of these compounds and their potential efficacy in inhibiting protozoan parasites. Anti-protozoal activity refers to the ability of a compound to inhibit the growth or survival of protozoan parasites, which are single-celled organisms that can cause various infectious diseases in humans and animals. Protozoan parasites can cause diseases such as malaria, leishmaniasis, trypanosomiasis, and toxoplasmosis, among others. The predicted and calculated activity values depicted on the graph represent the potency or effectiveness of the chlorinated steroids in terms of their anti-protozoal properties. With a confidence level of over 95%, the graph indicates a high degree of certainty in the accuracy of the predicted and calculated activity values. The exploration of chlorinated steroids for their anti-protozoal activity is of great interest in the field of parasitology and drug discovery. Identifying compounds that can effectively target and inhibit protozoan parasites can lead to the development of new treatments for various protozoal infections.

Two withanolides with a hemiketal bridge between what was originally ketone functions at C-12 and C-22 have also been discovered. Upon formation of the D-lactone, these compounds, known as **152** and **153**, were detected and identified from *Jaborosa rotacea* [188]. These compounds demonstrate a distinct structural arrangement, featuring a six-membered ring with a β-oriented hydroxy group at C-12 and a spiroketal at C-22. Figures 28 and 32 showcase the structures of various steroids, providing an overview of the diversity within the class. Furthermore, Table 1 presents the biological activities associated with plant chlorinated steroids, highlighting their cytostatic, anti-neoplastic, anti-eczematic, anti-diabetic, anti-bacterial, and other activities. These chlorinated steroids exhibit a range of characteristic biological activities, indicating their potential significance in various fields, including medicine, pharmacology, and agriculture. However, it is important to conduct further research, including in vitro and in vivo studies, to fully understand the mechanisms of action, therapeutic potential, and safety profile of these compounds.

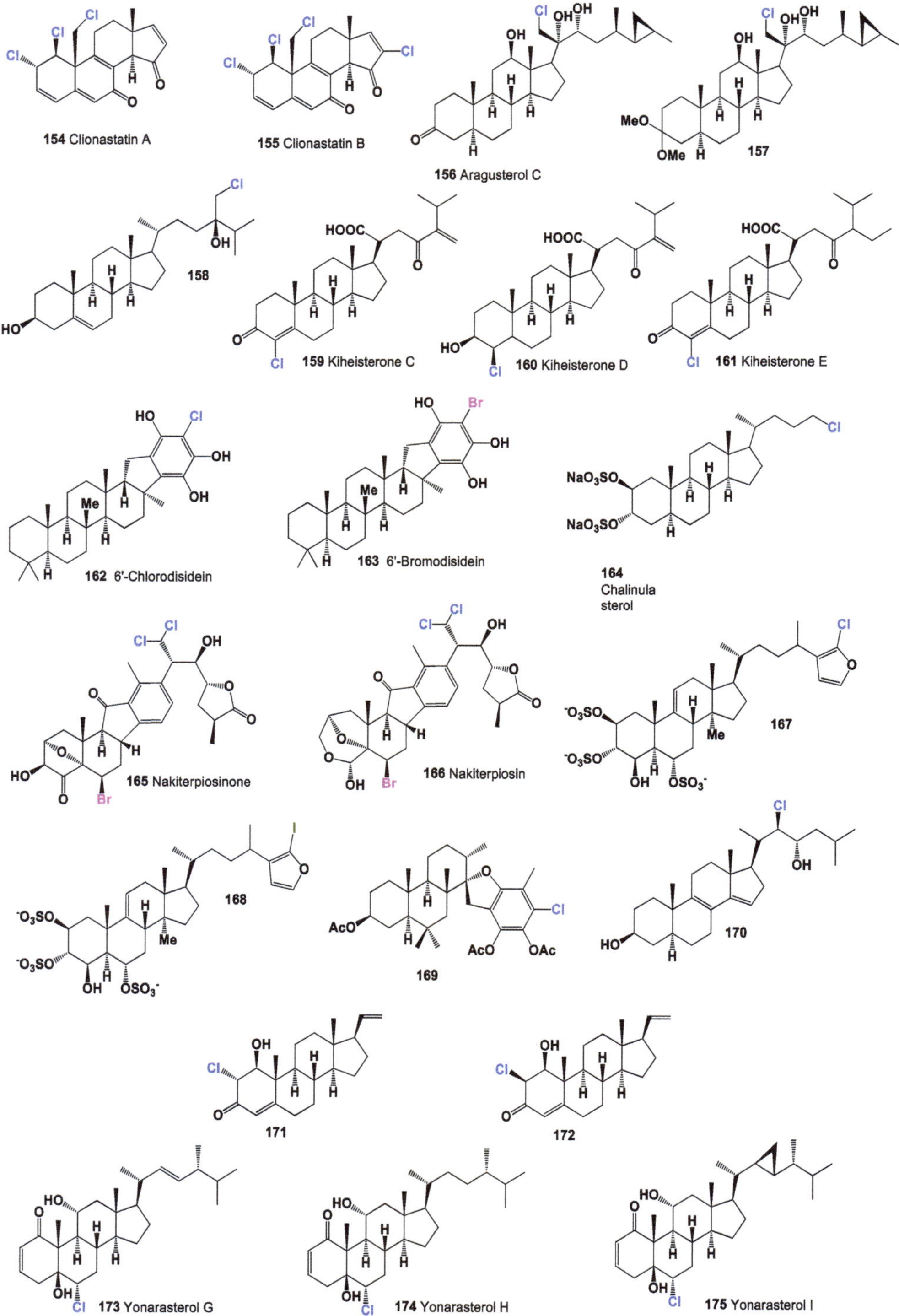

Figure 32. Halogenated steroids derived from marine sources.

4.2. Halogenated Steroids Derived from Marine Sources

Halogenated steroids derived from marine sources are natural compounds that contain halogen atoms (such as chlorine, bromine, or iodine) and are obtained from various marine organisms. These marine organisms can include algae, sponges, corals, mollusks, and other marine invertebrates. Halogenated steroids from marine sources exhibit diverse chemical structures and biological activities, making them of interest in the fields of pharmacology and drug discovery [143,145,147,154].

Strong cytotoxic chlorinated steroids known as clionastatins A (**154**) and B (**155**) have been discovered in the burrowing sponge *Cliona nigricans*. The structures of marine steroids can be observed in Figure 32 and activity see in Table 8. These remarkable compounds contain tri- and tetrachlorinated androstane derivatives, respectively. They are considered the first polyhalogenated steroids found in a living organism, whether marine or terrestrial, and represent the first instances of halogenated androstanes in nature [189]. Clionastatins A and B exhibit potent cytotoxic activity, making them of significant interest in the field of cancer research and drug development. These compounds have shown the ability to inhibit the growth of cancer cells in vitro and have demonstrated promising anti-cancer potential.

The discovery of clionastatins A and B highlights the unique chemistry and biodiversity found in marine organisms. These compounds contribute to our understanding of the natural products derived from marine sources and their potential therapeutic applications. Further research is needed to elucidate the precise mechanisms of action and therapeutic potential of clionastatins A and B, as well as to explore their structure–activity relationships. Investigating these compounds can provide insights into the development of novel anti-cancer agents and inspire the discovery of additional halogenated steroids derived from marine organisms.

Aragusterol C (**156**), a chlorinated steroid, was isolated from an Okinawan marine sponge of the genus *Xestospongia* sp. This compound exhibited strong inhibitory effects on the proliferation of KB cells in vitro. Furthermore, it demonstrated potent in vivo anti-tumor activity against L1210 cells in mice [190]. The distribution of biological activity percentages for aragusterol C is depicted in Figure 33. Another compound, aragusteroketal C (**157**), which is a steroid with a dimethylketal structure, was also isolated from the same sponge. This chlorinated steroid displayed cytotoxic activity against the KB tumor cell line, with an IC_{50} value of 4 ng/mL [191]. Additionally, a chlorinated steroid (**158**) was isolated from the soft coral *Sinularia brassica*. This coral-derived compound offers unique structural and chemical characteristics [192]. The coral sample associated with this compound is shown in Figure 34.

Table 8. Biological activities of halogenated steroids (**154–175**).

No.	Dominated Biological Activity (Pa) *	Additional Predicted Activities (Pa) *
154	Anti-neoplastic (0.860) Prostate disorders treatment (0.781)	Bone diseases treatment (0.722) Anti-inflammatory (0.639)
155	Anti-neoplastic (0.894) Prostate disorders treatment (0.799)	Bone diseases treatment (0.787) Anti-inflammatory (0.731)
156	Anti-neoplastic (0.934) Prostate cancer treatment (0.885) Anti-neoplastic (sarcoma) (0.875) Anti-neoplastic (renal cancer) (0.820)	Choleretic (0.879) Anti-hypercholesterolemic (0.828) Anti-fungal (0.781) Dermatologic (0.778)
157	Anti-neoplastic (0.922) Anti-neoplastic (sarcoma) (0.836)	Anti-osteoporotic (0.803) Bone diseases treatment (0.781)
158	Anti-hypercholesterolemic (0.937) Atherosclerosis treatment (0.831)	Respiratory analeptic (0.878) Anti-infertility, female (0.833)
159	Anti-neoplastic (0.881) Growth stimulant (0.751)	Dermatologic (0.771) Anti-fungal (0.696)

Table 8. *Cont.*

160	Anti-hypercholesterolemic (0.885)	Anesthetic general (0.823)
161	Anti-neoplastic (0.810) Apoptosis agonist (0.776)	Prostate disorders treatment (0.688) Acute neurologic disorders treatment (0.680)
162	Anti-neoplastic (0.805) Apoptosis agonist (0.744) Cytoprotectant (0.690) Prostate disorders treatment (0.681)	Dermatologic (0.750) Anti-viral (influenza) (0.738) Anti-bacterial (0.736) Anti-fungal (0.728)
163	Anti-neoplastic (0.805) Apoptosis agonist (0.744) Cytoprotectant (0.690) Prostate disorders treatment (0.681)	Dermatologic (0.750) Anti-viral (influenza) (0.738) Anti-bacterial (0.736) Anti-fungal (0.728)
164	Anti-neoplastic (0.851) Anti-carcinogenic (0.754)	Biliary tract disorders treatment (0.841) Bone diseases treatment (0.725)
165	Anti-neoplastic (0.882) Cytostatic (0.793)	Anti-bacterial (0.736) Anti-fungal (0.695)
166	Anti-neoplastic (0.822) Cytostatic (0.782)	Anti-parasitic (0.718) Anti-protozoal (0.714)
167	Glucan endo-1,3-b-D-glucosidase inhibitor (0.890)	Biliary tract disorders treatment (0.845)
168	Anti-neoplastic (0.884)	Anti-inflammatory (0.829)
169	Anti-inflammatory (0.829) Anti-viral (0.826)	Anti-neoplastic (0.784) Apoptosis agonist (0.763)
170	Anti-hypercholesterolemic (0.941) Atherosclerosis treatment (0.831)	Anti-infertility, female (0.833) Prostate disorders treatment (0.773)
171	Anti-neoplastic (0.912) Cytoprotectant (0.764) Prostate disorders treatment (0.767)	Respiratory analeptic (0.894) Erythropoiesis stimulant (0.776)
172	Anti-neoplastic (0.912) Cytoprotectant (0.764) Prostate disorders treatment (0.767)	Respiratory analeptic (0.894) Erythropoiesis stimulant (0.776) Apoptosis agonist (0.677)
173	Anti-hypercholesterolemic (0.911) Myocardial infarction treatment (0.900) Atherosclerosis treatment (0.811)	Apoptosis agonist (0.862) Anti-neoplastic (0.846) Prostate disorders treatment (0.823)
174	Respiratory analeptic (0.911) Myocardial infarction treatment (0.906)	Anti-hypercholesterolemic (0.845) Anti-diabetic (type 2) (0.669)
175	Myocardial infarction treatment (0.864) Immunosuppressant (0.734)	Dermatologic (0.785) Anti-psoriatic (0.728)

* Only activities with Pa > 0.7 are shown.

Cytotoxic chlorinated ketosteroids known as kiheisterones C (**159**), D (**160**), and E (**161**) were discovered in the extracts of the marine sponge *Strongylacedon* sp. from Maui [193]. These compounds exhibit cytotoxic activity and represent an intriguing class of chlorinated ketosteroids derived from a marine source. In addition, unique pentacyclic saturated sesterpenes condensed with a hydroxy-hydroquinone moiety, known as 6′-chlorodisidein (**162**) and 6′-bromodisidein (**163**), have been isolated from the marine sponge *Disidea pallescens* in the form of disulfate sodium calcium salts [194]. These compounds possess a distinct structural arrangement, incorporating both chlorine and bromine atoms. The discovery of these chlorinated compounds further highlights the chemical diversity and pharmacological potential of natural products derived from marine organisms. The cytotoxic and unique structural characteristics of kiheisterones and disideins offer promising avenues for further exploration in the fields of cancer research and drug development.

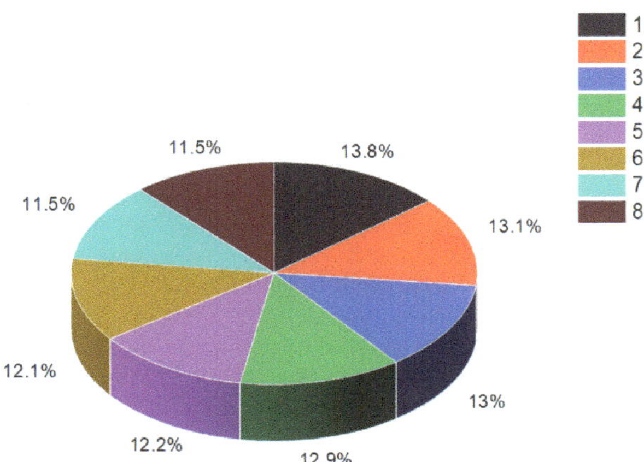

Figure 33. The percentage distribution of biological activities on the example of chlorinated steroid aragusterol C (**156**) from marine sponge *Xestospongia* sp., which has a wide range of pharmacological properties. The found activities are indicated under the numbers: 1, anti-neoplastic (13.8%); 2, prostate cancer treatment (13.1%); 3, anti-sarcoma cancer (13%); 4, anti-renal cancer (12.9%); 5, choleretic (12.2%); 6, anti-hypercholesterolemic (12.1%); 7, anti-fungal (11.5%); and 8, dermatologic (11.5%).

Figure 34. Some of the marine invertebrates and algae that produce chlorinated steroids. Chlorinated steroids (**154**) and B (**155**) were found in the sponge *Cliona nigricans* (**a**); chlorinated steroid (**158**) was isolated from the soft coral *Sinularia brassica* (**b**); the brown alga *Stypopodium flabelliforme* contained steroid (**169**) (**c**), and steroid (**170**) was found in starfish *Echilaster sepositus* (**d**).

Chalinulasterol (**164**), a chlorinated sterol disulfate, was isolated from the Caribbean sponge *Chalinula molitba* [195]. This compound represents a unique chlorinated sterol derivative found in a marine organism. Nakiterpiosinone (**165**) and nakiterpiosin (**166**), two related C-nor-D homosteroids, were identified in MeOH extracts of the sponge *Terpios hoshinota*. These compounds have shown potential as anti-cancer agents, particularly in tumors resistant to existing anti-mitotic agents and dependent on Hedgehog pathway responses for growth [196,197]. Their discovery highlights the importance of exploring marine sources for novel compounds with therapeutic potential.

The marine sponge *Topsentia* sp. yielded a chlorine-containing steroid sulfate (**167**) and the first natural iodinated steroid (**168**) [198]. These compounds showcase the chemical diversity of halogenated steroids derived from marine sources and contribute to our understanding of the unique natural products found in marine organisms. Chlorinated stypotriol triacetate (**169**) was detected in the dichloromethane extract of the brown alga *Stypopodium flabelliforme* [199]. This compound represents a chlorinated derivative of stypotriol, a sterol commonly found in brown algae. The identification of chlorinated derivatives expands our knowledge of the chemical variations within marine sterols. Furthermore, the (3β,5α,22R,23S)-22-chlorocholesta-8,14-diene-3,23-diol (**170**) was found in MeOH-CHCl$_3$ extracts of the starfish *Echinaster sepositus* [200]. This chlorinated steroid exhibits a unique structural arrangement and represents an interesting discovery in the field of marine natural products.

Two unique chloro-pregnane steroids (**171** and **172**) have been isolated from the eastern Pacific octocoral *Carijoa multiflora* [201]. These compounds exhibit distinct structures and represent novel chlorinated steroids found in the marine environment. The 3D graph depicting the predicted and calculated activity for compound **171** is shown in Figure 35. In addition, three chlorinated steroids, namely yonarasterols G (**173**), H (**174**), and I (**175**), were discovered in MeOH extracts of the Okinawan soft coral *Clavularia viridis* [202]. These compounds contribute to the growing repertoire of chlorinated steroids derived from marine sources. These compounds exhibit diverse chemical architectures and display unique halogenation patterns that contribute to their biological activities. The biological activities of marine halogenated steroids are varied, with anti-tumor, anti-fungal, anti-cancer, and anti-bacterial activities being characteristic among the compounds. Particularly, anti-cancer activity appears to be a common feature observed in the presented steroids.

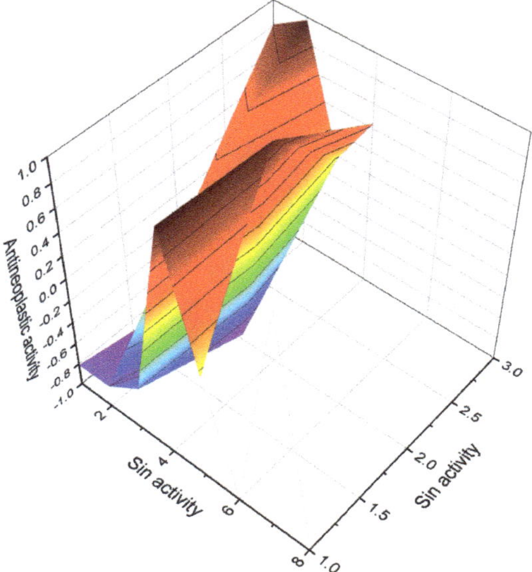

Figure 35. A 3D graph illustrating the predicted and calculated anti-neoplastic activity of halogenated

steroids (**156**, **157**, and **171**) derived from marine sources. Anti-neoplastic activity refers to the ability of a substance to inhibit or prevent the growth and spread of cancer cells. It is a crucial characteristic for potential cancer treatments. The graph provides valuable insights into the relationship between the structural features of these halogenated steroids and their potential effectiveness as anti-neoplastic agents. The predicted and calculated activity values depicted on the graph represent the potency or efficacy of these compounds for inhibiting the growth and proliferation of cancer cells. With a confidence level of over 92%, the graph indicates a high degree of certainty in the accuracy of the predicted and calculated anti-neoplastic activity values. This suggests that these halogenated steroids hold promise as potential candidates for further investigation as anti-cancer agents. The exploration of halogenated steroids derived from marine sources for their anti-neoplastic activity is of great interest in cancer research and drug development. Natural compounds with anti-neoplastic properties offer potential alternatives or adjuncts to traditional cancer therapies.

5. Conclusions

This comprehensive review has explored the diverse range of biological activity and structural variations found within steroids and related isoprenoid lipids. The analysis encompassed various natural compounds, including steroids with aromatic ring(s), steroid phosphate esters from marine invertebrates, and steroids bearing halogen atoms (I, Br, or Cl). These compounds are derived from sources such as fungi, fungal endophytes, plants, algae, and marine invertebrates. Through an examination of referenced literature sources, their biological activity was evaluated through in vivo and in vitro studies, as well as employing the QSAR method. The findings revealed a multitude of compounds exhibiting remarkable properties, including strong anti-neoplastic, anti-proliferative, anti-hypercholesterolemic, anti-Parkinsonian, anti-eczematic, anti-psoriatic, and various other activities. To enhance comprehension, the review incorporated visual aids such as 3D graphs illustrating the activity of individual steroids and images showcasing selected terrestrial or marine organisms. Furthermore, the review provided explanations elucidating certain types of biological activity associated with these compounds. Overall, the findings presented in this review not only contribute to the academic scientific knowledge in the field but also hold practical relevance for the development of pharmacological interventions and advancements in practical medicine. The review utilized data from various authors regarding the biological activity of natural steroids. To assess the potential activity of these steroids, the PASS program was employed. The PASS program utilizes structural features of compounds to predict their biological activity profiles. By inputting the structural information of the natural steroids into the program, their potential activity across multiple predefined activity classes was estimated. However, it is important to note that these predictions are based solely on structural information and should be validated through experimental studies.

Funding: This work did not receive any specific grant from funding agencies in the public, commercial, or not-for-profit sectors.

Institutional Review Board Statement: Not applicable.

Informed Consent Statement: Not applicable.

Data Availability Statement: Not applicable.

Conflicts of Interest: The author declares that he has no known competing financial interests or personal relationships that could affect the work described in this article.

References

1. Fahy, E.; Cotter, D.; Sud, M.; Subramaniam, S. Lipid classification, structures, and tools. *Biochim. Biophys. Acta* **2011**, *1811*, 637–647. [CrossRef] [PubMed]
2. Reszczyńska, E.; Hanaka, A. Lipids composition in plant membranes. *Cell Biochem. Biophys.* **2020**, *78*, 401–414. [CrossRef] [PubMed]
3. Bishop, G.J.; Koncz, C. Brassinosteroids and plant steroid hormone signaling. *Plant Cell* **2002**, *14*, S97–S110. [CrossRef] [PubMed]
4. Fernandes, D.; Loi, B.; Porte, C. Biosynthesis and metabolism of steroids in molluscs. *J. Steroid Biochem. Mol. Biol.* **2011**, *127*, 189–195. [CrossRef] [PubMed]
5. Sharifi, N.; Auchus, R.J. Steroid biosynthesis and prostate cancer. *Steroids* **2012**, *77*, 719–726. [CrossRef]
6. Li, J.; Papadopoulos, V.; Vihma, V. Steroid biosynthesis in adipose tissue. *Steroids* **2015**, *103*, 89–104. [CrossRef]
7. Moss, G.P. Nomenclature of steroids. *Pure Appl. Chem.* **1989**, *61*, 1783–1822. [CrossRef]
8. Russel, C.A. Organic chemistry: Natural products, steroids. In *Chemical History: Reviews of the Recent Literature*; Russell, C.A., Roberts, G.K., Eds.; RSC Publ.: Cambridge, UK, 2005.
9. Kirk, D.N.; Marples, B.A. The structure and nomenclature of steroids. In *Steroid Analysis*; Makin, H.L.J., Gower, D.B., Kirk, D.N., Eds.; Springer: Dordrecht, The Netherlands, 1995.
10. Li, H.M.; Chen, X.J.; Luo, D.; Fan, M.; Zhang, Z.J.; Peng, L.Y.; Wu, X.D.; Li, R.T.; Ji, X.; Zhao, Q.S. Protostane-type triterpenoids from *Alisma orientale*. *Chem. Biodivers.* **2017**, *14*, e1700452. [CrossRef]
11. Gribble, G.W. Biological activity of recently discovered halogenated marine natural products. *Mar. Drugs* **2015**, *13*, 4044–4136. [CrossRef]
12. Cabrita, M.T.; Vale, C.; Rauter, A.P. Halogenated compounds from marine algae. *Mar. Drugs* **2010**, *8*, 2301–2317. [CrossRef]
13. Ermolenko, E.V.; Imbs, A.B.; Gloriozova, T.A.; Poroikov, V.V.; Sikorskaya, T.V.; Dembitsky, V.M. Chemical diversity of soft coral steroids and their pharmacological activities. *Mar. Drugs* **2020**, *18*, 613. [CrossRef] [PubMed]
14. Mello, F.V.; Kasper, D.; Alonso, M.B.; Torres, J.P.M. Halogenated natural products in birds associated with the marine environment: A review. *Sci. Total Environ.* **2020**, *717*, 137000. [CrossRef] [PubMed]
15. Morais, T.; Cotas, J.; Pacheco, D.; Pereira, L. Seaweeds compounds: An ecosustainable source of cosmetic ingredients? *Cosmetics* **2021**, *8*, 8. [CrossRef]
16. Sanjeewa, K.A.; Jeon, Y.-J. Edible brown seaweeds: A review. *J. Food Bioact.* **2018**, *2*, 37–50. [CrossRef]
17. Sohn, S.-I.; Rathinapriya, P.; Balaji, S.; Jaya Balan, D.; Swetha, T.K.; Durgadevi, R.; Alagulakshmi, S.; Singaraj, P.; Pandian, S. Phytosterols in seaweeds: An overview on biosynthesis to biomedical applications. *Int. J. Mol. Sci.* **2021**, *22*, 12691. [CrossRef]
18. Gnanavel, V.; Roopan, S.M.; Rajeshkumar, S. Aquaculture: An overview of chemical ecology of seaweeds (food species) in natural products. *Aquaculture* **2019**, *507*, 1–6. [CrossRef]
19. Zhang, L.; Liao, W.; Huang, Y. Global seaweed farming and processing in the past 20 years. *Food Prod. Process. Nutr.* **2022**, *4*, 23. [CrossRef]
20. Sakthivel, R.; Devi, K.D. Antioxidant, anti-inflammatory and anticancer potential of natural bioactive compounds from seaweeds. *Stud. Nat. Prod. Chem.* **2019**, *63*, 113–160.
21. Dembitsky, V.M.; Savidov, N.; Poroikov, V.V. Naturally occurring aromatic steroids and their biological activities. *Appl. Microbiol. Biotechnol.* **2018**, *102*, 4663–4674. [CrossRef]
22. Kadis, B.M. Synthesis of Steroid Precursors. Ph.D. Thesis, Iowa State University, Ames, IA, USA, 1957.
23. Taub, D. Naturally occurring aromatic steroids. In *Total Synthesis of Natural Products*; John Wiley & Sons, Inc.: Hoboken, NJ, USA, 1973; Volume 2.
24. Rutherford, F.J. Ceric Oxidations of Aromatic Steroids and Related Compounds. Ph.D. Thesis, University of Edinburgh, Edinburgh, UK, 1972.
25. Niven, S.J. The Origins and Occurrence of Estrogenic A-Ring Aromatic Steroids in U.K. Sewage Treatment Works Effluents. Ph.D. Thesis, University of Plymouth, Plymouth, UK, 1999.
26. Gupta, R.R.; Jain, M. *Aliphatic and Aromatic Hydrocarbons, Steroids, Carbohydrates*; Springer: Berlin/Heidelberg, Germany, 2000.
27. Huang, H.; Yin, M.; Han, D. Novel parameters derived from alkylchrysenes to differentiate severe biodegradation influence on molecular compositions in crude oils. *Fuel* **2020**, *268*, 117366. [CrossRef]
28. Matyasik, I.; Bieleń, W. Aromatic steroids as a tool in geochemical interpretation. *Nafta-Gaz* **2015**, *71*, 376–383.
29. Yang, C.; Wang, Z.; Liu, Y.; Yang, Z.; Li, Y.; Shah, K. Aromatic steroids in crude oils and petroleum products and their applications in forensic oil spill identification. *Environ. Forensics* **2013**, *14*, 278–293. [CrossRef]
30. Barbanti, S.M.; Moldowan, J.M.; Watt, D.S.; Kolaczkowska, E. New aromatic steroids distinguish Paleozoic from Mesozoic oil. *Org. Geochem.* **2011**, *42*, 409–424. [CrossRef]
31. Li, L.; Jiang, L.; George, S.C.; Liu, Z. Aromatic compounds in lacustrine sediments from the Lower Cretaceous Jiufotang formation, Chaoyang basin (NE China). *Mar. Pet. Geol.* **2021**, *129*, 105111. [CrossRef]
32. Lednicer, D. *Steroid Chemistry at a Glance*; John Wiley & Sons, Inc.: Hoboken, NJ, USA, 2010; p. 152.
33. Pantoja, S.; Wakeham, S. Marine organic geochemistry: A general overview. In *Chemical Processes in Marine Environments*; Gianguzza, A., Pelizetti, E., Sammartano, S., Eds.; Environmental Science; Springer: Berlin/Heidelberg, Germany, 2000; pp. 43–74.
34. Killops, S.; Killops, V. FrontMatter. In *Front Matter, in Introduction to Organic Geochemistry*; Blackwell Publishing Ltd.: Malden, MA, USA, 2004.

35. Fluhmann, C.F. Estrogenic hormones: Their clinical usage. *Calif. West. Med.* **1938**, *49*, 362–366.
36. Edgar, A.; Doisy, E.A. An ovarian hormone: Preliminary report on its localization, extraction and partial purification, and action in test animals. *J. Am. Med. Assoc.* **1923**, *81*, 819–821.
37. Doisy, E.A.; Clement, D.V.; Sidney, T. Folliculin from urine of pregnant women. *Am. J. Phys.* **1929**, *90*, 329–330.
38. Butenandt, A. Über "Progynon" ein krystallisiertes weibliches Sexualhormon. *Naturwissenschaften* **1929**, *17*, 879. [CrossRef]
39. Butenandt, A. Über physikalische und chemische Eigenschaften des krystallisierten Follikelhormons. Untersuchungen über das weibliche Sexualhormon. *Hoppe-Seyler's Zeit. Physiol. Chem.* **1930**, *191*, 140–156. [CrossRef]
40. Dohrn, M.; Faure, W.; Poll, H.; Blotevogel, W. Tokokinine, Stoff mit sexualhormonartiger Wirkung aus Pflanzenzellen. *Med. Klin.* **1926**, *22*, 1417–1419.
41. Butenandt, A.; Jacobi, H. Über die Darstellung eines krystallisierten pflanzlichen Tokokinins (Thelykinins) und seine Identifizierung mit dem α-Follikelhormon. Untersuchungen über das weibliche Sexualhormon. *Hoppe Seyler's Z. Physiol. Chem.* **1933**, *218*, 104–112. [CrossRef]
42. Skarzynski, B. An oestrogenic substance from plant material. *Nature* **1933**, *131*, 766.
43. Janeczko, A.; Skoczowski, A. Mammalian sex hormones in plants. *Folia Histochem. Cytobiol.* **2005**, *43*, 71–79. [PubMed]
44. Zhang, J.S.; Yang, Z.H.; Tsao, T.H. The occurrence of estrogens in relation to reproductive processes in flowering plants. *Sex. Plant Reprod.* **1991**, *4*, 193–196. [CrossRef]
45. Zhong-han, Y.; Yin, T.; Zong-xun, C.; Tsao, T.H. The changes of steroidal sex hormone—Testosterone contents in reproductive organs of *Lilium davidii* Duch. *Acta Bot. Sin.* **1994**, *36*, 215–220.
46. Janot, M.M.; Devissaguet, P.; Khuong-Huu, Q.; Goutarel, R. Steroid alkaloids. LXVI. New alkaloids from the husks of *Holarrhena floribunda* (G. Don) Dur. and Schinz: Holarrheline, holadienine, holaromine and holaline. *Ann. Pharm. Fr.* **1967**, *25*, 733–748.
47. Cain, J.C. Miroestrol—An estrogen from the plant *Pueraria mirifica*. *Nature* **1960**, *188*, 774–777. [CrossRef]
48. Misico, R.I.; Veleiro, A.S.; Burton, G.; Oberti, J.C. Withanolides from *Jaborosa leucotricha*. *Phytochemistry* **1997**, *45*, 1045–1048. [CrossRef]
49. Cirigliano, A.M.; Veleiro, A.S.; Misico, R.I.; Tettamanzi, M.C.; Oberti, J.C.; Burton, G. Withanolides from *Jaborosa laciniata*. *J. Nat. Prod.* **2007**, *70*, 1644–1646. [CrossRef]
50. Valente, L.M.; Gunatilaka, A.A.; Glass, T.E.; Kingston, D.G.; Pinto, A.C. New norcucurbitacin and heptanorcucurbitacin glucosides from *Fevillea trilobata*. *J. Nat. Prod.* **1993**, *56*, 1772–1778. [CrossRef]
51. Igarashi, K. Studies on the steroidal components of domestic plants. XXXV. Structure of meteogenin. *Chem. Pharm. Bull.* **1961**, *9*, 722–729. [CrossRef]
52. Minato, H.; Shimaoka, A. Studies on the steroidal components of domestic plants. XLII. Narthogenin, isonarthogenin and neonogiragenin, three new sapogenins of metanarthecium luteoviride MAXIM. *Chem. Pharm. Bull.* **1961**, *9*, 729–734.
53. Pkheidze, T.A.; Gvazava, L.N.; Kemertelidze, É.P. Luvigenin and hecogenin from the leaves of *Yucca gloriosa*. *Chem. Nat. Compd.* **1991**, *27*, 376. [CrossRef]
54. Sobolewska, D.; Michalska, K.; Podolak, I.; Grabowska, K. Steroidal saponins from the genus Allium. *Phytochem. Rev.* **2016**, *15*, 1–35. [CrossRef] [PubMed]
55. Himeno, E.; Nagao, T.; Honda, J.; Okabe, H.; Irino, N.; Nakasumi, T. Structures of cayaponosides A, B, C and D, glucosides of new norcucurbitacins in the roots of *Cayaponia tayuya*. *Chem. Pharm. Bull.* **1992**, *40*, 2885–2887. [CrossRef] [PubMed]
56. Himeno, E.; Nagao, T.; Nonda, J.; Okabe, H.; Irino, N.; Nakasumi, T. Studies on the constituents of the root of Cayaponia tayuya (Vell) Cogn. I. Structures of cayaponosides, new 29-Nor-1,2,3,4,5,10-hexadehydro-cucurbitacin glucosides. *Chem. Pharm. Bull.* **1994**, *42*, 2295–2300. [CrossRef]
57. Konoshima, T.; Takasaki, M.; Kozuka, M.; Nagao, T.; Okabe, H. Inhibitory effects of cucurbitane triterpenoids on Epstein-Barr virus activation and two stage carcinogenesis of skin tumor. II. *Biol. Pharm. Bull.* **1995**, *18*, 284–287. [CrossRef]
58. Shirota, O.; Sekita, S.; Satake, M.; Morita, H.; Takeya, K.; Itokawa, H. Two cangorosin A type triterpene dimers from *Maytenus chuchuhuasca*. *Chem. Pharm. Bull.* **2004**, *52*, 1148–1150. [CrossRef]
59. Araújo Júnior, R.F.; Oliveira, A.L.; Pessoa, J.B.; Garcia, V.B. *Maytenus ilicifolia* dry extract protects normal cells, induces apoptosis, and regulates Bcl-2 in human cancer cells. *Exp. Biol. Med.* **2013**, *238*, 1251–1258. [CrossRef]
60. Vendruscolo, G.S.; Simoes, C.M.O.; Mentz, L.A. Etnobotanica no Rio Grande do Sul: Analise comparative entre o conhecimento original eatual sobre as plantas medicinais nativas. *Pesqui. Bot.* **2005**, *56*, 285–320.
61. Goncalves, M.I.A.; Martins, D.T.O. Plantas medicinais usadas pela populacao do municipio de Santo Antonio de Leverger, Mato Grosso, Brasil. *Rev. Bras. Farm.* **1998**, *79*, 56–61.
62. Si, Y.; Yao, X.H.; Zhang, C.K.; Tu, Z.B. C-32 triterpenes from *Taxodium ascendens*. *Biochem. Syst. Ecol.* **2005**, *33*, 211–214.
63. Otto, A.; White, J.D.; Simoneit, B.R.T. Natural product terpenoids in Eocene and Miocene conifer fossils. *Science* **2012**, *297*, 1543–1545. [CrossRef]
64. Guo, J.; Xue, J.; Hua, J.; Yin, Y.; Creech, D.L.; Han, J. Research status and trends of *Taxodium distichum*. *HortScience* **2023**, *58*, 317–326. [CrossRef]
65. Lu, Z.; Van Wagoner, R.M.; Harper, M.K.; Hooper, J.N.A.; Ireland, C.M. Two ring-A aromatized bile acids from the marine sponge *Sollasella moretonensis*. *Nat. Prod. Commun.* **2010**, *5*, 1571–1574. [PubMed]
66. Goddard, P.; Hill, M.J. The dehydrogenation of the steroid nucleus by human-gut bacteria. *Biochem. Soc. Trans.* **1973**, *1*, 1113–1115. [CrossRef]

67. Yeung, B.K.S.; Hamann, M.T.; Scheuer, P.J.; Kelly-Borges, M. Hapaioside: A 19-norpregnane glycoside from the sponge *Cribrochalina olemda*. *Tetrahedron* **1994**, *50*, 12593–12598. [CrossRef]
68. Di Girolamo, J.A.; Li, X.-C.; Jacob, M.R.; Clark, A.M.; Ferreira, D. Reversal of fluconazole resistance by sulfated sterols from the marine sponge *Topsentia* sp. *J. Nat. Prod.* **2009**, *72*, 1524–1528. [CrossRef]
69. Yan, X.-H.; Liu, H.-L.; Huang, H.; Li, X.-B.; Guo, Y.-W. Steroids with aromatic A rings from the Hainan soft coral *Dendronephthya studeri* Ridley. *J. Nat. Prod.* **2011**, *74*, 175–180. [CrossRef]
70. Poza, J.J.; Fernández, R.; Reyes, F.; Rodríguez, J.; Jiménez, C. Isolation, biological significance, synthesis, and cytotoxic evaluation of new natural parathiosteroids A-C and analogues from the soft coral *Paragorgia* sp. *J. Org. Chem.* **2008**, *73*, 7978–7984. [CrossRef]
71. Barrero, A.F.; Oltra, J.E.; Poyatos, J.A.; Jiménez, D.; Oliver, E. Phycomysterols and other sterols from the fungus *Phycomyces blakesleeanus*. *J. Nat. Prod.* **1998**, *61*, 1491–1496. [CrossRef] [PubMed]
72. Liu, X.H.; Tang, X.Z.; Miao, F.P.; Ji, N.Y. A new pyrrolidine derivative and steroids from an algicolous *Gibberella zeae* strain. *Nat. Prod. Commun.* **2011**, *6*, 1243–1246. [CrossRef] [PubMed]
73. Jiao, F.R.; Gu, B.B.; Zhu, H.R.; Zhang, Y.; Liu, K.C.; Zhang, W.; Shi-Hai, H.H.; Lin, H.W. Asperfloketals A and B, the first two ergostanes with rearranged A and D rings: From the sponge-associated *Aspergillus flocculosus*. *J. Org. Chem.* **2021**, *86*, 10954–10961. [CrossRef]
74. Luo, X.; Li, F.; Shinde, P.B.; Hong, J.; Lee, C.-O.; Im, K.S.; Jung, J.H. 26,27-Cyclosterols and other polyoxygenated sterols from a marine sponge *Topsentia* sp. *J. Nat. Prod.* **2006**, *69*, 1760–1768. [CrossRef] [PubMed]
75. Kim, E.L.; Li, J.L.; Hong, J.; Yoon, W.D.; Kim, H.S.; Liu, Y.; Wei, X.; Jung, J.H. An unusual 1(10→19)abeo steroid from a jellyfish-derived fungus. *Tetrahedron Lett.* **2016**, *57*, 2803–2806. [CrossRef]
76. Li, G.; Kusari, S.; Kusari, P.; Kayser, O.; Spiteller, M. Endophytic *Diaporthe* sp. LG23 produces a potent antibacterial tetracyclic triterpenoid. *J. Nat. Prod.* **2015**, *78*, 2128–2132. [CrossRef]
77. Rowland, S.J.; West, C.E.; Jones, D.; Scarlett, A.G.; Frank, R.A.; Hewitt, L.M. Steroidal aromatic 'naphthenic acids' in oil sands process affected water: Structural comparisons with environmental estrogens. *Environ. Sci. Technol.* **2011**, *45*, 9806–9815. [CrossRef]
78. Pounina, T.A.; Gloriozova, T.A.; Savidov, N.; Dembitsky, V.M. Sulfated and sulfur-containing steroids and their pharmacological profile. *Mar. Drugs* **2021**, *19*, 240. [CrossRef]
79. Wang, W.; Lee, Y.; Lee, T.G.; Mun, B.; Giri, A.G.; Lee, J.; Kim, H. Phorone A and isophorbasone A, sesterterpenoids isolated from the marine sponge *Phorbas* sp. *Org. Lett.* **2012**, *14*, 4486–4489. [CrossRef]
80. Gao, S.; Wang, Q.; Huang, L.J.S.; Lum, L.; Chen, C. Chemical, and biological studies of nakiterpiosin and nakiterpiosinone. *J. Am. Chem. Soc.* **2010**, *132*, 371–383. [CrossRef]
81. Venugopal, J.R.; Mukku, V.; Edrada, R.A.; Schmitz, F.J.; Shanks, M.K.; Chaudhuri, B.; Fabbro, D. New sesquiterpene quinols from a Micronesian sponge, *Aka* sp. *J. Nat. Prod.* **2003**, *66*, 686–689.
82. Misico, R.I.; Nicotra, V.E.; Oberti, J.C.; Barboza, G.; Gil, R.R.; Burton, G. Withanolides and related steroids. *Prog. Chem. Org. Nat. Prod.* **2011**, *94*, 127–229.
83. Crews, P.; Harrison, B. New triterpene-ketides (Merotriterpenes), haliclotriol A and B, from an Indo–Pacific *Haliclona* sponge. *Tetrahedron* **2000**, *56*, 9039–9046. [CrossRef]
84. Williams, D.E.; Steinø, A.; de Voogd, N.J.; Mauk, A.G.; Andersen, R.J. Halicloic acids A and B isolated from the marine sponge *Haliclona* sp. collected in the Philippines inhibit indoleamine 2,3-dioxygenase. *J. Nat. Prod.* **2012**, *75*, 1451–1458. [CrossRef] [PubMed]
85. Falk, H.; Wolkenstein, K. Natural product molecular fossils. In *Progress in the Chemistry of Organic Natural Products*; Kinghorn, A., Falk, H., Gibbons, S., Kobayashi, J., Eds.; Springer: Cham, Switzerland, 2017; Volume 104.
86. Oliveira, C.R.; Oliveira, C.J.F.; Ferreira, A.A.; Azevedo, D.A.; Neto, F.R.A. Characterization of aromatic steroids and hopanoids in marine and lacustrine crude oils using comprehensive two-dimensional gas chromatography coupled to time-of-flight mass spectrometry (GCxGC-TOFMS). *Org. Geochem.* **2012**, *53*, 131–136. [CrossRef]
87. Jacob, J.; Disnar, J.-R.; Boussafir, M.; Albuquerque, A.L.S.; Sifeddine, A. Contrasted distributions of triterpene derivatives in the sediments of Lake Caçó reflect paleoenvironmental changes during the last 20,000 yrs in NE Brazil. *Org. Geochem.* **2007**, *38*, 180–197. [CrossRef]
88. Nakanishi, K. Steroids. In *Natural Products Chemistry*; Nakanishi, K., Goto, T., Itô, S., Natori, S., Nozoe, S., Eds.; Academic Press: Cambridge, MA, USA, 1974; pp. 421–545.
89. Zuhrotun, A.; Suganda, A.G.; Nawawi, A. Phytochemical study of ketapang bark (*Terminalia catappa* L.). In Proceedings of the International Conference on Medicinal Plants, Surabaya, Indonesia, 21–22 July 2010.
90. Beall, D. Some notes on the isolation of oestrone and equilin from the urine of pregnant mares. *Biochem. J.* **1936**, *30*, 577–581. [CrossRef]
91. Schachter, B.; Marrian, G.F. Pregnant mares' sulfate from the urine of the isolation of estrone. *J. Biol. Chem.* **1938**, *126*, 663–669. [CrossRef]
92. Bachmann, W.E.; Cole, W.; Wilds, A.L. The total synthesis of the sex hormone equilenin. *J. Am. Chem. Soc.* **1939**, *61*, 974–975. [CrossRef]
93. Fritz, M.A.; Speroff, L. *Clinical Gynecologic Endocrinology and Infertility*; Lippincott Williams & Wilkin: Philadelphia, PA, USA, 2012; p. 751.

94. Toghueo, R.M.K.; Zabalgogeazco, I.; Vázquez de Aldana, B.R.; Boyoma, F.F. Enzymatic activity of endophytic fungi from the medicinal plants *Terminalia catappa, Terminalia mantaly* and *Cananga odorata. S. Afr. J. Bot.* **2017**, *109*, 146–153. [CrossRef]
95. Toghueo, R.M.K.; Ejiya, E.I.; Sahal, D.; Yazdani, S.S.; Boyom, F.F. Production of cellulolytic enzymes by endophytic fungi isolated from Cameroonian medicinal plants. *Int. J. Curr. Microbiol. Appl. Sci.* **2017**, *6*, 1264–1271. [CrossRef]
96. Parrish, S.M.; Yoshida, W.Y.; Williams, P.G. New diterpene isolated from a sponge of genus *Strongylophora. Planta Med.* **2016**, *82*, S1–S381. [CrossRef]
97. Qin, X.D.; Liu, J.K. Natural aromatic steroids as potential molecular fossils from the fruiting bodies of the ascomycete *Daldinia concentrica. J. Nat. Prod.* **2004**, *67*, 2133–2135. [CrossRef] [PubMed]
98. Brassell, S.C.; Eglinton, G.; Maxwell, J.R. The geochemistry of terpenoids and steroids. *Biochem. Soc. Trans.* **1983**, *11*, 575–586. [CrossRef]
99. Breger, I.A. Geochemistry of lipids. *J. Am. Oil Chem. Soc.* **1966**, *43*, 197–221. [CrossRef]
100. Huang, H.; Zhang, S.; Su, J. Palaeozoic oil–source correlation in the Tarim Basin, NW China: A review. *Org. Geochem.* **2016**, *94*, 32–46. [CrossRef]
101. Cheng, B.; Zhao, J.; Yang, C.; Tian, Y.; Liao, Z. Geochemical evolution of occluded hydrocarbons inside geomacromolecules: A review. *Energy Fuel* **2017**, *31*, 8823–8832. [CrossRef]
102. Machida, K.; Abe, T.; Arai, D.; Okamoto, M.; Shimizu, I.; de Voogd, N.J.; Fusetani, N.; Nakao, Y. Cinanthrenol A, an estrogenic steroid containing phenanthrene nucleus, from a marine sponge *Cinachyrella* sp. *Org. Lett.* **2014**, *16*, 1539–1541. [CrossRef]
103. Ludwig, B.; Gussler, G.; Wehrung, P.; Albrecht, P. C26-C29 triaromatic steroid derivatives in sediments and petroleum. *Tetrahedron Lett.* **1981**, *22*, 3313–3316. [CrossRef]
104. Mackenzie, A.S.; Brassell, S.C.; Eglinton, G.; Maxwell, J.R. Chemical fossils: The geological fate of steroids. *Science* **1982**, *217*, 491–504. [CrossRef]
105. Schnell, G.; Schaeffer, P.; Motscha, E.; Adam, P. Triterpenoids functionalized at C-2 as diagenetic transformation products of 2,3-dioxygenated triterpenoids from higher plants in buried wood. *Org. Biomol. Chem.* **2012**, *10*, 8276–8282. [CrossRef] [PubMed]
106. Le Milbeau, C.; Schaeffer, P.; Connan, J.; Albrecht, P.; Adam, P. Aromatized C-2 oxygenated triterpenoids as indicators for a new transformation pathway in the environment. *Org. Lett.* **2010**, *12*, 1504–1507. [CrossRef] [PubMed]
107. Su, Z.; Yuan, W.; Wang, P.; Li, S. Ethnobotany, phytochemistry, and biological activities of *Taxodium rich. Pharm. Crops* **2013**, *4*, 1–14.
108. Turekian, K.K.; Wedepohl, K.H. Distribution of the elements in some major units of the Earth's crust. *GSA Bull.* **1961**, *72*, 175–192. [CrossRef]
109. Worsfold, P.; McKelvie, I.; Monbet, P. Determination of phosphorus in natural waters: A historical review. *Anal. Chim. Acta* **2016**, *918*, 8–20. [CrossRef]
110. Dhuime, B.; Wuestefeld, A.; Hawkesworth, C.J. Emergence of modern continental crust about 3 billion years ago. *Nat. Geosci.* **2015**, *8*, 552–555. [CrossRef]
111. Krafft, F. Phosphorus: From elemental light to chemical element. *Angew. Chem. Int. Ed.* **1969**, *8*, 660–671. [CrossRef] [PubMed]
112. Holmes, R.R. Comparison of phosphorus and silicon: Hypervalency, stereochemistry, and reactivity. *Chem. Rev.* **1996**, *96*, 927–950. [CrossRef]
113. Su, J.; Dong, S.; Zhang, Y.; Li, Y.; Chen, X.; Li, J. Apatite fission track geochronology of the Southern Hunan province across the Shi-Hang Belt: Insights into the Cenozoic dynamic topography of South China. *Int. Geol. Rev.* **2017**, *59*, 981–995. [CrossRef]
114. Dorozhkin, S.V. Calcium orthophosphates: Occurrence, properties, biomineralization, pathological calcification and biomimetic applications. *Biomatter* **2011**, *1*, 121–164. [CrossRef]
115. Oelkers, E.H.; Montel, J.-M. Phosphates, and nuclear waste storage. *Elements* **2008**, *4*, 113–116. [CrossRef]
116. Ewing, R.C.; Wang, L. Phosphates as nuclear waste forms. *Rev. Mineral. Geochem.* **2002**, *48*, 67399. [CrossRef]
117. Portnov, A.M.; Gorobets, B.S. Luminescence of apatite from different rock types. *Dokl. Akad. Nauk SSSR* **1969**, *184*, 110–113.
118. Dorozhkin, S.V. Calcium orthophosphate cements for biomedical application. *J. Mater. Sci.* **2008**, *43*, 3028–3043. [CrossRef]
119. Dorozhkin, S.V. Bioceramics of calcium orthophosphates. *Biomaterials* **2010**, *31*, 1465–1485. [CrossRef] [PubMed]
120. Yeagle, P.L. *The Membranes of Cells*; Academic Press: Cambridge, MA, USA, 2016.
121. Dorozhkin, S.V.; Epple, M. Biological and medical significance of calcium phosphates. *Angew. Chem. Int. Ed.* **2002**, *41*, 3130–3146. [CrossRef]
122. De Riccardis, F.; Minale, L.; Riccio, R.; Giovannitti, B.; Iorizzi, M.; Debitus, C. Phosphated and sulfated marine polyhydroxylated steroids from the starfish *Tremaster novaecaledoniae. Gazz. Chim. Ital.* **1993**, *123*, 79–86.
123. Fujita, M.; Nakao, Y.; Matsunaga, S.; Seiki, M.; Itoh, Y.; van Soest, R.W.M.; Heubes, M.; Faulkner, D.J.; Fusetani, N. Isolation and structure elucidation of two phosphorylated sterol sulfates, MT1-MMP inhibitors from a marine sponge *Cribrochalina* sp.: Revision of the structures of haplosamates A and B. *Tetrahedron* **2001**, *57*, 3885–3890. [CrossRef]
124. Chianese, G.; Fattorusso, E.; Taglialatela-Scafati, O.; Bavestrello, G.; Calcinai, B.; Dien, H.A.; Ligresti, A.; Di Marzo, V. Desulfohaplosamate, a new phosphate-containing steroid from *Dasychalina* sp., is a selective cannabinoid CB2 receptor ligand. *Steroids* **2011**, *76*, 998–1002. [CrossRef]
125. Van Dullemen, H.M.; Tytgat, G.N.J. Colonoscopy in ileocolitis. In *Procedures in Hepatogastroenterology*; Tytgat, G.N.J., Mulder, C.J.J., Eds.; Developments in Gastroenterology; Springer: Dordrecht, The Netherlands, 1997; Volume 15.

126. Delrio, G.; Botte, V. Testosterone 17-phosphate and 19-nortestosterone 17-phosphate as substrate for rabbit prostate phosphatases. *Biochim. Biophys. Acta* **1970**, *218*, 327–332. [CrossRef]

127. Kokado, A.; Tsuji, A.; Maeda, M. Chemiluminescence assay of alkaline phosphatase using cortisol-21-phosphate as substrate and its application to enzyme immunoassays. *Anal. Chim. Acta* **1997**, *337*, 335–340. [CrossRef]

128. Ellam, T.J.; Chico, T.J. Phosphate: The new cholesterol? The role of the phosphate axis in non-uremic vascular disease. *Atherosclerosis* **2012**, *220*, 310–318. [CrossRef] [PubMed]

129. Davis, S.C.; Szoka, F.C., Jr. Cholesterol phosphate derivatives: Synthesis and incorporation into a phosphatase and calcium-sensitive triggered release liposome. *Bioconjug. Chem.* **1998**, *9*, 783–792. [CrossRef] [PubMed]

130. Sachs-Barrable, K.; Darlington, J.W.; Wasan, K.M. The effect of two novel cholesterol-lowering agents, disodium ascorbyl phytostanol phosphate and nanostructured aluminosilicate on the expression and activity of P-glycoprotein within Caco-2 cells. *Lipids Health Dis.* **2014**, *13*, 153–163. [CrossRef]

131. Kutney, J.P.; Pritchard, H.P.; Lukic, T. Novel Compounds and Compositions Comprising Sterols and/or Stanols and Cholesterol Biosynthesis Inhibitors and Use Thereof in Treating or Preventing a Variety of Diseases and Conditions. Japan Patent EP1644399A2, 7 September 2003.

132. Somogyi, G.; Nishitani, S.; Nomi, D.; Buchwald, P.; Prokai, L.; Bodor, N. Targeted drug delivery to the brain via phosphonate derivatives I Design, synthesis and evaluation of an anionic chemical delivery system for testosterone. *Int. J. Pharm.* **1998**, *166*, 15–26. [CrossRef]

133. Gunnarsson, P.O.; Norlén, B.J. Clinical pharmacology of polyestradiol phosphate. *Prostate* **1988**, *13*, 299–304. [CrossRef]

134. Zhang, Y.; Wu, X.; Lic, H.; Du, N.; Song, S.; Hou, W. Preparation and characterization of betamethasone sodium phosphate intercalated layered double hydroxide liposome nanocomposites. *Colloids Surf. A Physicochem. Eng. Asp.* **2017**, *529*, 824–831. [CrossRef]

135. Isaac, R.E.; Rees, H.H. Isolation, and identification of ecdysteroid phosphates and acetylecdysteroid phosphates from developing eggs of the locust, *Schistocerca gregaria*. *Biochem. J.* **1984**, *221*, 459–464. [CrossRef]

136. Chopra, A.; Doiphode, V.V. Ayurvedic medicine: Core concept, therapeutic principles, and current relevance. *Med. Clin.* **2002**, *86*, 75–89. [CrossRef]

137. Patwardhan, B.; Warude, D.; Pushpangadan, P.; Bhatt, N. Ayurveda, and traditional Chinese medicine: A comparative overview. *Evid.-Based Complement. Altern. Med.* **2005**, *2*, 465–473. [CrossRef]

138. Mishra, A. Traditional methods of food habits and dietary preparations in Ayurveda—The Indian system of medicine. *J. Ethn. Foods* **2019**, *6*, 14–24.

139. Viuda-Martos, M.; Ruiz-Navajas, Y.; Fernández-López, J.; Pérez-Alvarez, J.A. Functional properties of honey, propolis, and royal jelly. *J. Food Sci.* **2008**, *73*, R117–R124. [CrossRef] [PubMed]

140. Dembitsky, V.M. In silico prediction of steroids and triterpenoids as potential regulators of lipid metabolism. *Mar. Drugs* **2021**, *19*, 650. [CrossRef] [PubMed]

141. Dembitsky, V.M.; Gloriozova, T.A.; Poroikov, V.V.; Koola, M.M. QSAR study of some natural and synthetic platelet aggregation inhibitors and their pharmacological profile. *J. Appl. Pharm. Sci.* **2022**, *12*, 039–058. [CrossRef]

142. Ramadan, M.F.; Al-Ghamdi, A. Bioactive compounds and health-promoting properties of royal jelly: A review. *J. Funct. Foods* **2012**, *4*, 39–52. [CrossRef]

143. Gribble, G.W. Naturally occurring organohalogen compounds—A comprehensive survey. *Prog. Chem. Org. Nat. Prod.* **1996**, *68*, 1–496.

144. Gribble, G.W. Naturally occurring organohalogen compounds. *Acc. Chem. Res.* **1998**, *31*, 141–152. [CrossRef]

145. Gribble, G.W. The diversity of naturally occurring organobromine compounds. *Chem. Soc. Rev.* **1999**, *28*, 335–346. [CrossRef]

146. Gribble, G.W. Naturally occurring organohalogen compounds—A comprehensive update. *Prog. Chem. Org. Nat. Prod.* **2010**, *91*, 1–613.

147. Wang, C.; Du, W.; Lu, H.; Lan, J.; Liang, K.; Cao, S. A Review: Halogenated compounds from marine Actinomycetes. *Molecules* **2021**, *26*, 2754. [CrossRef]

148. Wang, C.; Lu, H.; Lan, J.; Ahammad Zaman, K.H.; Cao, S. A Review: Halogenated compounds from marine fungi. *Molecules* **2021**, *26*, 458. [CrossRef] [PubMed]

149. Dembitsky, V.M.; Tolstikov, G.A. Chlorine containing sesquiterpenes of higher plants. *Chem. Sustain. Dev.* **2002**, *10*, 363–370.

150. Dembitsky, V.M.; Tolstikov, G.A. *Natural Organic Halogenated Compounds*; Geo-Science: Novosibirsk, Russia, 2003; p. 367.

151. Dembitsky, V.M.; Tolstikov, G.A. Natural halogenated alkanes, cycloalkanes, and their derivatives. *Chem. Sustain. Dev.* **2003**, *11*, 803–810.

152. Dembitsky, V.M.; Tolstikov, G.A. Natural halogenated alkaloids. *Chem. Sustain. Dev.* **2003**, *11*, 451–466.

153. Dembitsky, V.M.; Tolstikov, G.A. Natural halogenated furanones, higher terpenes and steroids. *Chem. Sustain. Dev.* **2003**, *11*, 697–703.

154. Dembitsky, V.M.; Tolstikov, G.A. Natural halogenated non-terpenic C15-acetogenins of sea organisms. *Chem. Sustain. Dev.* **2003**, *11*, 329–339.

155. Dembitsky, V.M.; Tolstikov, G.A. Halogenated phenol compounds in lichens and fungi. *Chem. Sustain. Dev.* **2003**, *11*, 557–565.

156. Dembitsky, V.M.; Tolstikov, G.A. Natural halogenated mononuclear phenol compounds and their derivatives. *Chem. Sustain. Dev.* **2003**, *11*, 567–575.

157. Dembitsky, V.M.; Gloriozova, T.A.; Poroikov, V.V. Chlorinated plant steroids and their biological activities. *Int. J. Curr. Res. Biosci. Plant Biol.* **2017**, *4*, 70–85. [CrossRef]

158. Tschesche, R.; Baumgarth, M.; Welzel, P. Weitere inhaltsstoffe aus *Jaborosa integrifolia* Lam. III: Zur Struktur der Jaborosalactone C, D, und E. *Tetrahedron* **1968**, *24*, 5169–5179. [CrossRef]

159. Chen, L.X.; He, H.; Qiu, F. Natural withanolides: An overview. *Nat. Prod. Rep.* **2011**, *28*, 705–740. [CrossRef] [PubMed]

160. Ali, A.; Sahai, M.; Ray, A.B.; Slatkin, D.J. Physalolactone C, a new withanolide from *Physalis peruviana*. *J. Nat. Prod.* **1984**, *47*, 648–651. [CrossRef]

161. Frolow, F.; Ray, B.; Sahai, A.; Glotter, M.; Gottlieb, E.E.; Kirson, H.I. Withaperuvin and 4-deoxy-physalolactone, two new ergostane-type steroids from Physalis peruviana (Solanaceae). *J. Chem. Soc. Perkin Trans.* **1981**, *112*, 1029–1032. [CrossRef]

162. Shingu, K.; Yahara, S.; Okabe, H.; Nohara, T. Three new withanolides, physagulins E, F and G from *Physalis angulata* L. *Chem. Pharm. Bull.* **1992**, *40*, 2448–2451. [CrossRef]

163. Nittala, S.S.; Vande Velde, V.; Frolow, F.; Lavie, D. Chlorinated withanolides from *Withania somnifera* and *Acnistus breviflorus*. *Phytochemistry* **1981**, *20*, 2547–2552. [CrossRef]

164. Bessalle, R.; Lavie, D. Withanolide C, A chlorinated withanolide from *Withania somnifera*. *Phytochemistry* **1992**, *31*, 3648–3651. [CrossRef]

165. Kirson, I.; Glotter, E.E. Recent Developments in naturally occurring ergostane-type steroids. A Review. *J. Nat. Prod.* **1981**, *44*, 633–647. [CrossRef]

166. Fajardo, V.; Podesta, F.; Shamma, M.; Freyer, A.J. New withanolides from *Jaborosa magellanica*. *J. Nat. Prod.* **1991**, *54*, 554–563. [CrossRef]

167. Triana, J.; López, M.; Pérez, F.J.; Rico, M.; López, A.; Estévez, F.; Marrero, M.T.; Brouard, I.; León, F. Secondary metabolites from two species of Tolpis and their biological activities. *Molecules* **2012**, *17*, 12895–12909. [CrossRef]

168. Pramanick, S.; Roy, A.; Ghosh, S.; Majumder, H.K.; Mukhopadhyay, S. Withanolide Z, new chlorinated withanolide from *Withania somnifera*. *Planta Med.* **2008**, *74*, 1745–1753. [CrossRef]

169. Dinan, L.N.; Sarker, S.D.; Sik, V. 28-Hydroxywithanolide E from *Physalis peruviana*. *Phytochemistry* **1997**, *44*, 509–512. [CrossRef]

170. Lan, Y.-H.; Chang, F.-R.; Pan, M.-J.; Wu, C.-C.; Wu, S.-J.; Chen, S.-L. New cytotoxic withanolides from *Physalis peruviana*. *Food Chem.* **2009**, *116*, 462–471. [CrossRef]

171. Hsieh, P.-W.; Huang, Z.-Y.; Chen, J.-H.; Chang, F.-R.; Wu, C.-C. Cytotoxic withanolides from *Tubocapsicum anomalum*. *J. Nat. Prod.* **2007**, *70*, 747–756. [CrossRef] [PubMed]

172. Nagafuji, S.; Okabe, H.; Akahane, H.; Abe, F. Trypanocidal constituents in plants. 4. Withanolides from the aerial parts of *Physalis angulata*. *Biol. Pharm. Bull.* **2004**, *27*, 193–202. [CrossRef] [PubMed]

173. Nicotra, V.E.; Gil, R.R.; Vaccarini, C.; Oberti, J.C.; Burton, G. 15,21-Cyclowithanolides from *Jaborosa bergii*. *J. Nat. Prod.* **2003**, *66*, 1471–1475. [CrossRef] [PubMed]

174. Nicotra, V.E.; Gil, R.R.; Oberti, J.C.; Burton, G. Withanolides with phytotoxic activity from *Jaborosa caulescens* var. *caulescens* and *J. caulescens* var. *bipinnatifida*. *J. Nat. Prod.* **2007**, *70*, 808–813. [CrossRef]

175. Bonetto, G.M.; Gil, R.R.; Oberti, J.C.; Veleiro, A.S.; Burton, G. Novel withanolides from *Jaborosa sativa*. *J. Nat. Prod.* **1995**, *58*, 705–719. [CrossRef]

176. Kiyota, N.; Shingu, K.; Yamaguchi, K.; Yoshitake, Y.; Harano, K.; Yoshimitsu, H.; Ikeda, T.; Nohara, T. New C28 steroidal glycosides from *Tubocapsicum anomalum*. *Chem. Pharm. Bull.* **2007**, *55*, 34–39. [CrossRef]

177. Kiyota, N.; Shingu, K.; Yamaguchi, K.; Yoshitake, Y.; Harano, K.; Yoshimitsu, H. New C28 steroidal glycosides from *Tubocapsicum anomalum*. *Chem. Pharm. Bull.* **2008**, *56*, 1038–1044. [CrossRef]

178. Glotter, E.; Abraham, A.; Guenzberg, G.; Kirson, I. Naturally occurring steroidal lactones with a 17a-oriented side chain. Structure of withanolide E and related compounds. *J. Chem. Soc. Perkin Trans.* **1977**, *1*, 341–344. [CrossRef]

179. Shingu, K.; Marubayashi, N.; Ueda, I.; Yahara, S.; Nohara, T. Two new ergostane derivatives from *Tubocapsicum anomalum* (Solanaceae). *Chem. Pharm. Bull.* **1990**, *38*, 1107–1111. [CrossRef]

180. Cirigliano, A.M.; Veleiro, A.S.; Oberti, J.C.; Burton, G. Spiranoid withanolides from *Jaborosa odonelliana*. *J. Nat. Prod.* **2002**, *65*, 1049–1053. [CrossRef] [PubMed]

181. Cirigliano, A.M.; Misico, R.I. Spiranoid withanolides from *Jaborosa odonelliana* and *Jaborosa runcinata*. *Z. Naturforschung B Chem. Sci.* **2005**, *60*, 867–873. [CrossRef]

182. Garcia, M.E.; Navarro-Vazquez, S.P.A.; Phillips, D.D.; Gayathri, C.; Krakauer, H.; Stephens, P.W.; Nicotra, V.E.; Gil, R.R. Stereochemistry determination by powder X-ray diffraction analysis and NMR spectroscopy residual dipolar couplings. *Angew. Chem. Int. Ed.* **2009**, *48*, 5670–5676. [CrossRef] [PubMed]

183. Xia, G.; Cao, S.; Chen, L.; Qiu, F. Natural withanolides, an update. *Nat. Prod. Rep.* **2022**, *39*, 784–813. [CrossRef]

184. Moujir, L.M.; Llanos, G.G.; Araujo, L.; Amesty, A.; Bazzocchi, I.L.; Jiménez, I.A. Withanolide-type steroids from Withania aristata as potential anti-leukemic agents. *Molecules* **2020**, *25*, 5744. [CrossRef]

185. Ripperger, H.; Kamperdick, C. First isolation of physalins from the genus Saracha of Solanaceae. *Pharmazie* **1998**, *53*, 144–151.

186. Makino, B.; Kawai, M.; Ogura, T.; Nakanishi, M.; Yamamura, H.; Butsugan, Y. Structural revision of physalin H isolated from *Physalis angulata*. *J. Nat. Prod.* **1995**, *58*, 1668–1673. [CrossRef]

187. Kawai, M.; Makino, B.; Yamamura, H.; Butsugan, Y. Upon—Physalin L‖ isolated from *Physalis minima*. *Phytochemistry* **1996**, *43*, 661–667. [CrossRef]

188. Nicotra, V.E.; Ramacciotti, N.S.; Gil, R.R.; Oberti, J.C.; Feresin, G.E.; Guerrero, C.A.; Baggio, R.F.; Garland, M.T.; Burton, G. Phytotoxic withanolides from *Jaborosa rotacea*. *J. Nat. Prod.* **2006**, *69*, 783–801. [CrossRef]

189. Fattorusso, E.; Taglialatela-Scafati, O.; Petrucci, F.; Bavestrello, G.; Calcinai, B. Polychlorinated androstanes from the burrowing sponge *Cliona nigricans*. *Org. Lett.* **2004**, *6*, 1633–1635. [CrossRef] [PubMed]

190. Shimura, H.; Iguchi, K.; Yamada, Y.; Nakaike, S.; Yamagishi, T.; Matsumoto, K.; Yokoo, C. Aragusterol C: A novel halogenated marine steroid from an Okinawan sponge, *Xestospongia* sp., possessing potent antitumor activity. *Experientia* **1994**, *50*, 134–136. [CrossRef] [PubMed]

191. Kobayashi, M.; Chen, Y.J.; Higuchi, K.; Aoki, S.; Kitagawa, I. Marine natural products. XXXVII. Aragusteroketals A and C, two novel cytotoxic steroids from a marine sponge of *Xestospongia* sp. *Chem. Pharm. Bull.* **1996**, *44*, 1840–1842. [CrossRef]

192. Pham, G.N.; Kang, D.Y.; Kim, M.J.; Han, S.J.; Lee, J.H.; Na, M. Isolation of sesquiterpenoids and steroids from the soft coral *Sinularia brassica* and determination of their absolute configuration. *Mar. Drugs* **2021**, *19*, 523. [CrossRef] [PubMed]

193. Carney, J.R.; Scheuer, P.J.; Kelley-Borges, M. Three unprecedented chloro steroids from the Maui sponge *Strongylacidon* sp.: Kiheisterones C, D, and E. *J. Org. Chem.* **1993**, *58*, 3460–3462. [CrossRef]

194. Cimino, G.; De Luca, P.; De Stefano, S.; Minale, L. Disidein, a pentacyclic sesterterpene condensed with an hydroxyhydroquinone moiety, from the sponge *Disidea pallescens*. *Tetrahedron* **1975**, *31*, 271–275. [CrossRef]

195. Teta, R.; Della Sala, G.; Renga, B.; Mangoni, A.; Fiorucci, S.; Costantino, V. Chalinulasterol, a chlorinated steroid disulfate from the caribbean sponge *Chalinula molitba*. Evaluation of its role as PXR receptor modulator. *Mar. Drugs* **2012**, *10*, 1383–1390. [CrossRef]

196. Teruya, T.; Nakagawa, S.; Koyama, T.; Arimoto, H.; Kita, M.; Uemura, D. Nakiterpiosin and nakiterpiosinone, novel cytotoxic C-nor-D-homosteroids from the Okinawan sponge *Terpios hoshinota*. *Tetrahedron* **2004**, *60*, 6989–6993. [CrossRef]

197. Vil, V.; Gloriozova, T.A.; Zhukova, N.V.; Dembitsky, V.M. Highly oxygenated isoprenoid lipids derived from terrestrial and aquatic sources: Origin, structures and biological activities. *Vietnam J. Chem.* **2019**, *57*, 1–15. [CrossRef]

198. Guzii, A.G.; Makarieva, T.N.; Denisenko, V.A.; Dmitrenok, P.S.; Burtseva, Y.V.; Krasokhin, V.B.; Stonik, V.A. Topsentiasterol sulfates with novel iodinated and chlorinated side chains from the marine sponge *Topsentia* sp. *Tetrahedron Lett.* **2008**, *49*, 7191–7193. [CrossRef]

199. Areche, C.; Vaca, I.; Labbe, P.; Soto-Delgado, J.; Astudilloc, L.; Silva, M.; Rovirosa, J.; San-Martin, A. Biotransformation of stypotriol triacetate by *Aspergillus niger*. *J. Mol. Struct.* **2011**, *998*, 167–170. [CrossRef]

200. Minale, L.; Riccio, R.; De Simone, F.; Dini, A.; Pizza, C. Starfish saponins II. 22,23-epoxysteroids, minor genins from the starfish *Echinaster sepositus*. *Tetrahedron Lett.* **1979**, *20*, 645–648. [CrossRef]

201. Dort, E.; Díaz-Marrero, A.R.; Cueto, M.; D'Croz, L.; Maté, J.L.; San-Martín, A.; Darias, J. Unusual chlorinated pregnanes from the eastern Pacific octocoral *Carijoa multiflora*. *Tetrahedron Lett.* **2004**, *45*, 915–918. [CrossRef]

202. Iwashima, M.; Nara, K.; Nakamichi, Y.; Iguchi, K. Three new chlorinated marine steroids, yonarasterols G, H and I, isolated from the Okinawan soft coral, *Clavularia viridis*. *Steroids* **2001**, *66*, 25–32. [CrossRef] [PubMed]

Article

New Steroidal Selenides as Proapoptotic Factors

Izabella Jastrzebska [1,*], Natalia Wawrusiewicz-Kurylonek [2], Paweł A. Grześ [1], Artur Ratkiewicz [1], Ewa Grabowska [3], Magdalena Czerniecka [4], Urszula Czyżewska [4] and Adam Tylicki [4,*]

[1] Faculty of Chemistry, University of Białystok, Ciołkowskiego 1K, 15-245 Białystok, Poland; p.grzes@uwb.edu.pl (P.A.G.); artrat@uwb.edu.pl (A.R.)

[2] Department of Clinical Genetics, Medical University of Białystok, Waszyngtona 13, 15-089 Białystok, Poland; natalia.wawrusiewicz-kurylonek@umb.edu.pl

[3] Doctoral School of Exact and Natural Sciences, University of Bialystok, K. Ciolkowskiego 1K, 15-245 Bialystok, Poland; e.grabowska@uwb.edu.pl

[4] Faculty of Biology, University of Białystok, Ciołkowskiego 1J, 15-245 Białystok, Poland; m.siemieniuk@uwb.edu.pl (M.C.); urszula.czyzewska@uwb.edu.pl (U.C.)

* Correspondence: i.jastrzebska@uwb.edu.pl (I.J.); atyl@uwb.edu.pl (A.T.)

Abstract: Cytostatic and pro-apoptotic effects of selenium steroid derivatives against HeLa cells were determined. The highest cytostatic activity was shown by derivative **4** (GI_{50} 25.0 µM, almost complete growth inhibition after three days of culture, and over 97% of apoptotic and dead cells at 200 µM). The results of our study (cell number measurements, apoptosis profile, relative expression of apoptosis-related *APAF1*, *BID*, and mevalonate pathway-involved *HMGCR*, *SQLE*, *CYP51A1*, and *PDHB* genes, and computational chemistry data) support the hypothesis that tested selenosteroids induce the extrinsic pathway of apoptosis by affecting the cell membrane as cholesterol antimetabolites. An additional mechanism of action is possible through a direct action of derivative **4** to inhibit *PDHB* expression in a way similar to steroid hormones.

Keywords: antimetabolites; cell growth inhibition; gene expression; HeLa cells; in vitro culture

Citation: Jastrzebska, I.; Wawrusiewicz-Kurylonek, N.; Grześ, P.A.; Ratkiewicz, A.; Grabowska, E.; Czerniecka, M.; Czyżewska, U.; Tylicki, A. New Steroidal Selenides as Proapoptotic Factors. *Molecules* **2023**, *28*, 7528. https://doi.org/10.3390/molecules28227528

Academic Editors: Antal Csámpai, Erzsébet Mernyák, Marina Savić, Jovana Ajdukovic and Suzana Jovanović-Šanta

Received: 5 September 2023
Revised: 20 October 2023
Accepted: 6 November 2023
Published: 10 November 2023

1. Introduction

Selenosteroids (SeSt) are compounds formed by attaching a selenium-moiety to a steroid molecule [1]. Taking into consideration their structure, we can identify two groups of seleno-compounds: first with selenium directly connected to the steroid molecule to form a selenide and second where selenium is added to the steroid in the form of an organic (for example selenourea, benzsoselenazolones) [2,3] or inorganic (for example selenocyanate) [4] moiety. During the recent decades, much information about methods of SeSt synthesis has appeared [1]. Recently, selenocyano groups were directly introduced into pregnenolone, estradiol, and estrone [5] as well as using cholesterol and norcholesterol [6]. The results of that research indicate that some of the obtained SeSt exhibit antiproliferative properties against cancer cells. So far, little research has focused on the effects of SeSt on cells, and most studies have been limited to determining the effect of these compounds on the rate of cell proliferation with the determination of GI50 (growth inhibition—concentration of the compound resulting in a 50% reduction in the number of cells) or IC50 (inhibition concentration—concentration of compound causing 50% of metabolism inhibition) values in various in vitro tumor models [3,4,7,8] and bacteria [9]. Recent publications have also documented the inhibitory effects of selenosteroids on human tumors in animal models [5,6,10]. There is also information about the antioxidant actions of some SeSt [2,3]. Despite this prior knowledge about the effects of these compounds on cell cultures, there is still insufficient information to explain mechanism of their action at the cell biology level.

Last year a new method of selenosteroids synthesis was described [8]. This method employs a one-step, eco-friendly synthesis that bypasses some steps connected with the bad smell and the reactivity of commonly used seleno-reagents. Furthermore, the method demonstrated the potential to prepare libraries of steroids variously and selectively decorated with different organochalcogen moieties. Using this method, several seleno-steroidal derivatives were obtained based on the same steroid core containing selenium connected to the A ring of a steroid [11].

Our study aimed to compare the effects of obtained SeSt (Scheme 1) on HeLa cancer cell line in vitro in relation to other SeSt known from the literature and propose the mechanism of their action. In addition, having the above-mentioned model of SeSt structures, we were able to check their impact on cancer cells concerning chemical structure (the presence of a chlorine atom and the length of the linker between phenylene group and steroid core). Answers to these questions may guide further syntheses to optimize anti-proliferative effects. Knowing the previous data on the antiproliferative effect of SeSt on cells [1,4–6,10], we assumed that the tested compounds may have pro-apoptotic properties. From the point of view of their medical use, it is important which apoptotic pathway is activated [12]. In relation to the above, we consider the assumption that SeSt can act as cholesterol antimetabolites, so by interacting with the cell membrane, they can promote the extrinsic apoptosis pathway and can interact with the mevalonate pathway of steroid synthesis [13].

To test the above hypotheses, in addition to research on the HeLa cell growth rate under the influence of SeSt and their apoptosis profile, we also examined the expression of selected genes related to cholesterol synthesis and cell apoptosis as well as the possibility of SeSt interaction with the cell membrane using computational chemistry models.

To investigate the cholesterol synthesis pathway, we chose four genes: two enzymes that are key rat-limiting factors of sterol synthesis in human cells [11], *HMGCR* encoding 3-hydroxy-3-methylglutaryl-CoA reductase, which produces mevalonate, and *SQLE* encoding squalene monooxygenase, which catalyzes the oxidation of squalene to 2,3-epoxysqualene; the other two enzymes are *CYP51A1* encoding monooxygenase from the cytochrome P450 superfamily, which participates in cholesterol synthesis by removal of the methyl group at C-14 position from lanosterol [14], and *PDHB* encoding E1 beta subunit of pyruvate dehydrogenase complex which generates acetyl-CoA [15], a key substrate for cholesterol synthesis, by pyruvate oxidative decarboxylation.

Among the genes responsible for the apoptosis process, we chose two: *APAF1* encoding apoptotic protease activating factor-1, a cytoplasmic protein involved in apoptosome formation and activation of caspase 9 in the intrinsic apoptosis pathway [16,17], and *BID* encoding BH3 interacting-domain death agonist, which after activation participates in cytochrome C release from the mitochondrion [18]. Cytochrome C is necessary for apoptosome formation. BID protein is related to the extrinsic apoptotic pathway by the fact that caspase 8 is one of the BID activators. Caspase 8 in turn is a key protease activated in the extrinsic apoptotic pathway (Table 1).

Table 1. Characteristics of the genes for which expression was studied (based on www.ncbi.nlm.nih. gov (accessed on 28 October 2022); www.genecards.org (accessed on 30 October 2022)).

Types of Genes	Gene/Locus	Function of Encoded Protein
Related to cholesterol synthesis	*HMGCR*/5q13.3	3-hydroxy-3-methylglutaryl-CoA reductase—a transmembrane enzyme, bound to endoplasmic reticulum; catalyzes the conversion of (3*S*)-hydroxy-3-methylglutaryl-CoA to mevalonate, which is the rate-limiting step in the synthesis of cholesterol and other isoprenoids, playing a significant role in cellular cholesterol homeostasis; is regulated by a negative feedback mechanism through sterols and non-sterol metabolites derived from mevalonate.
	SQLE/8q24.13	squalene epoxidase—an enzyme with monooxygenase activity; catalyzes the stereospecific oxidation of squalene to (*S*)-2,3-epoxysqualene—the first oxygenation step in sterol biosynthesis.

Table 1. *Cont.*

Types of Genes	Gene/Locus	Function of Encoded Protein
	CYP51A1/7q21.2	lanosterol 14-alpha-demethylase, monooxygenase— a member of the cytochrome P450 superfamily (Cytochrome P450 Family 51, Subfamily A, Member 1); is an endoplasmic reticulum protein which catalyzes reactions involved in the synthesis of cholesterol, steroids, and other lipids; is an essential housekeeping enzyme and is evolutionarily highly conserved.
	PDHB/3p14.3	pyruvate dehydrogenase E1 subunit beta—is a nuclear-encoded mitochondrial multienzyme complex member that catalyzes the overall conversion of pyruvate to acetyl-CoA and carbon dioxide; provides a link between glycolysis and the tricarboxylic acid cycle.
Related to apoptosis	*APAF1/12q23.1*	apoptotic peptidase activating factor 1, a cytoplasmic protein that initiates apoptosis through the autocatalytic activation of procaspase 9 (dependent on ATP) and then caspase 3; is a component of the apoptosome
	BID/22q11.21	BH3 interacting domain death agonist, a proapoptotic protein that regulates apoptosis through binding to proteins BAX and /or BCL2; protein is activated by caspase-8, which is activated in the extrinsic apoptosis pathway; the BID protein is associated with cytochrome c release from mitochondria, thereby amplifying the apoptosis signal and contributing to the activation of the apoptosome.

2. Results and Discussion

For the purpose of this work, we synthesized selenosteroids **2–4** according to a procedure recently presented [8]. The functionalization of commercially available 1,4-androstadiene-3,17-dione (**1**) in a biphasic system was performed in a one-step protocol. The procedure as follows: biphasic system consisting of an equal volume of 10% HCl and ethyl acetate, diphenyl diselenide, and 10 equiv. zinc shaves was stirred until the organic layer was decolorized. Then the liquid phase was transferred under an argon atmosphere to a flask containing 1,4-androstadiene-3,17-dione (**1**). The resulting reaction mixture was stirred at room temperature for 3 h. Synthesis and the structure of seleno-Michael addition products **2–4** are shown in Scheme 1.

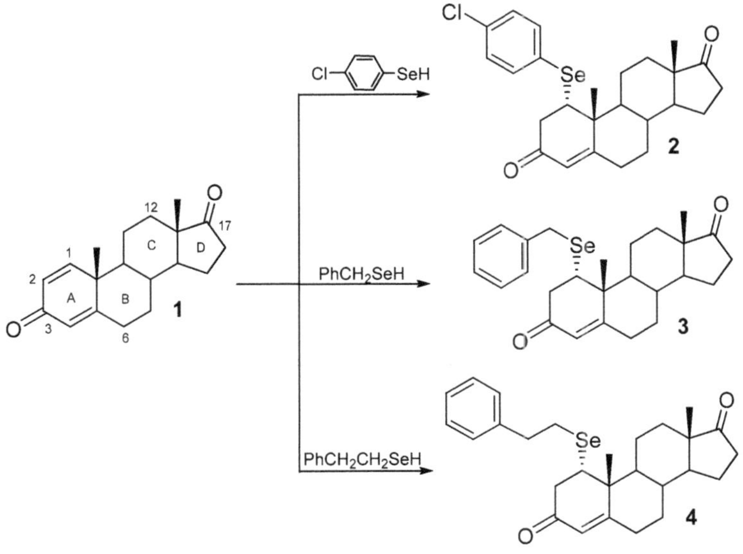

Scheme 1. Synthesis of selenosteroids **2–4**.

First, we explored the antiproliferative activity of SeSt **2–4** on the HeLa cancer cell line. Microscopic observation revealed a smaller amount as well as worse condition of cells in cultures with addition of tested SeSt compared with control (Figure 1). After 48 h of experiment on medium with SeSt, we noticed much fewer cells showed adhesion to the surface of the culture vessel and new cell division (compare microscope images, Figure 1). The worst condition of cells was observed on medium with compound **4**, where we found the fewest cells stuck to the culture vessel and the most cells in suspension that did not divide.

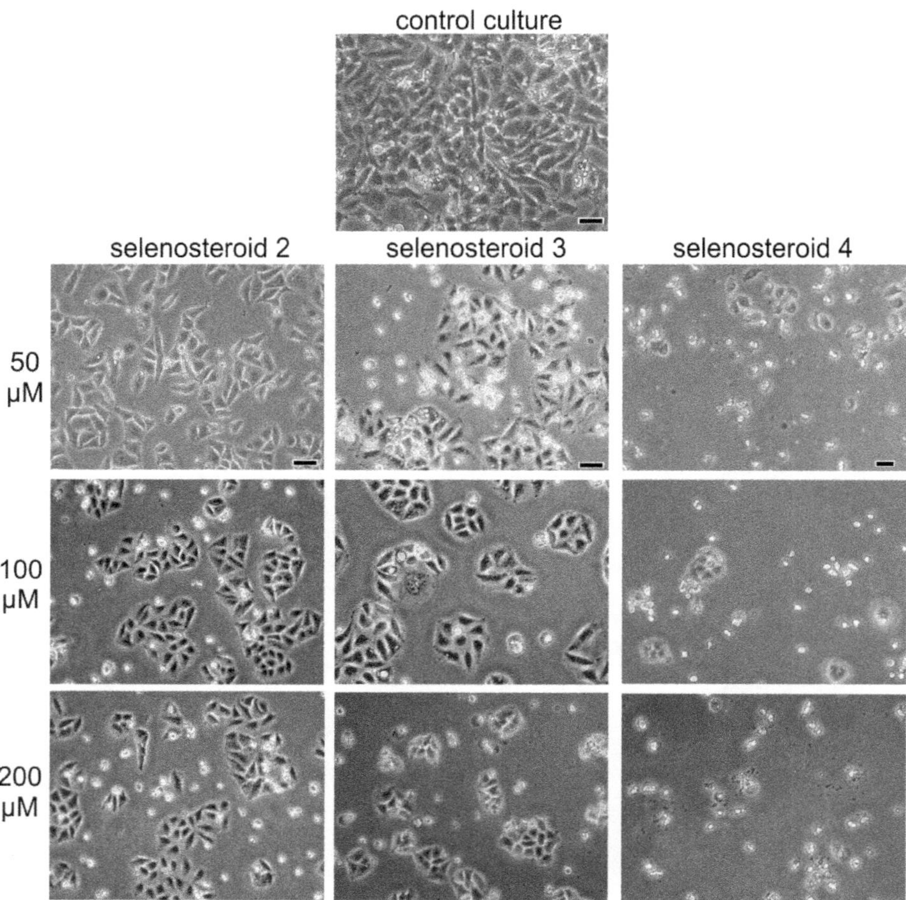

Figure 1. Microscopic observations of HeLa cells after 48 h of culture on a medium with selenosteroids (structure of selenosteroids according to Scheme 1), bar = 20 µm.

Analysis of the number of cells after three days of the experiment, when the control culture reached about 90% of confluence, showed significant, concentration-dependent inhibition of the growth rate in a culture with addition of all three tested SeSt (**2–4**) and substrate for synthesis (**1**) in the concentration range from 50 to 200 µM. Furthermore, a comparison between the growth rate of HeLa cells in the presence of the substrate (**1**) and the growth rate of cells under SeSt treatment shows that derivatives **3** and **4** exhibit significantly stronger cytostatic effects than substrate (**1**), while the effect of the substrate is comparable to that of SeSt **2** (compare Figure 2A–D). The most effective inhibitor of culture growth in comparison to control was selenosteroid **4** (Figure 2D), where already at a concentration of 50 µM we observed only 15.6% of cells relative to control, and the concentration of 100 and 200 µM almost completely inhibited culture growth. Less effect on cell number was found for selenosteroid **3** (at 50 µM 39.8%, at 100 µM 14.8% of control and at 200 µM almost complete inhibition, Figure 2C). Compound **2** as well as substrate (**1**) had the least influence on culture growth (at 200 µM, 39 0% of control; Figure 2A,B).

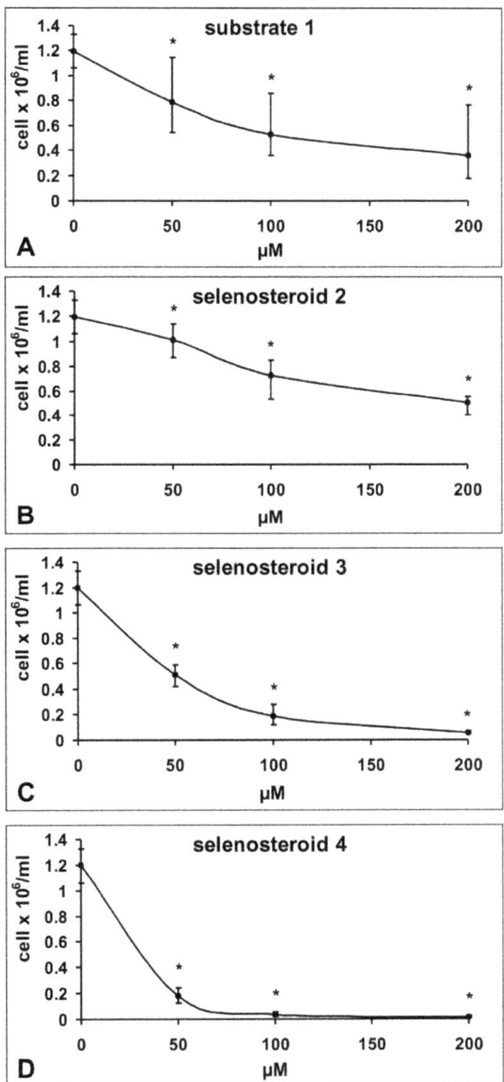

Figure 2. Changes in the number of HeLa cells in culture depending on the concentration of used compound after three days of culture (substrate **1**; 1,4-androstadiene-3,17-dione—substrate for synthesis of selenosteroid **2**, **3**, and **4**, according to Scheme 1). Data represent medians ± maximum and minimum values. Asterisks indicate statistically significant differences compared to the control (Mann–Whitney U-test, $p < 0.05$). The number of cells in the control culture corresponds to zero substrate and selenosteroid concentration.

Having the data concerning the dependence of the cells growth rate on the concentration of SeSt, we determined the GI_{50} values for each compound (Figure 3). The GI_{50} values were 25.0 µM for **4**, 46.3 µM for **3**, 128.1 µM for **2**, and 117 µM for substrate (**1**). Anova analysis showed that the difference between lowest GI_{50} value for **4** was statistically significant for substrate **1** and both SeSt **2** and **3**. SeSt **4** also acts stronger than **3**, and differences between substrate **1** and SeSt **2** were insignificant (Figure 3). The above results confirm the previous microscopic observations showing that the tested compounds affect HeLa cells with different strength, where the strongest cytostatic effect was that of compound **4**.

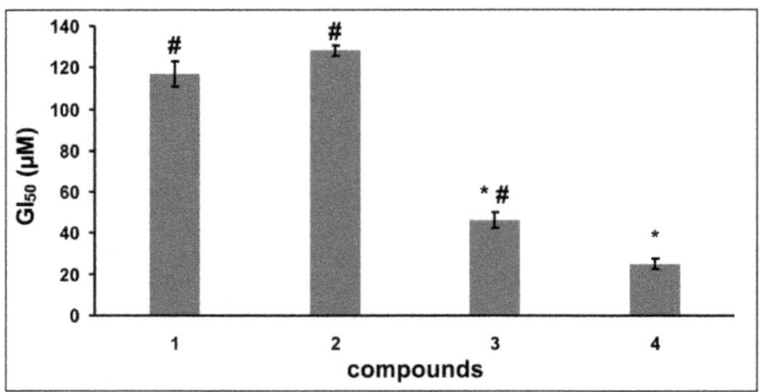

Figure 3. Comparison (means $\pm$ SE) of GI_{50} for substrate (**1**) and individual selenosteroids (**2–4**) based on data from Figure 2. Asterisks indicate values that differ from substrate **1**, and hashtags indicate values differ from selenosteroid **4** (Anova and RIR post-hoc test $p < 0.05$). The differences in GI_{50} values between selenosteroids **2** and substrate (**1**) were not statistically significant.

Currently, many selenium derivatives of steroids are known, some of which exhibit cytotoxic properties in relation to cancer cells (IC_{50} from a few to over 100 μM [1,3–6,10,19]); some of them also exhibit bacteriostatic properties [9]. Concerning HeLa cells, the lowest IC_{50} values are recorded for various A ring-modified selenocyanide and selenourea derivatives [3,4] (IC_{50} from a few to over 50 μM) and a phenylselenourea derivative in the C-3 position of the same ring [1] (IC_{50} approx. 2 μM). Recently, high cytotoxicity of selenocyanate derivatives in which the B ring of the steroid has been modified has been reported (IC_{50} about 6–30 μM) [6]. Among our phenylselenium derivatives at the C-1 position of ring A, the best cytostatic properties show derivative **4** (GI_{50} 25.0 μM) with an extended, two-carbon linker of the phenyl ring with selenium, compared to derivative **3** (almost twice higher GI_{50} in comparison with **4**). It should also be noted that the introduction of chloride into the SeSt significantly deteriorated the cytostatic properties (derivative **2** GI_{50} approx. 130 μM). Comparing literature data concerning the IC_{50} values of various SeSt in relation to breast cancer [2,4,10,19,20] (MCF-7) it can be concluded that these are more resistant to SeSt compared to HeLa cells and that modifications in various positions of the A ring of the steroid are generally most effective, which is also reflected in the results of our study. The exception is the last obtained selenocyanate derivatives at the B ring of the steroid, which in most cases show lower GI_{50} values for MCF-7 in comparison to HeLa [6]. Although the cytostatic properties (GI_{50}) of our derivative **4** are slightly weaker than (IC_{50}) those of the strongest SeSt known from the literature, it may prove to be prospective in the search for new cytostatic agents. In addition, our results may suggest the role of the length of the carbon chain between selenium and the phenyl ring concerning the cytostatic properties of the compound. Optimization of the length of this linker may be of key importance for the properties of the synthesized derivatives. In the case of our results, extending this linker by one carbon atom reduced the GI_{50} value twofold. The above observation is also reflected in other studies of selenocyanide steroid derivatives (Figure 4) [4,10].

The above-mentioned studies showed that the length of the carbon chain between the oxygen at the C-3 and NCSe groups is important for the cytotoxic properties of the derivative. The 2–8 carbon-linker derivatives had better cytotoxic properties compared to the shorter and longer linker derivatives. Therefore, optimization of the length and structure of the linker is a promising direction in the chemical synthesis of derivatives with the best cytostatic properties. If the length of the hydrophobic elements of SeSt derivatives has a significant impact on their cytotoxicity, it can be assumed that the mechanism of SeSt action on cells may be closely related to the interaction with the phospholipid component of cell membranes, leading to disturbances in their structure as cholesterol antimetabolites. This hypothesis is supported by the results of recent studies where it has been shown that some SeSt significantly increase the level of ROS in cells [19].

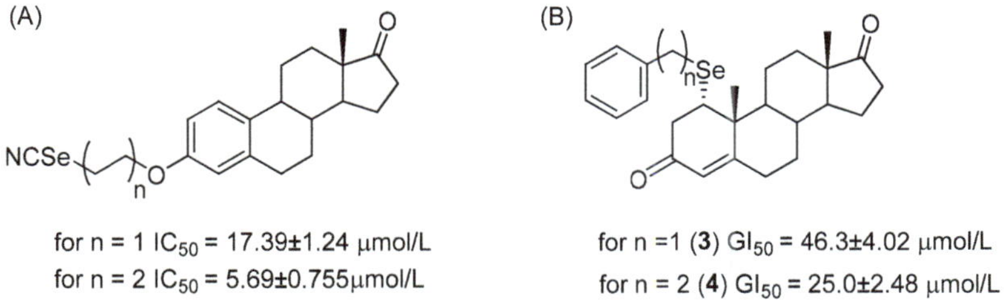

(A)   (B)

for n = 1 IC_{50} = 17.39±1.24 μmol/L   for n =1 (**3**) GI_{50} = 46.3±4.02 μmol/L

for n = 2 IC_{50} = 5.69±0.755 μmol/L   for n = 2 (**4**) GI_{50} = 25.0±2.48 μmol/L

Figure 4. Comparison of IC_{50}/GI_{50} values of different selenosteroids taking into account the length of the carbon linker between the selenium atom and the steroid core. (**A**) Structures described in ref. [4]: values given as mean ± SE; lengthening the chain from 2 to 4 carbon atoms resulted in an approximately 3-fold decrease in IC_{50}. (**B**) Our experimental data: values given as mean ± SE, where elongation of the carbon chain from 1 to 2 atoms resulted in an approximately 2-fold decrease in GI_{50}.

To propose the mechanism of action of the studied selenosteroids **2–4**, we analyzed the profile of cell viability and apoptosis under the influence of these SeSt as well as the expression of genes involved in the mevalonate pathway of steroid synthesis and the process of cell apoptosis.

Analyzing the survival and apoptosis profiles of cells under the influence of the tested SeSt, we found that all of them have a pro-apoptotic effect on HeLa cells to varying degrees. The observed effect was dependent on the concentration of SeSt (compare Figures 5–7).

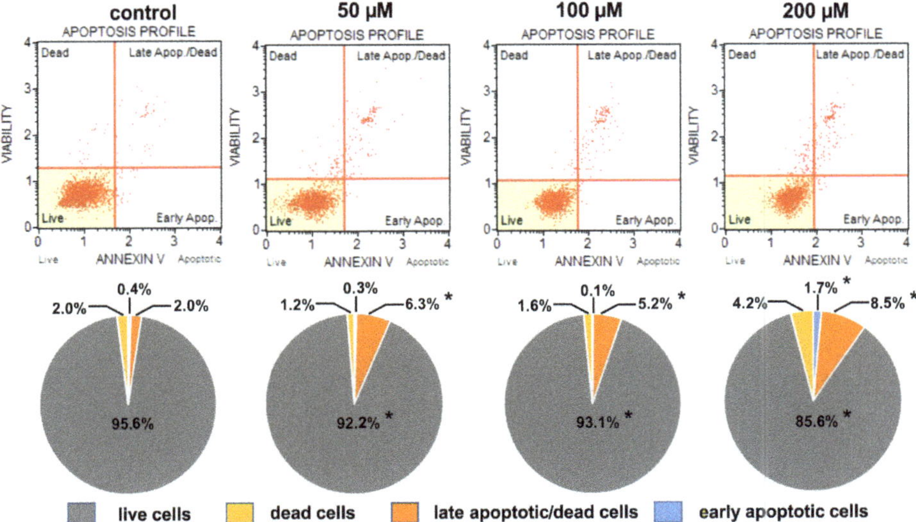

Figure 5. Comparison of the apoptotic profiles of HeLa control cultures and cultures on media with different concentrations of selenosteroid **2** after three days of culture. At the top are examples of histograms of individual cultures obtained as a result of cytometric analysis. At the bottom is the percentage of median data from a series of experiments. Asterisks indicate statistically significant differences concerning the control (Mann–Whitney U-test, $p < 0.05$).

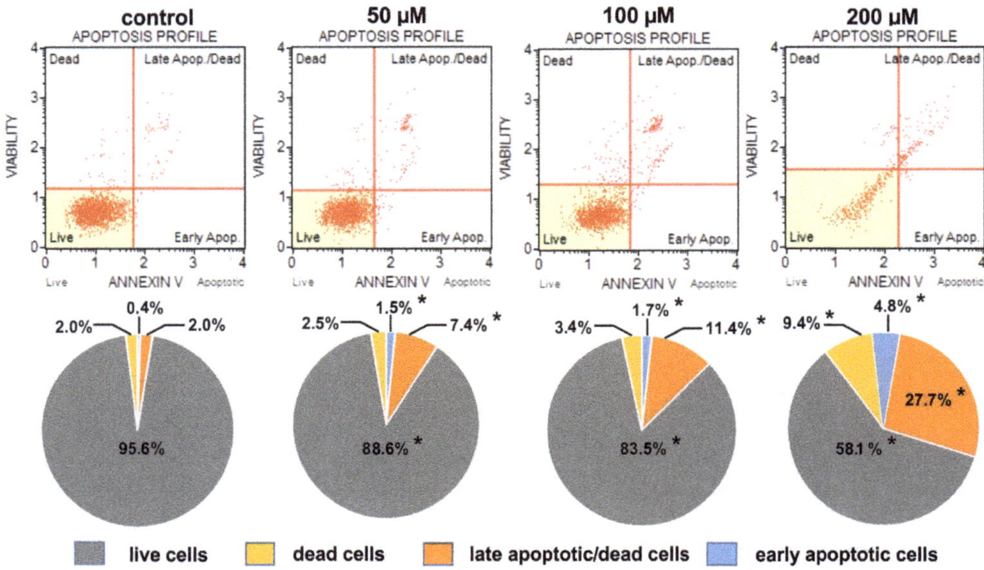

Figure 6. Comparison of the apoptotic profiles of HeLa control cultures and cultures on media with different concentrations of selenosteroid **3** after three days of culture. At the top are examples of histograms of individual cultures obtained as a result of cytometric analysis. At the bottom is the percentage of median data from a series of experiments. Asterisks indicate statistically significant differences concerning control (Mann–Whitney U-test, $p < 0.05$).

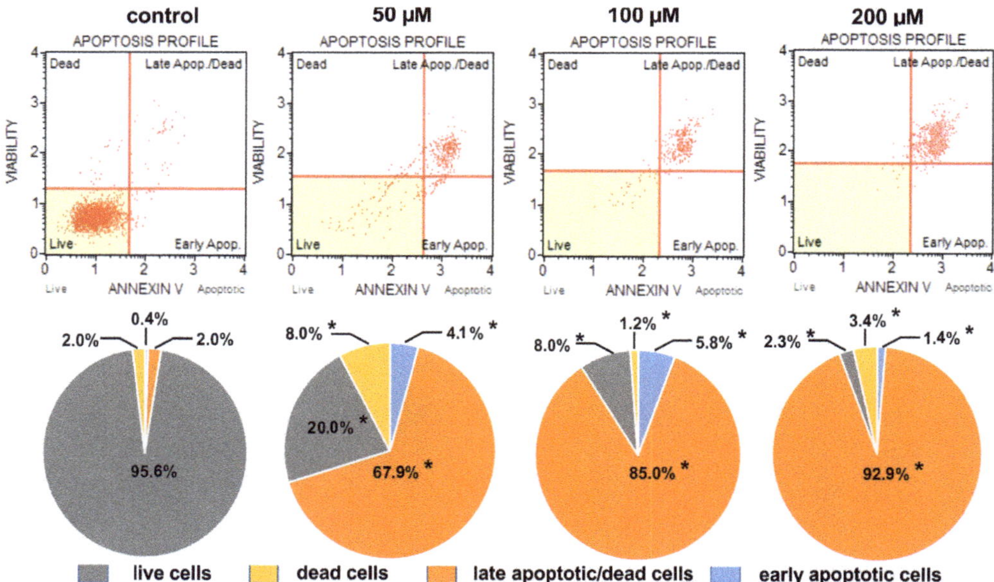

Figure 7. Comparison of the apoptotic profiles of HeLa control cultures and on media with different concentrations of selenosteroid **4** after three days of culture. At the top are examples of histograms of individual cultures obtained as a result of cytometric analysis. At the bottom is the percentage of median data from a series of experiments. Asterisks indicate statistically significant differences concerning control (Mann–Whitney U-test, $p < 0.05$).

The weakest pro-apoptotic effect was shown by **2**, where even at a concentration of 200 µM, live cells predominated (85.6%), but still there were fewer of them than in the control (95.6%); at the same time, the number of dead and apoptotic cells increased from 2.0% and 2.4% in control up to 4.2% and 10.2% in culture with **2**, respectively (Figure 5). The strongest pro-apoptotic effect was observed in the case of compound **4** (Figure 7), where at a concentration of 200 µM late apoptotic and dead cells definitely predominated (96.3%), and the number of viable cells decreased to 2.3%. In the case of this compound, a significant increase in apoptotic cell amount together with a strong reduction in the share of live cells was observed already at a concentration of 50 µM (Figure 7). A moderate pro-apoptotic effect was exerted by compound **3**, where substantial changes in the proportion of cells were observed at a concentration of 200 µM. In this case, we found 32.5% apoptotic cells (together early apoptotic and late apoptotic/dead), 9.4% dead cells, and 58.1% viable cells (Figure 6).

There are limited data in the literature on the effect of selenosteroids on the apoptosis process. The pro-apoptotic effect of selenosteroids on HeLa cells in vitro was found in the case of derivatives containing N-phenyl selenourea attached to the C3 position at the steroid molecule [2]. In these studies, approximately 60% apoptotic cells were noted at 25 µM concentration of N-phenyl selenourea steroid derivative, and this result was more than three times higher compared to diosgenin. During the same research, the authors did not find proapoptotic properties of thioureas steroid derivatives. More recently, information on the proapoptotic effect of selenosteroids has also appeared in the work of Huang Y. et al. [5,6,10]. In our studies, we found that selenosteroid **4** at a concentration of 50 µM caused apoptosis of about 72% of cells. In culture while compound **2** (organohalogen moiety linked to selenium) has a much lower proapoptotic effect. The intensity of the effect we observed was also correlated with the length of the carbon linker between the selenium attached to the C1 carbon of the A ring of the base steroid. A much stronger pro-apoptotic effect was noted in the case of derivative **4**, which has a longer linker compared to derivative **3**. The pro-apoptotic effect was also tested on steroidal ethynyl selenide derivatives where the selenium-containing residue was combined with the C-12 of the C ring, the C-6 of the B ring, or the C-17 of the D ring [17]. The last derivative has the highest proapoptotic effect (almost three-fold increase in the activity of caspases 3/7 in MCF-7 cells at a concentration of about 200 µM compared to the control). Due to the different methodology and structure of the studied SeSt, it is not easy to compare our results with the data from the literature; however, it can be concluded that the derivative **4** studied in this work has similar pro-apoptotic properties to other SeSt known from the literature. In addition, our results regarding the reduction in cell growth rate and promotion of the apoptosis process indicate that a good direction of chemical syntheses to improve the cytostatic properties of the tested SeSt will be the optimization of the structure and length of the carbon link between the selenium atom and phenol moiety in the described group of SeSt derivatives.

The in vitro cultures of HeLa cells with three tested SeSts enabled us to compare the relative expression profiles of genes (Figure 8) responsible for apoptosis induction with those related to cholesterol synthesis. Of the SeSt tested, the strongest effect for increase of cholesterol synthesis-involved genes expression (*HMGCR, SQLE, CYP51A1*) was observed for compounds **2** and **3** compared to cells cultured in the presence of compound **4** and the control culture. In contrast, *PDHB* gene expression was significantly decreased in culture with **4**, while the other compounds did not change their expression relative to the control. The presence of compound **2** in the culture medium increased the expression of the pro-apoptotic gene *BID* in HeLa cells compared to cells cultured in the presence of compounds **3** and **4** where the level of this transcript did not significantly change compared to the control. However, in the case of the *APAF1* gene, whose product is an essential factor of the apoptosome [16], we found a reduction in its expression under the influence of compounds **2** and **3**, as well as a tendency for reduction of its expression in the case of compound **4** in comparison with the control (Figure 8).

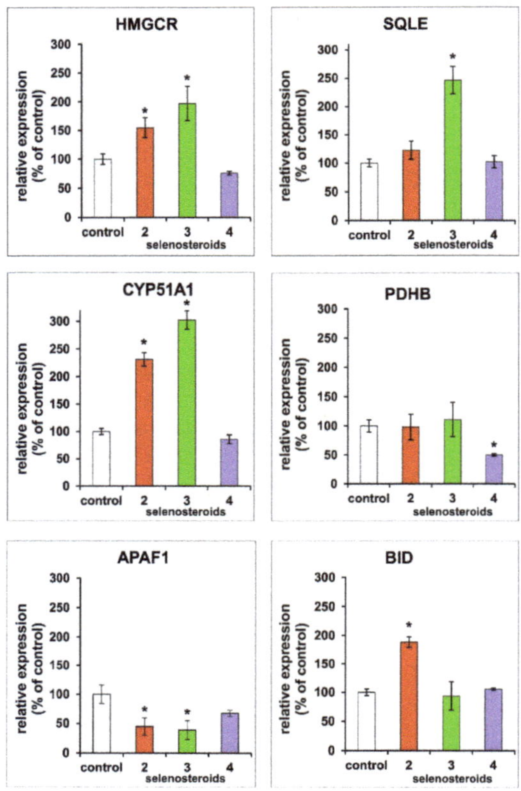

Figure 8. Comparison of the relative expression of selected genes after three days of HeLa cell culture in media supplemented with different selenosteroids (2—red color, 3—green color, 4—purple color) at a concentration equal to the GI50 value as shown in Figure 3. Values are shown as % of control ± standard deviation. Asterisks indicate statistically significant differences compared to the control (Student's *t*-test, $p < 0.05$).

The results of transcripts-level analysis of selected genes of the cholesterol synthesis pathway and those involved in the apoptosis process allowed us to make some observations that may contribute to verifying our hypothesis. Results obtained after three days of HeLa cell culture may suggest different mechanisms of action of tested SeSt. Compounds **2** and **3**, which have a milder effect on HeLa culture, act in the same way, while the most potent compound, **4**, may also have additional mechanisms of action.

The significant increase in *HMGCR*, *SQLE*, and *CYP51A1* gene expression in cultures with compounds **2** and **3** indicates the influence of SeSt on the synthetic rate of the mevalonate pathway in HeLa cells. Increased transcription levels of these key enzymes in the pathway indirectly support our hypothesis about the incorporation of these SeSt into the cell membrane and its destabilization. The key role of mevalonate pathway in tumor development is well known [13,21]. Although the activating effect of these compounds on cholesterol synthesis seems to be evident, the observed increase in expression of mevalonate pathway genes may be a specific response of the cell to membrane destabilization (increase in the rate of endogenous cholesterol synthesis should compensate for the membrane's destabilization). Overexpression of mevalonate pathway-involved genes (especially *SQLE*) was reported in different types of cancer cells [22–26].

The membrane destabilization may cause activation of the extrinsic apoptosis pathway, which may be indicated by overexpression of the *BID* gene. The BID protein is related to that pathway because it is activated by caspase 8, which is the key protease for the extrinsic apoptosis pathway [12]. Procaspase 8 may be part of two complexes related to the induction of apoptosis. The initial plasma membrane-bound complex (complex I) and cytoplasmic complex (complex II) are known as death-inducing signaling complexes [27]. Complex I is directly related to the induction of the extrinsic apoptotic pathway, while complex II triggers an amplified cascade of caspase activation via crosstalk with the intrinsic apoptosis pathway [28], probably by the release of proapoptotic factors from mitochondria. The observed reduction in *APAF1* expression, which is crucial for triggering the intrinsic apoptosis pathway (activation of caspase 9), may be a defensive reaction of cancer cells [29].

Taking into consideration the above-proposed mechanism, it can be assumed that the cytotoxic effect of the studied SeSt results from their interaction with the cell membrane and induction of the extrinsic apoptosis pathway.

SeSt **4** in the tested experimental system (concentration equal to GI_{50}) does not affect the expression of genes directly related to the mevalonate pathway as well as the proapoptotic *APAF1* and *BID* genes. Taking into account the strongest cytostatic effect of this SeSt, it can be concluded that in this case, the cells will not have time to activate defense mechanisms, as in the case of less efficient derivatives **2** and **3**. It is noteworthy that *PDHB* expression is reduced under the influence of SeSt **4**. The protein encoded by *PDHB* is a key component of the pyruvate dehydrogenase complex [30] that catalyzes the oxidative decarboxylation of pyruvate, which is one of the main sources of acetyl-CoA used in both catabolic (Krebs cycle) and anabolic reactions (e.g., steroid synthesis). Inhibition of the above-mentioned reaction leads to a decrease in the level of ATP and a reduction in lipid synthesis, which further also contributes to the disorganization of the cell membrane. The limitation of this reaction may be the main reason for the strongest anti-proliferative effect of derivative **4**. Comparing the data in Figure 8 and apoptotic profiles (Figures 5–7), it can be assumed that derivative **4**, in addition to affecting the membranes, acts inside the cell in a way similar to steroid hormones, limiting the expression of PDHB, which may additionally enhance the proapoptotic effect of this derivative.

To assess the effect of obtained SeSt on the stability of the membrane bilayer of the HeLa cells, a series of in silico experiments were performed, starting with molecular docking. As detailed below in the "Experimental" section, the procedure involved repeated dockings with the box shifting to capture all the diversity in the membrane structure. For every box position, docking calculations were performed to generate the best-scored (i.e., with the lowest value of ΔG) poses within. A number of poses, substantially differing in placement, were obtained as a result. The best docking scores for SeSt **2**, **3**, **4** are -10.4, -9.6, and -11.6 kcal/mol, respectively. This suggests the most efficient membrane penetration by SeSt **4**, which is in agreement with the in vitro experiments discussed above. Because of the best affinity, a visualization of the optimal pose for **4** is shown in Figure 9 below (in order to improve clarity, Figure 9 does not contain the SeSt **2** and **3**). It is seen that SeSt **4** is deeply buried in the membrane structure, filling the empty space between the layers. Similarly, the best poses for SeSt **2** and **3** are also located here. A significant variation (of about 5–6 kcal/mol) in docking scores between these favorable (between the layers) and other (inside the layers) positions is noted. The behavior of cholesterol is different. Sites comparable to the favorable ones for the new derivatives (i.e., those located between layers) are scored by poor ratings of the order of $-(5$–$7)$ kcal/mol. The best affinities (~-9 kcal/mol) are noted at the membrane edges. These results may suggest a different mechanism for the penetration of the membrane by cholesterol and proposed chemicals. Indeed, while cholesterol is embedded inside the phospholipid layers, the new compounds are rather deposited between these layers. The consequence may be the improper functioning of the penetrated membrane leading to its destabilization and, ultimately, cell apoptosis. Because of its best docking score, the most effective agent promoting this process appears to be SeSt **4**.

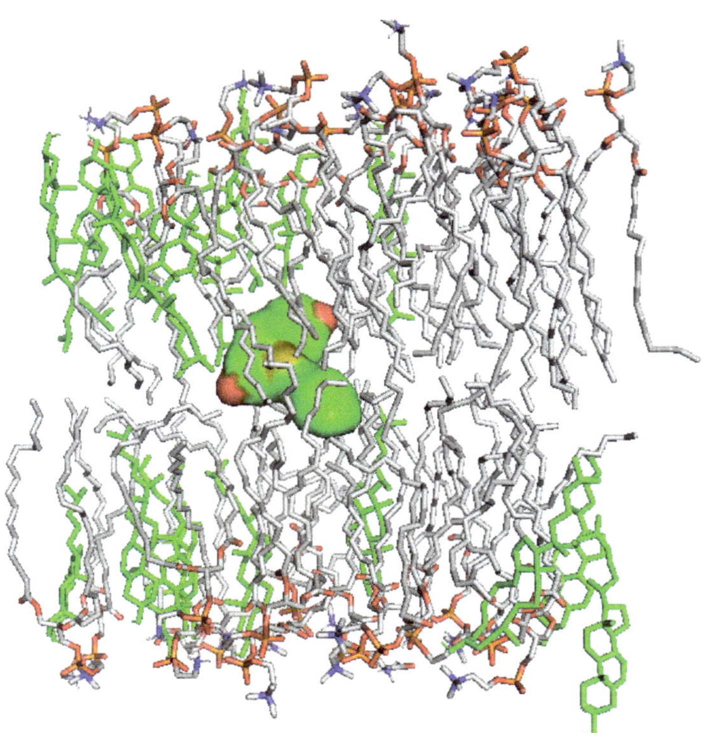

Figure 9. Position of selenosteroid **4** (surface view) in the lipid bilayer model of HeLa lineage cells. Cholesterol molecules forming the layers are highlighted in green. The water molecules and ions were removed for the sake of clarity.

In order to further understand the mechanism of interaction of the proposed derivatives with the HeLa cell membrane, 10 ns molecular dynamics simulations were carried out for both the membrane alone and the complexes with SeSt **2–4** embedded. The starting point, resulting from docking, was located between two monolayers of the membrane (see Figure 9 above). Visual inspection of the resulting trajectory indicates a movement of the ligand between the membrane-forming layers rather than their penetration. Due to the significant size of the ligand molecule, this is an expected result. In order to quantify the impact of the observed motion on the stability of the membrane, the change in RMSD (root mean square deviation) and Rg (radius of gyration) values along the trajectories were analyzed; results are plotted in Figure 10. RMSD analysis permits for assessing the system's stability during simulation. The averaged RMSD values of 8.2, 11., 11.1, and 10.2 Å for UNL and SeSt **2**, **3**, and **4**, respectively, are much higher than their counterparts for proteins, with typical average RMSDs of 2–5 Å [31,32]. This is not surprising since the cell membrane, consisting of noncovalently bonded phospholipids and sterols, is a much more labile structure than proteins. The interrelationships of the RMSDs for the different systems are important for assessing stability. From a comparison of the average values and from Figure 10a, it can be seen that ligand embedment in the membrane causes an observable increase in RMSD. The magnitude of this effect changes with time, although after 5 ns it appears to stabilize at about 2 Å. Although the influence of SeSt **4** appears to be smaller than that of the other two, there is a noticeable scattering of the interaction strength in the graph. In fact, the values of standard deviation (SD) and coefficient of variation (h = SD/mean) are by far the largest for SeSt **4** and the smallest for the ligand-free form. This may indicate that the actual influence of this compound on membrane stability is more significant than suggested by the RMSD analysis. On the other hand, the influence of all tested SeSt on membrane destabilization is apparent. The radius of gyration (Rg), defined as the root mean distance from the axis of gyration to a point where the total mass of the structure is concentrated, is considered a gauge of the compactness of the system. Its increase may indicate a destabilization of the whole structure—for example, the unfolding of the protein or disintegration of the lipid bilayer. The Rg for the structures considered here is plotted in Figure 10b. Its magnitude is initially utmost for the membrane without ligands. Over time, the relationship changes, and by the end of the simulation the unliganded form shows the smallest Rg. Except for SeSt **2**, the differences are not significant; however, Rg of SeSt **2** is surprisingly large and increases with time, which may indicate faster disintegration of the membrane. It seems to be consistent with the time evolution of RMSD, which reaches its highest value after the ninth second. A thorough clarification of this probably requires longer simulations to be carried out, however.

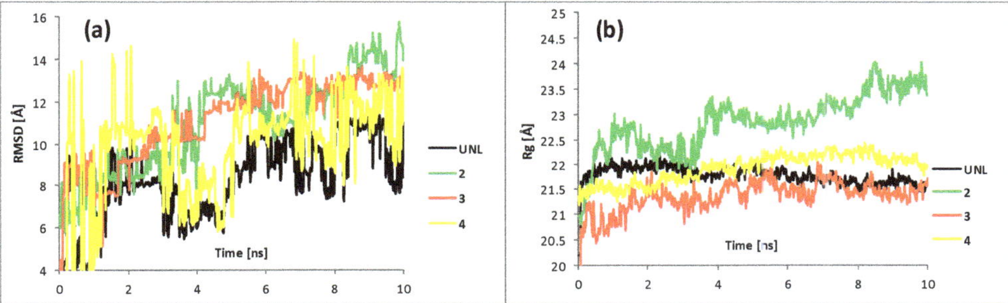

Figure 10. Analysis of RMSD and Rg of unliganded HeLa bilayer model (UNL) and three complexes during 10 ns of molecular dynamics simulation: (**a**) root mean square deviation (RMSD) for non-hydrogen atoms; (**b**) radius of gyration (Rg).

3. Conclusions

Our SeSt **4** has similar properties to other known selenosteroids and therefore can be considered as a prospective cytostatic agent. The results of our research, in contrast with the literature, allow us to conclude that the incorporation of selenium substituents into the A ring of steroid can give better cytostatic results against cancer cells, and the length of the carbon linker in the case of phenylselenium or selenocyanate derivatives is important for their action. The tested derivatives probably interact with the cell membrane as cholesterol antimetabolites, causing its disorganization and activation of the extrinsic apoptosis pathway, which may be evidenced by an increase in *BID* expression and reduction in *APAF1* expression (Figure 11). Moreover, our hypothesis has been confirmed by in silico experiments. Molecular docking proves that all SeSt penetrate the cell membrane by easily embedding between its layers. Among tested SeSt, the most favorable ΔG value is shown by **4**, which suggests its most efficient penetration of the bilayer structure. Molecular dynamics studies indicate a significant membrane destabilization (which may lead to cell apoptosis) after embedding of each of the compounds **2**–**4**. Derivative **4** penetrates the membrane most readily; for that reason, it appears to be the most potent in membrane disorganization of HeLa cells. In the case of the most potent derivative, **4**, an additional, intracellular mechanism is possible, causing a reduction in cell bioenergetics due to a decrease in *PDHB* expression.

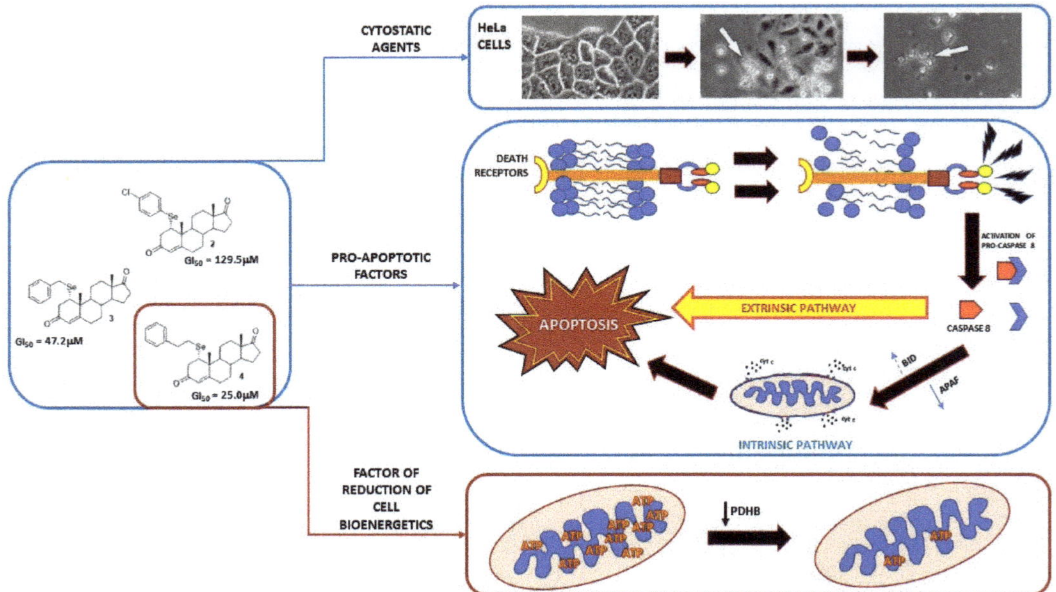

Figure 11. Proposed mechanism of action of the tested selenosteroids **2–4**. All tested compounds exhibit cytostatic activity with varying intensity depending on their chemical structure. The proapoptotic effect of selenosteroids may result from the destabilization of the cell membrane and activation of the extrinsic apoptotic pathway related to caspase 8. Elevated expression of BID activated by caspase 8 may enhance apoptosis by releasing cytochrome C from mitochondria. The most cytostatic derivative, **4**, additionally reduces PDHB expression, probably through a hormonal pathway, which leads to the disorganization of cell bioenergetics and inhibition of the mevalonate pathway by reducing acetyl-CoA synthesis.

4. Experimental Section

4.1. Synthesis

Compounds **2–4** were obtained according to the described protocol [11] as follows: Diselenide (0.13 mmol) was added to a flask with 2 mL of 10% HCl, 2 mL of ethyl acetate, and 10 equiv. of zinc shaves. The reaction was stirred vigorously until the discoloration of the organic layer occurred. Then the liquid was transferred into another flask under inert conditions (Ar) and the steroid (0.2 mmol) was added. The reaction mixture was stirred for 3 h. Then the reaction mixture was poured into water and extracted with ethyl acetate (3×20 mL). The organic layer was dried with Na_2SO_4, filtered, and the solvent removed under vacuum. The products were purified by flash chromatography. Obtained 1α-(4-chlorophenylselenyl)-androst-4-en-3,17-dione (**2**), 1α-benzylselenylandrost-4-en-3,17-dione (**3**), and 1α-phenylethylandrost-4-en-3,17-dione (**4**) were proved identical in all respects with the same compounds described in the literature [11].

4.2. Cells Culture

For evaluation of the impact of three tested SeSt (compound **2**, **3**, and **4**; Scheme 1) on cell growth and apoptosis, we used an in vitro culture of HeLa (ATCC-CCL-2) cells. Cultures were maintained in a NuAire NU-5820E incubator (37 °C, concentration of CO2: 5%, humidity: 95%) on medium MEM199 (Gibco ref. no. 31150-022), with 10% Fetal Bovine Serum and antibiotics (penicillin 50 U/mL, streptomycin 0,05 mg/mL). The initial culture density was approximately 1×10^5 cells/mL. Initially, all cultures were maintained for 24 h on control medium without SeSt. Subsequently, in control cultures the medium was replaced with fresh, whereas in experimental variants the medium was replaced with SeSt-containing in concentrations 50, 100, and 200 μM (first we prepared stock solutions with DMSO and added to the medium in volume not exceeding 100 μL). Final concentration of DMSO in the medium did not exceed 0.2%. According to our knowledge, this DMSO concentration is neutral to cells in vitro [33–35]). All cultures were grown in 6-well plates (one plate for each combination) until the control variant reached 90% of confluence (about 3 days). The cells were then harvested using trypsin-EDTA solution (Sigma ref. no. T3924) and suspended in the same volume of phosphate buffered saline (Sigma ref. no. D8537). For cell counting, we used automatic cell counter EVE-MT (NanoEnTec Inc., Seoul, Korea). Quantity assessment of live/dead with division of early and late apoptotic cells was performed with Merck Millipore Muse™ Cell Analyzer (0500-3115) using Muse™ Annexin V & Dead Cell assay kit (Cat. No. MCH100105), according to the manufacturer's instructions. During the experiment, observations of cultures were carried out using the Olympus inverted contrast phase microscope (CKX 41).

For the study of the relative expression of selected genes (see below), HeLa cultures were maintained as above, on media with SeSt concentrations equal to their GI_{50} values determined based on the cell growth curves plotted for individual SeSt.

4.3. Relative Gene Expression Analysis

Total RNA was isolated and purified using an RNeasy Mini Kit (Qiagen, Hilden, Germany) following the manufacturer's protocol. Total RNA integrity was verified by 1.5% agarose gel electrophoresis, identified by ethidium bromide staining and $OD_{260/280}$ absorbance ratio > 1.95 using a Nano-Drop ND-2000 spectrophotometer (ThermoFisher Scientific, Waltham, MA, USA). cDNA synthesis was performed using a High Capacity cDNA Reverse Transcription Kit (ThermoFisher Scientific, USA) on 1 μg of total RNA in the T-100 Thermal Cycler (Bio-Rad, Hercules, CA, USA).

The relative expression analysis of studied genes were measured by quantitative real-time PCR using commercially designed human QuantiTect Primer Assays (Qiagen): Hs_CYP51A1_1_SG (cytochrome P450 family 51 subfamily A member 1), Hs_SQLE_1_SG (squalene epoxidase), Hs_HMGCR_1_SG (3-hydroxy-3-methylglutaryl-CoA reductase), Hs_APAF1_1_SG (apoptotic peptidase activating factor 1), Hs_BID_1_SG (BH3 interacting domain death agonist), Hs_PDHB_1_SG (pyruvate dehydrogenase E1 subunit beta), and Hs_PRS18_1_SG (s18) as normalizer and an endogenous control. Real-time PCR reaction was performed in duplicate using the SYBR Green PCR Master Mix (Qiagen, Germany) carried out in the CFX OPUS Real-Time PCR System (Bio-Rad, Hercules, CA, USA). The thermal cycling conditions followed the manufacturer's instructions. At the end of the amplification phase, a melting curve analysis was carried out on the products formed to check their specificity. Based on the standard curve (which was generated employing a serial of five dilutions of cDNA derived from unstimulated cells in reaction with the house-keeping gene—s18), the levels of studied genes transcripts were calculated after normalization of amplification products to s18. We used the comparative C_T method for relative quantification (ΔC_T method) to calculate our data.

4.4. Statistical Analysis

Data represent the number of cells from 8–14 independent experiments. Data representing apoptotic profile and relative gene expression came from at last 4 independent experiments. The results were evaluated using the Shapiro–Wilk W-test to identify normal distribution of data and the Levene L-test for testing of homoscedastic variances. In the case of normal distribution of data and homoscedastic variances, t-Student test or Anova with RIR post-hoc test were used to compare the mean values. In the case of non-normal distribution of data, we used the nonparametric Mann–Whitney U-test to compare median values.

4.5. Molecular Docking

Molecular docking studies were conducted to better understand the mechanism of interaction of the proposed derivatives with the cell membrane using the latest (1.2.5) version of the

AutoDock Vina program [36]. The Pymol [37] molecular visualization package was utilized for the presentation of the results. The Membrane Builder [38] application from the CHARMM-GUI [39] online environment was used to create, solvate, and ionize the bilayer mimicking the membrane of the HeLa cell. This buildup was based on data from ref. [40], proposing the HeLa mimicking model (POPC:POPE:POPS:CHOL = 29:31:6:34, where POPC stands for phosphatidylcholine, POPE—phosphatidylethanolamine, POPS—phosphatidylserine, and CHOL—cholesterol). The prepared model was pretreated for docking by cleansing its structures from solvent and ions, adding polar H atoms and Kollman charges [41]. The cubic box of $20 \times 20 \times 20$ Å is set up to encapsulate the ligands during the docking procedure. The problem is to locate the optimal site to insert the ligand into the membrane. Typically, reference compounds are embedded in crystallographic structures; their location points out the active center. However, this is not the case here. Because of that, docking calculations were performed on a series of overlapping grid boxes covering the entire area of variation of the bilayer features. Since the lipid chains are oriented along the z-axis, the box moved both along this axis and in the xy-plane. The best-score poses were automatically detected. This was done with an in-house submitting script, repeating the calculations over the subsequent grid boxes, available from the authors upon request. Since there are no reliable crystallographic structures, validation of the docking protocol is problematic and must be approximated. To carry it out, we chose a membrane model similar to the HeLa mimicking models (POPE + POPC with 30 molar-% cholesterol) resulting from exhaustive (300 ns) molecular dynamics simulations [42,43] and validated against reliable experimental data. For that reason, it is believed that the arrangement of cholesterol moieties properly reflects the real structure. Validation involved removing one cholesterol molecule from the membrane and docking it again (redocking). A satisfactory agreement of the resulting conformation with the original one was found with RMSD = 2.045 Å (see Figure 12 below); thus, the identical methodology was subsequently used to investigate the new derivatives.

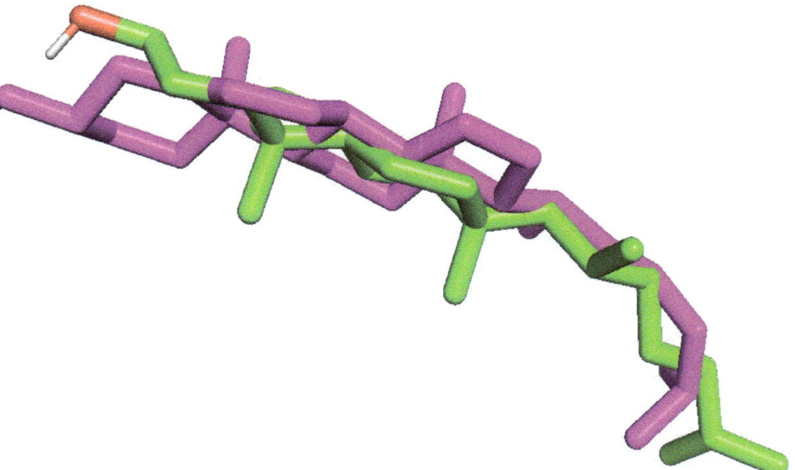

Figure 12. The alignment of the built-in cholesterol moiety from ref. [43] (green scaffold) superimposed on the best pose from the redocking procedure (magenta).

4.6. Molecular Dynamics

Molecular dynamics modeling was carried out using the NAMD 2.14 and VMD 1.9.4 programs [44,45]. The simulations were started with the best poses generated in the docking experiments. Every run began with an initial minimization for 50,000 steps, followed by a gradual heating from 0 to 310 K in 2 K steps and an additional 20,000 steps of equilibration. Preequilibrated structures were then subjected to a 10 ns production run with a timestep of 2 fs, and frames were written to the .dcd file for every 10,000 steps. Scripts provided by the NAMD/VMD developer were utilized to obtain the RMSD/Rg values for resulting trajectories. The Langevin piston methodology enforced a constant pressure of 1 atm with a decay period of 100 fs; a temperature of 310K was imposed by Langevin dynamics with a damping factor equal to 5 (1/s).

Author Contributions: Conceptualization, I.J., N.W.-K. and A.T.; Formal analysis, A.R., E.G. and U.C.; Investigation, P.A.G., M.C. and U.C.; Software, A.R., Writing—original draft, I.J. and A.T.; writing—review and editing, I.J., A.T. and A.R. All authors have read and agreed to the published version of the manuscript.

Funding: This study was supported by the Ministry of Education and Science, Poland, as a part of subsidies for maintaining the research potential granted to the Faculty of Biology (SWB-8) and Faculty of Chemistry of the University of Białystok. The equipment used for the cell culture at the Laboratory of Tissue Culture was funded by the Ministry of Science and Higher Education (grant no. 8636/E342/R/2014; Restructuration of Faculty of Biology and Chemistry). This work was financed from a scientific project by the subsidy of the Medical University of Bialystok (B.SUB.23.306).

Data Availability Statement: Data will be available on request.

Acknowledgments: We are grateful to Douglas F. Covey from Washington University in St. Louis, USA for editing the manuscript. The authors would like to thank the Computational Center of the University of Bialystok (Grant GO-008) for providing access to the supercomputer resources and the NAMD program.

Conflicts of Interest: The authors declare no conflict of interest.

References

1. Jastrzebska, I.; Grzes, P.A.; Niemirowicz-Laskowska, K.; Car, H. Selenosteroids—Promising hybrid compounds with pleiotropic biological activity: Synthesis and biological aspects. *J. Steroid Biochem. Mol. Biol.* **2021**, *213*, 105975. [CrossRef] [PubMed]
2. Romero-Hernández, L.L.; Merino-Montiel, P.; Montiel-Smith, S.; Meza-Reyes, S.; Vega-Báez, J.L.; Abasolo, I.; Schwartz, S.; López, Ó.; Fernández-Bolaños, J.G. Diosgenin-based thio(seleno)ureas and triazolyl glycoconjugates as hybrid drugs. Antioxidant and antiproliferative profile. *Eur. J. Med. Chem.* **2015**, *99*, 67–81. [CrossRef]
3. Fuentes-Aguilar, A.; Romero-Hernández, L.L.; Arenas-González, A.; Merino-Montiel, P.; Montiel-Smith, S.; Meza-Reyes, S.; Vega-Báez, J.L.; Plata, G.B.; Padrón, J.M.; López, Ó.; et al. New selenosteroids as antiproliferative agents. *Org. Biomol. Chem.* **2017**, *15*, 5041–5054. [CrossRef] [PubMed]
4. Huang, Y.; Peng, Z.; Wei, M.; Gan, C.; Zhang, Y.; Chen, S.; Xiao, J.; Cui, J. Synthesis and antiproliferative evaluation of some novel estradiol selenocyanates. *Steroids* **2022**, *181*, 108992. [CrossRef] [PubMed]
5. Huang, Y.; Peng, Z.; Wei, M.; Pang, L.; Cheng, Y.; Xiao, J.-A.; Gan, C.; Cui, J. Straightforward synthesis of steroidal selenocyanates through oxidative umpolung selenocyanation of steroids and their antitumor activity. *J. Steroid Biochem. Mol. Biol.* **2023**, *225*, 106203. [CrossRef]
6. Huang, Y.-M.; Cheng, Y.; Peng, Z.-N.; Pang, L.-P.; Li, J.-Y.; Xiao, J.-A.; Zhang, Y.-F.; Cui, J.-G. Synthesis and antitumor activity of some cholesterol-based selenocyanate compounds. *Steroids* **2023**, *194*, 109217. [CrossRef]
7. Chen, P.; Wang, P.; Song, N.; Li, M. Convergent synthesis and cytotoxic activities of 26-thio- and selenodioscin. *Steroids* **2013**, *78*, 959–966. [CrossRef]
8. Cui, J.; Pang, L.; Wei, M.; Gan, C.; Liu, D.; Yuan, H.; Huang, Y. Synthesis and antiproliferative activity of 17-[1′,2′,3′]-selenadiazolylpregnenolone compounds. *Steroids* **2018**, *140*, 151–158. [CrossRef]
9. Jastrzebska, I.; Mellea, S.; Salerno, V.; Grzes, P.A.; Siergiejczyk, L.; Niemirowicz-Laskowska, K.; Bucki, R.; Monti, B.; Santi, C. PhSeZnCl in the Synthesis of Steroidal β-Hydroxy-Phenylselenides Having Antibacterial Activity. *Int. J. Mol. Sci.* **2019**, *20*, 2121. [CrossRef]
10. Huang, Y.; Wei, M.; Peng, Z.; Cheng, Y.; Zhang, Y.; Li, J.; Xiao, J.; Gan, C.; Cui, J. Synthesis of estrone selenocyanate Compounds, anti-tumor activity evaluation and Structure-activity relationship analysis. *Bioorg. Med. Chem.* **2022**, *76*, 117086. [CrossRef]
11. Grześ, P.A.; Monti, B.; Wawrusiewicz-Kurylonek, N.; Bagnoli, L.; Sancineto, L.; Jastrzebska, I.; Santi, C. Simple Zn-Mediated Seleno- and Thio-Functionalization of Steroids at C-1 Position. *Int. J. Mol. Sci.* **2022**, *23*, 3022. [CrossRef] [PubMed]
12. Izadi, M.; Ali, T.A.; Pourkarimi, E. Over Fifty Years of Life, Death, and Cannibalism: A Historical Recollection of Apoptosis and Autophagy. *Int. J. Mol. Sci.* **2021**, *22*, 12466. [CrossRef] [PubMed]
13. Göbel, A.; Rauner, M.; Hofbauer, L.C.; Rachner, T.D. Cholesterol and beyond—The role of the mevalonate pathway in cancer biology. *Biochim. Biophys. Acta Rev. Cancer* **2020**, *1873*, 188351. [CrossRef]
14. Lepesheva, G.I.; Waterman, M.R. Sterol 14α-demethylase cytochrome P450 (CYP51), a P450 in all biological kingdoms. *Biochim. Biophys. Acta Gen. Subj.* **2007**, *1770*, 467–477. [CrossRef]

15. Saunier, E.; Benelli, C.; Bortoli, S. The pyruvate dehydrogenase complex in cancer: An old metabolic gatekeeper regulated by new pathways and pharmacological agents: Pyruvate dehydrogenase complex in cancer. *Int. J. Cancer* **2016**, *138*, 809–817. [CrossRef] [PubMed]

16. Dorstyn, L.; Akey, C.W.; Kumar, S. New insights into apoptosome structure and function. *Cell Death Differ.* **2018**, *25*, 1194–1208. [CrossRef]

17. Zou, H.; Henzel, W.J.; Liu, X.; Lutschg, A.; Wang, X. Apaf-1, a Human Protein Homologous to C. elegans CED-4, Participates in Cytochrome c–Dependent Activation of Caspase-3. *Cell* **1997**, *90*, 405–413. [CrossRef]

18. Billen, L.P.; Shamas-Din, A.; Andrews, D.W. Bid: A Bax-like BH3 protein. *Oncogene* **2008**, *27*, S93–S104. [CrossRef]

19. Grzes, P.A.; Sawicka, A.; Niemirowicz-Laskowska, K.; Wielgat, P.; Sawicka, D.; Car, H.; Jastrzebska, I. Metal-promoted synthesis of steroidal ethynyl selenides having anticancer activity. *J. Steroid Biochem. Mol. Biol.* **2023**, *227*, 106232. [CrossRef]

20. Cui, J.; Wei, M.; Pang, L.; Gan, C.; Xiao, J.; Shi, H.; Zhan, J.; Liu, Z.; Huang, Y. Synthesis and antiproliferative evaluation of novel steroid-benzisoselenazolone hybrids. *Steroids* **2019**, *152*, 108502. [CrossRef]

21. Juarez, D.; Fruman, D.A. Targeting the Mevalonate Pathway in Cancer. *Trends Cancer* **2021**, *7*, 525–540. [CrossRef]

22. Xu, Z.; Huang, L.; Dai, T.; Pei, X.; Xia, L.; Zeng, G.; Ye, M.; Liu, K.; Zeng, F.; Han, W.; et al. SQLE Mediates Metabolic Reprogramming to Promote LN Metastasis in Castration-Resistant Prostate Cancer. *OncoTargets Ther.* **2021**, *14*, 4285–4295. [CrossRef] [PubMed]

23. Xu, H.; Zhou, S.; Tang, Q.; Xia, H.; Bi, F. Cholesterol metabolism: New functions and therapeutic approaches in cancer. *Biochim. Biophys. Acta Rev. Cancer* **2020**, *1874*, 188394. [CrossRef] [PubMed]

24. Ge, H.; Zhao, Y.; Shi, X.; Tan, Z.; Chi, X.; He, M.; Jiang, G.; Ji, L.; Li, H. Squalene epoxidase promotes the proliferation and metastasis of lung squamous cell carcinoma cells though extracellular signal-regulated kinase signaling. *Thorac. Cancer* **2019**, *10*, 428–436. [CrossRef] [PubMed]

25. Brown, D.N.; Caffa, I.; Cirmena, G.; Piras, D.; Garuti, A.; Gallo, M.; Alberti, S.; Nencioni, A.; Ballestrero, A.; Zoppoli, G. Squalene epoxidase is a bona fide oncogene by amplification with clinical relevance in breast cancer. *Sci. Rep.* **2016**, *6*, 19435. [CrossRef]

26. Zhang, H.-Y.; Li, H.-M.; Yu, Z.; Yu, X.; Guo, K. Expression and significance of squalene epoxidase in squamous lung cancerous tissues and pericarcinoma tissues: Expression of SQLE mRNA and protein. *Thorac. Cancer* **2014**, *5*, 275–280. [CrossRef]

27. Micheau, O.; Tschopp, J. Induction of TNF Receptor I-Mediated Apoptosis via Two Sequential Signaling Complexes. *Cell* **2003**, *114*, 181–190. [CrossRef]

28. Berghe, T.V.; van Loo, G.; Saelens, X.; van Gurp, M.; Brouckaert, G.; Kalai, M.; Declercq, W.; Vandenabeele, P. Differential Signaling to Apoptotic and Necrotic Cell Death by Fas-associated Death Domain Protein FADD. *J. Biol. Chem.* **2004**, *279*, 7925–7933. [CrossRef]

29. Wu, C.-C.; Lee, S.; Malladi, S.; Chen, M.-D.; Mastrandrea, N.J.; Zhang, Z.; Bratton, S.B. The Apaf-1 apoptosome induces formation of caspase-9 homo- and heterodimers with distinct activities. *Nat Commun.* **2016**, *7*, 13565. [CrossRef]

30. Bunik, V.I.; Tylicki, A.; Lukashev, N.V. Thiamin diphosphate-dependent enzymes: From enzymology to metabolic regulation, drug design and disease models. *FEBS J.* **2013**, *280*, 6412–6442. [CrossRef]

31. Maliszewski, D.; Demirel, R.; Wróbel, A.; Baradyn, M.; Ratkiewicz, A.; Drozdowska, D. s-Triazine Derivatives Functionalized with Alkylating 2-Chloroethylamine Fragments as Promising Antimicrobial Agents: Inhibition of Bacterial DNA Gyrases, Molecular Docking Studies, and Antibacterial and Antifungal Activity. *Pharmaceuticals* **2023**, *16*, 1248. [CrossRef]

32. Wróbel, A.; Baradyn, M.; Ratkiewicz, A.; Drozdowska, D. Synthesis, Biological Activity, and Molecular Dynamics Study of Novel Series of a Trimethoprim Analogs as Multi-Targeted Compounds: Dihydrofolate Reductase (DHFR) Inhibitors and DNA-Binding Agents. *Int. J. Mol. Sci.* **2021**, *22*, 3685. [CrossRef]

33. Mahipal, S.; Kya, M.; Xiaoling, M. Effect of dimethyl sulfoxide on in vitro proliferation of skin fibroblast cells. *J. Biotech Res.* **2017**, *8*, 78–82.

34. Timm, M.; Saaby, L.; Moesby, L.; Hansen, E.W. Considerations regarding use of solvents in in vitro cell based assays. *Cytotechnology* **2013**, *65*, 887–894. [CrossRef] [PubMed]

35. Da Violante, G.; Zerrouk, N.; Richard, I.; Provot, G.; Chaumeil, J.C.; Arnaud, P. Evaluation of the Cytotoxicity Effect of Dimethyl Sulfoxide (DMSO) on Caco2/TC7 Colon Tumor Cell Cultures. *Biol. Pharm. Bull.* **2002**, *25*, 1600–1603. [CrossRef] [PubMed]

36. Trott, O.; Olson, A.J. AutoDock Vina: Improving the speed and accuracy of docking with a new scoring function, efficient optimization, and multithreading. *J. Comput. Chem.* **2010**, *31*, 455–461. [CrossRef]

37. The PyMOL Molecular Graphics System, Version 2.2.3. Copyright Schrodinger LLC. Available online: https://pymol.org (accessed on 30 October 2021).

38. Jo, S.; Kim, T.; Im, W. Automated Builder and Database of Protein/Membrane Complexes for Molecular Dynamics Simulations. *PLoS ONE* **2007**, *2*, e880. [CrossRef] [PubMed]

39. Jo, S.; Kim, T.; Iyer, V.G.; Im, W. CHARMM-GUI: A web-based graphical user interface for CHARMM. *J. Comput. Chem.* **2008**, *29*, 1859–1865. [CrossRef]

40. Botet-Carreras, A.; Montero, M.T.; Sot, J.; Domènech, Ò.; Borrell, J.H. Characterization of monolayers and liposomes that mimic lipid composition of HeLa cells. *Colloids Surfaces B Biointerfaces* **2020**, *196*, 111288. [CrossRef]

41. Singh, U.C.; Kollman, P.A. An approach to computing electrostatic charges for molecules. *J. Comput. Chem.* **1984**, *5*, 129–145. [CrossRef]

42. Hub, J.S.; Winkler, F.K.; Merrick, M.; De Groot, B.L. Potentials of Mean Force and Permeabilities for Carbon Dioxide, Ammonia, and Water Flux across a Rhesus Protein Channel and Lipid Membranes. *J. Am. Chem. Soc.* **2010**, *132*, 13251–13263. [CrossRef] [PubMed]
43. Wennberg, C.L.; Van Der Spoel, D.; Hub, J.S. Large Influence of Cholesterol on Solute Partitioning into Lipid Membranes. *J. Am. Chem. Soc.* **2012**, *134*, 5351–5361. [CrossRef]
44. Phillips, J.C.; Hardy, D.J.; Maia, J.D.C.; Stone, J.E.; Ribeiro, J.V.; Bernardi, R.C.; Buch, R.; Fiorin, G.; Hénin, J.; Jiang, W.; et al. Scalable molecular dynamics on CPU and GPU architectures with NAMD. *J. Chem. Phys.* **2020**, *153*, 044130. [CrossRef] [PubMed]
45. Humphrey, W.; Dalke, A.; Schulten, K. VMD: Visual molecular dynamics. *J. Mol. Graph.* **1996**, *14*, 33–38. [CrossRef] [PubMed]

Article

Aza Analogs of the TRPML1 Inhibitor Estradiol Methyl Ether (EDME)

Philipp Rühl and Franz Bracher *

Department of Pharmacy, Center for Drug Research, Ludwig-Maximilians University, 80539 Munich, Germany; philipp.ruehl@cup.lmu.de
* Correspondence: franz.bracher@cup.uni-muenchen.de

Abstract: Estradiol methyl ether (**EDME**) has recently been described by us as a very potent and subtype-specific inhibitor of the lysosomal cation channel TRPML1. Following the principle of bioisosteres, we worked out efficient synthetic approaches to ring-A aza-analogs of EDME, namely a methoxypyridine and a methoxypyrimidine analog. Both target compounds were obtained in good overall yields in six and eight steps starting from 19-nortestosterone via the oxidative cleavage of ring A followed over several intermediates and with the use of well-selected protective groups by re-cyclization to provide the desired hetero-analogs. The methoxypyridine analog largely retained its TRPML1-inhibitory activity, whereas the methoxypyrimidine analog significantly lost activity.

Keywords: cation channels; TRPML1; bioisostere; steroids; estrogens; ring cleavage; protective groups; pyridines; pyrimidines; cyclization

Citation: Rühl, P.; Bracher, F. Aza Analogs of the TRPML1 Inhibitor Estradiol Methyl Ether (EDME). *Molecules* **2023**, *28*, 7428. https://doi.org/10.3390/molecules28217428

Academic Editors: Marina Savić, Erzsébet Mernyák, Jovana Ajdukovic and Suzana Jovanović-Šanta

Received: 14 October 2023
Revised: 27 October 2023
Accepted: 31 October 2023
Published: 4 November 2023

1. Introduction

TRPML1 is one of three members (TRPML1-3) of the TRPML cation channels group, a subfamily within the transient receptor potential (TRP) superfamily. As a non-selective lysosomal channel permeable to Ca^{2+}, Na^+, Fe^{2+}, Zn^{2+} and other cations, it plays an important role in multiple physiological processes but also in several human diseases. A mutation with loss of function of TRPML1 causes Mucolipidosis Type IV, a neurodegenerative lysosomal storage disorder [1]. Furthermore, TRPML1 has gained interest as it is associated to be involved in various processes in different cancers, e.g., melanoma [2] and non-small lung cancer [3], and its influence on cardiovascular [4] and neurodegenerative diseases has been discussed [5].

Therefore, obtaining access to inhibitors and activators for this target as pharmacological tools or even as possible future therapeutic options is of great interest.

Phosphatidylinositol 3,5-bisphosphate (PI(3,5)P2), a major constituent of the lysosomal membrane, has been described as an endogenous activator of all TRPML channels, while phosphatidylinositol 4-5-bisphosphate (PI(4,5)P2), which is mainly found in the plasma membrane, has been identified as endogenous inhibitor of TRPML1 and TRPML3 [6]. Due to their structural characteristics (polarity), these two molecules are not suitable as pharmacological tools as they cannot permeate cell membranes.

As a consequence, several low-molecular activators and inhibitors of TRPML1 with suitable pharmacokinetic properties have been developed in recent years. While MK6-83 [7], SF-51 and ML-SA1 [8] are examples of unselective TRPML activators, the only known selective activator of TRPML1 is ML1-SA1 [9]. Despite the evident need for potent TRPML1 inhibitors, the number of these is still limited.

While the indoline derivative **ML-SI1** and the 1,2-diaminocyclohexane derivative **ML-SI3** are known examples in the literature for isoform-unselective inhibitors [10–12], we described the steroidal compound estradiol methyl ether (**EDME**) as the first highly potent (IC_{50} 0.6 μM) and subtype-selective TRPML1 inhibitor in our previous work (Figure 1) [13].

This inhibitor was identified by random screening of 2,430 compounds on hTRPML1ΔNC-YFP, a plasma membrane variant of wild-type TRPML1. Subsequently, we tested other pharmacologically relevant steroidal compounds and found that natural and synthetic steroids lacking aromaticity in ring A (the typical structure motif of estrogens) are virtually inactive (cholesterol, phytosterols, glucocorticoids, mineralocorticoids, antiestrogens, antiandrogens, 5α-reductase inhibitors). In the class of estrogens, the native hormone 17β-estradiol showed significantly reduced inhibition (IC$_{50}$ 5.3 μM) and only mestranol, a congener of EDME bearing an additional ethinyl group at C-17, showed considerable activity. Inversion of the configuration of the 17α-hydroxy group eliminated inhibitory activity in all cases. Finally, we synthesized ten modified versions of EDME, most of which have a replacement of the methoxy group at C-3 in common with a lipophilic residue. Out of these, the 3-vinylestrane **PRU-10** (IC$_{50}$ 0.41 μM) and the 3-acetyl derivative **PRU-12** (IC$_{50}$ 0.28 μM) (Figure 1) showed stronger TRPML1 inhibition, an improved selectivity profile compared to **EDME**, and reduced estrogenic activity.

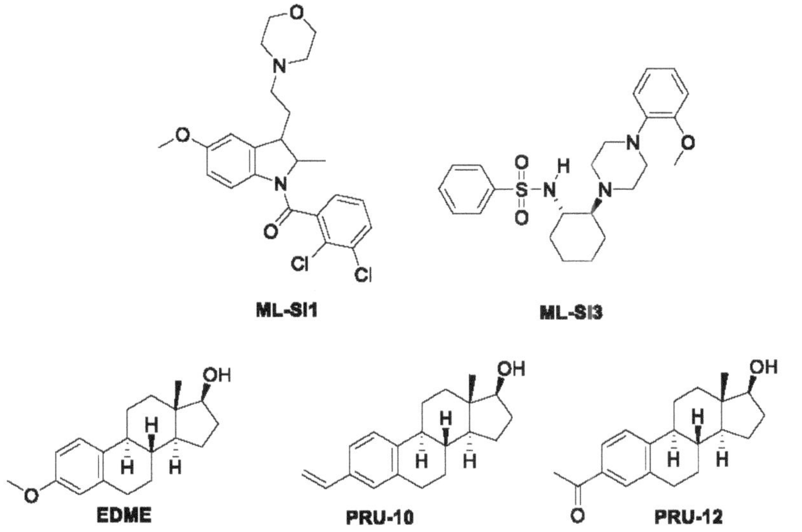

Figure 1. Structures of published TRPML inhibitors.

Based on this preliminary evidence on structure–activity relationships, we focussed on additional modifications of ring A of **EDME**. Due to our own positive experience with the synthesis of aza analogs of steroidal lead structures for improving or modulating biological activities [14,15], we decided to investigate a pyridine-type 4-aza analog (**1**) and a pyrimidine-type 2,4-diaza analog (**2**) of **EDME**.

Following the well-established principle of "bioisosteres" [16], single functional groups in a bioactive molecule can be replaced by other, more or less similar groups in order to extend or improve potency, enhance selectivity, alter the physicochemical properties or metabolism, or improve pharmacokinetics or toxicity. The bioisosteric replacement of phenyl rings can be performed in a classical manner with the introduction of neutral aromatic rings (thiophene, furan) or azaarenes (pyridine, pyrimidine, pyrazine) [17], further "nonclassical" biosisosteres (acetylene, bridged aliphatic ring systems) have been developed [18]. The azaarene bioisosteres have gained significant interest since they can introduce basic properties as well as H-bond acceptor and/or H-bond donor properties and thus improve (or reduce) the interaction with the target protein.

2. Results and Discussion

2.1. Chemistry

As we intended to synthesize both target compounds in an enantiomerically pure form, we selected a "chiral pool" approach [19] for our syntheses. Estradiol or **EDME** were not suitable starting materials for this approach due to the lack of feasible synthetic methods for the conversion of phenols/phenol ethers into pyridines and pyrimidines. The same holds for non-estrogenic sterols bearing a methyl group (C-19) at C-10 since this residue would prevent aromatization of ring A. For our purposes, 19-nortestosterone (nandrolone; **3**) was identified as the best precursor for a couple of reasons: this (commercially available and affordable) homochiral compound already has the required configurations at the stereocenters in rings C and D, its ring A is a cyclohexenone that can be cleaved by oxidation, and, last but not least, there is no methyl group at C-10. The published oxidative degradation of 19-nortestosterone (**3**) under the cleavage of the C-4,C-5 bond and decarboxylation yields a ketocarboxylic acid of type **A** [20]. The formal integration of ammonia and oxidative aromatization should provide a ring A pyridone, which was then to be *O*-methylated to provide the desired 4-aza analog **1** of **EDME**.

Oxidative degradation of the propionate side chain in **A** [20] would provide a ketoaldehyde of type **B**, which, upon treatment with *O*-methylisourea, should provide the methoxypyrimidine target compound **2**. In both series, temporary protection of the 17-OH group had to be considered (Figure 2).

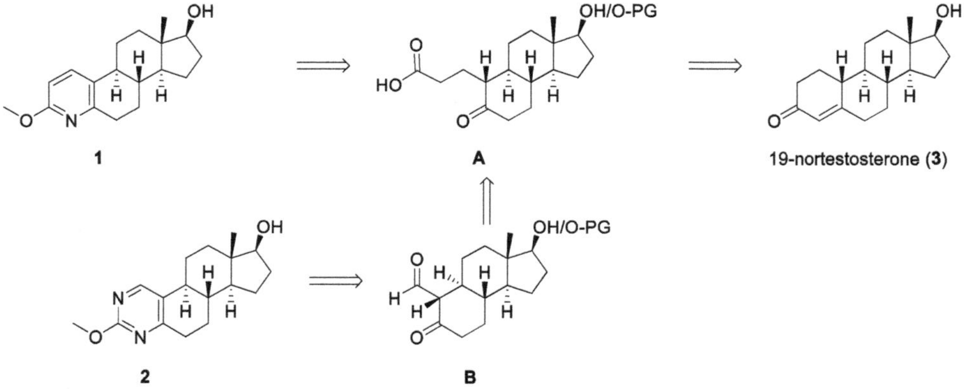

Figure 2. Retrosynthesis of the target compounds **1** and **2** (PG—protective group).

2.1.1. 4-Aza Analog of EDME

Our chiral pool approach started with the oxidative cleavage of ring A of 19-nortestosterone (**3**) to provide ketocarboxylic acid **4**. While Holt et al. [20] described an ozonolysis protocol with a yield of 50%, we obtained **4** in a yield of 94.5% by using $NaIO_4/KMnO_4$ as the oxidant, a method established for a related degradation of a 19-methyl steroid in the course of the synthesis of the drug finasteride [21]. Subsequent treatment with ammonium acetate in acetic acid under reflux [22] resulted in ring closure to two poorly separable unsaturated lactams, **5a** with a Δ5,10- and **5b** with a Δ5,6 double bond (yield: 87.4%, ratio **5a:5b**: 15:85). Unfortunately, we could not find a suitable oxidant for direct dehydrogenation of these lactams to the ring A pyridone **6** (Scheme 1). Most likely, the unprotected 17-OH group interfered with the examined oxidants (DDQ, iodine-based reagents, and others). As a consequence, we examined protective groups for 17-OH.

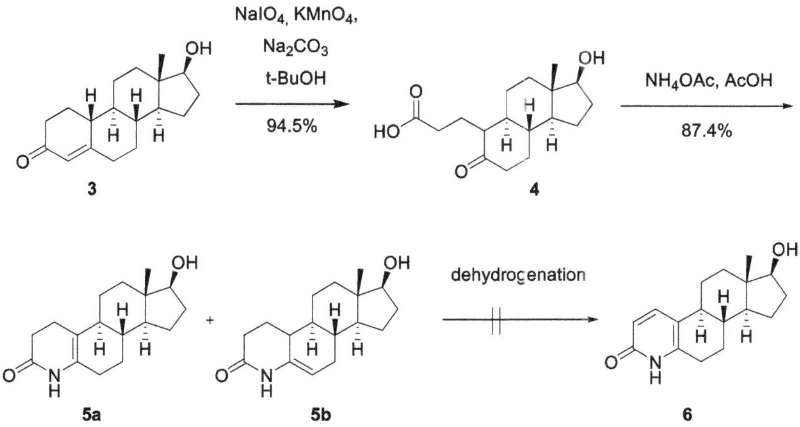

Scheme 1. First attempt for the synthesis of the 4-aza analog 1 of EDME.

Our first attempts to utilize MOM protection of the starting material 19-nortestosterone (**3**) failed early due to problems with introducing this protective group. The following experiments using TBDMS protection gave promising results in the early steps (for details, see Supporting Information) but failed due to the instability of the TBDMS ether as soon as experiments were performed under acidic conditions (no details shown).

Finally, we turned to benzyl protection of 17-OH. Surprisingly, standard protocols for the protection of this secondary alcohol under alkaline (NaH/benzyl halides) or acidic conditions [23] (benzyl trichloroacetimidate/TFA) failed to provide acceptable yields. However, Dudley's protocol [24] utilizing 2-benzyloxy-1-methylpyridinium triflate/MgO gave the desired benzyl ether **7** in 82.1% yield. Following the general strategy described above, oxidative ring A degradation with $NaIO_4$/$KMnO_4$ yielded ketocarboxylic acid **8** in a 99.2% yield. Subsequent treatment with ammonium acetate in acetic acid under reflux resulted in ring closure to two (still poorly separable) unsaturated lactams, **9a** with a Δ5,10- and **9b** with a Δ5,6 double bond (73.4% yield of the mixture, ratio **9a**:**9b**: 15:85).

With this mixture of isomers in hand, we again investigated numerous reagents used in previous publications to dehydrogenate dihydropyridines. These included treatment with MnO_2 [25] (result: no conversion), $Pb(OAc)_4$ [26] (result: decomposition), air oxidation [27] (result: no conversion), and treatment with $KMnO_4$ (result: decomposition).

As all of these experiments failed, we attempted formal dehydrogenation via halogenation at the methylene group next to the lactam carbonyl, followed by dehydrohalogenation utilizing published reagents from the pyridone and related fields (SO_2Cl_2 [28], iodotrimethylsilane [29], $CuBr_2$ [30]). Neither of these reagents gave noteworthy amounts of the dehydrogenation product. Finally, treating **9a**/**9b** with a combination of reagents ($CuBr_2$, LiBr, 1,3-dimethoxybenzene, trifluoromethanesulfonic acid in acetonitrile) that was originally used for a cyclohexenone-to-phenol conversion in 19-norandrost-4-en-3-ones [31] gave the desired pyridone **10** in 33.8% yield. This product was accompanied by small amounts (7.8%) of the 17-*O*-deprotected pyridone **6** (see Scheme 1).

O-Methylation of **10** to provide the methoxypyridine derivative **11** was achieved in 47.2% yield using iodomethane/Ag_2CO_3 [32]. As the final step, the benzyl protective group at 17-OH had to be removed without affecting the methoxypyridine unit. This step turned out to be more difficult than expected. Under standard *O*-debenzylation conditions (hydrogenolysis under Pd catalysis), no conversion was achieved; an alternative Pd-catalyzed method using Et_3SiH as the reductant [33] failed as well. A published method for the selective cleavage of benzyl ethers utilizing $CrCl_2$/LiI [34] surprisingly led to the cleavage of the methyl ether at the pyridine ring and left the benzyl ether untouched. Pyridone **10** (the precursor of **11**) was obtained in a 90% yield. Finally, the desired *O*-debenzylation was achieved by means of BCl$_3$ [35]. The carbinol **1** was obtained in a 90.5% yield, and the methoxy group at C-3 was not affected (Scheme 2).

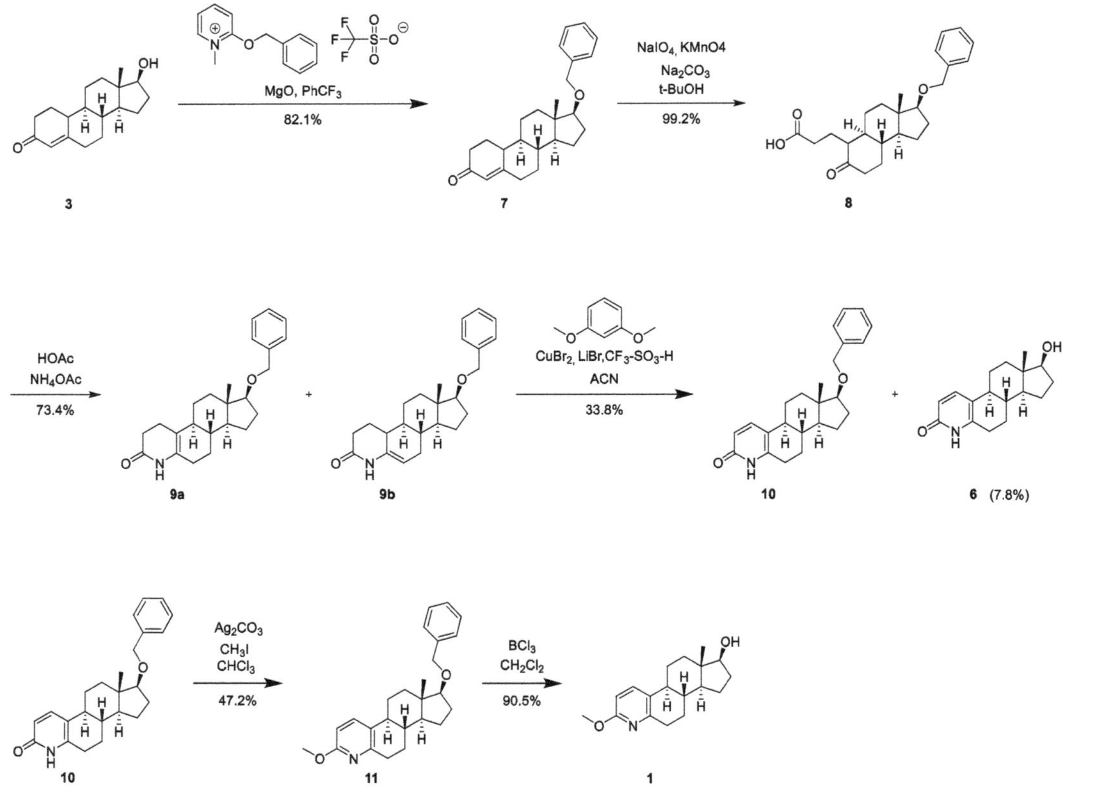

Scheme 2. Successful approach to the 4-aza analog **1** utilizing benzyl protection at 17-OH.

2.1.2. 2,4-Diaza (pyrimidine) Analog of **EDME**

As mentioned above (Figure 2), the methoxypyrimidine motif of the target compound **2** should be built up by cyclocondensation of a ketoaldehyde of type **B** with *O*-methylisourea. This approach has, in principle, been published before in a French patent claimed by Roussel Uclaf in 1967 [36]; however, this route started with a fully synthetic precursor [37] of undefined stereochemistry (most likely racemic), and neither full experimental details nor acceptable spectroscopic data on the characterization of intermediates and the final product were presented.

Our chiral pool approach started once again with 19-nortestosterone (**3**). For this new purpose, the ketocarboxylic acid **4** obtained by oxidative cleavage of ring A (Scheme 1) needed to be degraded further in order to convert the propionate side-chain into a formyl group (see Figure 2) following, in general, a poorly detailed protocol published by Holt et al. [20]. For this purpose, ketocarboxylic acid **4** was first converted into its methyl ester **12** via a higher-yielding protocol utilizing iodomethane/Cs$_2$CO$_3$ (96.9% yield), followed by conversion of the keto group into the dioxolane **13** (67.5% yield). Next, and distinct from the Holt protocol, the 17-OH group was protected by conversion into the TBDS ether **14** (74.4% yield), in order to circumvent interference of the acidic 17-OH group with the strong base LDA required for the following step. Then, the methyl propionate side chain was converted into the α,β-unsaturated ester **15** in 72.2% yield by a selenation-selenoxide elimination protocol including treatment with LDA/diphenyldiselenide and oxidation with H$_2$O$_2$ [38], followed by spontaneous elimination. Two-carbon degradation was then performed by ozonolysis followed by work-up with dimethyl sulfide to provide the aldehyde **16** in 75.8% yield. The treatment of **16** with acetic acid in THF-water resulted in simultaneous deprotection of the dioxolane and the TBDS ether to provide the ketoaldehyde **17** in 79.2% yield. Finally, treatment with *O*-methylisourea gave the target methoxypyrimidine **2** in 37.7% yield (Scheme 3).

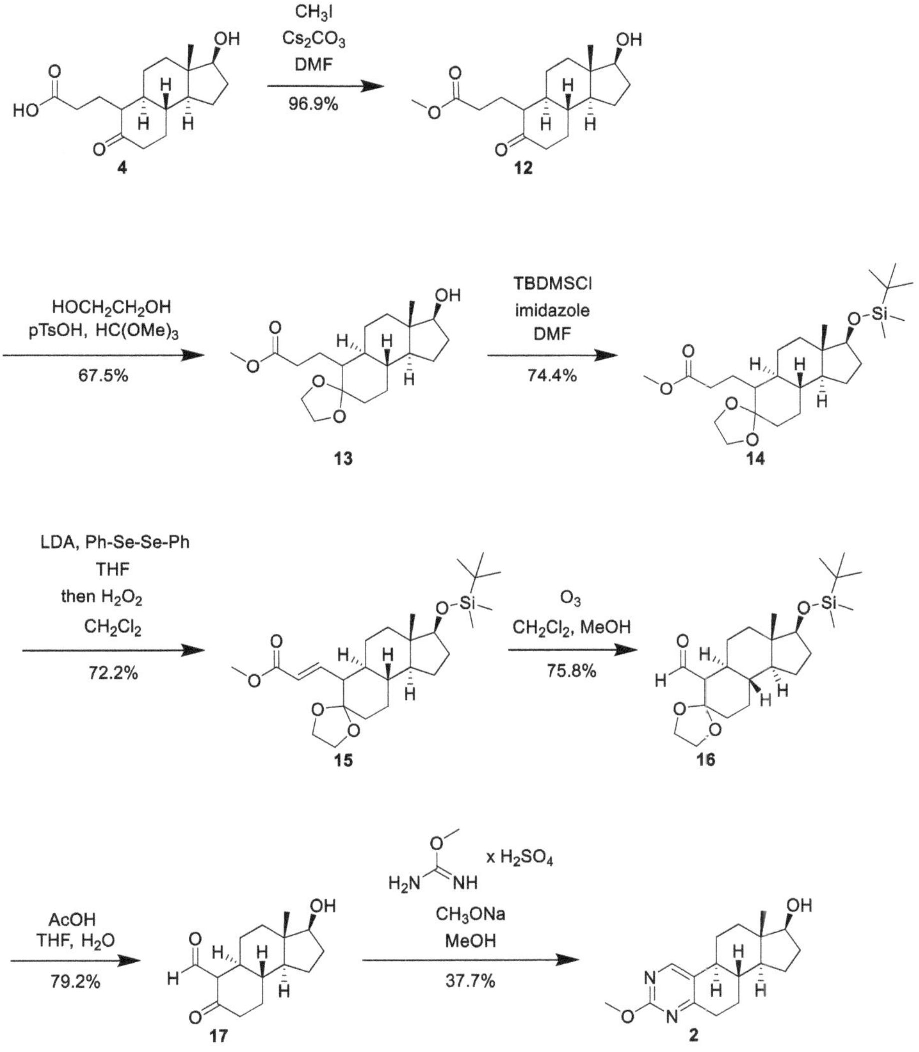

Scheme 3. Synthesis of the 2,4-diaza analog **2** of **EDME**.

2.2. Biological Testing

The two target compounds, pyridine analog **1** and pyrimidine analog **2**, as well as the inadvertently obtained pyridone analog **6** were submitted to our previously described [13] test for inhibition of TRPML1 on hTRPML1ΔNC-YFP, a plasma membrane variant of wild-type TRPML1 by means of a fluorimetric Ca^{2+} influx assay. The results are shown in Table 1.

Table 1. TRPML1-inhibitory activities of lead structure **EDME** and the three synthesized aza analogs.

Compound	IC$_{50}$ for TRPML1 Inhibition
EDME	0.60 µM
1	1.0 µM
2	8.8 µM
6	40 µM

Compared to EDME, the 4-aza analog **1** showed slightly reduced TRPML1-inhibitory activity (factor <2 less potent), the 2,4-diaza analog **2**; however, it is only a very weak inhibitor, and the pyridone analog **6** is virtually inactive.

3. Materials and Methods

3.1. Chemistry

All NMR spectra (^{1}H, ^{13}C, DEPT, H-H-COSY, HSQC, HMBC) were recorded at 23 °C on an Avance III 400 MHz Bruker BioSpin or Avance III 500 MHz Bruker BioSpin instrument (Bruker, Billerica, MA, USA) unless otherwise specified. Chemical shifts δ are stated in parts per million (ppm) and are calibrated using residual protic solvents as an internal reference for proton (CD$_2$Cl$_2$: δ = 5.32 ppm, MeOD δ = 3.31 ppm, DMSO: δ = 2.50 ppm) and for carbon the central carbon resonance of the solvent (CD$_2$Cl$_2$: δ = 53.84 ppm, MeOD δ = 49.00 ppm, DMSO: δ = 39.52 ppm). Multiplicity is defined as s—singlet, d—doublet, t—triplet, q—quartet, and m—multiplet. NMR spectra were analyzed with the NMR software MestReNova, version 12.0.1-20560 (Mestrelab Research S.L., Santiago de Compostela, Spain). Numbering of the carbon atoms in seco-steroids: For the sake of comparability, we kept using the numbering the single carbon atoms had in the intact steroids, since, in the following, the seco-steroidal intermediates were cyclized to the azasteroids later. High-resolution mass spectra were performed by the LMU Mass Spectrometry Service applying a Thermo Finnigan MAT 95 (Thermo Fisher Scientific, Waltham, MA, USA) or Joel MStation Sektorfeld instrument (Peabody, MA, USA) at a core temperature of 250 °C and 70 eV for EI or a Thermo Finnigan LTQ FT Ultra Fourier Transform Ion Cyclotron Resonance device (Thermo Fisher Scientific, Waltham, MA, USA) at 250 °C for ESI. IR spectra were recorded on a Perkin Elmer FT-IR Paragon 1000 instrument (Perkin Elmer, Hong Kong, China) as neat materials. The absorption bands were reported in wave number (cm^{-1}) with ATR PRO450-S. Melting points were determined by the open tube capillary method on a Büchi melting point B-540 apparatus and are uncorrected. The HPLC purities were determined using an HP Agilent 1100 HPLC (Agilent, Santa Clara, CA, USA) with a diode array detector at 210 nm and an Agilent Poroshell column (120 EC-C18; 3.0 × 100 mm; 2.7 micron) with acetonitrile/water as eluent. Values for specific rotation (α) were measured at 23 °C at a wavelength of λ = 589 nm (Na-D-line) using a Perkin Elmer 241 Polarimeter instrument (Perkin Elmer, Hong Kong, China). All samples were dissolved in dichloromethane (layer thickness l = 10 cm, concentration c = 0.1 mg/100 mL). All chemicals used were of analytical grade. Isohexane, ethyl acetate and methylene chloride were purified by distillation. All reactions were monitored by thin-layer chromatography (TLC) using pre-coated plastic sheets, POLYGRAM® SIL G/UV254 from Macherey-Nagel (Düren, Germany). Flash column chromatography was performed on Merck silica gel Si 60 (0.015–0.040 mm). Ozonolysis was performed on an Ozonova Type OG700-10WC (Jeske Ozontechnik, Ruderserg, Germany).

3-((3S,3aS,5aS,6R,9aR,9bS)-3-Hydroxy-3a-methyl-7-oxododecahydro-1H-cyclopenta[a]naphthalen-6-yl)propanoic acid (**4**): To a solution of 19-nortestosterone (**3**; 3.01 g, 10.9 mmol, 1.00 eq) in 60 mL of *tert*-butanol were added 19.5 mL of a saturated aqueous Na$_2$CO$_3$ solution. The mixture was heated at reflux, and a solution of NaIO$_4$ (23.4 g, 110 mmol, 10.00 eq) and KMnO$_4$ (0.130 g, 0.821 mmol, 7.50 mol%) in water (66 mL), preheated to 80 °C, was added via a dropping funnel over a time period of 30 min. After cooling, the reaction mixture was filtered, and the filter cake was washed with 10 mL of water. The filtrate was acidified with 6M HCl to pH 2 and then extracted with dichloromethane (4 × 20 mL). The organic phase was washed with water (20 mL) and dried over anhydrous sodium sulfate. After filtration and removal of the solvent, the crude product was purified by flash column chromatography (isohexane/ethyl acetate 1:1) to yield a colorless oil (3.08 g, 10.4 mmol, 95.4%). ^{1}H NMR (400 MHz, DMSO-d_6) δ/ppm = 11.95 (s, 1H, COOH), 4.48 (d, J = 4.8 Hz, 1H, OH), 3.46 (td, J = 8,5 Hz, 4,7 Hz, 1H, 17-H), 2.42 (m, 1H, 6-H$_a$), 2.29 (ddd, J = 11.2 Hz, 7.8 Hz, 2.5 Hz, 1H, 10-H), 2.21 (m, 1H, 6-H$_b$), 2.17 (m, 1H, 2-H$_a$), 2.09 (m, 1H, 2-H$_b$), 1.88 (m, 1H, 7-H$_a$), 1.75 (m, 1H, 1-H$_a$), 1.72 (m, 1H, 12-H$_a$), 1.69 (m, 1H, 15-H$_a$), 1.63 (m, 1H, 1-H$_b$), 1.57 (m, 1H, 8-H), 1.52 (m, 1H, 11-H$_a$), 1.36 (m, 1H, 16-H$_a$), 1.30 (m, 1H, 15-H$_b$), 1.24 (m, 1H, 11-H$_b$ or 16-H$_b$), 1.22 (m, 1H, 11-H$_b$ or 16-H$_b$), 1.15 (m, 1H, 7-H$_b$), 1.04 (m, 1H, 9-H), 1.00 (m, 1H, 12-H$_b$), 0.95 (m, 1H, 14-H), 0.72 (s, 3H, 18-H) ^{13}C NMR (101 MHz, DMSO-d_6) δ/ppm = 211.41 (C-5), 174.6 (C-3), 79.84 (C-17), 52.85 (C-10), 49.08 (C-14), 47.50 (C-9), 42.69 (C-13), 41.19 (C-6), 40.19 (C-8), 36.16 (C-12), 30.96 (C-2), 30.77 (C-7), 29.73 (C-16), 26.59 (C-15), 22.96 (C-11), 20.57 (C-1), 11.25 (C-18) IR (ATR): ν_{max}/cm^{-1} = 2921, 2359, 1698, 1636, 1455, 1385, 1261, 1127, 1055, 805, 696 HRMS (EI): m/z = [M$^{\bullet+}$] calculated for C$_{17}$H$_{26}$O$_4$$^{\bullet+}$: 294.1826; found: 294.1825.

(4bS,6aS,7S,9aS,9bR)-7-Hydroxy-6a-methyl-1,3,4,4a,4b,5,6,6a,7,8,9,9a,9b,10-tetradecahydro-2H-indeno[5,4-f]quinolin-2-one (**5a**) and *(4bS,6aS,7S,9aS,9bR)-7-hydroxy-6a-methyl-1,3,4,4b,5,6,6a,7,8,9,9a,9b,10,11-tetradecahydro-2H-indeno[5,4-f]quinolin-2-one* (**5b**): A mixture of compound **4** (0.795 g, 2.70 mmol, 1.00 eq) and ammonium acetate (0.728 g, 9.45 mmol, 3.50 eq) in glacial acetic acid (20 mL) was stirred and heated at reflux for 4 h. After cooling, the mixture was concentrated under reduced pressure and the residue was poured into water. The precipitate was collected by filtration, washed with water (10 mL) and dissolved in dichloromethane (20 mL). The resulting solution was washed with NaOH (1M, 3 × 10 mL), water (10 mL) and brine (10 mL), filtered over a hydrophobic filter, and concentrated in vacuo. The crude product was purified by flash column chromatography (isohexane/ethyl acetate 3:1) to yield 0.650 g (2.36 mmol, 87.4%) of a mixture of lactams **5a** and **5b** (ratio **5a/5b**: 15:85) as a beige solid.

5a: m.p.: 209 °C ^{1}H NMR (500 MHz, CD$_2$Cl$_2$) δ/ppm = 5.88 (s, 1H, NH), 3.65 (t, J = 8.5 Hz, 1H, 17-H), 2.38 (m, 2H, 2-H), 2.32 (m, 1H, 1-H$_a$), 2.20 (m, 1H, 1-H$_b$), 2.15 (m, 1H, 6-H$_a$), 2.04 (m, 1H, 16-H$_a$), 1.95 (m, 1H, 6-H$_b$), 1.87 (m, 1H, 11-H$_a$ or 15-H$_a$), 1.81 (dt, J = 12.3 Hz, 3.1 Hz, 1H, 12-H$_a$), 1.75 (m, 1H, 7-H$_a$), 1.71 (m, 1H, 9-H), 1.60 (m, 1H, 11-H$_a$ or 15-H$_a$), 1.43 (m, 1H, 16-H$_b$), 1.34 (m, 1H, 8-H or 11-H$_b$ or 15-H$_b$), 1.31 (m, 1H, 8-H or 11-H$_b$ or 15-H$_b$), 1.26 (m, 1H, 11-H$_b$ or 15-H$_b$), 1.22 (m, 1H, 7-H$_b$), 1.17 (m, 1H, 12-H$_b$), 1.11 (m, 1H, 14-H$_b$), 0.75 (s, 3H, 18-H) ^{13}C NMR (126 MHz, CD$_2$Cl$_2$) δ/ppm = 171.03 (C-3), 128.63 (C-5), 112.88 (C-10), 81.99 (C-17), 49.70 (C-14), 44.31 (C-9), 44.00 (C-13), 39.50 (C-8), 37.16 (C-12), 31.11 (C-2), 30.92 (C-16), 27.34 (C-6), 26.33 (C-7), 25.68 (C-11 or C-15), 23.30 (C-11 or C-15), 22.21 (C-1), 11.46 (C-18) IR (ATR): ν_{max}/cm^{-1} = 3465, 2913, 2868, 1683, 1668, 1542, 1507, 1473, 1456, 1388, 1319, 1284, 1224, 1186, 1133, 1055, 1027, 894, 842 HRMS (EI): m/z = [M$^{\bullet+}$] calculated for C$_{17}$H$_{25}$NO$_2$$^{\bullet+}$: 275.1880; found: 275.1880.

5b: m.p.: 218 °C ^{1}H NMR (500 MHz, CD$_2$Cl$_2$) δ/ppm = 7.34 (s, 1H, NH), 4.86 (dt, J = 5.1 Hz, 2.3 Hz, 1H, 6-H), 3.63 (t, J = 8.6 Hz, 1H, 17-H), 2.47 (m, 1H, 2-H$_a$), 2.37 (m, 1H, 2-H$_b$), 2.10 (m, 1H, 7-H$_a$), 2.05 (m, 1H, 16-H$_a$), 2.02 (m, 1H, 10-H), 1.92 (m, 1H, 1-H$_a$), 1.81 (dt, J = 12.6 Hz, 3.4 Hz, 1H, 12-H$_a$), 1.65 (m, 1H, 11-H$_a$), 1.60 (m, 1H, 15-H$_a$), 1.46 (m, 1H, 7-H$_b$ or 11-H$_b$), 1.43 (m, 1H, 16-H$_b$), 1.40 (m, 1H, 8-H), 1.31 (m, 1H, 1-H$_b$), 1.29 (m, 1H, 15-H$_b$), 1.26 (m, 1H, 7-H$_b$ or 11-H$_b$), 1.15 (m, 1H, 12-H$_b$), 1.04 (m, 1H, 14-H), 1.00 (m, 1H, 9-H), 0.76 (s, 3H, 18-H) ^{13}C NMR (126 MHz, CD$_2$Cl$_2$) δ/ppm = 169.77 (C-3), 136.46 (C-5), 102.48 (C-6), 82.05 (C-17), 50.71 (C-14), 43.79 (C-9), 43.35 (C-13), 39.98 (C-10), 37.08 (C-8), 36.66 (C-12), 32.36 (C-2), 30.78 (C-16), 29.18 (C-11), 26.60 (C-1), 25.29 (C-7), 23.47 (C-15), 11.10 (C-18) IR (ATR): ν_{max}/cm^{-1} = 2920, 2308, 1636, 1558, 1541, 1507, 1457, 1386, 1055, 735 HRMS (EI): m/z = [M$^{\bullet+}$] calculated for C$_{17}$H$_{25}$NO$_2$$^{\bullet+}$: 275.1880; found: 275.1880.

(8R,9S,13S,14S,17S)-17-(Benzyloxy)-13-methyl-1,2,6,7,8,9,10,11,12,13,14,15,16,17-tetradecahydro-3H-cyclopenta[a]phenanthren-3-one (**7**): 19-Nortestosterone (**3**; 2.47 g, 9.00 mmol, 1.00 eq), 2-benzyloxy-1-methylpyridinium triflate (6.29 g, 18.0 mmol, 2.00 eq) and magnesium oxide (vacuum-dried, 0.744 g, 18.0 mmol, 2.00 eq) were combined in a round bottom flask. Benzotrifluoride (20 mL) was added, and the resulting suspension was heated under stirring at 83 °C for 24 h. After cooling to room temperature, the reaction mixture was diluted with dichloromethane (20 mL) and filtered through Celite. After evaporation of the solvent, the crude product was purified by flash column chromatography (isohexane/ethyl acetate 10:1) to yield 2.69 g (7.39 mmol, 82.1%) of compound **7** as a white solid. m.p.: 176 °C ^{1}H NMR (400 MHz, DMSO-d_6) δ/ppm = 7.31 (m, 4H, benzyl aromatic ortho and meta Hs), 7.26 (m, 1H, benzyl aromatic para H), 5.72 (s, 1H, 4-H), 4.50 (s, 2H, benzyl CH$_2$), 3.41 (t, J = 8.2 Hz, 1H, 17-H), 2.42 (m, 6-H$_a$), 2.26 (m, 6-H$_b$), 2.22 (m, 1H, 1-H$_a$), 2.20 (m, 1H, 2-H$_a$), 2.16 (m, 1H, 10-H), 1.97 (m, 1H, 16-H$_a$), 1.87 (dt, J = 12.2 Hz, 3.2 Hz, 1H, 2-H$_b$), 1.78 (m, 1H, 15-H$_a$), 1.73 (m, 1H, 7-H$_a$), 1.55 (m, 1H, 11-H$_a$), 1.48 (m, 1H, 16-H$_b$), 1.43 (m, 1H, 12-H$_a$), 1.37 (m, 1H, 1-H$_b$ or 15-H$_b$), 1.30 (m, 1H, 8-H), 1.26 (m, 1H, 11-H$_b$), 1.21 (m, 1H, 1-H$_b$ or 15-H$_b$), 1.12 (m, 1H, 12-H$_b$), 1.01 (m, 1H, 14-H), 0.95 (m, 1H, 7-H$_b$), 0.81 (s, 3H, 18-H), 0.77 (m, 1H, 9-H) ^{13}C NMR (101 MHz, DMSO-d_6) δ/ppm = 198.41 (C-3), 166.76 (C-5), 139.18 (benzyl, qu$_a$ternary carbon), 128.15 (benzyl, aromatic para), 127.15 (4C, benzyl aromatic ortho and meta), 123.75 (C-4), 87.59 (C-17), 70.74 (benzyl CH$_2$), 49.27 (C-14), 48.95 (C-9), 42.69 (C-13), 41.67 (C-10), 37.09 (C-12), 36.16 (C-2), 34.60 (C-6), 30.28 (C-7), 27.49 (C-16), 26.13 (C-1 or C-15), 25.64 (C-1 or C-15), 22.81 (C-11), 11.68 (C-18). IR (ATR): ν_{max}/cm^{-1} = 2927, 2870, 2350, 2307, 1717, 1653, 1558, 1541, 1507, 1489, 1473, 1456, 1388, 1339, 1067 HRMS (EI): m/z = [M$^{•+}$] calculated for C$_{25}$H$_{32}$O$_2$$^{•+}$ 364.2397; found: 364.2396.

3-((3S,3aS,5aS,9aR,9bS)-3-(Benzyloxy)-3a-methyl-7-oxododecahydro-1H-cyclopenta[a]naphthalen-6-yl)propanoic acid (**8**): To a solution of compound **7** (2.63 g, 7.20 mmol, 1.00 eq) in 75 mL of *tert*-butanol were added 13.5 mL of a saturated aqueous Na$_2$CO$_3$ solution. The mixture was heated at reflux, and a solution of NaIO$_4$ (15.4 g, 72.0 mmol, 10.00 eq) and KMnO$_4$ (85.3 mg, 0.540 mmol, 7.50 mol%) in water (45 mL), preheated to 80 °C, was added via a dropping funnel over a time period of 30 min. After cooling, the reaction mixture was filtered, and the filter cake was washed with 10 mL of water. The filtrate was acidified with 6M HCl to pH 2 and then extracted with dichloromethane (4 × 20 mL). The organic phase was washed with water (20 mL) and dried over anhydrous sodium sulfate. After filtration and removal of the solvent, the crude product was purified by flash column chromatography (isohexane/ethyl acetate 3:1) to yield compound **8** as a colorless oil (2.86 g, 7.44 mmol, 99.2%) ^{1}H NMR (500 MHz, CD$_2$Cl$_2$) δ/ppm = 7.33 (2s, 4H, benzyl aromatic ortho and meta Hs) 7.26 (hept, J = 3.8 Hz, 1H, benzyl aromatic para H), 4.51 (s, 2H, benzyl CH$_2$), 3.45 (td, J = 7.5 Hz, 6.8 Hz, 2.5 Hz, 1H, 17-H), 2.41 (m, 1H, 2-H$_a$), 2.37 (m, 1H, 6-H$_a$), 2.35 (m, 1H, 6-H$_b$), 2.30 (m, 1H, 2-H$_b$), 2.26 (m, 1H, 7-H$_a$), 2.23 (m, 1H, 10-H), 2.05 (m, 1H, 16-H$_a$), 2.01 (m, 1H, 7-H$_b$), 1.96 (m, 1H, 12-H$_a$), 1.91 (m, 1H, 1-H$_a$), 1.80 (m, 1H, 15-H$_a$), 1.77 (m, 1H, 1-H$_b$), 1.62 (m, 1H, 11-H$_a$), 1.59 (m, 1H, 8-H), 1.55 (m, 1H, 16-H$_b$), 1.39 (m, 1H, 15-H$_b$), 1.35 (m, 1H, 11-H$_b$), 1.20 (m, 1H, 12-H$_b$), 1.08 (m, 1H, 9-H), 1.03 (m, 1H, 14-H), 0.90 (s, 3H, 18-H) ^{13}C NMR (126 MHz, CD$_2$Cl$_2$) δ/ppm = 212.32 (C-5), 178.14 (C-3), 139.84 (benzyl, quaternary carbon), 128.57 (benzyl, aromatic para), 127.74 (4C, benzyl aromatic ortho and meta), 88.66 (C-17), 72.00 (benzyl CH$_2$), 54.17 (C-10), 50.17 (C-14), 48.71 (C-9), 43.59 (C-13), 42.21 (C-6), 40.72 (C-8), 37.81 (C-12), 31.76 (C-7), 31.34 (C-2), 28.17 (C-16), 27.51 (C-15), 23.64 (C-11), 21.08 (C-1), 11.98 (C-18). IR (ATR): ν_{max}/cm^{-1} = 2927, 2871, 2349, 2307, 1868, 1705, 1653, 1558, 1541, 1521, 1507, 1497, 1456, 1418, 1362, 1279, 869 HRMS (EI): m/z = [M$^{•+}$] calculated for C$_{24}$H$_{32}$O$_4$$^{•+}$: 384.2295; found: 384.2294.

(4bS,6aS,7S,9aS,9bR)-7-(Benzyloxy)-6a-methyl-1,3,4,4a,4b,5,6,6a,7,8,9,9a,9b,10-tetradecahydro-2H-indeno[5,4-f]quinolin-2-one (**9a**) and *(4bS,6aS,7S,9aS,9bR)-7-(benzyloxy)-6a-methyl-1,3,4,4b,5,6,6a,7,8,9,9a,9b,10,11-tetradecahydro-2H-indeno[5,4-f]quinolin-2-one* (**9b**): A mixture of compound **8** (2.54 g, 6.60 mmol, 1.00 eq) and ammonium acetate (1.78 g, 23.1 mmol, 3.50 eq)

in glacial acetic acid (60 mL) was stirred and heated at reflux for 4 h. After cooling, it was concentrated under reduced pressure and the remaining residue was poured into water. The precipitate was filtered, washed with water (20 mL) and dissolved in dichloromethane (40 mL). The resulting solution was washed with NaOH (1M, 3 × 20 mL), water (20 mL) and brine (20 mL), filtered over a hydrophobic filter and concentrated in vacuo. The crude product was purified by flash column chromatography (isohexane/ethyl acetate 5:1) to provide a total of 1.77 g (4.85 mmol, 73.4%) of fractions containing compounds **9a/9b** (ratio **9a:9b**: ca. 15:85) as beige solids (pure **9a**: 0.150 g, 0.420 mmol, 6.3%, pure **9b**: 0.870 g, 2.37 mmol, 35.9%, mixed fraction: 0.750 g, 2.05 mmol, 31.1%; ratio **9a:9b**: ca. 15:85).

9a: m.p.: 207 °C ^{1}H NMR (400 MHz, CD_2Cl_2) δ/ppm = 7.32 (m, 4H, benzyl aromatic ortho and meta Hs), 7.26 (m, 1H, benzyl aromatic para H), 6.69 (s, 1H, NH), 4.51 (s, 2H, benzyl CH_2), 3.47 (m, 1H, 17-H), 2.39 (m, 2H, 2-H), 2.32 (m, 1H, 1-H_a), 2.20 (m, 1H, 1-H_b), 2.14 (m, 1H, 6-H_a), 2.04 (m, 1H, 16-H_a), 1.98 (m, 1H, 12-H_a), 1.92 (m, 1H, 6-H_b), 1.85 (m, 1H, 7-H_a), 1.75 (m, 1H, 11-H_a), 1.69 (m, 1H, 9-H), 1.63 (m, 1H, 15-H_a), 1.55 (m, 1H, 16-H_b), 1.34 (m, 1H, 15-H_b), 1.30 (m, 1H, 8-H), 1.26 (m, 1H, 12-H_b), 1.24 (m, 1H, 11-H_b), 1.20 (m, 1H, 7-H_b), 1.14 (m, 1H, 14-H), 0.84 (s, 3H, 18-H) ^{13}C NMR (101 MHz, CD_2Cl_2) δ/ppm = 170.75 (C-3), 139.95 (benzyl, quaternary carbon), 128.61 (C-5), 128.56 (2C, benzyl aromatic ortho or meta), 127.72 (2C, benzyl aromatic ortho or meta), 127.60 (benzyl, aromatic para), 112.82 (C-10), 88.88 (C-17), 72.00 (benzyl CH_2), 49.88 (C-14), 44.31 (C-9), 44.18 (C-13), 39.29 (C-8), 38.36 (C-12), 31.17 (C-2), 28.39 (C-16), 27.38 (C-6), 26.33 (C-11), 25.81 (C-7), 23.32 (C-15), 22.24 (C-1), 12.21 (C-18). IR (ATR): ν_{max}/cm^{-1} = 3087, 2925, 2870, 2348, 2307, 1868, 1698, 1558, 1542, 1521, 1507, 1490, 1455, 1387, 1338 HRMS (EI): m/z = [M$^{\bullet+}$] calculated for $C_{24}H_{31}NO_2$ $^{\bullet+}$: 365.2349; found: 365.2354.

9b: 207 °C ^{1}H NMR (400 MHz, CD_2Cl_2) δ 7.52 (s, 1H, NH), 7.33 (2s, 4H, benzyl aromatic ortho and meta Hs), 7.26 (ddt, J = 5.7 Hz, 3.7 Hz, 2.2 Hz, 1H, benzyl aromatic para H), 4.87 (dt, J = 5.1 Hz, 2.3 Hz, 1H, 6-H), 4.52 (s. 2H, benzyl CH_2), 3.45 (dd, J = 8.7 Hz, 7.6 Hz, 1H, 17-H), 2.47 (ddd, J = 17.8 Hz, 5.2 Hz, 2.0 Hz, 1H, 2-H_a), 2.35 (ddd, J = 18.0 Hz, 13.0 Hz, 5.9 Hz, 1H, 2-H_b), 2.12 (m, 1H, 7-H_a), 2.07 (m, 1H, 7-H_b), 2.03 (m, 1H, 16-H_a), 2.00 (m, 1H, 10-H), 1.97 (m, 1H, 12-H_a), 1.92 (m, 1H, 11-H_a or 15-H_a), 1.58 (m, 1H, 1-H_a), 1.55 (m, 1H, 16-H_b), 1.48 (m, 1H, 8-H), 1.42 (m, 1H, 11-H_a or 15-H_a), 1.32 (m, 1H, 1-H_b), 1.29 (m, 1H, 12-H_b), 1.25 (m, 1H, 11-H_b or 15-H_b), 1.22 (m, 1H, 11-H_b or 15-H_b), 1.05 (m, 1H, 14-H), 1.01 (m, 1H, 9-H), 0.85 (s, 3H, 18-H) ^{13}C NMR (101 MHz, CD_2Cl_2) δ/ppm = 169.91 (C-3), 139.93 (benzyl, quaternary carbon), 136.45 (C-5), 128.56 (2C, benzyl aromatic ortho or meta), 127.73 (2C, benzyl aromatic ortho or meta), 127.61 (benzyl, aromatic para), 102.56 (C-6), 88.92 (C-17), 72.02 (benzyl CH_2), 50.89 (C-14), 43.81 (C-9), 43.53 (C-13), 39.97 (C-10), 37.86 (C-12), 36.87 (C-8), 32.35 (C-2), 29.20 (C-7), 28.23 (C-16), 26.70 (C-15), 25.27 (C-11), 23.51 (C-1), 11.85 (C-18). IR (ATR): ν_{max}/cm^{-1} = 3195, 3062, 2920, 2872, 1716, 1569, 1355, 1332, 1317, 1190, 1139, 1070, 1045, 843, 800, 737, 695, 647 HRMS (EI): m/z = [M$^{\bullet+}$] calculated For $C_{24}H_{31}NO_2$$^{\bullet+}$: 365.2349; found: 365.2349.

(4bS,6aS,7S,9aS,9bR)-7-(Benzyloxy)-6a-methyl-1,4b,5,6,6a,7,8,9,9a,9b,10,11-dodecahydro-2H-indeno[5,4-f]quinolin-2-one (**10**) and *(4bS,6aS,7S,9aS,9bR)-7-hydroxy-6a-methyl-1,4b,5,6,6a,7,8, 9,9a,9b,10,11-dodecahydro-2H-indeno[5,4-f]quinolin-2-one* (**6**): A mixture of compounds **9a/9b** (0.256 g, 0.700 mmol, 1.00 eq) was suspended in acetonitrile (2 mL). 1.3-Dimethoxybenzene (0.2 mL), a suspension of copper (II) bromide (87.6 mg, 0.392 mmol, 0.560 eq) and lithium bromide (78.6 mg, 0.896 mmol, 1.28 eq) in acetonitrile (2 mL) and a solution of methane-sulfonic acid (23.2 μL, 0.350 mmol, 0.500 eq in 1.2 mL acetonitrile) were added. The resulting mixture was stirred and heated at reflux for 5 h. Then water (5 mL) was added, and the mixture was extracted with dichloromethane (3 × 10 mL). The organic layers were combined, washed with brine (10 mL), and filtered through a hydropho-bic filter. After evaporation of the solvent, the crude product was purified by flash column chromatography to yield 86.0 mg (0.237 mmol, 33.8%) of **10** (eluated first with

dichloromethane/methanol 100:3) as a beige solid and 14.9 mg (0.0546 mmol, 7.8%) of **6** (eluated second with dichloromethane/methanol 10:1) as a beige solid.

10: m.p.: 302 °C ^{1}H NMR (400 MHz, CD$_2$Cl$_2$) δ/ppm = 12.78 (br s, 1H, NH), 7.44 (d, J = 9.4 Hz, 1H, 1-H), 7.33 (m, 4H, benzyl aromatic ortho and meta Hs), 7.27 (dq, J = 7.4, 2.8 Hz, 1H, benzyl aromatic para H), 6.28 (d, J = 9.3 Hz, 1H, 2-H), 4.53 (s, 2H, benzyl CH$_2$), 3.50 (t, J = 8.2 Hz, 1H, 17-H), 2.70 (m, 2H, 6-H), 2.14 (m, 1H, 11-H$_a$), 2.08 (m, 1H, 16-H$_a$), 2.05 (m, 1H, 12-H$_a$), 2.02 (m, 1H, 9-H), 1.89 (m, 1H, 7-H$_a$), 1.68 (ddd, J = 12.4 Hz, 6.8 Hz, 2.8 Hz, 1H, 15-H$_a$), 1.57 (ddd, J = 13.4 Hz, 7.8 Hz, 3.0 Hz, 1H, 16-H$_b$), 1.43 (m, 1H, 8-H), 1.39 (m, 1H, 11-H$_b$), 1.36 (m, 1H, 15-H$_b$), 1.34 (m, 1H, 12-H$_b$), 1.29 (m, 1H, 7-H$_b$), 1.19 (td, J = 11.4 Hz, 6.9 Hz, 1H, 14-H), 0.85 (s, 3H, 18-H). ^{13}C NMR (101 MHz, CD$_2$Cl$_2$) δ/ppm = 164.69 (C-3), 143.70 (C-5), 140.66 (C-1), 139.91 (benzyl, quaternary carbon), 128.57 (2C, benzyl aromatic ortho or meta), 127.73 (2C, benzyl aromatic ortho or meta), 127.62 (benzyl, aromatic para), 118.06 (C-10), 116.73 (C-2), 88.82 (C-17), 72.02 (benzyl CH$_2$), 49.85 (C-14), 43.91 (C-13), 42.35 (C-9), 38.74 (C-8), 38.01 (C-12), 28.30 (C-16), 27.46 (C-6), 26.53 (C-11), 26.07 (C-7), 23.35 (C-15), 12.05 (C-18) IR (ATR): ν_{max}/cm^{-1} = 3087, 2933, 2869, 2348, 2307, 1869, 1845, 1716, 1614, 1542, 1522, 1508, 1496, 1456, 1420 HRMS (EI): m/z = [M$^{\bullet+}$] calculated for C$_{24}$H$_{29}$NO$_2$ $^{\bullet+}$: 363.2193; found: 363.2192.

6: m.p.: 306 °C ^{1}H NMR (500 MHz, MeOD-d$_4$) δ/ppm = 7.61 (d, J = 9.4 Hz, 1H, 1-H), 6.36 (d, J = 9.3 Hz, 1H, 2-H), 3.66 (m, 1H, 17-H), 2.70 (m, 2H, 6-H), 2.24 (dq, J = 12.7 Hz, 3.7 Hz, 1H, 11-H$_a$), 2.09 (m, 1H, 9-H), 2.04 (m, 1H, 16-H$_a$), 1.97 (m, 1H, 12-H$_a$), 1.93 (m, 1H, 7-H$_a$), 1.69 (dddd, J = 12.3 Hz, 9.6 Hz, 7.1 Hz, 3.3 Hz, 1H, 15-H$_a$), 1.53 (dddd, J = 13.1 Hz, 11.6 Hz, 8.2 Hz, 3.3 Hz, 1H, 16-H$_b$), 1.44 (m, 1H, 8-H), 1.40 (m, 1H, 11-H$_b$ or 15-H$_b$), 1.38 (m, 1H, 11-H$_b$ or 15-H$_b$), 1.35 (m, 1H, 7-H$_b$), 1.28 (m, 1H, 12-H$_b$), 1.22 (m, 1H, 14-H), 0.79 (s, 3H, 18-H) ^{13}C NMR (126 MHz, MeOD-d$_4$) δ/ppm = 165.20 (C-3), 144.55 (C-5), 142.44 (C-1), 120.32 (C-10), 117.12 (C-2), 82.28 (C-17), 50.50 (C-14), 44.51 (C-13), 43.27 (C-9), 39.87 (C-8), 37.72 (C-12), 30.65 (C-16), 27.92 (C-6), 27.13 (C-11), 26.77 (C-7), 23.86 (C-15), 11.70 (C-18) IR (ATR): ν_{max}/cm^{-1} = 3399, 2929, 2869, 1651, 1606, 1550, 1507, 1449, 1375, 1338, 1293, 1253, 1196, 1136, 1100, 1081, 1057, 1022, 960 HRMS (EI): m/z = [M$^{\bullet+}$] calculated for C$_{17}$H$_{23}$NO$_2$ $^{\bullet+}$: 273.1723; found: 273.1724.

(4bS,6aS,7S,9aS,9bR)-7-(Benzyloxy)-2-methoxy-6a-methyl-4b,6,6a,7,8,9,9a,9b,10,11-decahydro-5H-indeno[5,4-f]quinoline (**11**): To a solution of compound **10** (83.6 mg, 0.230 mmol, 1.00 eq) in chloroform (3.5 mL), silver carbonate (320 mg, 1.15 mmol, 5.00 eq) and iodomethane (0.859 mL, 13.8 mmol, 60.0 eq) were added and the mixture was stirred for 40 h at ambient temperature under exclusion of light. Thereafter, the mixture was filtered through Celite, which was washed with chloroform (5 mL), and the filtrate was concentrated in vacuo. The crude product was purified by flash column chromatography (isohexane/ethyl acetate 4:1) to yield 41.0 mg (0.109 mmol, 47.2%) of compound **11** as a colorless solid. m.p.: 102 °C ^{1}H NMR (500 MHz, CD$_2$Cl$_2$) δ/ppm = 7.48 (d, J = 8.5 Hz, 1H, 1-H), 7.34 (m, 4H, benzyl aromatic ortho and meta Hs), 7.26 (ddt, J = 8.6 Hz, 5.5 Hz, 2.5 Hz, 1H, benzyl aromatic para H), 6.49 (d, J = 8.5 Hz, 1H, 2-H), 4.54 (benzyl CH$_2$), 3.84 (OCH$_3$), 3.52 (t, J = 8.3 Hz, 1H, 17-H), 2.84 (m, 2H, 6-H), 2.23 (m, 1H, 11-H$_a$), 2.18 (m, 1H, 9-H), 2.10 (m, 1H, 16-H$_a$), 2.06 (m, 1H, 12-H$_a$), 1.95 (dtd, J = 10.6 Hz, 4.6 Hz, 2.2 Hz, 1H, 7-H$_a$), 1.71 (dddd, J = 12.4 Hz, 9.7 Hz, 7.0 Hz, 3.3 Hz, 1H, 15-H$_a$), 1.59 (dddd, J = 13.2 Hz, 11.5 Hz, 7.9 Hz, 3.4 Hz, 1H, 16-H$_b$), 1.49 (m, 1H, 11-H$_b$), 1.44 (m, 1H, 8-H), 1.41 (m, 1H, 15-H$_b$), 1.39 (m, 1H, 7-H$_b$), 1.36 (m, 1H, 12-H$_b$), 1.22 (m, 1H, 14-H), 0.85 (s, 3H, 18-H) ^{13}C NMR (126 MHz, CD$_2$Cl$_2$) δ/ppm = 161.97 (C-3), 154.62 (C-5), 139.96 (benzyl, quaternary carbon), 136.57 (C-1), 128.57 (2C, benzyl aromatic ortho or meta), 128.21 (C-10), 127.74 (2C, benzyl aromatic ortho or meta), 127.61 62 (benzyl, aromatic para), 107.59 (C-2), 88.93 (C-17), 72.02 (benzyl CH$_2$), 53.32 (OCH$_3$), 50.36 (C-14), 43.81 (C-13), 43.67 (C-9), 38.70 (C-8), 38.17 (C-12), 32.94 (C-6), 28.33 (C-16), 27.42 (C-7), 26.77 (C-11), 23.45 (C-15), 12.00 (C-18). IR (ATR): ν_{max}/cm^{-1} = 2928, 2871, 2348, 2306, 1869, 1716,

1698, 1670, 1654, 1596, 1558, 1541, 1507, 1474, 1457, 1419 HRMS (EI): m/z = [M$^{\bullet+}$] calculated for $C_{25}H_{31}NO_2{}^{\bullet+}$: 377.2349; found: 377.2354.

(4bS,6aS,7S,9aS,9bR)-2-Methoxy-6a-methyl-4b,6,6a,7,8,9,9a,9b,10,11-decahydro-5H-indeno[5, 4-f]quinolin-7-ol (**1**): Under a nitrogen atmosphere compound **11** (18.9 mg, 0.0500 mmol, 1.00 eq) was dissolved in dichloromethane (1.0 mL) and cooled to −78 °C. Then, boron trichloride solution (1M in dichloromethane, 0.15 mL, 0.150 mmol, 3.00 eq) was added dropwise, and the resulting solution was allowed to warm to 0 °C and stirred at this temperature for 2 h. Thereafter, the mixture was quenched with methanol (1 mL) and filtered through Celite. After evaporation of the solvent, the crude product was purified by flash column chromatography (isohexane/ethyl acetate 3:1 with 1% triethylamine) to yield 13.0 mg (0.0452 mmol, 90.5%) of compound **1** as a white solid m.p. 157 °C $[\alpha]_D^{23}$ = 2.5° (CH$_2$Cl$_2$) ^{1}H NMR (400 MHz, CD$_2$Cl$_2$) δ/ppm = 7.48 (d, J = 8.5 Hz, 1H, 1-H), 6.49 (d, J = 8.5 Hz, 1H, 2-H), 3.84 (s, 3H, OCH$_3$), 3.69 (t, J = 8.4 Hz, 1H, 17-H), 2.85 (td, J = 7.0 Hz, 5.8 Hz, 2.6 Hz, 2H, 6-H), 2.25 (m, 1H, 11-H$_a$), 2.20 (m, 1H, 9-H), 2.08 (m, 1H, 16-H$_a$), 1.95 (m, 1H, 7-H$_a$), 1.92 (m, 1H, 12-H$_a$), 1.71 (m, 1H, 15-H$_a$), 1.49 (m, 1H, 11-H$_b$), 1.44 (m, 1H, 1H, 16-H$_b$), 1.42 (m, 1H, 1H, 8-H), 1.38 (m, 1H, 7-H$_b$), 1.34 (m, 1H, 15-H$_b$), 1.26 (m, 1H, 12-H$_b$), 1.19 (m, 1H, 14-H), 0.76 (s, 3H, 18-H) ^{13}C NMR (101 MHz, CD$_2$Cl$_2$) δ/ppm = 161.98 (C-3), 154.63 (C-5), 136.59 (C-1), 128.20 (C-10), 107.59 (C-2), 82.08 (C-17), 53.35 (OCH$_3$), 50.20 (C-14), 43.68 (C-13), 43.65 (C-9), 38.93 (C-8), 37.00 (C-12), 32.93 (C-6), 30.92 (C-16), 27.43 (C-7), 26.67 (C-11), 23.42 (C-15), 11.25 (C-18). IR (ATR): ν_{max}/cm^{-1} = 2928, 2870, 2349, 2307, 1715, 1654, 1596, 1542, 1507, 1475, 1420, 1385, 1309, 1286, 1257, 1309, 1257, 1080 HRMS (EI): m/z = [M$^{\bullet+}$] calculated for $C_{18}H_{25}NO_2{}^{\bullet+}$: 287.1880; found: 287.1888 Purity (HPLC, acetonitrile/water 70:30): >95% (λ = 210 nm), >95% (λ = 254 nm).

Methyl 3-((3S,3aS,5aS,9aR,9bS)-3-Hydroxy-3a-methyl-7-oxododecahydro-1H-cyclopenta[a]naphthalen-6-yl)propanoate (**12**): Ketocarboxylic acid **4** (3.06 g, 10.4 mmol, 1.00 eq), Cs$_2$CO$_3$ (6.78 g, 20.8 mmol, 2.00 eq) and dry DMF (34 mL) were added to an oven-dried round-bottom flask and the mixture was stirred for 30 min at room temperature. Then, iodomethane (2.37 mL, 15.6 mmol, 1.50 eq) was added, and the reaction mixture was stirred overnight at room temperature. After quenching with H$_2$O (30 mL), the mixture was extracted with diethyl ether (3 × 30 mL). The combined organic phases were washed with water (30 mL) and brine (30 mL), dried over anhydrous sodium sulfate, filtered and concentrated in vacuo. The residue was purified by flash column chromatography (isohexane/ethyl acetate 2:1) to obtain methyl ester **12** as a colorless oil (3.11 g, 10.1 mmol, 96.9%) ^{1}H NMR (400 MHz, DMSO-d_6) δ/ppm = 4.49 (d, 1H, J = 4.9 Hz, OH), 3.45 (td, 1H, J = 8.5 Hz, 4.9 Hz, 17-H), 2.41 (m, H, 6-H$_a$), 2.29 (m, 1H, 10-H), 2.24 (m, 1H, 6-H$_b$), 2.20 (m, 1H, 2-H$_a$), 2.13 (m, 1H, 2-H$_b$), 1.88 (m, 1H, 7-H$_a$), 1.81 (m, 1H, 1-H$_a$), 1.73 (m, 1H, 12-H$_a$), 1.68 (m, 1H, 15-H$_a$), 1.64 (m, 1H, 1-H$_b$), 1.58 (m, 1H, 8-H), 1.51 (m, 1H, 11-H$_a$), 1.37 (m, 1H, 16-H$_a$), 1.30 (m, 1H, 15-H$_b$), 1.24 (m, 1H, 16-H$_b$), 1.19 (m, 1H, 11-H$_b$), 1.12 (m, 1H, 7-H$_b$), 1.04 (m, 1H, 9-H), 0.99 (m, 1H, 12-H$_b$), 0.93 (m, 1H, 14-H), 0.71 (s, 3H, 18-H) ^{13}C NMR (101 MHz, DMSO-d_6): δ/ppm = 211.39 (C-5), 173.49 (C-3), 79.86 (C-17), 52.75 (C-10), 51.22 (OCH$_3$), 49.09 (C-14), 47.41 (C-9), 42.70 (C-13), 41.17 (C-6), 40.19 (C-8), 36.17 (C-12), 30.75 (C-2), 30.66 (C-7), 29.73 (C-16), 26.58 (C-15), 22.97 (C-11), 20.58 (C-1), 11.25 (C-18). IR (ATR): ν_{max}/cm^{-1} = 2943, 2308, 1733, 1715, 1647, 1542, 1457, 1387, 1055 HRMS (EI): m/z = [M$^{\bullet+}$] calculated for $C_{18}H_{28}O_4{}^{\bullet+}$: 308.1982; found: 308.1982.

Methyl 3-((3S,3aS,5aS,6R,9aR,9bS)-3-Hydroxy-3a-methyldodecahydrospiro-[cyclopenta[a]naphthalene-7,2'-[1,3]dioxolan]-6-yl)propanoate (**13**): A mixture of ketone **12** (3.09 g, 10.0 mmol, 1.00 eq), trimethyl orthoformate (24.1 mL, 220.0 mmol, 22.0 eq), ethylene glycol (24 mL, 430 mmol, 43.0 eq) and p-toluenesulfonic acid (0.194 g, 1.00 mmol, 0.100 eq) in a round bottom flask was stirred overnight at room temperature. The mixture was diluted with ethyl acetate, and the solution was washed with saturated aqueous sodium bicarbonate solution. The organic phase was dried over anhydrous sodium sulfate, filtered, and concentrated. The residue was purified by flash column chromatography (isohexane/ethyl acetate 2:1) to obtain the dioxolane **13** as a colorless solid (2.38 g, 6.75 mmol, 67.5%). m.p.: 84 °C ^{1}H NMR (400 MHz, DMSO-d_6) δ/ppm = 4.44 (d, 1H, J = 4.8 Hz, OH), 3.88 (m, 2H, ethylene), 3.81 (m, 2H, ethylene), 3.56 (2s, 3H, OCH$_3$) 3.43 (td, 1H, J = 8.4 Hz, 4.9 Hz, 17-H), 2.38 (ddd, J = 16.7 Hz, 10.0 Hz, 6.6 Hz, 1H, 2-H$_a$), 2.24 (ddd, J = 16.0 Hz, 10.0 Hz, 6.1 Hz, 1H, 2-H$_b$), 1.82 (m, 1H, 16-H$_a$), 1.76 (m, 1H, 6-H$_a$), 1.70 (m, 1H, 12-H$_a$), 1.64 (m, 1H, 15-H$_a$), 1.61 (m, 1H, 1-H$_a$), 1.57 (m, 1H, 1-H$_b$), 1.47 (m, 1H, 11-H$_a$ or 7-H$_a$), 1.44 (m, 1H, 11-H$_a$ or 7-H$_a$), 1.38 (m, 1H, 10-H), 1.31 (m, 1H, 16-H$_b$), 1.21 (m, 1H, 6-H$_b$), 1.16 (m, 1H, 11-H$_b$ or 15-H$_b$), 1.13 (m, 1H, 11-H$_b$ or 15-H$_b$), 1.06 (m, 1H, 8-H), 1.01 (m, 1H, 9-H or 12-H$_b$), 0.97 (m, 1H, 9-H or 12-H$_b$), 0.95 (m, 1H, 7-H$_b$), 0.91 (m, 1H, 14-H), 0.63 (s, 3H, 18-H) ^{13}C NMR (101 MHz, DMSO-d_6) δ/ppm = 173.76 (C-3), 110.55 (C-5), 79.99 (C-17), 63.88 (ethylene), 63.86 (ethylene), 51.14 (OCH$_3$), 49.41 (C-14), 47.32 (C-10), 44.90 (C-9), 42.71 (C-13), 40.22 (C-8), 36.58 (C-12), 33.91 (C-6), 32.68 (C-2), 29.82 (C-16), 27.17 (C-7), 25.92 (C-15), 22.93 (C-11), 21.13 (C-1), 11.35 (C-18). IR (ATR): ν_{max}/cm^{-1} = 1868, 2307, 1732, 1698, 1647, 1635, 1321 HRMS (EI): m/z = [M$^{\bullet+}$] calculated for C$_{20}$H$_{32}$O$_5$$^{\bullet+}$: 352.2244; found: 352.2244.

Methyl 3-((3S,3aS,5aS,6R,9aR,9bS)-3-((tert-butyldimethylsilyl)oxy)-3a-methyl dodecahydro spiro[cyclopenta[a]naphthalene-7,2'-[1,3]dioxolan]-6-yl)propanoate (**14**): Compound **13** (2.38 g, 6.75 mmol, 1.00 eq) was dissolved in dimethylformamide (14 mL). Then imidazole (0.957 g, 14.1 mmol, 3.80 eq) and tert-butyldimethylsilyl chloride (1.06 g, 7.03 mmol, 2.00 eq) were added, and the resulting mixture was stirred overnight at room temperature. After addition of water (10 mL), the mixture was extracted with ethyl acetate (3 × 10 mL). The combined organic layers were washed with 1M hydrochloric acid (30 mL), water (20 mL) and brine (20 mL), dried over anhydrous sodium sulfate, filtered and concentrated in vacuo. The crude product was purified by flash column chromatography (isohexane/ethyl acetate 9:1) to obtain a colorless solid (2.35 g, 5.02 mmol, 74.4%). m.p.: 86 °C ^{1}H NMR (400 MHz, CD$_2$Cl$_2$) δ/ppm = 3.94 (m, 2H, ethylene), 3.89 (m, 2H, ethylene), 3.61 (s, 3H, ester CH$_3$), 3.58 (m, 1H, 17-H), 2.43 (m, 1H, 2-H$_a$), 2.31 (m, 1H, 2-H$_b$), 1.88 (m, 1H, 12-H$_a$), 1.80 (m, 1H, 6-H$_a$), 1.74 (m, 1H, 7-H$_a$), 1.71 (m, 1H, 1-H$_a$), 1.69 (m, 1H, 10-H), 1.56 (m, 1H, 16-H$_a$), 1.54 (m, 1H, 11-H$_a$ or 15-H$_a$), 1.51 (m, 1H, 11-H$_a$ or 15-H$_a$), 1.44 (m, 1H, 12-H$_b$), 1.39 (m, 1H, 16-H$_b$), 1.29 (m, 1H, 6-H$_b$), 1.25 (m, 1H, 1-H$_b$), 1.21 (m, 1H, 11-H$_b$), 1.14 (m, 1H, 15-H$_b$), 1.09 (m, 1H, 8-H or 9-H), 1.06 (m, 1H, 8-H or 9-H), 1.02 (m, 1H, 7-H$_b$), 0.97 (m, 1H, 14-H), 0.87 (s, 9H, tert-butyl), 0.72 (s, 3H, 18-H), 0.01 (s, 3H, dimethylsilyl), 0.01 (s, 3H, dimethylsilyl). ^{13}C NMR (101 MHz, CD$_2$Cl$_2$) δ/ppm = 175.01 (ester carbonyl), 111.91 (C-5), 82.36 (C-17), 64.89 (ethylene), 51.69 (methyl ester), 49.96 (C-14), 48.53 (C-10), 45.93 (C-9), 43.99 (C-13), 41.33 (C-8), 37.70 (C-12), 34.93 (C-6), 33.69 (C-2), 31.44 (C-16), 27.97 (C-7), 27.04 (C-15), 26.18 (tert-butyl CH$_3$), 23.89 (C-11), 21.96 (C-1), 18.54 (tert-butyl C), 11.77 (C-18), −4.25 (dimethylsilyl), −4.54 (dimethylsilyl). IR (ATR): ν_{max}/cm^{-1} = 2926, 2885, 2854, 2307, 1735, 1472, 1162, 1093, 899, 885 HRMS (EI): m/z = [M$^{\bullet+}$] calculated for C$_{26}$H$_{46}$O$_5$Si$^{\bullet+}$: 466.3109; found: 466.3102.

Methyl (E)-3-((3S,3aS,5aS,6R,9aR,9bS)-3-((tert-butyldimethylsilyl)oxy)-3a-methyldodecah ydrospiro[cyclopenta[a]naphthalene-7,2'-[1,3]dioxolan]-6 yl)acrylate (**15**): Dry THF (1.2 mL) and lithium diisoproylamide (2M in in THF, 3.76 mL, 7.53 mmol, 1.25 eq) were added to a flame-dried Schlenk flask under nitrogen. The solution was cooled down to -78 °C, and after 10 min, a solution of compound **14** (2.34 g, 5.02 mmol, 1.00 eq) in 8.5 mL of dry THF was added dropwise via a syringe. After stirring for 25 min, a solution of diphenyldiselenide (0.888 g, 2.84 mmol, 1.25 eq) in 8.8 mL dry THF was added quickly. The mixture was stirred at −78 °C for 30 min and then gradually warmed up to room temperature over a 2 h period. The reaction mixture was then quenched by adding a saturated ammonium chloride solution (50 mL). After extraction with ethyl acetate (3 × 5 mL), the combined organic layers were washed with 1M hydrochloric acid (50 mL), water (50 mL), saturated aqueous sodium bicarbonate solution (50 mL) and brine (50 mL), dried over anhydrous sodium sulfate and filtered. After the evaporation of the solvent, an orange solid was obtained.

Dichloromethane (15 mL) was added to this solid, and the resulting solution was cooled to 0 °C. The temperature of the solution was monitored throughout the whole reaction. Then, a solution of hydrogen peroxide (30%, 4.4 mL, 131 mmol, 26.0 eq) in water (4.4 mL) was added dropwise. After the addition was complete, the temperature of the reaction mixture rose quickly to about 30 °C, dropping thereafter. The mixture was allowed to come to room temperature and stirred until the reaction was complete (TLC control). The reaction mixture was transferred to a separation funnel containing saturated aqueous sodium bicarbonate solution (50 mL) and extracted with dichloromethane (3 × 25 mL). The combined organic layers were dried over anhydrous sodium sulfate, filtered and concentrated in vacuo. The crude product was purified by flash column chromatography (isohexane/ethyl acetate 9:1) to yield compound **15** (1.69 g, 3.63 mmol, 72.2%) as a colorless solid. m.p.: 96 °C
^{1}H NMR (400 MHz, CD_2Cl_2) δ/ppm = 6.72 (dd, 1H, J = 15.7 Hz, 10.0 Hz, 1-H), 5.82 (d, 1H, J = 15.7 Hz, 2-H), 3.90-3.70 (m, 4H, ethylene), 3.68 (s, 3H, ester CH_3), 3.58 (dd, 1H, J = 8.8 Hz, 7.8 Hz, 17-H), 2.21 (t, J = 10.6 Hz, 1H, 10-H), 1.89 (dtd, J = 13.1 Hz, 9.1 Hz, 5.7 Hz, 1H, 16-H_a), 1.79 (dt, J = 13.5 Hz, 3.0 Hz, 1H, 6-H_a), 1.66 (dt, J = 12.2 Hz, 2.9 Hz, 1H, 12-H_a), 1.60 (m, 1H, 7-H or 11-H), 1.56 (m, 1H, 15-H_a), 1.45 (m, 1H, 6-H_b), 1.41 (m, 1H, 16-H_b), 1.38 (m, 1H, 7-H or 11-H), 1.30 (m, 1H, 7-H or 11-H), 1.26 (m, 1H, 15-H_b), 1.24 (m, 1H, 9-H), 1.12 (m, 1H, 8-H), 1.08 (m, 1H, 7-H or 11-H), 1.02 (m, 1H, 12-H_b), 0.98 (m, 1H, 14-H), 0.87 (s, 9H, tert-butyl), 0.71 (s, 3H, 18-H), 0.00 (s, 6H, dimethylsilyl) ^{13}C NMR (101 MHz, CD_2Cl_2) δ/ppm = 167.04 (ester carbonyl), 148.73 (C-1), 124.24 (C-2), 110.53 (C-5), 82.26 (C-17), 65.78 (Ethylen), 65.50 (ethylene), 55.39 (C-10), 51.72 (OCH_3), 49.86 (C-14), 45.00 (C-9), 44.15 (C-13), 40.48 (C-8), 37.38 (C-12), 35.68 (C-6), 31.37 (C-16), 28.03 (C-7 or C-11), 28.02 (C-7 or C-11), 26.17 (tert-butyl CH_3), 23.83 (C-15), 18.52 (tert-butyl C), 11.72 (C-18), −4.25 (dimethylsilyl), −4.25 (dimethylsilyl). IR (ATR): v_{max}/cm^{-1} = 2952, 2928, 2858, 1718, 1652, 1472, 1435, 1163, 900, 772 HRMS (EI): m/z = [M$^{\bullet+}$] calculated for $C_{26}H_{44}O_5Si^{\bullet+}$: 464.2953; found: 464.2951.

(3S,3aS,5aS,9aR,9bS)-3-((tert-Butyldimethylsilyl)oxy)-3a-methyldodecahydrospiro[cyclopenta[a] naphthalene-7,2'-[1,3]dioxolane]-6-carbaldehyde (**16**): Compound **15** (1.44 g, 3.10 mmol) was dissolved in a mixture of dichloromethane (18 mL) and methanol (10 mL). The solution was cooled to −78 °C and then treated with ozone (5 min, flow: 50 L/h, 55 W). Progress of the reaction was monitored via TLC. After excess ozone had been removed by a stream of nitrogen, dimethyl sulfide (18.0 mL, 243 mmol) was added, and the reaction mixture was allowed to warm gradually. It was then stirred overnight at room temperature. The mixture was diluted with dichloromethane (18 mL) and washed with saturated aqueous sodium bicarbonate solution (2 × 50 mL) and brine (50 mL). The organic layer was dried over anhydrous sodium sulfate filtered, and the solvent was evaporated to obtain a colorless solid (0.960 g, 2.35 mmol, 75.8%), which was used as such in the next step. m.p.: 135 °C
^{1}H NMR (500 MHz, CD_2Cl_2; pure compound obtained by tedious flash chromatography) δ/ppm = 4.01(m, 4H, ethylene), 3.61 (s, 3H, OCH_3), 3.59 (t, J = 8.3 Hz, 17-H), 2.54 (d, J = 11.5 Hz, 1H, 10-H), 1.92 (m, 1H, 16-H_a), 1.89 (m, 1H, 6-H_a), 1.72 (m, 1H, 12-H_a), 1.62 (m, 1H, 7-H_a or 11-H_a), 1.56 (m, 1H, 15-H_a), 1.49 (m, 1H, 7-H_a or 11-H_a), 1.45 (m, 1H, 16-H_b), 1,41 (m, 1H, 8-H), 1.39 (m, 1H, 6-H_b), 1.36 (m, 1H, 7-H_b or 11-H_b), 1.28 (m, 1H, 15-H_b), 1.15 (m, 1H, 9-H), 1.14 (m, 1H, 7-H_b or 11-H_b), 1.05 (m, 1H, 12-H_b), 1.01 (m, 1H, 14-H), 0.87 (s, 9H, $(CH_3)_3$), 0.73 (s, 3H, 18-H), 0.01 (s, 3H, dimethylsilyl), 0.01 (s, 3H, dimethylsilyl) ^{13}C NMR (126 MHz, CD_2Cl_2) δ/ppm = 172.18 (C-1), 109.88 (C-5), 82.11 (C-17), 64.98 (ethylen), 64.95 (ethylen), 57.16 (C-10), 49.55 (C-14), 44.20 (C-13), 44.20 (C-8) 40.33 (C-9), 37.06 (C-12), 34.11 (C-6), 31.29 (C-16), 27.58 (C-7 or C-11), 27.34 (C-7 or C-11), 26.15 (tert-butyl CH_3), 23.75 (C-15), 18.51 (tert-butyl quaternary carbon), 11.68 (C-18), −4.26 (dimethylsilyl), −4.57 (dimethylsilyl). IR (ATR): v_{max}/cm^{-1} = 2927, 2308, 1733, 1717, 1653, 1558, 1261, 900 HRMS (EI): m/z = [M$^{\bullet+}$] calculated for $C_{23}H_{40}O_4Si^{\bullet+}$: 408.2690; found: 408.2658.

(3S,3aS,5aS,9aR,9bS)-3-Hydroxy-3a-methyl-7-oxododecahydro-1H-cyclopenta[a]naphthalene-6-carbaldehyde (**17**): Crude compound **16** (0.899 g, about 2.20 mmol) was suspended in a mixture of glacial acetic acid (22.0 mL), THF (7.5 mL) and water (7.5 mL) and stirred

overnight at room temperature. After the addition of 16.4 mL of a 50% solution of acetic acid in water, the mixture was refluxed for 1 h. After cooling to room temperature, brine (20 mL) was added, and the mixture was extracted with ethyl acetate (4 × 50 mL). The combined organic extracts were washed with saturated aqueous sodium bicarbonate solution (50 mL), dried over anhydrous sodium sulfate and filtered. After evaporation of the solvent, compound **17** was obtained as a colorless oil (0.436 g, 1.74 mmol, about 79.2%), which was used as such in the next step. ^{1}H NMR (400 MHz, CD$_2$Cl$_2$; pure compound obtained by tedious flash chromatography) δ/ppm = 15.45 (d, J = 6.2 Hz, 0.41H, enol OH), 9.60 (dd, J = 4.6 Hz, 2.2 Hz, 1H, 0.26H, aldehyde, keto tautomer), 8.28 (d, J = 5.7 Hz, 0.46H, aldehyde, enol tautomer), 3.90 (m, 1H), 3.66 (m, 1H, 17-H), 2.46 (m, 1H), 2.04 (m, 1H), 1.97 (m, 1H), 1.84 (m, 1H), 1.68 (m, 1H), 1.61 (m, 1H), 1.57 (m, 1H), 1.45 (m, 1H), 1.40 (m, 1H), 1.36 (m, 1H), 1.31 (m, 1H), 1.27 (m, 1H), 1.20 (m, 1H), 1.15 (m, 1H), 1.09 (m, 1H), 0.80-0.73 (3s, 3H, 18-H). ^{13}C NMR (101 MHz, CD$_2$Cl$_2$) δ/ppm = 204.59 (aldehyde, keto tautomer), 200.93 (C-5, keto tautomer) 194.50 (C-5, enol tautomer), 178.27 (aldehyde, enol tautomer), 113.83 (C-10, enol tautomer), 81.98 (C-17), 62.51 (C-10, keto tautomer), 49.56 (C-14), 43.58 (C-13), 41.19 (CH), 39.04 (CH), 36.52 (CH$_2$), 33.95 (CH$_2$), 30.69 (CH$_2$), 26.25 (CH$_2$), 26.17 (CH$_2$), 23.38 (CH$_2$), 11.20 (C-18). IR (ATR): ν_{max}/cm^{-1} = 2927, 2307, 1733, 1716, 1636, 1457, 1082 HRMS (EI): m/z = [M$^{\bullet+}$] calculated for C$_{15}$H$_{22}$O$_3$$^{\bullet+}$: 250.1563; found: 250.1563.

(4bS,6aS,7S,9aS,9bR)-2-Methoxy-6a-methyl-4b,6,6a,7,8,9,9a,9b,10,11-decahydro-5H-indeno[5,4-f]quinazolin-7-ol (**2**): Crude compound **17** (0.401 g, 1.60 mmol, 1.00 eq) was added to a round bottom flask and dissolved in 10 mL of dry methanol. Then, methyl carbamimidate sulfate (0.826 g, 4.80 mmol, 3.00 eq) and 3.2 mL of a freshly prepared solution of sodium methanolate in methanol (0.11 g of sodium in 3.2 mL of dry methanol, 4.80 mmol, 3.00 eq) were added to the flask and the mixture was refluxed for 8 h under a nitrogen atmosphere. After cooling to room temperature, water (100 mL) was added, and the mixture was extracted with ethyl acetate (3 × 50 mL). The combined organic layers were washed with water and brine and dried over anhydrous sodium sulfate. After filtration and evaporation of the solvent, the residue was purified by flash column chromatography (isohexane/ethyl acetate 2:1 with 1% triethylamine) to obtain compound **2** as a white solid (0.174 g, 0.603 mmol, 37.7%). m.p.: 179 °C $[\alpha]_D^{23}$ = 2.1° (CH$_2$Cl$_2$) ^{1}H NMR (400 MHz, CD$_2$Cl$_2$) δ/ppm = 8.31 (s, 1H, 1-H), 3.91 (s, 3H, OCH$_3$), 3.70 (t, 1H, J = 8.5 Hz, 17-H), 2.83 (m, 2H, 6-H), 2.29 (m, 1H, 11-H$_a$), 2.20 (m, 1H, 9-H), 2.08 (m, 1H, 16-H$_a$), 1.96 (m, 1H, 7-H$_a$), 1.92 (m, 1H, 12-H$_a$), 1.70 (m, 1H, 15-H$_a$), 1.48 (m, 1H, 11-H$_b$), 1.45 (m, 1H, 16-H$_b$), 1.42 (m, 1H, 8-H), 1.39 (m, 1H, 7-H$_b$), 1.36 (m, 1H, 15-H$_b$), 1.29 (m, 1H, 12-H$_b$), 1.18 (m, 1H, 14-H), 0.76 (s, 3H, 18-H) ^{13}C NMR (101 MHz, CD$_2$Cl$_2$) δ/ppm = 168.87 (C-5), 164.32 (C-3), 156.38 (C-1), 127.06 (C-10), 82.14 (C-17), 54.86 (OCH$_3$), 50.12 (C-14), 43.78 (C-13), 42.11 (C-9), 38.76 (C-8), 36.85 (C-12), 32.74 (C-6), 30.91 (C-16), 26.92 (C-7), 26.10 (C-11), 23.54 (C-15), 11.36 (C-18). IR (ATR): ν_{max}/cm^{-1} = 2943, 2866, 2307, 1734, 1654, 1587, 1546, 1467, 1389, 1323, 1034, 749 HRMS (ESI): m/z = [M + H]$^+$ calculated for C$_{17}$H$_{25}$N$_2$O$_2$$^+$: 289.1911; found: 289.1912 Purity (HPLC, acetonitrile/water 50:50): >96% (λ = 210 nm), >97% (λ = 254 nm).

3.2. Biological Testing

hTRPML1ΔNC-YFP, a plasma membrane variant of wild-type TRPML1, was obtained from HEK293 cells stably expressing plasma membrane-targeted TRPML1 by trypsination and after resuspending in HEPES buffered solution. IC$_{50}$ values for TRPML1 inhibition were determined on a fluorescence imaging plate reader built into a robotic liquid handling station (Freedom Evo 150, Tecan, Mannedorf, Switzerland) using the calcium dye Fluo-4/AM (Invitrogen, Thermo Fisher Scientific, Waltham, MA, USA) according to the test procedure described in our previous work [13] in the section Compound screening and generation of concentration–response curves.

4. Conclusions

In conclusion, we have worked out straightforward chiral pool syntheses of the 4-aza (**1**) and 2,4-diaza analog (**2**) of the TRPML1 inhibitor estradiol methyl ether (**EDME**) starting with oxidative cleavage of ring A of the readily available steroid 19-nortestosterone (**3**) to provide ketocarboxylic acid **4** as the central intermediate for both target compounds. By utilizing carefully selected protective groups for the 17-OH group (benzyl in the pyridine synthesis, TBDS in the pyrimidine synthesis) and oxidants for dehydrogenation (CuBr$_2$/LiBr/methanesulfonic acid in the pyridine synthesis) and chain degradation (selenylation/selenoxide elimination/ozonolysis in the pyrimidine synthesis) both target compounds were obtained in 6 and 8 steps, respectively, and in acceptable overall yields (8.6%, 7.5%).

While the 4-aza analog **1** showed significant TRPML1-inhibitory activity (only factor <2 less potent than the gold standard **EDME**), the 2,4-diaza analog **2** significantly lost inhibitory potency, and the pyridone analog, obtained as an unexpected side product, was completely inactive. This leads to the conclusion that for the cation channel TRPML1, aza analogs are not promising bioisosteres of **EDME**.

Supplementary Materials: The following supporting information can be downloaded at: https://www.mdpi.com/article/10.3390/molecules28217428/s1, Additional synthetic procedures; Figure S1: Numbering of the compounds (for assignment of NMR signals); ^{1}H- and ^{13}C-NMR spectra of the compounds.

Author Contributions: Conceptualization, F.B.; methodology, F.B. and P.R.; investigation, P.R.; resources, F.B.; data curation, F.B. and P.R.; writing—original draft preparation, F.B. and P.R.; writing—review and editing, F.B. and P.R.; visualization, F.B. and P.R.; supervision, F.B. All authors have read and agreed to the published version of the manuscript.

Funding: This research received no external funding.

Data Availability Statement: Experimental data, spectra, and protocols are stored in an electronic lab journal by the authors.

Acknowledgments: The authors thank Nicole Urban and Michael Schäfer, University of Leipzig, Germany, for performing the screenings for TRPML1-inhibitory activity and Lars Almendinger for support with NMR measurements.

Conflicts of Interest: The authors declare no conflict of interest.

References

1. Bargal, R.; Avidan, N.; Ben-Asher, E.; Olender, Z.; Zeigler, M.; Frumkin, A.; Raas-Rothschild, A.; Glusman, G.; Lancet, D.; Bach, G. Identification of the gene causing mucolipidosis type IV. *Nat. Genet.* **2000**, *26*, 118–122. [CrossRef]
2. Kasitinon, S.Y.; Eskiocak, U.; Martin, M.; Bezwada, D.; Khivansara, V.; Tasdogan, A.; Zhao, Z.; Mathews, T.; Aurora, A.B.; Morrison, S.J. TRPML1 Promotes Protein Homeostasis in Melanoma Cells by Negatively Regulating MAPK and mTORC1 Signaling. *Cell Rep.* **2019**, *28*, 2293–2305. [CrossRef] [PubMed]
3. Yin, C.; Zhang, H.; Liu, X.; Zhang, H.; Zhang, Y.; Bai, X.; Wang, L.; Li, H.; Li, X.; Zhang, S.; et al. Downregulated MCOLN1 Attenuates The Progression Of Non-Small-Cell Lung Cancer By Inhibiting Lysosome-Autophagy. *Cancer Manag. Res.* **2019**, *11*, 8607–8617. [CrossRef] [PubMed]
4. Xing, Y.; Sui, Z.; Liu, Y.; Wang, M.-m.; Wei, X.; Lu, Q.; Wang, X.; Liu, N.; Lu, C.; Chen, R.; et al. Blunting TRPML1 channels protects myocardial ischemia/reperfusion injury by restoring impaired cardiomyocyte autophagy. *Basic Res. Cardiol.* **2022**, *117*, 20. [CrossRef]
5. Santoni, G.; Maggi, F.; Amantini, C.; Marinelli, O.; Nabissi, M.; Morelli, M.B. Pathophysiological Role of Transient Receptor Potential Mucolipin Channel 1 in Calcium-Mediated Stress-Induced Neurodegenerative Diseases. *Front. Physiol.* **2020**, *11*, 251. [CrossRef] [PubMed]
6. Zhang, X.; Li, X.; Xu, H. Phosphoinositide isoforms determine compartment-specific ion channel activity. *Proc. Natl. Acad. Sci. USA* **2012**, *109*, 11384–11389. [CrossRef]
7. Chen, C.C.; Keller, M.; Hess, M.; Schiffmann, R.; Urban, N.; Wolfgardt, A.; Schaefer, M.; Bracher, F.; Biel, M.; Wahl-Schott, C.; et al. A small molecule restores function to TRPML1 mutant isoforms responsible for mucolipidosis type IV. *Nat. Commun.* **2014**, *5*, 4681. [CrossRef]

8. Shen, D.; Wang, X.; Li, X.; Zhang, X.; Yao, Z.; Dibble, S.; Dong, X.P.; Yu, T.; Lieberman, A.P.; Showalter, H.D.; et al. Lipid storage disorders block lysosomal trafficking by inhibiting a TRP channel and lysosomal calcium release. *Nat. Commun.* **2012**, *3*, 731. [CrossRef] [PubMed]

9. Spix, B.; Butz, E.; Chen, C.-C.; Scotto Rosato, A.; Tang, R.; Jeridi, A.; Kudrina, V.; Melzergeb Plesch, E.; Wartenberg, P.; Arlt, E.; et al. Lung emphysema and impaired macrophage elastase clearance in mucolipin 3 deficient mice. *Nat. Commun.* **2022**, *13*, 318. [CrossRef] [PubMed]

10. Wang, W.; Gao, Q.; Yang, M.; Zhang, X.; Yu, L.; Lawas, M.; Li, X.; Bryant-Genevier, M.; Southall, N.T.; Marugan, J.; et al. Up-regulation of lysosomal TRPML1 channels is essential for lysosomal adaptation to nutrient starvation. *Proc. Natl. Acad. Sci. USA* **2015**, *112*, E1373–E1381. [CrossRef]

11. Leser, C.; Keller, M.; Gerndt, S.; Urban, N.; Chen, C.C.; Schaefer, M.; Grimm, C.; Bracher, F. Chemical and pharmacological characterization of the TRPML calcium channel blockers ML-SI1 and ML-SI3. *Eur. J. Med. Chem.* **2021**, *210*, 112966. [CrossRef] [PubMed]

12. Kriegler, K.; Leser, C.; Mayer, P.; Bracher, F. Effective chiral pool synthesis of both enantiomers of the TRPML inhibitor trans-ML-SI3. *Arch. Pharm.* **2022**, *355*, e2100362. [CrossRef] [PubMed]

13. Rühl, P.; Rosato, A.S.; Urban, N.; Gerndt, S.; Tang, R.; Abrahamian, C.; Leser, C.; Sheng, J.; Jha, A.; Vollmer, G.; et al. Estradiol analogs attenuate autophagy, cell migration and invasion by direct and selective inhibition of TRPML1, independent of estrogen receptors. *Sci. Rep.* **2021**, *11*, 8313. [CrossRef]

14. Mayer, C.D.; Bracher, F. Cytotoxic ring A-modified steroid analogues derived from Grundmann's ketone. *Eur. J. Med. Chem.* **2011**, *46*, 3227–3236. [CrossRef] [PubMed]

15. Renard, D.; Perruchon, J.; Giera, M.; Müller, J.; Bracher, F. Side chain azasteroids and thiasteroids as sterol methyltransferase inhibitors in ergosterol biosynthesis. *Bioorg. Med. Chem.* **2009**, *17*, 8123–8137. [CrossRef]

16. Meanwell, N.A. Synopsis of some recent tactical application of bioisosteres in drug design. *J. Med. Chem.* **2011**, *54*, 2529–2591. [CrossRef] [PubMed]

17. Subbaiah, M.A.M.; Meanwell, N.A. Bioisosteres of the Phenyl Ring: Recent Strategic Applications in Lead Optimization and Drug Design. *J. Med. Chem.* **2021**, *64*, 14046–14128. [CrossRef]

18. Tse, E.G.; Houston, S.D.; Williams, C.M.; Savage, G.P.; Rendina, L.M.; Hallyburton, I.; Anderson, M.; Sharma, R.; Walker, G.S.; Obach, R.S.; et al. Nonclassical Phenyl Bioisosteres as Effective Replacements in a Series of Novel Open-Source Antimalarials. *J. Med. Chem.* **2020**, *63*, 11585–11601. [CrossRef]

19. Brill, Z.G.; Condakes, M.L.; Ting, C.P.; Maimone, T.J. Navigating the Chiral Pool in the Total Synthesis of Complex Terpene Natural Products. *Chem. Rev.* **2017**, *117*, 11753–11795. [CrossRef]

20. Holt, D.A.; Levy, M.A.; Brandt, M.; Metcalf, B.W. Inhibition of pyridine-nucleotide-dependent enzymes by pyrazoles. Synthesis and enzymology of A novel A-ring pyrazole steroid. *Steroids* **1986**, *48*, 213–222. [CrossRef]

21. Rasmusson, G.H.; Reynolds, G.F.; Utne, T.; Jobson, R.B.; Primka, R.L.; Berman, C.; Brooks, J.R. Azasteroids as inhibitors of rat prostatic 5 alpha-reductase. *J. Med. Chem.* **1984**, *27*, 1690–1701. [CrossRef] [PubMed]

22. Xia, P.; Yang, Z.-Y.; Xia, Y.; Zhang, H.-B.; Zhang, K.-H.; Sun, X.; Chen, Y.; Zheng, Y.-Q. Synthesis of N-Substituted 3-Oxo-17β-carboxamide-4-aza-5α-androstanes and the Tautomerism of 3-Oxo-4-aza-5-androstenes. *Heterocycles* **1998**, *47*, 703–716. [CrossRef]

23. Skaanderup, P.R.; Poulsen, C.S.; Hyldtoft, L.; Jørgensen, M.R.; Madsen, R. Regioselective Conversion of Primary Alcohols into Iodides in Unprotected Methyl Furanosides and Pyranosides. *Synthesis* **2002**, *2002*, 1721–1727. [CrossRef]

24. Poon, K.W.C.; Dudley, G.B. Mix-and-Heat Benzylation of Alcohols Using a Bench-Stable Pyridinium Salt. *J. Org. Chem.* **2006**, *71*, 3923–3927. [CrossRef] [PubMed]

25. Zhang, J.; Yan, Y.; Hu, R.; Li, T.; Bai, W.J.; Yang, Y. Enantioselective Total Syntheses of Lyconadins A-E through a Palladium-Catalyzed Heck-Type Reaction. *Angew. Chem. Int. Ed.* **2020**, *59*, 2860–2866. [CrossRef]

26. Fischer, D.F.; Sarpong, R. Total Synthesis of (+)-Complanadine A Using an Iridium-Catalyzed Pyridine C–H Functionalization. *J. Am. Chem. Soc.* **2010**, *132*, 5926–5927. [CrossRef]

27. Lee, A.S.; Liau, B.B.; Shair, M.D. A unified strategy for the synthesis of 7-membered-ring-containing Lycopodium alkaloids. *J. Am. Chem. Soc.* **2014**, *136*, 13442–13452. [CrossRef]

28. Wu, B.; Bai, D. The First Total Synthesis of (±)-Huperzine B. *J. Org. Chem.* **1997**, *62*, 5978–5981. [CrossRef]

29. King, A.O.; Anderson, R.K.; Shuman, R.F.; Karady, S.; Abramson, N.L.; Douglas, A.W. Iodotrimethylsilane-mediated 2-monohalogenation of 4-aza-5.alpha.-androstan-3-one steroids. *J. Org. Chem.* **1993**, *58*, 3384–3386. [CrossRef]

30. King, L.C.; Ostrum, G.K. Selective Bromination with Copper(II) Bromide1. *J. Org. Chem.* **1964**, *29*, 3459–3461. [CrossRef]

31. Sander, M.; Gries, J.; Schuetz, A. Method for Aromatizing 19-Norandrost-4-en-3-Ones to Estra-1,3,5(10)-Trienes. U.S. Patent Application WO 2009/074313 A1, 18 June 2009.

32. White, J.D.; Li, Y.; Kim, J.; Terinek, M. Cyclobutane Synthesis and Fragmentation. A Cascade Route to the Lycopodium Alkaloid (−)-Huperzine A. *J. Org. Chem.* **2015**, *80*, 11806–11817. [CrossRef] [PubMed]

33. Mandal, P.K.; McMurray, J.S. Pd–C-Induced Catalytic Transfer Hydrogenation with Triethylsilane. *J. Org. Chem.* **2007**, *72*, 6599–6601. [CrossRef] [PubMed]

34. Falck, J.R.; Barma, D.K.; Baati, R.; Mioskowski, C. Differential Cleavage of Arylmethyl Ethers: Reactivity of 2,6-Dimethoxybenzyl Ethers. *Angew. Chem. Int. Ed.* **2001**, *40*, 1281–1283. [CrossRef]

35. Aversa, R.J.; Burger, M.T.; Dillon, M.P.; Dineen, T.A., Jr.; Lou, Y.; Nishiguchi, G.A.; Ramurthy, S.; Rico, A.C.; Rauniyar, V.; Sendzik, M.; et al. Compounds and Compositions as Raf Kinase Inhibitors. U.S. Patent Application WO 2016/038582 2016, 11 September 2014.
36. Bertin, D.; Nedelec, L.; Pierdet, A. Diaza Steroids. FR 1481182, 19 May 1967.
37. Nomine, G. Derivatives of 7-oxo-1,2,3,4,5,6,7,8-Octahydronaphthalene. FR 1234734, 19 October 1960.
38. de Avellar, I.G.J.; Vierhapper, F.W. Novel Partial Synthetic Approaches to Replace Carbons 2,3,4 of Steroids. A Methodology to Label Testosterone and Progesterone with 13C in the Steroid A Ring. Part 1. *Tetrahedron* **2000**, *56*, 9957–9965. [CrossRef]

Article

Montmorillonite Catalyzed Synthesis of Novel Steroid Dimers

Aneta M. Tomkiel [1],*, Adam D. Majewski [1,2], Leszek Siergiejczyk [1] and Jacek W. Morzycki [1],*

[1] Faculty of Chemistry, University of Bialystok, Ciołkowskiego 1K, 15-245 Bialystok, Poland; a.majewski@uwb.edu.pl (A.D.M.); nmrbial@uwb.edu.pl (L.S.)

[2] Doctoral School of Exact and Natural Sciences, University of Bialystok, Ciołkowskiego 1K, 15-245 Bialystok, Poland

* Correspondence: a.tomkiel@uwb.edu.pl (A.M.T.); morzycki@uwb.edu.pl (J.W.M.); Tel.: +48-85-738-80-44 (A.M.T.); +48-85-738-82-60 (J.W.M.)

Abstract: The reactions of sterols (androst-5-en-3β-ol-17-one, diosgenin, and cholesterol) and their tosylates with hydroquinone aimed at the synthesis of *O,O*-1,4-phenylene-linked steroid dimers were studied. The reaction course strongly depended on the conditions used. The study has shown that the major reaction products are the elimination products and unusual steroid dimers resulting from the nucleophilic attack of the hydroquinone C2 carbon atom on the steroid C3 position, followed by an intramolecular addition to the C5–C6 double bond. A different reaction course was observed when montmorillonite K10 was used as a catalyst. The reaction of androst-5-en-3β-ol-17-one under the promotion of this catalyst afforded the *O,O*-1,4-phenylene-linked steroid dimer in addition to the disteroidal ether. The formation of the latter compound was suppressed by using 3-tosylate as a substrate instead of the free sterol. The reactions of androst-5-en-3β-ol-17-one tosylate and cholesteryl tosylate with hydroquinone catalyzed by montmorillonite K10 carried out under optimized conditions afforded the desired dimers in 31% and 67% yield, respectively.

Keywords: steroid dimers; 1,4-phenylene-linked dimers; montmorillonite; DHEA; cholesterol; steroids

Citation: Tomkiel, A.M.; Majewski, A.D.; Siergiejczyk, L.; Morzycki, J.W. Montmorillonite Catalyzed Synthesis of Novel Steroid Dimers. *Molecules* **2023**, *28*, 7068. https://doi.org/ 10.3390/molecules28207068

Academic Editors: Marina Savić, Erzsébet Mernyák, Jovana Ajdukovic and Suzana Jovanović-Šanta

Received: 28 September 2023
Revised: 10 October 2023
Accepted: 10 October 2023
Published: 13 October 2023

1. Introduction

During the last few decades steroid dimers have emerged as an interesting new area of steroid chemistry. The first review article on steroid dimers was published by Li and Dias in 1997 [1]. Since then, numerous reports have appeared on these compounds, which have been reviewed in the book [2] and in several review articles by Nahar et al. [3,4]. In the beginning, steroid dimers, which were formed as byproducts of certain reactions, were considered mere curiosities. Later, steroid dimers were found in natural sources, e.g., japindine isolated from the root bark of *Chonemorpha macrophylla* [5], crellastatins from marine sponge *Crella* sp. [6], cephalostatins from tiny marine worm *Cephalodiscus gilchristi* [7], and ritterazines from tunicate *Ritterella tokioka* (Figure 1) [8]. The highly cytotoxic cephalostatins and ritterazines inspired chemists to attempt their chemical synthesis [9–11]. The synthesis of these dimeric steroidal pyrazine alkaloids and their analogs was particularly important because the natural sources of these compounds are extremely limited. These intensive studies were covered by several review articles [12–15]. Oligomeric steroids with or without spacer groups can be used as chiral building blocks to construct artificial receptors and as architectural components in biomimetic/molecular recognition chemistry. Davis has briefly reviewed his work in this area, directed toward the construction of enzyme mimics [16]. The study on dimerization and oligomerization of the cholic acid skeleton as an architectural building block was particularly intensive [17–19]. Various methods of synthesis of steroid dimers with linear [20] or cyclic [21] structures, e.g., Sonogashira coupling, Yamaguchi esterification, Wurtz reaction, ring closing metathesis, etc., were employed [22]. In the last decades, the interesting chemical, biological, and physical properties of the growing number of steroid dimers synthesized or isolated from living organisms have triggered increased activity in this field. The large number of steroid dimers described so far include compounds that have shown properties as catalysts [23], artificial receptors [24], molecular umbrellas [25], as well as activity as sulfatase inhibitors [26], antimalarials [27], and cytotoxic and antiproliferative agents [28,29], amongst others. Recent solid-state studies revealed that several crystalline steroid dimers act as molecular rotors [30].

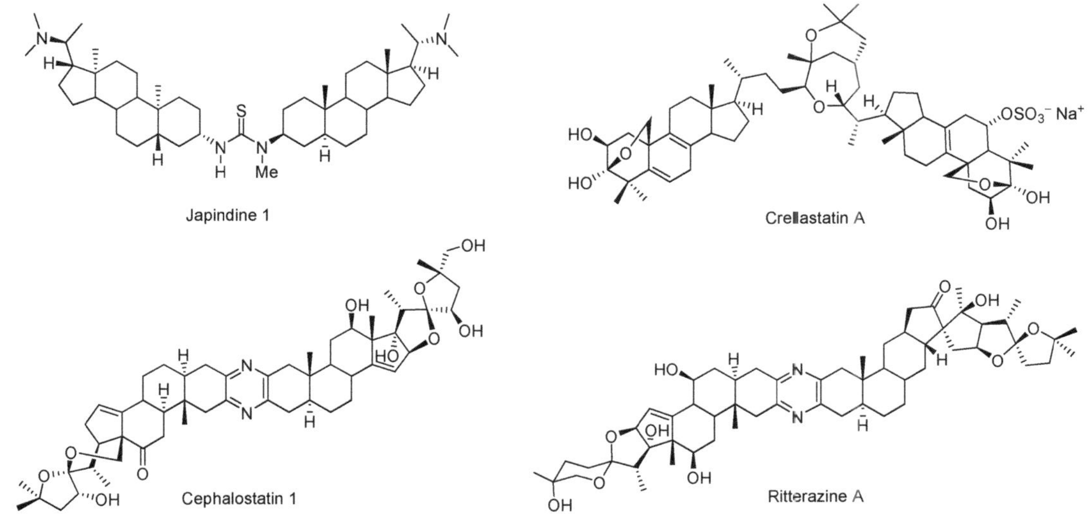

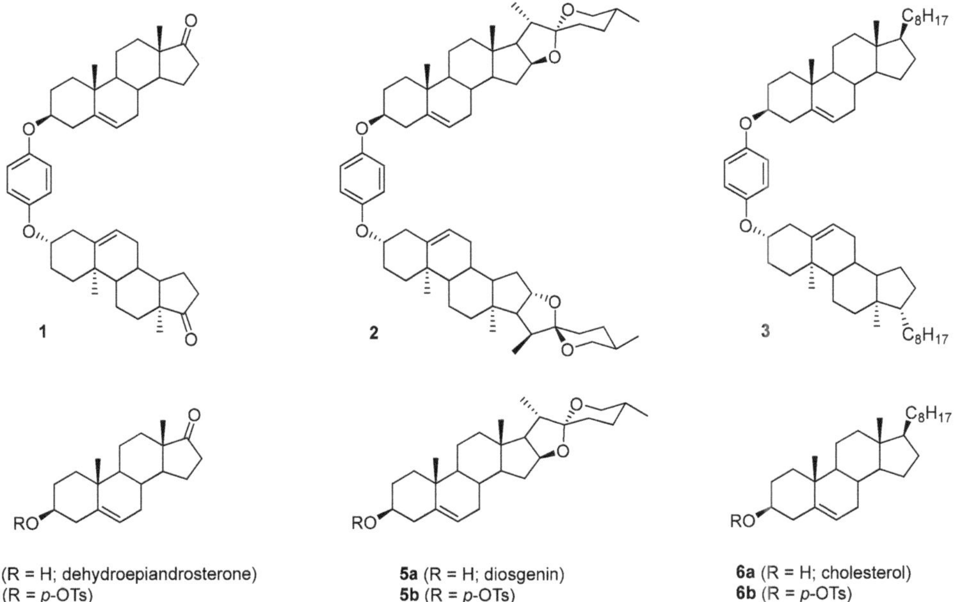

Figure 1. Naturally occurring steroid dimers.

4a (R = H; dehydroepiandrosterone)
4b (R = *p*-OTs)

5a (R = H; diosgenin)
5b (R = *p*-OTs)

6a (R = H; cholesterol)
6b (R = *p*-OTs)

Figure 2. Target steroid dimers (**1–3**) and starting materials (**4–6**).

The aim of this research was the synthesis of steroid dimers, in which steroid units are connected by an *O*,*O*-1,4-phenylene linker (Figure 2). These compounds (**1**−**3**) were selected as potential molecular rotors. Molecular motors are an important class of nanomolar-sized molecular machines that use different energy sources to generate unidirectional mechanical motion. In the crystal lattice, they contain both a static part and a fragment capable of free rotation [31]. Compound **1** has a carbonyl group at C17, which allows modification of the structure by functional group manipulation. Compounds **1** and **2** have not been described before, while compound **3** was previously obtained in low yield from cholesterol hydroquinone ether as an undesired product of an electrochemical reaction [32]. In the present study, a chemical procedure was elaborated for the efficient synthesis of dimers **1–3** from readily available sterols, androst-5-en-3β-ol-17-one **4a** (dehydroepiandrosterone; DHEA), (25*R*)-spirost-5-en-3β-ol **5a** (diosgenin), and cholesterol **6a**, respectively.

2. Results and Discussion

The synthetic plan of dimers **1–3** assumed the conversion of starting sterols **4a–6a** into the corresponding *p*-tosylates **4b–6b** and their reaction with hydroquinone. It is well known that upon leaving *p*-TsO⁻ the mesomerically stabilized homoallylic carbocation (Scheme 1) is generated, which can react with a nucleophilic reagent, e.g., hydroquinone, in the 3β position. For stereoelectronic reasons, only 3β-substituted steroid products can be formed. In fact, the nucleophile may also attack the 6β position, but the resulting product (so-called *i*-steroid) is thermodynamically less stable and is not formed as a product under acidic conditions. However, the reaction of androst-5-en-3β-ol-17-one tosylate **4b** (2 equiv.) with hydroquinone (1 equiv.) did not afford the expected dimer **1**. There was no reaction in boiling acetone (even in the presence of *p*-TsOH), while the analogous reaction carried out in dioxane (24 h at reflux) yielded mostly the nonpolar elimination products **7** and **8** (Table 1; run 1). The fast solvent−free reaction at 120 °C (run 2) gave a mixture of several products, which were carefully analyzed.

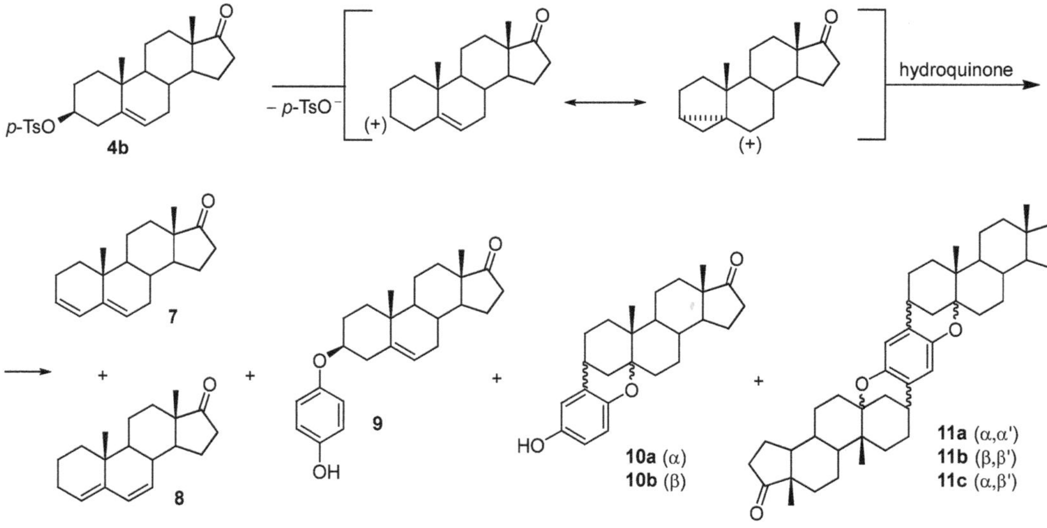

Scheme 1. The reactions of androst-5-en-3β-ol-17-one tosylate with hydroquinone.

In addition to a mixture of dienes **7** and **8**, androst-5-en-3β-ol-17-one hydroquinone ether **9** and unexpected products **10** and **11** as stereomeric mixtures were obtained. The compound **9** could be an intermediate in the synthesis of dimer **1**, but the desired compound **1** was not found among the reaction products. Instead, the different dimers were isolated, **11a–11c**, and their likely precursors, **10a** and **10b**. The similar reaction products, but in a different ratio (more elimination products), were formed in xylenes at reflux for 3 h (run 3). The structures of the products prove that both nucleophilic sites of hydroquinone, the oxygen atom and the carbon atom C2 of the aromatic ring, take part in the reaction. Interestingly, during the reaction, the hydroquinone carbon atom binds to the steroid C3 position, while the oxygen atom binds to C5. The direction of the reaction was deduced from the analysis of ^{1}H NMR spectra. The chemical shifts of protons at C3 were shifted upfield (compared to δ 4.34 in tosylate **4b**) and appeared at δ below 3 ppm. Therefore, the oxygen atom in this compound cannot be attached to C3. It seems that compounds **10b** and **11b** resulted from an attack of the carbon atom of hydroquinone on the mesomeric carbocation, which is formed upon the leaving of p-TsO$^-$ from **4b** (Scheme 2).

Table 1. The results of tosylate **4b** (2 equiv.) reaction with hydroquinone (1 equiv.) under different conditions.

Run No./ Reaction Conditions	Structure of Products				Conversion
	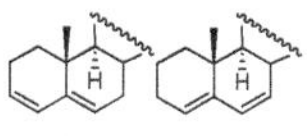 7	8	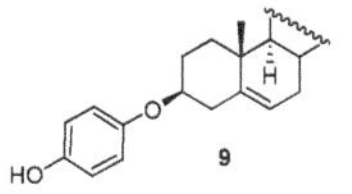 9	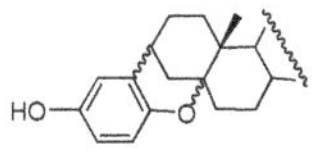 10a, 10b	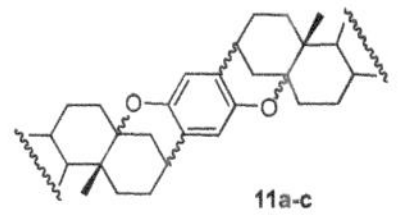 11a-c
1: Dioxane, reflux, 24 h	16%	<5%	-	-	30%
2: 120 °C, 5 min	23%	12%	10a: <5%, 10b: <5%	11a: 7%, 11b: 5%, 11c: 9%	100%
3: Xylenes, reflux, 3 h	48%	13%	10a: <1%, 10b: <1%	11a: <5%, 11b: 5%, 11c: 5%	100%
4: Ball mill, 48 h	7%	19%	10a: <1%, 10b: <1%	11a: <5%, 11b: <5%, 11c: 5%	50%
5: Acetone, MW, 110 °C, 6 h	47%	8%	10a: <1%, 10b: <1%	11a: <1%, 11b: <5%, 11c: 5%	90%
6: Dioxane, US, 45 °C, 12 h	5%	-	-	-	15%

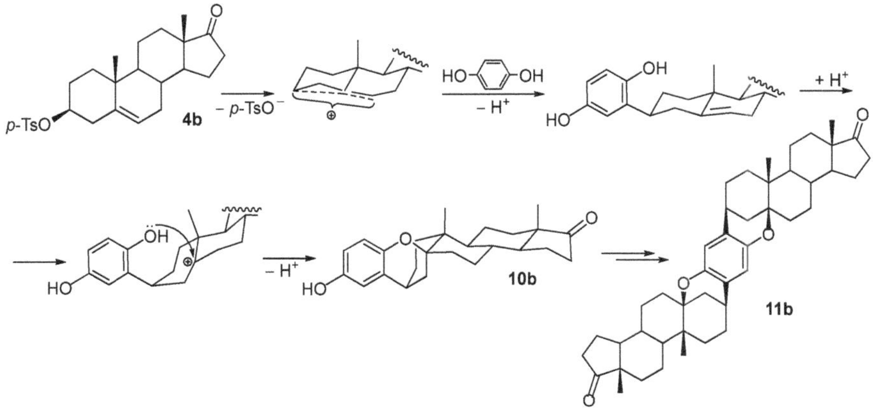

Scheme 2. A presumable mechanism of **11b** formation.

This attack is possible from the steroid β side only due to stereoelectronic reasons. In the next step, an intramolecular addition of phenol to the steroid C5–C6 double bond occurred, leading to a six-membered ring formation. Compound **10b** obtained this way may undergo the same reaction sequence to provide the dimer **11b**. The mechanism of the dimer **11a** formation (via **10a**) is less obvious (Scheme 3). Apparently, the attack of hydroquinone on steroid tosylate **4b** occurred from the less hindered α side in this case. However, the reaction timing is not clear. It could be either a nucleophilic substitution of 3-tosylate with the carbon C2 of hydroquinone followed by an intramolecular addition to the C5–C6 double bond or a concerted attack of hydroquinone on both 3-tosylate and the double bond.

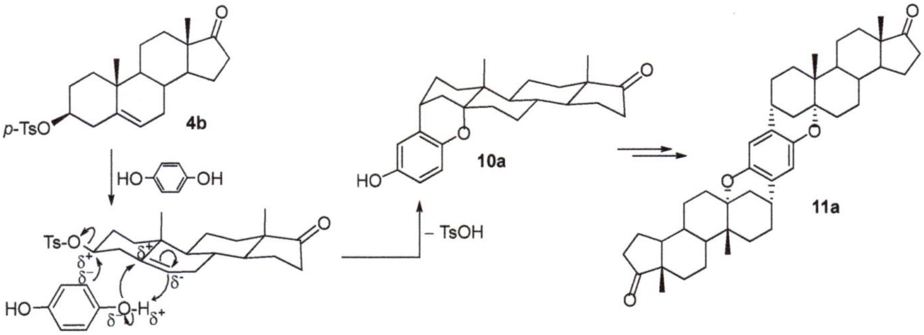

Scheme 3. A presumable mechanism of **11a** formation.

The dimers **11a**, **11b**, and **11c** were separated by column chromatography since they differ in polarity. The Rf values of dimers determined by their migration on TLC plates three times developed in the solvent system hexane—ethyl acetate (74:26) were 0.77, 0.66, and 0.71, respectively. The C2 symmetrical dimers **11a** and **11b** were isolated in similar amounts, while the mixed unsymmetrical dimer **11c** was obtained in larger quantities. Optimized structures of dimers **11a** and **11b** are presented in Figure 3. The ^{1}H NMR spectra of **11a** and **11b** were very similar, except for protons in the A-ring region. Particularly, the chemical shifts of protons at C3 and C3′ in these dimers proved different. The two-proton signal in dimer **11a** came out at δ 2.90 ppm, while an analogous two-proton signal in **11b** appeared at δ 2.85 ppm. Of course, the ^{1}H NMR spectrum of mixed dimer **11c**, which consists of two different steroid units, showed both proton signals (δ 2.85 and 2.90 ppm) as presented in Figure 4. The difference of 0.05 ppm between the chemical shifts of protons at the C3 and C3′ positions in dimers **11a** and **11b** is comparable to the literature values described for analogous A/B ring systems. For example, the 3β-H signal in 3α-hydroxy-5α-androstane-17-one is located at 4.05 ppm [33], while the analogous signal (3α-H) in 3β-hydroxy-5β-androstane-17-one shows up at 4.11 ppm [34]. In addition, the chemical shifts of the C4 protons in dimer **11a** appear at δ 1.40 and 2.08 ppm, compared to 1.43 and 2.48 ppm for the corresponding proton signals in compound **11b**. The signal of the 4α proton in **11b** is significantly (0.40 ppm) shifted downfield compared to the analogous signal in **11a**. A similar difference (Δδ 0.43 ppm) between the chemical shifts of protons at C4 was reported in the literature for 5α- and 5β-androstan-17-ones [35]. The full assignment of signals in ^{1}H and ^{13}C NMR spectra was carried out based on different NMR techniques (DEPT, COSY, HMQC, and HMBC) and is presented in Table S1 (Supplementary Material).

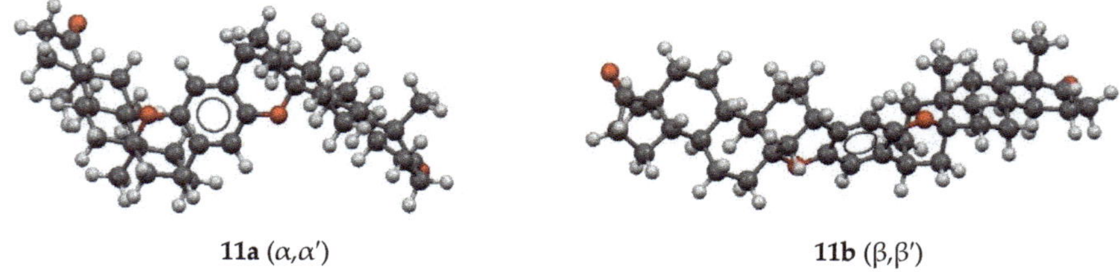

Figure 3. Ball and stick presentation of optimized structures of dimers **11a** and **11b**.

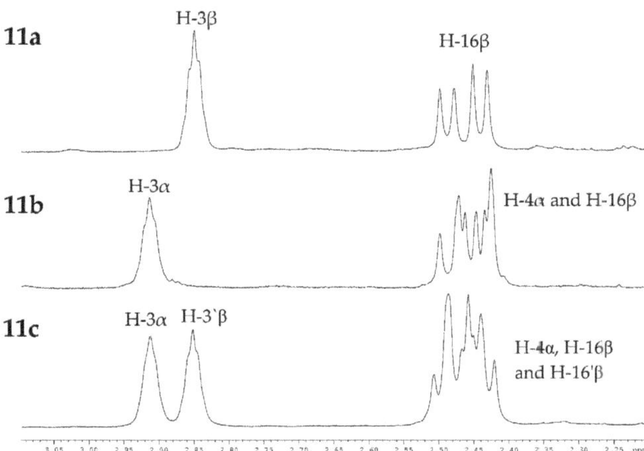

Figure 4. The diagnostic fragment of 1H NMR spectra of **11a**, **11b**, and **11c** dimers.

A series of experiments were performed using mechanochemistry (Table 1; run 4), microwave (run 5), or ultrasound irradiation (run 6), but none of them produced reasonable yields of tosylate **4b** with hydroquinone reaction products.

In the next series of experiments, montmorillonite was used as a catalyst. Montmorillonite is a very soft phyllosilicate group of minerals that form when they precipitate from water solution as microscopic crystals, known as clay. It is named after Montmorillon in France. Montmorillonite, a member of the smectite group, is a 2:1 clay, meaning that it has two tetrahedral sheets of silica sandwiching a central octahedral sheet of alumina. The acid-treated montmorillonite clays have been extensively used in various catalytic processes [36]. Other heterogeneous Brønsted and Lewis acid catalysts could be developed using the cation-exchange ability of the montmorillonite interlayer. The organic syntheses using the above acid catalysts have several advantages: ease of preparing the solid catalysts, high catalytic activities, wide applicability to large molecules, simple workup procedure, and reusability of the catalyst [37]. Montmorillonite was also used for the preparation of ethers from alcohols and phenols. A comprehensive review of the catalytic *O*-alkylation of phenol and hydroquinone has recently appeared [38]. Also, the preparation of cholesteryl ethers from cholesterol and various alcohols or phenols has been described [39].

Before we started studying reactions catalyzed with montmorillonite, the reaction of androst-5-en-3β-ol-17-one tosylate (**4b**) with hydroquinone in the presence of basic Al_2O_3 was carried out. Unfortunately, the reaction was dirty and led to *i*-steroid products (e.g., 3α,5α-cyclo-androst-6-en-17-one) predominately. Therefore, a conclusion has been drawn that basic conditions should be avoided, and further reactions were carried out with K10 (commercial montmorillonite) as a catalyst under different conditions (Table 2). For these reactions, a non-activated androst-5-en-3β-ol-17-one (**4a**) was used instead of tosylate **4b**. In addition to the previously described products, new dimeric products, i.e., disteroidal ether **12** and the desired hydroquinone disteroidal diether **1**, appeared among the reaction products. The synthesis of disteroidal ethers was previously described. They can be efficiently prepared from Δ^5-steroids by treating them with montmorillonite in a dichloromethane solution [40] or by an electrochemical method [41]. In both cases, a mesomeric homoallylic carbocation (Scheme 1) is formed as an intermediate, which finally reacts with the starting sterol.

Table 2. The results of androst-5-en-3β-ol-17-one **4a** reaction with hydroquinone catalyzed with montmorillonite K10 under different conditions.

Run No./ Reaction Conditions	Structure of Products 7, 8	9	11a-c	12	1	Conversion
1 [a]: K10 (120 °C; 0.5 g)	14%	17%	1%	26%	10%	80%
2 [a]: K10 (120 °C; 1 g)	19%	11%	1%	26%	9%	80%
3 [b]: K10 (120 °C; 0.5 g)	15%	13%	1%	27%	12%	95%
4 [b]: K10 (200 °C; 0.5 g)	36%	8%	<5%	12%	16%	80%
5 [b]: K10 (280 °C; 0.5 g)	28%	12%	<5%	7%	**23%**	90%
6 [b]: K10 (400 °C; 0.5 g)	19%	17%	1%	22%	6%	80%
7 [b]: H⁺-K10 (280 °C; 0.5 g)	26%	16%	<5%	12%	13%	95%
8 [c]: H⁺-K10 (280 °C; 0.5 g)	13%	7%	-	57%	1%	99%
9 [b]: Ti⁴⁺-K10 (280 °C; 0.5 g)	40%	18%	-	22%	1%	95%
10 [b]: Cu²⁺-K10 (120 °C; 0.5 g)	33%	12%	-	9%	12%	95%

Reaction conditions: [a] DHEA (100 mg; 0.35 mmol), hydroquinone (19 mg; 0.175 mmol), $CHCl_3$, reflux 4 h. [b] DHEA (100 mg; 0.35 mmol), hydroquinone (15.4 mg; 0.14 mmol), $CHCl_3$, reflux 4 h. [c] DHEA (100 mg; 0.35 mmol), hydroquinone (15.4 mg; 0.14 mmol), $CHCl_3$, 50 °C, 24 h.

The reaction was optimized against the quantity of catalyst used. With 500 mg of K10 activated at 120 °C, a 10% yield of dimer **1** was achieved (Table 2; run 1); higher amounts of catalyst deteriorated (run 2) the reaction result. The temperature of the montmorillonite activation is very important. Montmorillonite changes its properties during activation due to the desorption of water, dehydration, and alteration of crystalline structure. The activation at 120 °C causes water desorption, which exposes acid groups. This increases the number and strength of acid centers. A higher temperature of activation results in partial dehydration of Si-OH and Al-OH groups. Then the number of Brønsted acid sites decreases, while the number of Lewis acid (e.g., Al^{3+}) sites increases. The reactions of hydroquinone with 2.5 equiv. of DHEA (**4a**) in $CHCl_3$ at reflux catalyzed with K10 (500 mg) activated at different temperatures were carried out (runs 3–6). The best result was obtained with K10 activated at 280 °C (run 5). The desired dimer **1** was obtained in 23% under optimized conditions. Further attempts to increase the yield of dimer **1** failed. The experiments presented in Table 2 were conducted in 5 mL of $CHCl_3$ at reflux. The reaction carried out as in the case of run 5, but in a more diluted solution (10 mL of solvent), gave essentially the same result (23% of **1**). The reaction performed in 1,2-dichloroethane at reflux (83 °C) provided only 13% of **1**, while in other solvents at reflux (cyclohexane, toluene, xylenes, acetonitrile, acetone, THF, and dioxane), dimer **1** was not formed at all. Also, the attempts to carry out the reaction at room temperature were completely unsuccessful. In the next series of experiments, montmorillonite activated with hydrochloric acid was used. With this catalyst dried at 280 °C (H⁺-K10), up to a 57% yield of disteroidal ether **12** (Table 2; runs 7 and 8) could be achieved, but the yield of dimer **1** was negligible. It seems that increasing the number of Brønsted reactive sites deteriorates the yield of dimer **1**. Some experiments were carried out with montmorillonite activated with the ions Ti^{4+} or Cu^{2+}. However, this modification of the catalyst promoted only the elimination processes (runs 9 and 10).

The optimal reaction conditions (as in run 5) were applied to analogous reactions of other sterols—diosgenin **5a** and cholesterol **6a**. In the latter case, the yield of desired dimer **3** was slightly better (28%) than that achieved for DHEA (**4a**). The dimer **3** was accompanied by dicholesteryl ether (22%), hydroquinone mono cholesteryl ether (10%), and the elimination products (35%). The yield of diosgenin-derived dimer **2** was lower (12%) under the same conditions. The corresponding disteroidal ether was isolated in tiny amounts [42].

In the next experiments, androst-5-en-3β-ol-17-one tosylate (**4b**) and cholesteryl tosylate (**6b**) served as substrates. The advantage of using tosylates instead of free sterols is that disteroidal ethers cannot be formed under these conditions, which is important from the point of view of reaction product separation. Also, dimers **11a–c** (or analogous compounds derived from cholesterol) were formed only in negligible amounts. In addition, much milder reaction conditions were used, resulting in fewer by-products. The results of tosylate reactions are shown in Table 3. It should be noted that all reactions occurred at ambient temperature and required less catalyst. In the case of the androstane series (runs 1–3), the best yield of the desired dimer **1** (31%) was observed for the reaction catalyzed by Ti^{4+} activated montmorillonite K10 (run 3). Without Ti^{4+} activation, the yield of **1** was 23%, provided that K10 was calcined at 500 °C (run 2). The reactions of cholesteryl tosylate **6b** with hydroquinone (runs 4–6) worked much better and provided dimer **3** in excellent yields. Especially the catalyst dried at 500 °C (run 5) was found to be very active in promoting the reaction, even in low quantities (50 mg). The likely explanation for the high activity of montmorillonite K10 against cholesterol **6a** and its tosylate **6b** is the lack of additional groups in these substrates capable of binding to the active catalyst sites.

Table 3. Reactions of tosylates **4b** and **6b** with hydroquinone catalyzed with montmorillonite K10 under different conditions.

Run No./ Reaction Conditions	Structure of Products				Conversion
1 [a]: **4b**, K10 (280 °C; 300 mg), r.t., 4 days	22%	19%	1%	15%	80%
2 [a]: **4b**, K10 (500 °C; 300 mg), r.t., 3 days	21%	14%	1%	23%	100%
3 [a]: **4b**, Ti^{4+}-K10 (280 °C; 300 mg), r.t., 3 days	28%	19%	-	31%	100%
4 [b]: **6b**, K10 (280 °C; 50 mg), r.t., 5 days	17%	-	-	32%	70%
5 [b]: **6b**, K10 (500 °C; 50 mg), r.t., 3 days	10%	-	-	67%	100%
6 [b]: **6b**, Ti^{4+}-K10 (500 °C; 50 mg), r.t., 3 days	23%	-	-	42%	80%

Reaction conditions: [a] DHEA tosylate (88.5 mg; 0.2 mmol), hydroquinone (8.4 mg; 0.076 mmol), $CHCl_3$ (5 mL). [b] cholesteryl tosylate (100 mg; 0.2 mmol), hydroquinone (8.4 mg; 0.076 mmol), $CHCl_3$ (5 mL).

3. Materials and Methods

3.1. General Experimental Data

All solvents were freshly distilled prior to use. Anhydrous solvents were prepared by distillation over appropriate drying agents under an argon atmosphere. The stabilizer (ethanol) contained in commercially available chloroform was removed before use. The reactions were monitored by TLC on silica gel plates 60 F254, and spots were visualized either by a UV hand lamp or by charring with molybdophosphoric acid/cerium(IV) sulfate in H_2SO_4. The reaction products were isolated by chromatographic methods using silica gel pore size 40 Å (70–230 mesh). ^{1}H and ^{13}C NMR (400 and 100 MHz, respectively) spectra of all compounds were recorded using a Bruker Avance II spectrometer in a CDCl$_3$ or CDCl$_3$/MeOD mixture and referenced to TMS (0.0 ppm) and CDCl$_3$ (77.0 ppm), respectively. Infrared spectra were recorded using Attenuated Total Reflectance (ATR) as solid samples with a Nicolet 6700 FT-IR spectrometer. Mass spectra were obtained with an Accurate-Mass Q-TOFLC/MS 6530 spectrometer with electrospray ionization (ESI). Melting points were determined on a Kofler bench melting point apparatus.

Tosylates **4b** [43] and **6b** [44] were prepared according to literature procedures.

Unmodified montmorillonite K10 or metal cation-exchanged montmorillonites: H^+-K10, Ti^{4+}-K10, and Cu^{2+}-K10 were used as catalysts. Modified forms: H^+-K10 [45], Ti^{4+}-K10 [46], and Cu^{2+}-K10 [47] were obtained according to the literature procedures. All catalysts were activated prior to use by heating at high temperatures (unmodified montmorillonite K10 at 120 °C, 200 °C, 280 °C, 400 °C, or 500 °C; H^+-K10 at 280 °C; Ti^{4+}-K10 at 280 °C or 500 °C; and Cu^{2+}-K10 at 120 °C). Characterization data and NMR spectra are presented in the supporting information.

3.2. General Experimental Procedure for the Formation of All New Compounds

3.2.1. Solvent-Free Reaction of Tosylate **4b** with Hydroquinone (Table 1, Run No. 2)

The mixture of tosylate **4b** (100 mg. 0.2 mmol) and hydroquinone (11 mg. 0.1 mmol) was stirred and heated to 120 °C for 5 min under an argon atmosphere. Then, the reaction mixture was cooled to room temperature and subjected to silica gel column chromatography, which resulted in the separation of compounds: **7** and **8** (eluted with hexane in 23% yield), **11a** (eluted with hexane/ethyl acetate 97:3 mixture in 7% yield), **11c** (eluted with hexane/ethyl acetate 95:5 mixture in 9% yield), **10a** (eluted with hexane/ethyl acetate 93:7 mixture in <5% yield), **11b** (eluted with hexane/ethyl acetate 93:7 mixture in 5% yield), **9** (eluted with hexane/ethyl acetate 87:13 mixture in 12% yield), and **10b** (eluted with hexane/ethyl acetate 87:13 mixture in <5% yield).

Compound 7: white solid (hexane/CH$_2$Cl$_2$); mp 85–87 °C; Rf = 0.32 (hexane/ethyl acetate 9:1); IR (ATR) v_{max} 2912, 2856, 1739 cm^{-1}; ^{1}H NMR (CDCl$_3$, 400 MHz) δ 5.95 (1H, d, J = 9.9 Hz, H-4), 5.63 (1H, m, H-3), 5.42 (1H, m, H-6), 2.48 (1H, dd, J = 19.2 Hz, J = 8.8 Hz, H-16β), 0.99 (3H, s, H-19), 0.93 (3H, s, H-18); ^{13}C NMR (CDCl$_3$, 100 MHz) δ 221.1 (C), 141.6 (C), 128.7 (CH), 125.4 (CH), 122.1 (CH), 51.9 (CH), 48.5 (CH), 47.7 (C), 35.8 (CH$_2$), 35.3 (C), 33.7 (CH$_2$), 31.44 (CH$_2$), 31.41 (CH), 30.6 (CH$_2$), 23.0 (CH$_2$), 21.8 (CH$_2$), 20.3 (CH$_2$), 18.8 (CH$_3$), 13.7 (CH$_3$); HRMS *m/z* 271.2054 (calcd for $C_{19}H_{27}O^+$, 271.2056).

Compound 8: white solid; Rf = 0.36 (hexane/ethyl acetate 9:1); IR (ATR) v_{max} 2916, 1736, 1506 cm^{-1}; ^{1}H NMR (CDCl$_3$, 400 MHz) δ 5.99 (1H, dd, J = 9.8 Hz, J = 2.5 Hz, H-4), 5.56 (1H, d, J = 9.8 Hz, H-7), 5.47 (1H, m, H-3), 0.98 (3H, s, H-19), 0.96 (3H, s, H-18); ^{13}C NMR (CDCl$_3$, 100 MHz) δ 220.7 (C), 142.2 (C), 129.9 (CH), 125.6 (CH), 124.7 (CH), 51.6 (CH), 49.6 (CH), 48.3 (C), 36.5 (CH), 35.8 (CH$_2$), 34.8 (C), 34.5 (CH$_2$), 31.6 (CH$_2$), 25.3 (CH$_2$), 21.5 (CH$_2$), 20.2 (CH$_2$), 18.4 (CH$_2$), 18.3 (CH$_3$), 13.8 (CH$_3$); HRMS *m/z* 271.2061 (calcd for $C_{19}H_{27}O^+$, 271.2056).

Compound 10a: white solid; Rf = 0.66 (3 × hexane/ethyl acetate 74:26); IR (ATR) v_{max} 3325, 2927, 1737, 1241, 1206, 1190, 1153, 1077, 1053, 813, 785 cm^{-1}; ^{1}H NMR (CDCl$_3$, 400 MHz) δ 6.67 (1H, d, J = 8.6 Hz, H-Ar), 6.60 (1H, dd, J = 8.6 Hz, J = 3.0 Hz, H-Ar), 6.48 (1H, d, J = 3.0 Hz, H-Ar), 4.26 (1H, bs, -OH), 2.87 (1H, m, H-3β), 2.46 (1H, dd, J = 18.9 Hz, J = 8.3 Hz, H-16β), 1.05 (3H, s, H-19), 0.88 (3H, s, H-18); ^{13}C NMR (CDCl$_3$, 100 MHz) δ 221.4 (C), 150.1 (C), 147.9 (C), 128.0 (C), 115.3 (CH), 114.2 (CH), 113.9 (CH), 78.5 (C), 51.2 (CH), 47.8 (C), 46.2 (CH), 42.1 (C), 35.9 (CH$_2$), 34.4 (CH), 33.6 (CH$_2$), 32.7 (CH), 31.8 (CH$_2$), 31.5 (CH$_2$), 29.41 (CH$_2$), 29.36 (CH$_2$), 24.8 (CH$_2$), 21.7 (CH$_2$), 20.0 (CH$_2$), 16.2 (CH$_3$), 13.8 (CH$_3$); HRMS *m/z* 381.2428 (calcd for $C_{25}H_{33}O_3^+$, 381.2424).

Compound 11a: white solid; Rf = 0.77 (3 × hexane/ethyl acetate 74:26); IR (ATR) v_{max} 2915, 2853, 1737, 1242, 1205, 1151, 1001, 813, 785 cm^{-1}; ^{1}H NMR (CDCl$_3$, 400 MHz) δ 6.41

(2H, s, H-Ar), 2.85 (2H, m, H-3β), 2.46 (2H, dd, J = 18.9 Hz, J = 8.2 Hz, H-16β), 1.04 (6H, s, H-19), 0.88 (6H, s, H-18); ^{13}C NMR (CDCl$_3$, 100 MHz) δ 221.5 (2 × C), 148.4 (2 × C), 126.1 (2 × C), 112.8 (2 × CH), 78.1 (2 × C), 51.2 (2 × CH), 47.8 (2 × C), 46.1 (2 × CH), 42.1 (2 × C), 35.9 (2 × CH$_2$), 34.4 (2 × CH), 33.7 (2 × CH$_2$), 32.6 (2 × CH), 32.2 (2 × CH$_2$), 31.6 (2 × CH$_2$), 29.5 (2 × CH$_2$), 29.3 (2 × CH$_2$), 24.8 (2 × CH$_2$), 21.7 (2 × CH$_2$), 20.0 (2 × CH$_2$), 16.1 (2 × CH$_3$), 13.8 (2 × CH$_3$); HRMS *m/z* 651.4408 (calcd for C$_{44}$H$_{59}$O$_4$$^+$, 651.4408).

Compound 11b: white solid; Rf = 0.66 (3 × hexane/ethyl acetate 74:26); IR (ATR) ν$_{max}$ 2914, 2853, 1737, 1242, 1205, 813, 785 cm^{-1}; ^{1}H NMR (CDCl$_3$, 400 MHz) δ 6.43 (2H, s, H-Ar), 2.90 (2H, m, H-3α), 2.51–2.43 (4H, m, H-4α and H-16β), 1.03 (6H, s, H-19), 0.88 (6H, s, H-18); ^{13}C NMR (CDCl$_3$, 100 MHz) δ 221.0 (2 × C), 149.1 (2 × C), 125.4 (2 × C), 113.1 (2 × CH), 78.4 (2 × C), 51.6 (2 × CH), 47.9 (2 × C), 43.8 (2 × CH), 42.9 (2 × C), 35.9 (2 × CH$_2$), 34.6 (2 × CH), 34.4 (2 × CH$_2$), 32.6 (2 × CH), 31.6 (2 × CH$_2$), 30.8 (2 × CH$_2$), 28.7 (2 × CH$_2$), 27.9 (2 × CH$_2$), 27.1 (2 × CH$_2$), 21.8 (2 × CH$_2$), 20.6 (2 × CH$_2$), 17.6 (2 × CH$_3$), 13.8 (2 × CH$_3$); HRMS *m/z* 651.4401 (calcd for C$_{44}$H$_{59}$O$_4$$^+$, 651.4408).

Compound 11c: white solid; Rf = 0.71 (3 × hexane:ethyl acetate 74/26); IR (ATR) ν$_{max}$ 2929, 2870, 1737, 1241, 1194, 1152, 785 cm^{-1}; ^{1}H NMR (CDCl$_3$, 400 MHz) δ 6.43 (2H, s, H-Ar), 2.91 (1H, m, H-3α), 2.85 (1H, m, H-3β), 1.03 (3H, s, H-19), 1.01 (3H, s, H-19′), 0.88 (s, 6H, H-18 and H-18′); ^{13}C NMR (CDCl$_3$, 100 MHz) δ 221.4 (C), 220.9 (C), 149.0 (C), 148.7 (C), 126.1 (C), 125.2 (C), 113.0 (CH), 112.7 (CH), 78.3 (C), 78.1 (C), 51.6 (CH), 51.2 (CH), 47.8 (2 × C), 46.1 (CH), 43.9 (CH), 42.9 (C), 42.1 (C), 35.9 (2 × CH$_2$), 34.5 (CH), 34.4 (CH), 34.4 (CH$_2$), 33.6 (CH$_2$), 32.5 (2 × CH), 32.0 (CH$_2$), 31.6 (CH$_2$), 31.5 (CH$_2$), 30.7 (CH$_2$), 29.5 (CH$_2$), 29.4 (CH$_2$), 28.8 (CH$_2$), 27.9 (CH$_2$), 27.0 (CH$_2$), 24.8 (CH$_2$), 21.8 (CH$_2$), 21.7 (CH$_2$), 20.6 (CH$_2$), 20.0 (CH$_2$), 17.6 (CH$_3$), 16.2 (CH$_3$), 13.8 (2 × CH$_3$); HRMS *m/z* 651.4403 (calcd for C$_{44}$H$_{59}$O$_4$$^+$, 651.4408).

3.2.2. Optimal Procedure for Preparation of Dimer 1 in the Montmorillonite K10 Catalyzed Reaction of Androst-5-en-3β-ol-17-one (4a) with Hydroquinone (Table 2, Run No. 5)

A stirred mixture of androst-5-en-3β-ol-17-one (**4a**) (100 mg, 0.35 mmol), hydroquinone (15.4 mg, 0.14 mmol), and unmodified montmorillonite K10 activated at 280 °C (500 mg) in dry chloroform (5 mL) was gently refluxed under argon, and the reaction progress was monitored by TLC. After completion of the reaction (4 h), the suspension was filtered through a sintered glass funnel, and the precipitate was washed with a methanol/chloroform 2:8 mixture (3 × 50 mL). The filtrate was evaporated in a vacuum. The residue was subjected to column chromatography on silica gel, which resulted in the separation of compounds **7** and **8** (eluted with hexane in 28% yield), **12** (eluted with hexane/ethyl acetate 95:5 mixture in 7% yield), **1** (eluted with hexane/ethyl acetate 94:6 mixture in 23% yield), **9** (eluted with hexane/ethyl acetate 87:13 mixture in 12% yield), and small amounts of compounds **11a**, **11b**, and **11c** (total <5%).

Compound 1: colorless crystals (CH$_2$Cl$_2$/ethyl acetate); mp 252–254 °C; Rf = 0.50 (3 × benzene/ethyl acetate 94:6); IR (ATR) ν$_{max}$ 2938, 2907, 1744, 1731, 1501, 1214, 1043, 1029, 815 cm^{-1}; ^{1}H NMR (CDCl$_3$, 400 MHz) δ 6.82 (4H, s, H-Ar), 5.42 (2H, m, H-6), 4.00 (2H, m, H-3α), 1.09 (6H, s, H-19), 0.91 (6H, s, H-18); ^{13}C NMR (CDCl$_3$, 100 MHz) δ 221.0 (2 × C), 151.8 (2 × C), 140.7 (2 × C), 121.4 (2 × CH), 117.4 (4 × CH), 77.9 (2 × CH), 51.8 (2 × CH), 50.3 (2 × CH), 47.5 (2 × C), 38.8 (2 × CH$_2$), 37.1 (2 × CH$_2$), 37.0 (2 × C), 35.8 (2 × CH$_2$), 31.5 (2 × CH), 31.4 (2 × CH$_2$), 30.8 (2 × CH$_2$), 28.3 (2 × CH$_2$), 21.9 (2 × CH$_2$), 20.4 (2 × CH$_2$), 19.4 (2 × CH$_3$), 13.5 (2 × CH$_3$); HRMS *m/z* 651.4409 (calcd for C$_{44}$H$_{59}$O$_4$$^+$, 651.4408).

Compound 9: colorless crystals (CH$_2$Cl$_2$/ethyl acetate); mp 277–279 °C; Rf = 0.48 (3 × hexane/ethyl acetate 74:26); IR (ATR) ν$_{max}$ 3311, 2948, 2864, 1710, 1505, 1211, 1028, 819 cm^{-1}; ^{1}H NMR (CDCl$_3$/MeOD, 400 MHz) δ 6.76 (2H, d, J = 19.2 Hz, H-Ar), 6.71 (2H, d, J = 19.1 Hz, H-Ar), 5.37 (1H, m, H-6), 3.93 (1H, m, H-3α), 1.05 (3H, s, H-19), 0.87 (3H, s, H-18); ^{13}C NMR (CDCl$_3$/MeOD, 100 MHz) δ 221.9 (C), 150.8 (C), 150.7 (C), 140.7 (C), 121.3 (CH), 117.9 (2 × CH), 115.8 (2 × CH), 78.3 (CH), 51.7 (CH), 50.2 (CH), 47.6 (C), 38.8 (CH$_2$),

37.0 (CH_2), 36.9 (C), 35.8 (CH_2), 31.4 (CH), 31.3 (CH_2), 30.7 (CH_2), 28.2 (CH_2), 21.8 (CH_2), 20.3 (CH_2), 19.4 (CH_3), 13.5 (CH_3); HRMS *m/z* 381.2438 (calcd for $C_{25}H_{33}O_3{}^+$, 381.2424).

Compound 12: colorless crystals (hexane/CH_2Cl_2); mp 268–269 °C; Rf = 0.45 (3 × benzene/ethyl acetate 94:6); IR (ATR) ν_{max} 2931, 2895, 1731, 1094, 1058, 1005 cm^{-1}; ^{1}H NMR ($CDCl_3$, 400 MHz) δ 5.38 (2H, m, H-6), 3.30 (2H, m, H-3α), 2.47 (2H, dd, *J* = 19.2 Hz, *J* = 8.6 Hz, H-16β), 1.04 (6H, s, H-19), 0.89 (6H, s, H-18); ^{13}C NMR ($CDCl_3$, 100 MHz) δ 221.2 (2 × C), 141.5 (2 × C), 120.6 (2 × CH), 76.2 (2 × CH), 51.8 (2 × CH), 50.3 (2 × CH), 47.5 (2 × C), 40.0 (2 × CH_2), 37.3 (2 × CH_2), 37.0 (2 × C), 35.8 (2 × CH_2), 31.5 (2 × CH), 31.4 (2 × CH_2), 30.8 (2 × CH_2), 29.3 (2 × CH_2), 21.9 (2 × CH_2), 20.3 (2 × CH_2), 19.4 (2 × CH_3), 13.5 (2 × CH_3).

Analogously, steroid dimers **2** (12%) and **3** (28%) were obtained from diosgenin (**5a**) and cholesterol (**6a**), respectively, according to the procedure described above for androst-5-en-3β-ol-17-one (**4a**). The elimination products (~35% in both cases), disteroidal ethers (12% and 22%, respectively), and hydroquinone mono steroidal ethers (12% and 10%, respectively) were also formed.

Compound 2: colorless crystals (hexane/CH_2Cl_2); mp 307–309 °C; Rf = 0.38 (3 × benzene/ethyl acetate 94:6); IR (ATR) ν_{max} 2925, 1502, 1225, 1050, 1016, 809 cm^{-1}; ^{1}H NMR ($CDCl_3$, 400 MHz) δ 6.82 (4H, s, H-Ar), 5.38 (2H, m, H-6), 4.43 (2H, m, H-16), 3.98 (2H, m, H-3α), 3.49 (2H, m, H-26β), 3.39 (2H, t, *J* = 10.9 Hz, H-26α), 1.08 (6H, s, H-19), 0.99 (6H, d, *J* = 6.9 Hz, H-21), 0.81 (6H, s, H-18), 0.80 (6H, d, *J* = 5.0 Hz, H-27); ^{13}C NMR ($CDCl_3$, 100 MHz) δ 151.8 (2 × C), 140.5 (2 × C), 121.9 (2 × CH), 117.4 (4 × CH), 109.3 (2 × C), 80.8 (2 × CH), 78.0 (2 × CH), 66.8 (2 × CH_2), 62.1 (2 × CH), 56.5 (2 × CH), 50.1 (2 × CH), 41.6 (2 × CH), 40.3 (2 × C), 39.8 (2 × CH_2), 38.8 (2 × CH_2), 37.2 (2 × CH_2), 37.0 (2 × C), 32.1 (2 × CH_2), 31.8 (2 × CH_2), 31.43 (2 × CH), 31.38 (2 × CH_2), 30.3 (2 × CH), 28.8 (2 × CH_2), 28.3 (2 × CH_2), 20.8 (2 × CH_2), 19.4 (2 × CH_3), 17.1 (2 × CH_3), 16.3 (2 × CH_3), 14.5 (2 × CH_3); HRMS *m/z* 903.6502 (calcd for $C_{60}H_{87}O_6{}^+$, 903.6497).

Diosgenin derived hydroquinone mono steroidal ether: colorless crystals (hexane/CH_2Cl_2); mp 165–166 °C; Rf = 0.35 (hexane/ethyl acetate 88:12); IR (ATR) ν_{max} 3309, 2929, 1507, 1210, 1046, 1012, 830 cm^{-1}; ^{1}H NMR ($CDCl_3$, 400 MHz) δ 6.80 (2H, d, *J* = 9.0 Hz, H-Ar), 6.75 (2H, d, *J* = 9.0 Hz, H-Ar), 5.38 (1H, m, H-6), 4.76 (1H, bs, -OH), 4.43 (1H, m, H-16), 3.97 (1H, m, H-3α), 3.49 (1H, m, H-26β), 3.39 (1H, t, *J* = 10.9 Hz, H-26α), 1.07 (3H, s, H-19), 0.99 (3H, d, *J* = 6.9 Hz, H-21), 0.81 (3H, s, H-18), 0.80 (3H, d, *J* = 6.4 Hz, H-27); ^{13}C NMR ($CDCl_3$, 100 MHz) δ 151.6 (C), 149.8 (C), 140.5 (C), 121.9 (CH), 117.8 (CH), 116.0 (CH), 109.4 (C), 80.8 (CH), 78.3 (CH), 66.9 (CH_2), 62.1 (CH), 56.5 (CH), 50.1 (CH), 41.6 (CH), 40.3 (C), 39.8 (CH_2), 38.8 (CH_2), 37.1 (CH_2), 37.0 (C), 32.1 (CH_2), 31.8 (CH_2), 31.42 (CH), 31.37 (CH_2), 30.3 (CH), 28.8 (CH_2), 28.3 (CH_2), 20.9 (CH_2), 19.4 (CH_3), 17.1 (CH_3), 16.3 (CH_3), 14.5 (CH_3); HRMS *m/z* 507.3476 (calcd for $C_{33}H_{47}O_4{}^+$, 507.3469).

3.2.3. Optimal Procedure for Preparation of Dimer **12** in the Montmorillonite K10 Catalyzed Reaction of Androst-5-en-3β-ol-17-one (**4a**) (Table 2, Run No. 8)

A stirred mixture of androst-5-en-3β-ol-17-one (**4a**) (100 mg, 0.35 mmol), hydroquinone (15.4 mg, 0.14 mmol), and modified montmorillonite H^+-K10 activated at 280 °C (500 mg) in dry chloroform (5 mL) was heated to 50 °C for 24 h under argon, and the reaction progress was monitored by TLC. After completion of the reaction, the suspension was filtered through a sintered glass funnel, and the precipitate was washed with a methanol/chloroform 2:8 mixture (3 × 50 mL). The filtrate was evaporated in a vacuum. The residue was subjected to silica gel column chromatography, which resulted in separation of compounds **7** and **8** (eluted with hexane in 13% yield), **12** (eluted with hexane/ethyl acetate 95:5 mixture in 57% yield), **1** (eluted with hexane/ethyl acetate 94:6 mixture in 1% yield), and **9** (eluted with hexane/ethyl acetate 87:13 mixture in 7% yield).

3.2.4. Optimal Procedure for Preparation of Dimer **1** in the Montmorillonite K10 Catalyzed Reaction of Tosylate **4b** with Hydroquinone (Table 3, Run No. 3)

A mixture of tosylate **4b** (88.5 mg; 0.2 mmol), hydroquinone (8.4 mg; 0.076 mmol), and modified montmorillonite Ti^{4+}-K10 activated at 280 °C (300 mg) in dry chloroform (5 mL) was stirred under argon, and the reaction progress was monitored by TLC. After completion of the reaction (3 days), the suspension was filtered through a sintered glass funnel, and the precipitate was washed with a methanol/chloroform 2:8 mixture (3 × 50 mL). The filtrate was evaporated in a vacuum. The residue was subjected to column chromatography on silica gel, which resulted in the separation of compounds **7** and **8** (eluted with hexane in 28% yield), **1** (eluted with hexane/ethyl acetate 94:6 mixture in 31% yield), and **9** (eluted with hexane/ethyl acetate 87:13 mixture in 19% yield).

The reaction of tosylate **6b** with hydroquinone was carried out according to the procedure described above for tosylate **4b,** but using montmorillonite K10 activated at 500 °C (50 mg) as a catalyst (Table 3, Run No. 5). Dimer **3** was obtained with a 67% yield. Small amounts of elimination products (10%) were also formed.

4. Conclusions

The reactions of sterols and their tosylates with hydroquinone were studied. They provided different steroid dimers depending on the reaction conditions. The solvolytic reactions of DHEA tosylate afforded the elimination products in addition to three stereoisomeric steroid dimers **11a–11c**, which resulted from the nucleophilic attack of the C2 carbon atom of hydroquinone on the C3 position of the steroid, followed by an intramolecular addition to the C5–C6 double bond. The cause of reactions catalyzed by montmorillonite was different. The major reaction products were the steroid dimers with a 3,3′-*O,O*-1,4-phenylene linker **1–3** and the disteroidal ethers **12**. The formation of the latter compounds was suppressed by using sterol tosylates for the montmorillonite-catalyzed reactions with hydroquinone. As a result, an excellent yield (67%) of hydroquinone dicholesteryl diether (dimer **3**) was achieved.

Supplementary Materials: The following supporting information can be downloaded at: https://www. mdpi.com/article/10.3390/molecules28207068/s1, Table S1: ^{1}H NMR and ^{13}C NMR signal assignments for dimers **11a** and **11b**, ^{1}H and ^{13}C NMR spectra of all synthesized compounds.

Author Contributions: Conceptualization, A.M.T. and J.W.M.; methodology, A.M.T. and J.W.M.; validation, J.W.M.; formal analysis, A.M.T. and L.S.; investigation, A.D.M., A.M.T. and L.S.; resources, J.W.M.; data curation, A.M.T. and A.D.M.; writing—original draft preparation, J.W.M.; writing—review and editing, J.W.M. and A.M.T.; visualization, A.M.T.; supervision, J.W.M.; project administration, A.M.T.; funding acquisition, J.W.M. All authors have read and agreed to the published version of the manuscript.

Funding: This research received no external funding.

Data Availability Statement: All generated data will be available from the authors upon request.

Acknowledgments: The authors thank the University of Bialystok for its continuous support of our research programs. The authors are grateful to Jadwiga Maj for her skillful technical assistance.

Conflicts of Interest: The authors declare no conflict of interest.

Sample Availability: Not available.

References

1. Li, Y.; Dias, J.R. Dimeric and oligomeric steroids. *Chem. Rev.* **1997**, *97*, 283–304. [CrossRef] [PubMed]
2. Nahar, L.; Sarker, S.D. *Steroid Dimers—Chemistry and Applications in Drug Design and Delivery*; John Wiley & Sons: Chichester, UK, 2012; ISBN 978-1-119-97284-6.
3. Nahar, L.; Sarker, S.D.; Turner, A.B. A review on synthetic and natural steroid dimers: 1997–2006. *Curr. Med. Chem.* **2007**, *14*, 1349–1370. [CrossRef] [PubMed]
4. Nahar, L.; Sarker, S.D. A review on steroid dimers: 2011–2019. *Steroids* **2020**, *164*, 108736. [CrossRef] [PubMed]

5.	Banerji, J.; Chatterjee, A.; Itoh, Y.; Kikuchi, T. New steroid alkaloids from *Chonemorpha macrophylla* G. don (Chonemorpha fragrans Moon Alston). *Indian J. Chem.* **1973**, *11*, 1056–1057, ISSN/ISBN 0019-5103.
6.	D'Auria, M.V.; Giannini, C.; Zampella, A.; Minale, L.; Debitus, C.; Roussakis, C. Crellastatin A: A Cytotoxic Bis-Steroid Sulfate from the Vanuatu Marine Sponge *Crella* sp. *J. Org. Chem.* **1998**, *63*, 7382–7388. [CrossRef]
7.	Pettit, G.R.; Kamano, Y.; Dufresne, G.; Inoue, M.; Christie, N.; Schmidt, K.M.; Doubek, D.L. Isolation and structure of the unusual Indian Ocean *Cephalodiscus gilchristi* components, cephalostatins 5 and 6. *Can. J. Chem.* **1989**, *67*, 1509–1513. [CrossRef]
8.	Fuzukawa, S.; Matsunaga, S.; Fusetani, N. Ritterazine A, a highly cytotoxic dimeric steroidal alkaloid, from the tunicate Ritterella tokioka. *J. Org. Chem.* **1994**, *59*, 6164–6166. [CrossRef]
9.	LaCour, T.G.; Guo, C.; Bhandaru, S.; Boyd, M.R.; Fuchs, P.L. Interphylal product splicing: The first total syntheses of cephalostatin 1, the North hemisphere of ritterazine G, and the highly active hybrid analogue, ritterostatin $G_N 1_N$[1]. *J. Am. Chem. Soc.* **1998**, *120*, 692–697. [CrossRef]
10.	Shi, Y.; Jia, L.; Xiao, Q.; Lan, Q.; Tang, X.; Wang, D.; Li, M.; Ji, Y.; Zhou, T.; Tian, W. A Practical Synthesis of Cephalostatin 1. *Chem. Asian J.* **2011**, *6*, 786–790. [CrossRef]
11.	Jautelat, R.; Müller-Fahrnow, A.; Winterfeldt, E. A novel oxidative cleavage of the steroidal skeleton. *Chem. Eur. J.* **1999**, *5*, 1226–1233. [CrossRef]
12.	Iglesias-Arteaga, M.A.; Morzycki, J.W. Cephalostatins and ritterazines. In *The Alkaloids, Chemistry and Biology*; Knölker, H.-J., Ed.; Academic Press: Cambridge, MA, USA, 2013; Volume 72, pp. 153–279. [CrossRef]
13.	Ganesan, A. The dimeric steroid-pyrazine marine alkaloids: Challenges for isolation, synthesis, and biological studies. *Angew. Chem. Int. Ed. Engl.* **1996**, *35*, 611–615. [CrossRef]
14.	Gryszkiewicz-Wojtkielewicz, A.; Jastrzębska, I.; Morzycki, J.W.; Romanowska, D.B. Approaches towards the synthesis of cephalostatins, ritterazines and saponins from Ornithogalum saundersiae—New natural products with cytostatic activity. *Curr. Org. Chem.* **2003**, *7*, 1257–1277. [CrossRef]
15.	Lee, S.; LaCour, T.G.; Fuchs, P.L. Chemistry of trisdecacyclic pyrazine antineoplastics: The cephalostatins and ritterazines. *Chem. Rev.* **2009**, *109*, 2275–2314. [CrossRef] [PubMed]
16.	Davis, A.P. Cholaphanes et al.; steroids as structural components in molecular engineering. *Chem. Soc. Rev.* **1993**, *22*, 243–253. [CrossRef]
17.	Bonar-Law, R.P.; Davis, A.P. Synthesis of steroidal cyclodimers from cholic acid; a molecular framework with potential for recognition and catalysis. *J. Chem. Soc. Chem. Commun.* **1989**, *15*, 1050–1052. [CrossRef]
18.	Joachimiak, R.; Paryzek, Z. Synthesis and alkaline metal ion binding ability of new steroid dimers derived from cholic and lithocholic acids. *J. Inclusion Phenom. Macrocycl. Chem.* **2004**, *49*, 127–132. [CrossRef]
19.	Virtanen, E.; Kolehmainen, E. Use of bile acids in pharmacological and supramolecular applications. *Eur. J. Org. Chem.* **2004**, *16*, 3385–3399. [CrossRef]
20.	Vazquez-Chavez, J.; Aguilar-Granda, A.; Iglesias Arteaga, M.A. Synthesis and characterization of a fluorescent steroid dimer linked through C-19 by a 1,4-Bis(phenylethynyl)phenylene fragment. *Steroids* **2022**, *187*, 109098. [CrossRef]
21.	Czajkowska, D.; Morzycki, J.W. Synthesis of cholaphanes by ring closing metathesis. *Tetrahedron Lett.* **2007**, *48*, 2851–2855. [CrossRef]
22.	Jurášek, M.; Džubák, P.; Sedlák, D.; Dvořáková, H.; Hajdúch, M.; Bartůněk, P.; Drašar, P. Preparation, preliminary screening of new types of steroid conjugates and their activities on steroid receptors. *Steroids* **2013**, *78*, 356–361. [CrossRef]
23.	Guthrie, J.P.; Cossar, J.; Darson, B.A. A water soluble dimeric steroid with catalytic properties. Rate enhancements from hydrophobic binding. *Can. J. Chem.* **1986**, *64*, 2456–2469. [CrossRef]
24.	Chattopadhyay, P.; Pandey, P.S. Synthesis and binding ability of bile acid-based receptors for recognition of flavin analogues. *Tetrahedron* **2006**, *62*, 8620–8624. [CrossRef]
25.	Janout, V.; Staina, I.V.; Bandyopadhyay, P.; Regen, S.L. Evidence for an umbrella mechanism of bilayer transport. *J. Am. Chem. Soc.* **2001**, *123*, 9926–9927. [CrossRef]
26.	Fournier, D.; Poirier, D. Estradiol dimers as a new class of steroid sulfatase reversible inhibitors. *Bioorg. Med. Chem. Lett.* **2009**, *19*, 693–696. [CrossRef] [PubMed]
27.	Opsenica, D.; Pocsfalvi, G.; Juranić, Z.; Tinant, B.; Declercq, J.-P.; Kyle, D.E.; Milhous, W.K.; Šolaja, B.A. Cholic acid derivatives as 1,2,4,5-tetraoxane carriers: Structure and antimalarial and antiproliferative activity. *J. Med. Chem.* **2000**, *43*, 3274–3282. [CrossRef]
28.	Moser, B.R. Review of cytotoxic cephalostatins and ritterazines: Isolation and synthesis. *J. Nat. Prod.* **2008**, *71*, 487–491. [CrossRef] [PubMed]
29.	Jurášek, M.; Černohorská, M.; Řehulka, J.; Spiwok, V.; Sulimenko, T.; Dráberová, E.; Darmostuk, M.; Gurská, S.; Frydrych, I.; Buriánová, R.; et al. Estradiol dimer inhibits tubulin polymerization and microtubule dynamics. *J. Steroid Biochem. Mol. Biol.* **2018**, *183*, 68–79. [CrossRef]
30.	Czajkowska-Szczykowska, D.; Rodríguez-Molina, B.; Magaña-Vergara, N.E.; Santillan, R.; Morzycki, J.W.; Garcia-Garibay, M.A. Macrocyclic Molecular Rotors with Bridged Steroidal Frameworks. *J. Org. Chem.* **2012**, *77*, 9970–9978. [CrossRef] [PubMed]
31.	Lino, R.; Kinbara, K.; Bryant, Z. Introduction: Molecular Rotors. *Chem. Rev.* **2020**, *120*, 1–4. [CrossRef]
32.	Tomkiel, A.M.; Kowalski, J.; Płoszyńska, J.; Siergiejczyk, L.; Łotowski, Z.; Sobkowiak, A.; Morzycki, J.W. Electrochemical synthesis of glycoconjugates from activated sterol derivatives. *Steroids* **2014**, *82*, 60–67. [CrossRef]

33. Hunter, A.C.; Collins, C.; Dodd, H.T.; Dedi, C.; Koussoroplis, S.-J. Transformation of a series of saturated isomeric steroidal diols by Aspergillus tamarii KITA reveals a precise stereochemical requirement for entrance into the lactonization pathway. *J. Steroid Biochem. Mol. Biol.* **2010**, *122*, 352–358. [CrossRef]

34. Shimizu, S.; Hagiwara, K.; Itoh, H.; Inoue, M. Unified total synthesis of five bufadienolides. *Org. Lett.* **2020**, *22*, 8652–8657. [CrossRef] [PubMed]

35. Yang, Y.; Haino, T.; Usui, S.; Fukazawa, Y. Shielding effect of ether C-O bond obtained from proton chemical shifts of 4-oxa-5α- and 4-oxa-5β-androstan-17-ones. *Tetrahedron* **1996**, *52*, 2325–2336. [CrossRef]

36. Lloyd, L. *Handbook of Industrial Catalysts*; Springer: New York, NY, USA, 2011; pp. 181–182. [CrossRef]

37. Kaneda, K. Cation-exchanged montmorillonites as solid acid catalysts for organic synthesis. *Synlett* **2007**, *7*, 0999–1015. [CrossRef]

38. Bhongale, P.; Joshi, S.; Mali, N. A comprehensive review on catalytic *O*-alkylation of phenol and hydroquinone. *Catal. Rev.* **2023**, *65*, 455–500. [CrossRef]

39. Lu, B.; Li, L.-J.; Li, T.-S.; Li, J.-L. Montmorillonite clay catalysis. Part 13. 1 Etherification of cholesterol catalysed by montmorillonite K-10. *J. Chem. Res. Synop.* **1998**, *9*, 604–605. [CrossRef]

40. Li, T.; Li, H.; Guo, J.; Jin, T. Montmorillonite clay catalysis I: An efficient and convenient procedure for preparation of 5(6)/5′(6′)-unsaturated 3β-disteryl ethers. *Synth. Commun.* **1996**, *26*, 2497–2502. [CrossRef]

41. Kowalski, J.; Morzycki, J.W.; Sobkowiak, A.; Wilczewska, A.Z. Unusual electrochemical oxidation of cholesterol. *Steroids* **2008**, *73*, 543–548. [CrossRef]

42. Zmysłowski, A.; Sitkowski, J.; Bus, K.; Ofiara, K.; Szterk, A. Synthesis and search for 3β,3′β-disteryl ethers after high-temperature treatment of sterol-rich samples. *Food Chem.* **2020**, *329*, 127132. [CrossRef]

43. Dhingra, N.; Bhardwaj, T.R.; Mehta, N.; Mukhopadhyay, T.; Kumar, A.; Kumar, M. Synthesis, antiproliferative activity, acute toxicity and assessment of the antiandrogenic activities of new androstane derivatives. *Arch. Pharm. Res.* **2011**, *34*, 1055–1063. [CrossRef]

44. Shuping, W.; Zhiqin, J.; Heting, L.; Li, Y.; Daixun, Z. Sensitized photooxygenation of cholesterol and pseudocholesterol derivatives via singlet oxygen. *Molecules* **2001**, *6*, 52–60. [CrossRef]

45. Wallis, P.J.; Gates, W.P.; Patti, A.F.; Scott, J.L.; Teoh, E. Assessing and improving the catalytic activity of K-10 montmorillonite. *Green Chem.* **2007**, *9*, 980–986. [CrossRef]

46. Ebitani, K.; Kawabata, T.; Nagashima, K.; Mizugaki, T.; Kaneda, K. Simple and clean synthesis of 9,9-bis[4-(2-hydroxyethoxy)phenyl]fluorene from the aromatic alkylation of phenoxyethanol with fluoren-9-one catalysed by titanium cation-exchanged montmorillonite. *Green Chem.* **2000**, *2*, 157–160. [CrossRef]

47. Joseph, T.; Shanbhag, G.V.; Halligudi, S.B. Copper(II) ion-exchanged montmorillonite as catalyst for the direct addition of N-H bond to CC triple bond. *J. Mol. Catal. A Chem.* **2005**, *236*, 139–144. [CrossRef]

Article

Exploration of Binding Affinities of a 3β,6β-Diacetoxy-5α-cholestan-5-ol with Human Serum Albumin: Insights from Synthesis, Characterization, Crystal Structure, Antioxidant and Molecular Docking

Mahboob Alam 

Department of Safety Engineering, Dongguk University Wise, 123 Dongdae-ro, Gyeongju-si 780714, Gyeongbuk, Republic of Korea; mahboobchem@gmail.com

Abstract: The present study describes the synthesis, characterization, and in vitro molecular interactions of a steroid 3β,6β-diacetoxy-5α-cholestan-5-ol. Through conventional and solid-state methods, a cholestane derivative was successfully synthesized, and a variety of analytical techniques were employed to confirm its identity, including high-resolution mass spectrometry (HRMS), Fourier transforms infrared (FT-IR), nuclear magnetic resonance (NMR), elemental analysis, and X-ray single-crystal diffraction. Optimizing the geometry of the steroid was undertaken using density functional theory (DFT), and the results showed great concordance with the data from the experiments. Fluorescence spectral methods and ultraviolet–vis absorption titration were employed to study the in vitro molecular interaction of the steroid regarding human serum albumin (HSA). The Stern-Volmer, modified Stern-Volmer, and thermodynamic parameters' findings showed that steroids had a significant binding affinity to HSA and were further investigated by molecular docking studies to understand the participation of active amino acids in forming non-bonding interactions with steroids. Fluorescence studies have shown that compound **3** interacts with human serum albumin (HSA) through a static quenching mechanism. The binding affinity of compound **3** for HSA was found to be 3.18×10^4 mol^{-1}, and the Gibbs free energy change (ΔG) for the binding reaction was -9.86 kcal mol^{-1} at 298 K. This indicates that the binding of compound **3** to HSA is thermodynamically favorable. The thermodynamic parameters as well as the binding score obtained from molecular docking at various Sudlow's sites was -8.2, -8.5, and -8.6 kcal/mol for Sites I, II, and III, respectively, supporting the system's spontaneity. Aside from its structural properties, the steroid demonstrated noteworthy antioxidant activity, as evidenced by its IC$_{50}$ value of 58.5 µM, which is comparable to that of ascorbic acid. The findings presented here contribute to a better understanding of the pharmacodynamics of steroids.

Keywords: diacetoxy-5α-cholestane; crystal structure HSA binding; multi-spectroscopy; DFT

Citation: Alam, M. Exploration of Binding Affinities of a 3β,6β-Diacetoxy-5α-cholestan-5-ol with Human Serum Albumin: Insights from Synthesis, Characterization, Crystal Structure, Antioxidant and Molecular Docking. *Molecules* **2023**, *28*, 5942. https://doi.org/10.3390/molecules28165942

Academic Editors: Marina Savić, Erzsébet Mernyák, Jovana Ajdukovic and Suzana Jovanović-Šanta

Received: 3 July 2023
Revised: 3 August 2023
Accepted: 4 August 2023
Published: 8 August 2023

1. Introduction

Cholestane derivatives are important steroid molecules that have several biological functions in the human body. Cholesterol, the primary precursor for cholestane derivatives, is an essential component of cell membranes and is required for the synthesis of hormones and bile acids [1]. Some cholestane derivatives have demonstrated positive effects in the treatment of different diseases [1–5]. However, it is worth noting that elevated blood cholesterol levels could potentially heighten the risk of heart-related illnesses [6–8]. Statins stand out among the extensively researched cholesterol derivatives [3,9], given their remarkable efficacy in lowering cholesterol levels in individuals with hypercholesterolemia [10,11] as well as their application in various other treatments [12]. The mechanism of action of statins involves the inhibition of HMG-CoA reductase [10], an enzyme responsible for cholesterol synthesis in the liver. By reducing cholesterol in the blood, statins can reduce the risk of heart disease and stroke [13]. The cholestane skeleton is a rigid, four-ring structure found in cholesterol and other steroid molecules. Introducing heteroatoms, such as nitrogen, oxygen, and sulfur, into the cholestane skeleton can alter the biological activity of these molecules. Estradiol [14,15] is one example of a steroid hormone that is synthesized from cholesterol and contains three hydroxyl groups. Estradiol is crucial for reproductive health and has other effects such as anti-inflammatory and neuroprotective properties. As a prominent protein in blood plasma, human serum albumin (HSA) is essential for the transport of numerous molecules throughout the body [16–18]. HSA has different binding sites that interact with various compounds, including fatty acids, hormones, and drugs [19,20]. Its interaction with drugs is crucial, as it affects the distribution, metabolism, and bioavailability. HSA interacts with small-molecule drugs, peptides, and proteins, and its binding affects drug pharmacokinetics and pharmacodynamics [21,22]. It can also alter drugs' activity and potency by causing conformational changes in the protein that affect its stability [23]. Furthermore, HSA plays a vital role influencing their pharmacokinetics and pharmacodynamics [24]. The investigation of human serum albumin (HSA) and drug interactions has been greatly aided by molecular docking [19,25]. The technique of molecular docking enables the prediction of binding modes and affinity between drugs and HAS [22]. Its related tools facilitate the exploration of stability and conformational alterations within the protein–drug complex. These tools are essential in drug discovery, as they help researchers understand the mechanisms of drug action, improve drug efficacy, and reduce the risk of adverse effects. Researchers have extensively studied the binding affinities of steroidal and nonsteroidal compounds with human serum albumin (HSA) using various techniques such as synthesis, characterization, crystal structure analysis, and molecular docking [26–29]. These studies have provided valuable insights into the interactions and structural features of these compounds with HSA and have opened up possibilities for therapeutic applications. A novel preclinical approach to drug optimization has been developed using HSA binding affinity as an indicator of changes in the serum half-life (T1/2). In this strategy, scaffold sites of drug candidates not involved in target interactions are altered to proactively identify longer or shorter dosing regimens for humans to complement existing medicinal chemistry efforts. HSA, a major plasma protein, has a significant impact on drug binding, which in turn affects drug delivery, efficacy, and pharmacokinetics. It is also used in clinical settings as a drug delivery system. In this review [21], the properties of drug binding sites within the HSA structure are described, and an overview of drug–HSA interactions is provided. Cholestane derivatives are being investigated by researchers for their ability to bind to HSA in the body, which may provide insight into molecular mechanisms governing drug metabolism and activity [23,30]. The results of these studies contribute to the development of novel drugs with targeted biological activities, potentially leading to therapeutic advances. Microwave irradiation (MWI) was investigated as a means of introducing acetyl functionality to the OH group [31–33]. Acetic acid and acetyl chloride were used as a cheaper acetylating reagent in place of acetic anhydride, which has been banned in some countries due to safety concerns. In conventional processes using acetic anhydride, the reaction takes about 1–4 h at high temperatures and requires a distillation step to separate acetic acid formed during the reaction. The MWI method eliminates excess solvent usage during the reaction, leading to safer and more energy-efficient operation. 3β,6β-diacetoxy-5α-cholestan-5-ol was synthesized using both conventional and MWI methods as part of a green synthesis strategy [34]. Both techniques are equally useful; the former is used for macro-level synthesis, and the latter is used for micro-level synthesis. As part of this study, a comprehensive analysis was conducted to reveal the mechanism of binding between synthesized 3β,6β-diacetoxy-5α-cholestan-5-ol (3) and human serum albumin (HSA). To gain a deeper understanding of the interaction between the compound and the protein, spectroscopic methods and molecular docking were used in this study. The synthesized compound **3** obtained by both conventional and green methods was identified by applying physicochemical techniques, Single-X-ray diffraction, and mixed melting points. This study provided new insights into the binding mechanism of 3β,6β-diacetoxy-

5α-cholestan-5-ol and HSA. The investigation involved both in-silico and experimental analysis, which provided comprehensive and supportive information regarding the steroid binding process with HSA. Additionally, the DPPH radical scavenging potential of the compound was determined using the 1,1-diphenyl-2-picrylhydrazine (DPPH) assay and compared to standard ascorbic acid.

2. Results and Discussion

2.1. Chemistry

The synthesis of 3β,6β-diacetoxy-5α-cholestan-5-ol was achieved through two protocols: conventional and nonconventional. The conventional method is more suitable for macro-level synthesis, likely for larger-scale production, while the nonconventional protocol is better suited for micro-level synthesis in the laboratory setting. The novelty of the nonconventional protocol synthesis lies in utilizing a microwave to speed up the reaction, resulting in a shorter reaction time and higher yield. The solid-state procedure streamlines the reaction workup and improves its environmental friendliness. Notably, the melting points of the products obtained through conventional and solid-state methods are comparable, indicating no product degradation with the solid-state approach. Overall, the solid-state synthesis of 3β,6β-diacetoxy-5α-cholestan-5-ol is a novel and efficient method, made valuable by the combination of microwave acceleration and a simplified workup.

An analysis of the synthesized steroidal compounds obtained by solid-state synthesis as well as conventional methods with a melting point of 167–168 °C revealed that it has the molecular composition $C_{31}H_{52}O_5$. A sharp peak with a medium band at 3480 cm^{-1} corresponds to a tertiary OH group undergoing asymmetric stretching vibration and two sets of strong bands in the ranges of 1238–1266 cm^{-1} and 1736–1741 cm^{-1}, indicating the presence of two acetyloxy groups in symmetric and asymmetric stretching vibrations, respectively. A band at 1031 cm^{-1}, which indicates the presence of a C-O-C bond with asymmetric stretching vibration, was among the distinctive features that could be seen in the infrared (IR) spectrum of acetoxy-steroids. Different signals could be seen in the nuclear magnetic resonance (NMR) spectrum, including a broad multiplet centered at δ 5.37 (1H, dd, C6-αH, equatorial, J = 9.0, 6.0 Hz Hz, 4.9 (1H, m, W1/2 = 18 Hz) for C3-H, axial, A/B ring junction trans), and a sharp singlet at δ 2.24 for the OH group, which vanished when D_2O was added. Additionally, two distinct singlets were seen at δ 2.07 and δ 2.0, which stood for two acetoxy groups (CH_3-CO-O-). The C10-CH_3 and C13-CH_3 protons were responsible for the signals at δ 1.03 and δ 0.73, respectively, while other methyl protons showed up at δ 0.97 and δ 0.87. ^{13}C NMR spectroscopy established the existence of two acetoxy groups in steroid 3. The signals at δ 171.0 and δ 171.3 correspond to the two acetoxy groups, which could be located at the 3-position and 6-position of the steroidal skeleton. Based on the elemental and physical data provided, which correspond to a melting point of 168°, it is possible to positively identify the substance mentioned in the document, which is known as 3β,6β-diacetoxy-5-hydroxy-5α-cholestane and has a reported melting point of 167°. Finally, the single-crystal X-ray crystallography technique was used for confirming the synthesized compound **3** geometry.

2.2. X-ray Crystal Structure and Molecular Geometry of the Synthesized 3β,6β-Diacetoxy-5-hydroxy-5α-cholestane (3)

X-ray analysis of the synthesized steroid compound confirmed its structure and stereochemistry. 3β,6β-diacetoxy-5-hydroxy-5α-cholestane (3) crystallizes in the space group C2 monoclinic crystal system. The unit cell contained four asymmetric molecules, each with similar geometric parameters. The dimensions of the unit cell are a = 31.634(3), b = 9.9420(11), and c = 9.7325(11). The steroid backbone adopts a chair conformation in the A, B, and C rings [35]. Ring A and ring B contain acetoxy groups at C3 and C6, respectively, while there is a hydroxyl group at C5 at the α position, which connects the junction of the two rings. The anisotropic displacement tensor of the terminal atoms of the 3β-acetoxy group, present at the 3-position of the cholestane skeleton, is strongly anisotropic, indicating that these atoms have a large amplitude of vibration perpendicular to the mean plane of the group, as has been reported in the literature [36]. The five-membered D ring is in an envelope conformation with C13 at the flap. The A/B and C/D ring in the steroid core exhibit trans junctions, which means that the C-C bonds connecting these rings are oriented in a trans configuration with the two carbon atoms on opposite sides of the ring. All of the carbon atoms in the lateral alkyl chain are in a straight line because the chain is fully extended. Steroids are characterized by the presence of trans junctions and a fully extended lateral alkyl chain. These characteristics affect how steroids interact with other molecules, which is crucial for the biological activity of steroids. The average bond lengths are: C(sp3)–C(sp3) = 1.53 Å, C(sp2)=O = 1.23 Å, Csp3–O = 1.45 Å, and Csp3–H = 0.959 Å. These bond lengths are very close to their theoretical values. There are intermolecular hydrogen bond interactions between the C5-O-H···O=C-O-C6 and C3-O-C=O···H-O-C5 groups of the nucleus of steroids (Table 1). These interactions are similar in distance, 2.077 Å, and enrich the stability of the crystal structure by keeping the molecules in a regular array, as shown in packing Figure 1. Some bond distances, especially the acetyl group at C3, were found to be slightly altered. This may be because the crystal quality was not standard quality for diffraction analysis. The endocyclic and torsional bond angles in the steroid skeleton are shown in Table 1. These angles were determined by X-ray crystallography and compared with values obtained by density functional theory (DFT) calculations. The experimental result closely resembles the predicted result, indicating that the synthesized steroid compound's structure is accurate (Figure 1).

Table 1. Comparison of some important geometrical parameters of optimized compound **3** with XRD data parameters.

Bond Length (Å)	DFT	XRD	Bond Angle (°)	DFT	XRD	Bond Dihedral (°)	DFT	XRD
C9-O82	1.458	1.434	C11C9O82	109.7313	115.22	C85C84O82C9	178.52	−168.28
O82-C84	1.350	1.430	C6C9O82	107.2475	96.2	C9O82C84O83	−112.4	−46.56
C84-O82	1.208	1.51	C9O82C84	117.6730	102.24	C6C3C24C22	−173.09	−172.70
C84-C85	1.508	1.665	O82C84C85	110.9093	97.0	C6C3C24C46	65.46	66.37
C14-O1	1.449	1.425	C3C24C22	111.0458	109.35	C45C24C22C25	61.91	58.17
O1-H2	0.964	0.821	C3C24C46	108.2078	109.77	C11C14O1H2	48.470	73.57
C9-H10	1.089	0.980	C46C24C22	110.5004	110.14	C11C14C15C17	179.67	178.89
C15-H16	1.092	0.979	C28C31C40	116.4791	118.42	C1C14C15O75	173.30	177.62
C15-O75	1.457	1.456	C28C31C42	110.6404	109.22	C15C14C15O75	−178.32	178.60
O75-C76	1.354	1.314	C42C31C40	110.1862	109.14	C15O75C76O81	1.046	0.81
C76-O81	1.206	1.234	C37C40C50	112.5386	112.93	C25C28C31C40	−165.88	−164.94
C76-C77	1.508	1.506	C40C50C52	113.2645	113.19	C25C28C31C42	67.276	69.37
C77-H80	1.093	0.959	C40C50C56	110.1896	111.27	C42C31C40C58	46.779	48.17
C24-C45	1.548	1.563	C52C50C56	110.3467	110.83	C42C31C40C37	−79.231	−79.66
C31-C42	1.544	1.563	C14O1H2	108.4482	109.44	C37C40C50C52	177.46	176.77
C40-C50	1.549	1.540	C11C14C15	112.0757	113.42	C40C50C56C59	168.60	175.86
C50-C52	1.537	1.523	C14C15O75	109.2216	109.22	C52C50C56C59	−65.573	−55.65
C40-H41	1.099	0.980	C17C15O75	110.9283	109.83	O1C14C24C46	−178.66	−179.90
C50-H51	1.098	0.980	C15O75C76	117.5999	119.15	H33C32C31C42	−177.83	−179.52
C65-C71	1.535	1.523	O75C76C77	110.8554	111.16	C59C62C65C71	−171.31	−169.70
C65-C67	1.535	1.471	O75C76O81	124.0454	122.62	C59C62C65C67	64.6696	69.04
C65-C66	1.099	0.980	C67C75C71	110.4441	109.45	C15C17C28C32	173.10	171.49

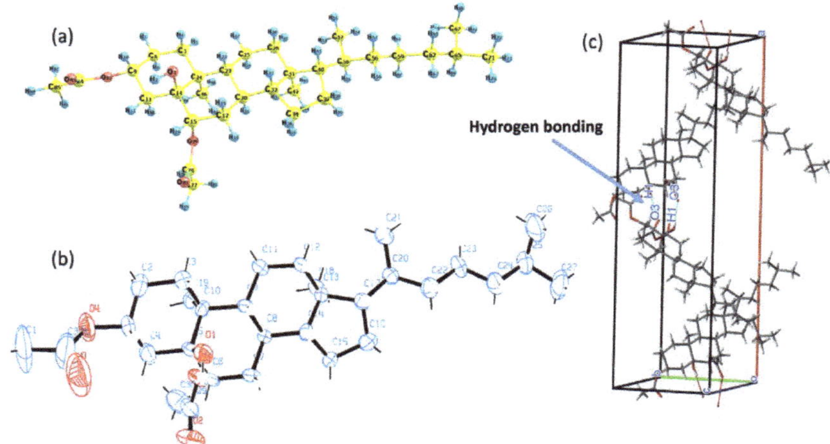

Figure 1. (**a**) Geometry optimization using a B3LYP/6-311++G(d,p) basis set, (**b**) An ORTEP plot of molecule 3, with the atoms numbered at the 50% probability level, and, (**c**) A visualization of the molecular packing of 3, with hydrogen bonds (O-H···O, H···O-H) shown as dashed lines.

Computational methods are increasingly used to study heterocyclic molecules. DFT can predict structures, energetics, and electronic properties. FMOs play a key role in reactivity, optical properties, and biological activity. Reduced density gradients (RDGs) are a measure of the electron density in a molecule. They are significant because they can be used to identify regions of high electron density, which are often associated with chemical bonds. RDGs can also be used to study the interactions between molecules. To study the electronic structure of the steroid molecule (3), a Gaussian software package was utilized to conduct calculations employing density functional theory (DFT). The initial coordinate source for theoretical calculations was the Crystallographic Information File (CIF). Density functional theory (DFT) was employed to optimize the molecular structure of 3β,6β-diacetoxy-5-hydroxy-5α-cholestane, utilizing the B3LYP functional and the 6-311++G(d,p) basis set. The structures obtained from the DFT calculations are shown in Figure 1a along with the corresponding numbering scheme. The calculated values by DFT calculation were matched with the values achieved by X-ray diffraction analysis. The results demonstrate a correlation between the two sets of data, showing agreement between the electronic structure determined by DFT, the experimental results from X-ray diffraction (Table 1), and the literature [36]. Certain bond lengths and torsional angles, however, showed deviations because the DFT calculations were performed in the gas phase, whereas X-ray diffraction experiments are typically performed on solid-state samples. Furthermore, the molecule's crystal must be capable of producing accurate diffraction patterns.

The crystal packing of the molecules reveals that they exist as O-H···O hydrogen-bonded dimers. The hydrogen bonding occurs between the hydroxyl group at C5 of one molecule and the carbonyl O atom of the acetoxy moiety of the other molecule. The O–H···O hydrogen bond distances are 2.81 Å. In addition to the hydrogen bonds, there are also three short intramolecular distances: H30B···O4 2.56 Å, H6···O3 2.65 Å, and H30A···O4 2.56 Å. These short distances suggest that there is a significant interaction between the polar hydrogen and the carbonyl O atoms. The interaction between these atoms helps stabilize the molecule by keeping the atoms in close proximity. This interaction also helps determine the structure of the molecule by affecting the orientation of the atoms (Table 2).

Table 2. Hydrogen-bond geometry (Å, °).

D—H···A	D—H	H···A	D···A	D—H···A	Symmetry Code
O1—H1···O3	0.78(9)	2.11(9)	2.881(9)	173(12)	x, y, 1−z

The six-membered rings in a molecule (3) adopt conformations that closely resemble a chair form. This is supported by the puckering parameters determined by Cremer and Pople (1975) [37]. The puckering parameters for ring A are Q = 0.584 Å, θ = 3.6°, and φ = 212°. For ring B, the parameters are Q = 0.549 Å, θ = 104°, and φ = 173.6°. And for ring C, the values are Q = 0.576 Å, θ = 176.7°, and φ = 216°. The D-ring exhibits a twisted conformation around the C13—C14 bond with the puckering parameters Q = 0.452 Å and φ = 113.6°, as shown in Table 3. All rings in the molecule are in a trans-fused configuration. The acetoxy group at C3 is positioned 3β with respect to ring A, while the substituents at ring B are in the 5α and 6β orientations, respectively.

Table 3. Puckering Parameters of Steroid (3) with Five and Three Six-Membered Rings in its Skeleton.

Cholestane-Skeleton	Puckering Parameters		
	Q (Å)	Θ (°)	Φ (°)
Ring A (Six-Membered)	0.584	3.6	212
Ring B (Six-Membered)	0.549	104	173.6
Ring C (Six-Membered)	0.576	176.7	216
Ring D (Five-Membered)	0.452	-	13.6

2.3. Reduced Density Gradients and Frontier Molecular Orbitals (FMOs)

Apart from the Hirshfeld surfaces mentioned in Section 2.4, the color-filled reduced density gradient (RDG) of 3β,6β-diacetoxy-5-hydroxy-5α-cholestane was also examined in real space using electron density analysis. This study was performed to further understand the influence of intermolecular and intramolecular interactions within the crystal structure. As shown in Figure 2, different color regions were analyzed to better understand these interactions. The surfaces are color-coded based on the magnitude of sign(λ_2)p on a scale of blue–green–red. Blue signifies powerful attractive forces, green represents weak attractive interactions, and a strong non-bonded overlap is represented by red in this color scheme. The strength of interactions can be visually assessed by examining the scatter plot and filled isosurface plots. Denser points in the scatter diagram (Figure 2b) indicate a higher electron density and stronger interactions. A dense peak and scattered dots of sign(λ_2)p within the purview of −0.02∼0 a.u in the provided figure indicate the presence of interactions within the molecule with varying intensities, ranging from weak to strong van der Waals interactions (−0.02 to −0.01 a.u.). The green peak observed only in this region confirms the absence of intramolecular hydrogen bonding within the molecule. Furthermore, the predominance of van der Waals interactions suggests that they are the main force in the system. Positive values of sign(λ_2)p indicate the presence of steric effects, resulting in red regions observed in the molecule.

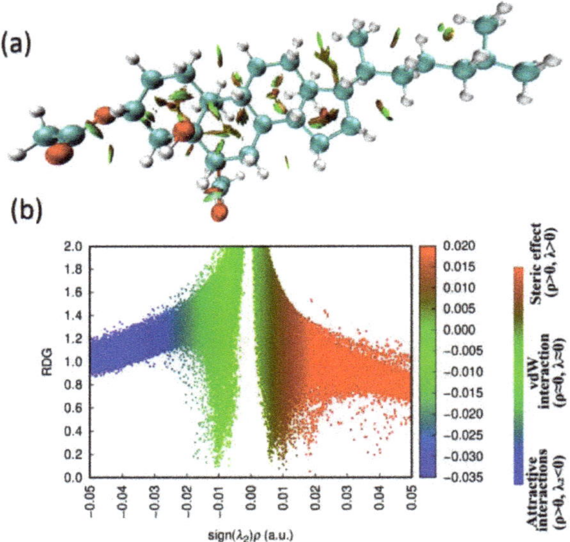

Figure 2. RDG map and NCI plot of the crystal's gradient isosurface: (**a**) analysis of noncovalent interactions; (**b**) scatter diagram.

The presented data on the frontier molecular orbitals (FMOs) of compound (**3**) are significant, as they reveal key insights into its reactivity, stability, and receptor binding affinity. The molecule's HOMO and LUMO distribution indicates important binding regions for receptor interactions. Molecular properties are determined by the properties of the frontier molecular orbitals (FMOs), which are the highest occupied molecular orbitals (HOMO) and the lowest unoccupied molecular orbitals (LUMOs). The low-energy LUMO of the acceptor molecule and the higher HOMO value of the donor molecule are both necessary for a molecule to be able to grant electrons to a proper acceptor [38,39]. Molecule reactivity is contingent upon the energy difference between the HOMO and the LUMO, a measurement of the energy disparity between the HOMO and LUMO. A larger HOMO–LUMO gap, which electrons find challenging in transitioning between distant energy levels, indicates lower reactivity. A larger HOMO–LUMO gap means that it is chemically hard because it is difficult to modify it by adding or removing electrons [40]. The spatial arrangement of the HOMO and the LUMO is critical for the recognition of receptor binding sites in ligands (molecules). Figure 3 illustrates that the HOMO and LUMO of 3β,6β-diacetoxy-5-hydroxy-5α-cholestane (**3**) are mainly distributed on the hydroxyl and acetoxy groups and the steroid ring. This suggests that these specific regions of the ligand are more likely to bind to the receptor, suggesting their importance in receptor interactions. The HOMO and LUMO plots of compound (**3**) are shown in Figure 2, with positive phases highlighted in red and negative phases highlighted in green. The energies HOMO and LUMO are 7.38 and 0.44 eV, respectively, resulting in a 6.94 eV HOMO–LUMO gap. This significant gap reflects the robustness and stability of 3β,6β-diacetoxy-5-hydroxy-5α-cholestane (**3**), which corresponds to the characteristics of its skeletal structure. The atomic orbitals (AOs) associated with the oxygen atoms of the hydroxy and acetoxy groups located at the 5, 3, and 6-positions of the steroid skeleton (-O-H and -C=O in Figure 1) contribute to the formation of a mixed n- and p-type HOMO and a p-LUMO* when the HOMO and LUMO orbitals of compound (**3**) are surface-analyzed. Consequently, due to the aforementioned interactions, a delocalization effect was noticed in the acetoxy groups when switching from HOMO to LUMO. This means that the electron density spreads out and becomes more distributed over the acetoxy groups in LUMO orbitals compared to HOMO orbitals. A relatively large HOMO–LUMO gap indicates that the compound is chemically hard, which can enhance its stability and receptor binding ability. The delocalized nature of the orbitals may also contribute to these properties. This information can be used to guide the design and optimization of the compound for specific applications, such as receptor binding or drug delivery.

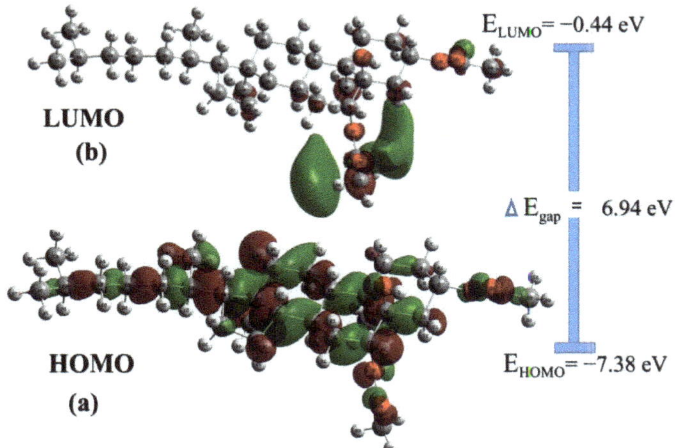

Figure 3. Spatial plots of (a) HOMO and (b) LUMO showing the energy gap.

2.4. Hirshfeld Surface Analysis

The 3D Hirshfeld surface analysis is crucial for understanding the stability of organic crystals. It quantitatively identifies hydrogen–hydrogen (H-H) interactions as the most significant contributors, comprising 83.8% of the total interactions in the present molecule 3. Specific intra- and intermolecular interactions, which are regarded as the crystal's support structure, control the stability of organic crystals. Therefore, a crucial component of crystal engineering is the qualitative and quantitative measurement of these interactions. An effective method for assessing the intramolecular and intermolecular interactions that take place during crystal packing is the 3D Hirshfeld surface analysis. Based on a Hirshfeld surface study, two-dimensional fingerprint plots were created using CrystalExplorer 21.5.

The Hirshfeld surface of steroid (3) was generated with an isovalue of 0.5 using the electron density and mapped onto the dnorm descriptor. The color scheme used ranges from red (shorter distances than the sum of van der Waals radii) to white and blue (longer distances than the sum of van der Waals radii). The 3D mapping was carried out in the range of −0.5293 to 2.1132 dnorm values, where negative values are shown in red, positive values are shown in blue, and zero values are shown in white. These colors represent intermolecular contacts that are short, long, or weak. Intermolecular forces were depicted as crimson circular marks on the Hirshfeld surface. The presence of red circle spots on the pentacyclic ring of steroid (3) in Figure 4a,b indicates intermolecular interactions with a second steroid molecule in a reciprocal manner. Figure 4c–f show a 2D fingerprint plot of 3β,6β-diacetoxy-5-hydroxy-5α-cholestane, which reveals two distinct spikes that display various interactions between neighboring molecules in the crystal lattice. Reciprocal hydrogen–hydrogen (H-H), oxygen–hydrogen (O-H), and carbon–hydrogen (C-H) intermolecular interactions appear in the 2D fingerprint plots as single, double, and unidentified spikes, accounting for 83.8%, 14.4%, and 1.0% of the total plot, respectively. Figure 4d, in particular, shows that H-H interactions are important in maintaining the stability of the crystal structure, as evidenced by their central location among the scattered points. These H-H contacts contribute a significant 83.8% of the total Hirshfeld surfaces, as shown in Figure 4. These findings offer valuable insights in designing new crystals with enhanced properties and predicting crystal packing. The technique also provides qualitative information about the short range and relatively strong nature of H-H interactions, influencing the crystal's properties. Overall, this analysis aids in comprehending the relationship between the crystal structure and stability, enabling improved crystal engineering.

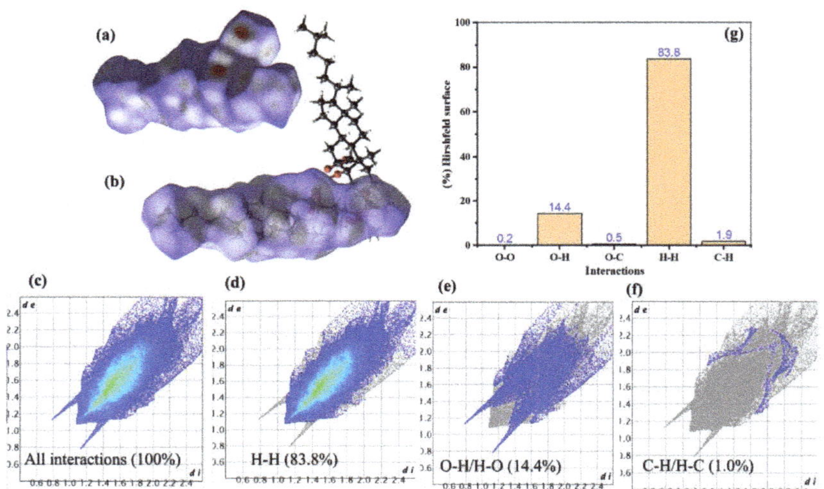

Figure 4. The 2D fingerprint plot shows the distribution of intermolecular interactions in the crystal. The shape index (**a,b**) displays the different types of interactions, while the total surface area (**c**) shows the overall interaction network. The proportions of H/H (**d**), O/H/H/O (**e**), C/H/C/H (**f**), and other interactions are also shown in the bar diagram (**g**).

2.5. Human Serum Albumin (HSA)–Steroid Binding Studies

HSA–steroid binding studies investigate the interplays involving human serum albumin and steroids in order to better understand their binding mechanisms. HSA functions as a steroid carrier protein, influencing their bioavailability and pharmacokinetics. Various spectroscopic tools including UV–vis absorption spectroscopy and fluorescence spectroscopy have been employed to investigate the binding interactions between ligands and HSA, allowing for the molecular evaluation of their biological activities. These tools have improved researchers' understanding of the molecular mechanisms underlying steroid biology and their engagements with human serum albumin (HSA). Moreover, these approaches can be used to identify the participating amino acid residues in steroid binding and judge the binding strength between steroids and HSA.

2.6. UV–Vis Absorption Analysis

UV–visible absorption is a highly responsive method used to observe and study the formation of protein–ligand complexes and gain insight into how complex formation affects protein structural integrity [41]. The procedure is mainly due to the enhanced absorption of UV–vis radiation caused by samples and protein molecules. Tryptophan, tyrosine, and phenylalanine constitute the primary sources of aromatic moieties that are present and absorb light. The maximum wavelength of human serum albumin (HSA) is 280 nm, which is characteristic of the three aromatic residues. Particularly responsive to alterations in the microenvironment, tryptophan absorbs primarily at 280 nm. In general, modifications to the aromatic residues' microenvironment that follow ligand binding offer insights into the phenomenon of binding. Figure 5 displays the HSA absorption spectra in the presence and absence of 3β,6β-diacetoxy-5-hydroxy-5α-cholestane. The peak in absorbance observed at 278 nm is focused on the microenvironment surrounding the tryptophan and tyrosine residues of HSA. The magnitude of the peaks decreased as the various concentrations of compound **3** were introduced to the constant amount of HSA. Overall, the peak of the magnitude confirms that HSA causes significant structural changes when it interacts with steroids.

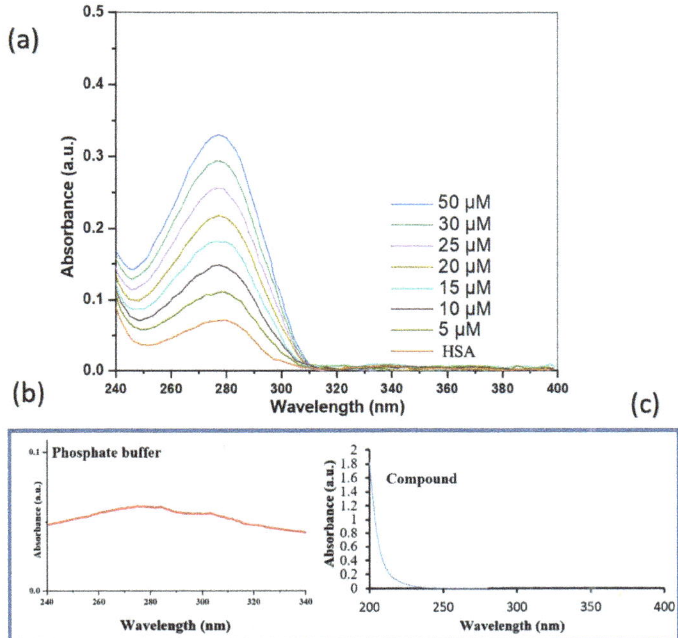

Figure 5. (a) HSA–steroid absorption spectra at increasing concentrations of compound (**3**) (5–30 and 50 μM), with the HSA concentration held constant at 5 μM, (b) UV–vis spectrum of the buffer, and (c) UV spectrum of compound **3**.

To check for interference from the buffer solution, a control was performed with increasing concentrations of the compound ranging from 5 to 50 μM, in the absence of HSA (Figure 5b). The results showed that the buffer solution had a very low absorbance with a flat spectrum at all wavelengths, so it did not interfere with the measurement of the absorbance of other substances. Additionally, no significant change was observed in the UV–vis spectrum of the control as the concentration of steroidal compound **3** was increased. Finally, the wavelength of the compound in the UV–vis spectrum (Figure 5c) was found to be approximately 212 nm, which is outside of the wavelength range of HSA, so there was no interference in the results.

2.7. Fluorescence Studies

Fluorescence spectroscopy has demonstrated its utility as a valuable scientific tool in the investigation of a variety of effects on biomacromolecules, including protein–drug interactions, conformational changes, and the mechanism of quenching in protein–drug interactions [42]. Three amino acids—tryptophan, tyrosine, and phenylalanine—are authorized to regulate the fluorescence of proteins. Tryptophan is the most important fluorophore and fluoresces most strongly in subdomain IIA of the active cavity of HSA. Tyrosine and phenylalanine fluoresce as well, but much more weakly than tryptophan does. Tyrosine fluorescence can be quenched by ionization or by proximity to other amino acids, such as amino groups, carboxyl groups, or tryptophan. In this study, the binding mechanism and formation of complexes involving HSA were explored by titrating it against 3β,6β-diacetoxy-5-hydroxy-5α-cholestane. The concentration of HSA was kept constant at 5 μM, while the concentration of 3β,6β-diacetoxy-5-hydroxy-5α-cholestane was varied from 5 to 50 μM. The presence and absence of the steroid were compared by obtaining HSA spectra, as shown in Figure 6a. The findings indicated that HSA exhibits a fluorescence emission peak at approximately 338 nm. The peak suppression increased as the steroid concentration increased, indicating the formation of a complex that engaged with HSA. The intensity of the fluorescence of HSA dropped by about 41% at the highest steroid dose (50 µM), while the wavelength (λem) and peak morphology remained unaffected. This suggests that steroid 3 binds to HSA in a non-covalent manner. 3β,6β-diacetoxy-5-hydroxy-5α-cholestane (steroid) induced a slow fluorescence quenching in HSA. The quenching of HSA fluorescence by the drug was concentration-dependent, with more prominent quenching observed at 50 µM. The mechanism of fluorescence quenching can take place statically or dynamically. A stable complex between the protein and the quencher results in the quenching of static fluorescence. Various studies consistently demonstrate an inverse relationship between the quenching constant (Kq) and temperature. This relationship is attributed to the fact that the frequency of diffusion decreases with increasing temperature. Dynamic fluorescence quenching occurs due to transient collisions between the quencher and protein. As the frequency of collisions rises, the quenching rate constant (Kq) also escalates in a temperature-dependent manner. Additionally, without actual ligand binding, dynamic fluorescence quenching can take place, which means that HSA's shape and function are unaffected, providing consistency with the literature. The mechanism of fluorescence quenching was determined using the Stern–Volmer equation establishing a mathematical relationship between the quencher concentration and protein fluorescence quenching. The Stern–Volmer plot demonstrating the interaction of HSA with steroid (3) is shown in Figure 6b. The Stern–Volmer quenching constant (Ksv) can be used to estimate the quenching efficiency of a drug. A higher Ksv value indicates that the drug is better at quenching the fluorescence of a fluorophore (any macromolecule or protein). The Ksv coefficient of the present steroid 3 was determined to be 1.45×10^4 M^{-1}. These values indicated that, with a certain amount of the quencher added, the fluorescence intensity decreases by 1%. The bimolecular quenching constant, Kq, can be calculated from the quenching constant, Ksv, and the average time spent in the excited state, τ0, using the following Equation (1):

$$kq = Ksv/\tau 0 \tag{1}$$

where Ksv is the Stern–Volmer constant and τ0 is the average lifetime of the protein in the absence of a quencher. For biopolymers, the value of τ0 is usually 10^{-9} s.

The Stern–Volmer equation was used to estimate the rate constant at which HSA and steroid (3) bind, also known as the bimolecular quenching rate constant (Kq). The value of Kq was found to be approximately 1.45×10^{12} M^{-1}s^{-1}. This finding supports the existence of a static quenching process that contributes to the formation of a stable steroid–HSA complex in its resting state. The bimolecular quenching constant, Kq, is necessary to comprehend how HSA and drugs interact. The interaction between HSA and steroid (3) at ambient temperature was studied using a linear regression plot of F0/F against [Q]. The KSV and Kq values were determined to be in the order of 10^4. This finding suggests that the quenching mechanism between HSA and 3β,6β-diacetoxy-5-hydroxy-5α-cholestane likely began with a static quenching mechanism, as documented in the literature [43]. Additionally, employing a modified Stern–Volmer equation relies on a double logarithmic graph plotting log (F0 F)/F vs. log [Q], the binding constant (Kb) and binding sites (n) on HSA can be ascertained. The slope of the resulting line indicates the number of binding sites (n), while the intercept (ordinate of the origin, shown in Figure 6c) corresponds to the logarithmic value of K_b (logK_b). Fluorescence quenching experiments yielded the HSA–steroid (3) binding and thermodynamic parameters, which are also shown in Table 4. In general, drugs with higher binding constants exhibit stronger binding to receptors than drugs with lower binding constants. The regression equation produced a slope close to one, indicating that, under specific conditions, a single binding site with a high affinity for steroids on HSA was found. The steroid 3 displayed a binding constant of 3.18×10^4 M^{-1}, ascertained through experimentation. This high value indicates that 3β,6β-diacetoxy-5-hydroxy-5α-cholestane binds well to HSA. In addition, the calculated Gibbs free energy (ΔG_o) for the binding of HSA to steroids is -9.86 kcal/mol, and these findings show that the binding process happens spontaneously in nature. The molecular docking resulted in a binding score of -8.6 kcal/mol, suggesting a significant interaction between HSA and steroid 3 because it falls within the range of binding affinities deemed biologically relevant. The experimental and predicted binding fractions agree with each other, further supporting the existence of this interaction.

Table 4. In the temperature range from 298 to 318 K, fluorescence quenching experiments were carried out to ascertain the binding and thermodynamic properties of the interaction between has and steroid (3). For this, two equations—the Stern–Volmer equation and the modified Stern–Volmer equation—were used. The steroid concentration ranged from 0 to 30 and 50 μM, while the HSA concentration remained constant at 5 μM.

Temp	Ksv ($\times 10^4$ M^{-1})	Kq ($\times 10^{12}$ M^{-1}s^{-1})	R^2	K_b ($\times 10^5$ M^{-1})	n	$\Delta G°$ (Kcal/mol)	R^2
298 K	1.45	1.45	0.9846	3 18	1	−9.86	0.9193
308 K	1.49	1.49	0.9853	2 72	1	−9.53	0.962
318 K	1.5	1.5	0.9855	2 42	1	−9.20	0.985

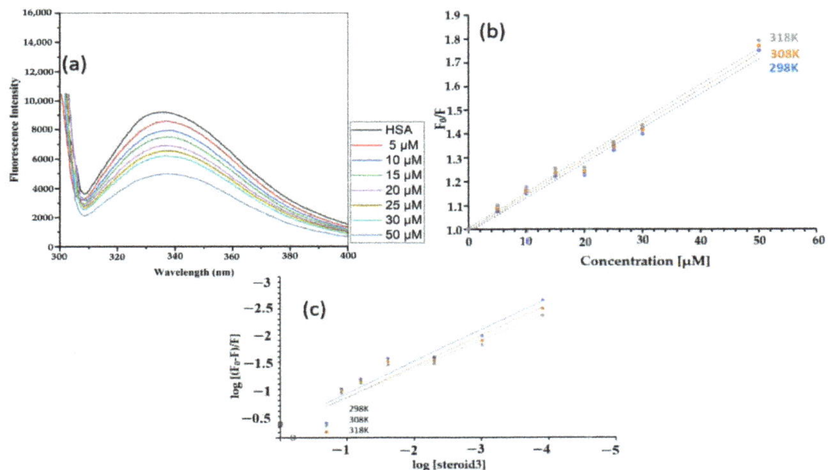

Figure 6. (a) Fluorescence emission spectra of HSA–steroid 3 with an increasing steroid concentration (0–30 and 50 μM), while the concentration of HSA was kept fixed to 5 μM in physiological pH 7.4 sodium phosphate buffer at 298 K; (b) Stern–Volmer plot of F0/F versus steroid 3 for HSA with steroid 3 interactions at various temperature such as 298 K, 308 K, and 318 K; and (c) Modified Stern–Volmer graph plotting log [(F0 − F)/F] vs. log [steroid 3] for HSA with steroid 3 interactions.

The effect of temperature on the HSA–Stern–Volmer quenching constants with compound (3) was investigated. The studies were performed at 298, 308, and 318 K. As shown in Table 4 and Figure 6b, the K$_{SV}$ values decreased as the temperature increased. This indicates that the quenching mechanism is static, as the interaction between HSA and compound (3) is not affected by the increased thermal energy. Similarly, the bimolecular quenching constant (Kq) decreased as the temperature increased from 298 to 318 K, which is consistent with a static quenching mechanism. Ligands bind to proteins through non-covalent interactions, including hydrogen bonding, van der Waals forces, hydrophobic interactions, and electrostatic interactions [43]. The dimensions and signs suggested by Ross and Subramanian are employed for the computation of the thermodynamic parameters (ΔH and ΔS) connected to these interactions [44]. The binding forces are controlled by the thermodynamic variables ΔH and ΔS in protein–ligand binding. While ΔS denotes changes in disorder, ΔH represents the reaction's heat production or absorption. Positive ΔS denotes increased disorder, and negative ΔH denotes heat release, both of which favor binding. The sum of ΔH and ΔS determines the driving force of the binding, ΔG, with a negative value indicating favorable binding. The following Equations (2)–(4) are used to calculate the thermodynamic parameters:

$$\Delta G^\circ = -RT \ln(K) \qquad (2)$$

$$\Delta H^\circ = -T\Delta S^\circ + \Delta G^\circ \qquad (3)$$

$$\Delta S^\circ = \Delta G^\circ / T \qquad (4)$$

In this context, ΔG° represents the alteration in free energy, ΔH° indicates a change in enthalpy, ΔS° denotes the change in entropy, R stands for the gas constant (8.314 J/mol/K), T represents the temperature (in Kelvin), and K is the constant.

The slope and intercept of the van't Hoff plot can be used to calculate the enthalpy change (ΔH°) and entropy change (ΔS°) of a binding reaction, respectively [45]. The values for ΔH°, ΔS°, and the binding free energy (ΔG°) are shown in Table 5. A negative ΔH° indicates that the binding process is exothermic and favors the interaction, indicating that the hydrophobic and hydrogen bond interactions between the drug and human serum albumin are exothermic, releasing heat upon formation. A positive ΔS° indicates that there is more disorder during binding, which is advantageous because it suggests that the interactions between the drug and human serum albumin are not very strong, allowing them to bind in more disordered conformations. The presence of a negative ΔH° and positive ΔS° indicates that hydrogen bonds and hydrophobic interactions are the main binding forces between the drug and human serum albumin. These interactions, despite being weak, are advantageous because they dissipate heat and permit more disordered conformations. Analyzing the data in Table 6 shows that the magnitude of the negative values decreases with an increasing temperature. This observation suggests that as the temperature increases, the spontaneity of the reaction decreases.

Table 5. Thermodynamic parameters obtained from studying the interaction between a steroid and HSA.

Temperature (K)	ΔG° (kcal/mol)	ΔH° (kJ/mol)	ΔS° (J/mol/K)
298	−9.86	−11.87	20.12
308	−9.53	−11.54	19.72
318	−9.20	−11.21	19.32

Table 6. Protein–ligand interaction profiler for hydrophobic, salt bridges, and hydrogen bonding interactions of active amino residues at various binding sites of the Human Serum Albumin (HSA) with steroid **3** were obtained using the PLIP web tool.

Binding Site I						
Hydrophobic Interactions						
Index	Residue	Amino Acid	Distance	Ligand Atom	Protein Atom	Binding Energy (kcal/mol)
1	115A	LEU	3.93	9098	1736	
2	115A	LEU	3.82	9103	1739	
3	115A	LEU	3.64	9119	1737	
4	138A	TYR	3.76	9124	2106	
5	138A	TYR	3.71	9125	2107	−8.2
6	142A	ILE	3.30	9103	2183	
7	161A	TYR	3.70	9123	2527	
8	165A	PHE	3.71	9125	2591	
9	186A	ARG	3.70	9109	2896	

Table 6. *Cont.*

Hydrogen Bonds								
Index	Residue	Amino Acid	Distance H-A	Distance D-A	Donor Angle	Protein Donor?	Side Chain	Donor Atom	Acceptor Atom
1	115A	LEU	2.35	2.98	116.15	√	X	1080 [Nam]	5609 [O3]
2	115A	LEU	2.19	2.79	118.50	X	X	5609 [O3]	1083 [O2]

Salt Bridges						
Index	Residue	Amino Acid	Distance Protein	Protein Positive	Ligand Group	Ligand Atoms
1	146A	HIS	5.36	√	Carboxylate	9126, 9128
2	186A	ARG	5.00	√	Carboxylate	9130, 9132
3	190A	LYS	4.21	√	Carboxylate	9130, 9132

Binding site II

Hydrophobic Interactions					
Index	Residue	Amino Acid	Distance	Ligand Atom	Protein Atom	Binding Energy (kcal/mol)
1	214A	TRP	3.28	9108	3327	
2	218A	ARG	3.52	9108	3380	
3	219A	LEU	3.45	9125	3405	
4	223A	PHE	3.73	9122	3480	
5	223A	PHE	3.78	9125	3479	
6	234A	LEU	3.53	9125	3651	−8.5
7	238A	LEU	3.50	9123	3712	
8	238A	LEU	3.71	9124	3713	
9	260A	LEU	3.44	9124	4020	
10	264A	ILE	3.34	9124	4092	
11	290A	ILE	3.58	9122	4487	
12	291A	ALA	3.38	9119	4504	

Hydrogen Bonds								
Index	Residue	Amino Acid	Distance H-A	Distance D-A	Donor Angle	Protein Donor?	Side Chain	Donor Atom	Acceptor Atom
1	222A	ARG	3.45	4.03	115.90	√	√	3456 [Ng+]	5646 [O3]

Salt Bridges						
Index	Residue	Amino Acid	Distance Protein	Protein Positive	Ligand Group	Ligand Atoms
1	195A	LYS	4.68	√	Carboxylate	9130, 9132
2	195A	LYS	4.38	√	Carboxylate	9126, 9128

Table 6. *Cont.*

				Binding Site III		
				Hydrophobic Interactions		
Index	Residue	Amino Acid	Distance	Ligand Atom	Protein Atom	Binding Energy (kcal/mol)
1	344A	VAL	3.39	9124	5311	
2	387A	LEU	2.92	9105	5991	
3	387A	LEU	3.14	9103	5993	
4	388A	ILE	3.78	9105	6011	
5	407A	LEU	3.63	9133	6325	
6	430A	LEU	3.54	9133	6711	
7	430A	LEU	3.81	9114	6712	−8.6
8	449A	ALA	3.26	9107	7000	
9	450A	GLU	3.73	9123	7011	
10	453A	LEU	3.06	9106	7061	
11	453A	LEU	3.39	9125	7058	
12	485A	ARG	3.53	9124	7547	
13	488A	PHE	3.82	9129	7597	
14	488A	PHE	3.37	9097	7595	

				Salt Bridges		
Index	Residue	Amino Acid	Distance Protein	Protein Positive	Ligand Group	Ligand Atoms
1	410A	ARG	4.77	√	Carboxylate	9130, 9132
2	414A	LYS	5.26	√	Carboxylate	9126, 9128

2.8. Molecular Docking Analysis

HSA has three major binding sites referred to as subdomains I, II, and III. These subdomains are allosterically coupled, meaning that the binding of a ligand to one subdomain can affect the binding of ligands to the other subdomains The binding intensity of a compound can vary at different binding sites, such as sites I, II, and III [46]. The best target site in the protein can be predicted based on the binding score obtained from various binding sites in a docking analysis. Molecular docking is a useful computational method for investigating the interactions between macromolecules (like proteins) and small molecules, ligands, or drugs. It plays a crucial role in the drug discovery and development processes by analyzing and forecasting the binding affinities and modes of ligands to target proteins. A popular piece of software called AutoDock Vina uses an algorithm to explore a ligand's conformational space and forecast the binding pose and affinities it will have for a target protein. By accounting for electrostatic interactions, hydrogen bonds, and van der Waals forces, it evaluates the energy of the interaction between the ligand and the protein. The human serum albumin (HSA) molecule is composed of three structurally related domains: domain I, which contains amino acid residues 1–195, domain II, which contains amino acid residues 196–383, and domain III, which contains amino acid residues 385–585, as shown in Figure 7a. 3β,6β-diacetoxy-5-hydroxy-5α-cholestan, a steroidal molecule (Figure 7b), was individually docked with the active pockets of HSA in domains I, II, and III to examine its binding preferences (Figure 7c).

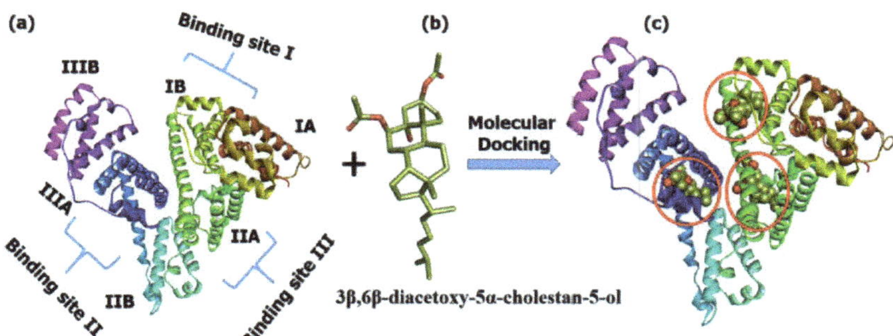

Figure 7. General representation of molecular docking. (**a**) A receptor with three binding sites, labeled I, II, and III. (**b**) Steroid 3. (**c**) Steroid 3 docked in each of the three binding sites.

The docked pose of HSA with the steroid (3) in binding sites I, II, and III with the highest binding energy (negative) is shown in Figure 8 (binding sites I, II, and III). 3β,6β-diacetoxy-5-hydroxy-5α-cholestan showed the highest binding energy at three different sites within the HSA. Specifically, the energy of site I was determined to be −8.2 kcal/mol, while it was −8.5 kcal/mol for site II and −8.6 kcal/mol for site III. In the docking model, the presence of actively participating amino acids was observed, contributing to non-bonding interactions such as hydrogen bonding and van der Waals forces at different sites within HSA. To further investigate these interactions, specific docking poses were chosen to analyze non-bonded interactions between steroids and active amino acid residues responsible for forming the active pocket of HSA. At binding site 1, steroid (3) had a binding score of −8.2 kcal/mol. In this model, the oxygen atom of the acetoxy group forms a hydrogen bond with LYS199 of HSA. In addition, several amino acids including ASP451, TRP214, ARG222, GLU153, SER192, and ILE290 were found to surround steroids (2), resulting in van der Waals interactions. This information is depicted in Figure 8. Additionally, salt bridges and hydrophobic interactions between various amino acids and various steroid atoms at particular distances were seen (Table 6). The amino acids LEU115, TYR138, ILE142, TYR161, PHE165, ARG186, HIS146, ARG186, and LYS190 were involved in these interactions. These interactions, along with the hydrophobic interactions and salt bridges including alkyl and pi-alky, collectively contributed to the proper orientation and stabilization of the ligand within the active pocket.

As a result, the close interactions between the ligand and the active amino acid residues yielded a high docking score. The details of these interactions can be found in Table 6.

At binding site II, the best-docked pose of the steroid (3)–HSA obtained by various interactions was found to have the highest binding score of −8.5 kcal/mol. An analysis of Figure 8 (binding site II) shows that the oxygen atom and keto-oxygen atom of the acetoxy group at the 6-position of 3β,6β-diacetoxy-5-hydroxy-5α-cholestane (3) form conventional hydrogen bonds with the amino acids TYR411 and ARG410, respectively. These hydrogen bonds play an important role in close interactions with amino acids present at the active sites of the protein. van der Waals forces were also observed, surrounding the molecule with different types of amino acids, including LEU430, ARG348, LEU460, LEU407, PHE403, ASN391, SER489, PHE488, LEU491, and GLU450. These amino acids, which form non-bonding interactions, assist the molecule and protein in close host–guest interactions. In addition to these interactions, some amino acids, including VAL344, LEU387, PRO384, ILE388, MET446, ALA449, LEU457, ARG453, and LEU453, participate in the formation of alkyl and pi-alkyl interactions of the steroid (3) skeleton, as shown in the figure. Hydrophobic amino acids, such as TRP214, ARG218, LEU219, PHE223, LEU234, LEU238, LEU238, LEU260, ILE264, ILE290, and ALA291, are buried inside the protein such that they are shielded from water. These hydrophobic amino acids make a protein fold stable even during interactions with different atoms of a molecule at specific distances, as shown in Table 6. It was discovered that LYS195 of these amino acids played a role in building a salt bridge with the carboxylate of steroid (3) to increase the stability of the docking model of steroid (3)–HSA.

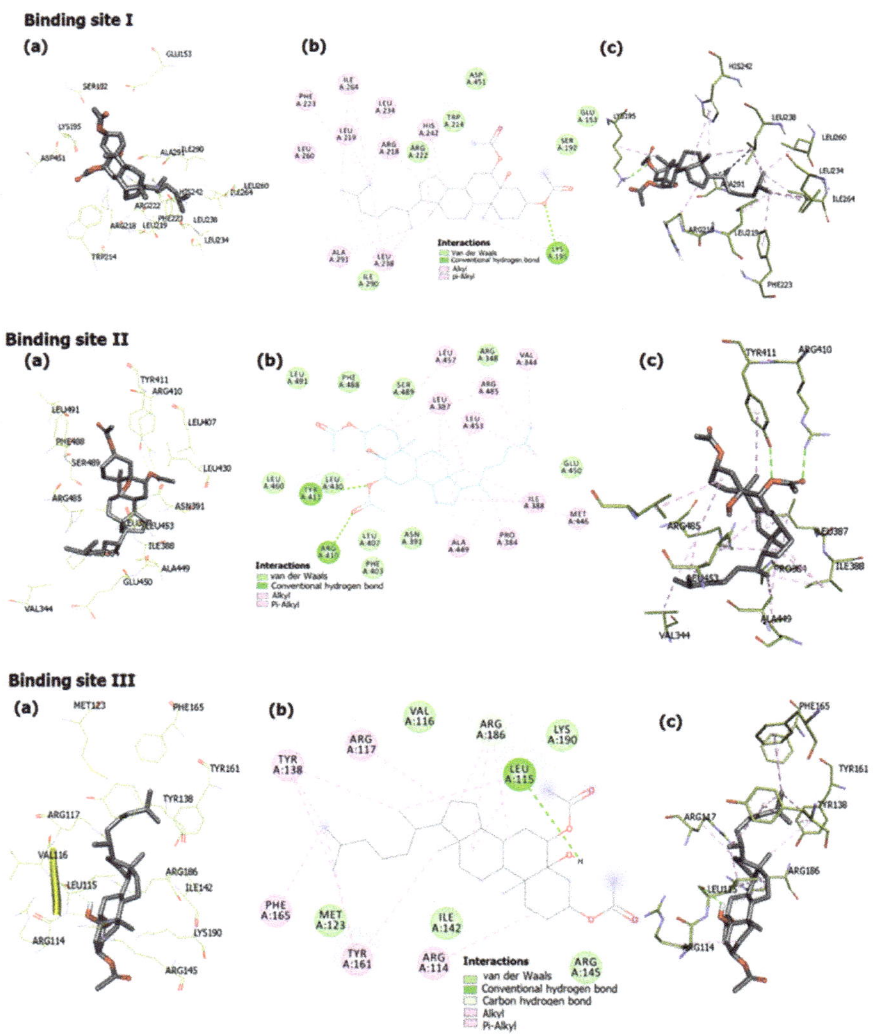

Figure 8. Binding site I, binding site II and binding site III show the best-docked poses with various interactions in (**a**) 3D, (**b**) a 2D-diagram, and (**c**) ligand interactions with active amino acids.

The best docking model for the steroid (3)–HSA had a binding score of −8.6 kcal/mol, according to the docking simulation. This suggests a strong interaction between site III and the steroid. The OH group of the steroid and the amino acid LEU115 most likely form a hydrogen bond, which accounts for the strong interaction. The steroid's binding to the site is probably going to be stabilized by this hydrogen bond. In addition to hydrogen bond formation, other non-bonding forces were observed in the best docking model. At binding site III, specific amino acids, namely, VAL116, LYS190, ILE142, MET123, and ARG146, were observed in close proximity to the steroid molecule. These amino acids interacted with each other through van der Waals forces, which are weak attractions between molecules or atoms. ARG186 was also involved in the formation of a carbon–hydrogen bond, indicating that it interacted with a carbon atom in the steroid molecule.

Furthermore, ARG117, TYR138, PHE165, TYR161, and ARG114 displayed alkyl and pi-alkyl forces with the skeleton of the steroid, indicating their contribution to the complex's stabilization, as depicted in Figure 8. As shown in Table 6, several hydrophobic amino acids, including VAL344, LEU387, ILE388, LEU407, LEU430, ALA499, GLU450, LEU453, ARG485, and PHE488, interacted with different atoms of the steroidal skeleton to improve the docking model's stability with the protein's secondary structure. Furthermore, at binding site III, ARG410 and LYS414 formed salt bridges with the steroid's carboxylate group. This suggests that these amino acids are important for enhancing the stability of the protein–steroid complex. The combined effect of these non-bonding forces results in the overall interaction between the steroid molecule and site III. The strength of this interaction is most likely determined by the number and type of non-bonding forces present. HSA–steroid interactions were examined and compared to previously published interaction profiles for HSA–ligand interactions. The comparison includes various binding sites composed of specific amino acids, such as Asp, Leu, Val, Lys, Phe, Tyr, Ser, Asn, Ala, Glu, Gln, and Gly. Some of these amino acids were also found to be involved in HSA–steroid interactions. These findings are supported by studies conducted by Shen et al. (2017) [47], and Veeralakshmi et al. (2017) [48]. Following a review of the literature, it was observed that a significant number of amino acid residues surrounding drug molecules exhibited hydrophobic properties. This finding suggests that the presence of hydrophobic effects might contribute to the stabilization of complexes formed between drugs and HSA. Likewise, most of the amino acid residues neighboring compound **3** exhibited hydrophobic characteristics. This suggests that the stability of the steroid–HSA complex could be influenced by the hydrophobic effects as well. Molecular docking analyses give a useful framework for observing and investigating on a microscopic level. They enable the quantitative study of the binding affinity score, which can provide insights into ligand–protein interactions [47,48]. The affinity of a compound for binding can exhibit variability based on the structural characteristics of the compound itself, the particular amino acids engaged in the binding process, and the specific binding site involved. As illustrated in this study, compound **3** demonstrates a higher binding affinity towards subdomain IIA compared to its affinities towards subdomains I, II, and III.

2.9. Antioxidant Potential Analysis

The ability of steroid 3 to reduce DPPH radicals was used to assess its antioxidant potential. The reduction potential of steroid 3 was measured by a decrease in the absorbance of DPPH radicals at 517 nm. The free radical scavenging activity of steroid 3 may be due to its ability to donate electrons to DPPH free radicals. This steroid has two acetoxy groups at the 3 and 6 positions and an OH group at the 5 position. These functional groups are either electron-rich or readily transfer hydrogen, and they can donate electrons to the unpaired electrons of the DPPH radical. This leads to the formation of hydrazine, a stable molecule with paired electrons. The findings indicated that steroid 3 could lower DPPH radicals, and the IC_{50} value—the amount of steroid 3 needed to lower 50% of the DPPH radicals—was determined to be 58.5 µM. This is shown in Figure 9a,b, which represent the percentage of scavenging versus the concentration. This demonstrates the significant free radical scavenging activity of steroid 3. Although the IC_{50} value of steroid 3 was notably higher than the standard (ascorbic acid IC_{50} = 47.7 mM) [49], careful examination revealed that, in terms of the antioxidant study of samples, the IC_{50} values of the sample and the standard were relatively close. This suggests that steroid 3, albeit to a slightly lesser extent, is also effective at scavenging free radicals. Scavenging free radicals is an important antioxidant activity, so this suggests that compound **3**, a steroid, has the potential to be a promising candidate for the development of novel antioxidant therapies.

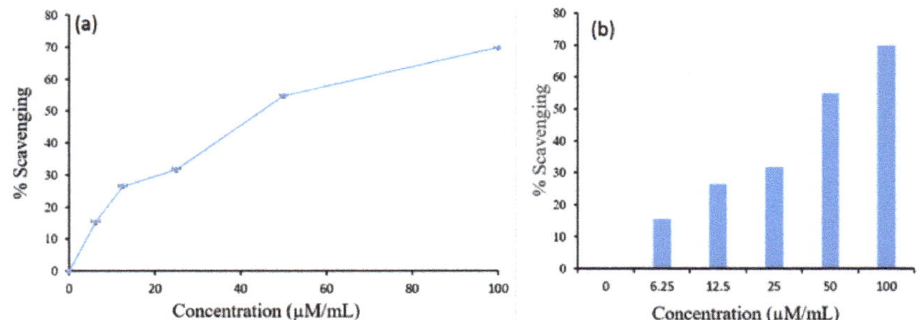

Figure 9. (**a**) DPPH radical scavenging activity (%) versus concentration (μM) and (**b**) bar diagram showing the antioxidant activity.

3. Materials and Method

Melting points are reported uncorrected. Infrared (IR) spectra were obtained by a Perkin Elmer 1600 FTIR spectrophotometer with KBr. Proton and carbon nuclear magnetic resonance (^{1}H & ^{13}C-NMR) spectra were recorded using a Varian VXR-300s instrument with Me4Si and CDCI$_3$ as the internal standard. Chemical shifts are reported in parts per million (ppm), and the abbreviations are used to describe the signals observed (s, singlet; br, broad; m, multiplet centered at; d, doublet; t, triplet). When deuterium (D$_2$O) was added, the O-H signals disappeared without significant alterations in other regions of the NMR spectra. Ultraviolet (UV) spectra were captured in chloroform using a Perkin Elmer 1800 FTIR instrument. To visualize spots and track the advancement of reactions, silica gel G-coated TLC plates were employed, along with iodine. Chromatograms were developed using mixtures of light petroleum-benzene and CHC1$_3$-acetone. Acetone was used to crystallize the product for further purification, and anhydrous sodium sulfate was employed as a desiccant to dry a solution.

3.1. Preparation of 3β,6β-Diacetoxy-5α-cholestan-5-ol

3β,6β-Diacetoxy-5α-cholestan-5-ol can be synthesized using two different methods: conventional and solid-state techniques.

Conventional procedure [50–53]: A mixture of cholestane-3β,5α,6β-triol (500 mg) (or 3β-acetoxy-5α-cholestane-5, 6β-diol as an alternative), anhydrous pyridine (3 mL), and acetic anhydride (2 mL) was subjected to heating in a water bath for a duration of 3 h. Pouring the reaction mixture into 50 g of crushed ice allowed it to cool. The solid was washed thoroughly with water and diluted hydrochloric acid (5%, 2 × 20 mL) to remove any residual pyridine. Further washing with water was performed, then filtered under suction, followed by air-drying, resulting in a yellow solid with a 90% yield. The recrystallization of the solid product using acetone resulted in the formation of crystals, which were identified as 3β,6β-diacetoxy-5α-cholestan-5-ol. The melting point of the crystals was determined to be 167–168 °C, which closely aligns with the reported m.p. of 167 °C, as specified in the literature [54].

Solid-state procedure: Cholestane-3β,5α,6β-triol (**1**) (250 mg) (or 3β-acetoxy-5α-cholestane-5, 6β-diol (2) as an alternative starting material) (250 mg), acetic anhydride (1.0 mL) or acetyl chloride, and basic alumina (2.0 g) were thoroughly ground using a mortar and pestle. The reaction mixture contained in a 50 mL flask (or vial) was irradiated in a microwave for a duration of 30 s to 2.0 min. The progress of the reaction was monitored using thin-layer chromatography (TLC). The flask was cooled once the reaction was judged to be finished, and the organic materials were then diluted with water and filtered. The precipitate was dissolved in ether, followed by washing with water and drying using anhydrous sodium sulfate. Subsequently, the solvent evaporated, and the resulting residue crystallized from acetone afforded the compound 3β,6β-diacetoxy-5α-cholestan-5-ol (Scheme 1), which exhibited a melting point of 167–168 °C (Scheme 1). Notably, the melting points of the mixtures obtained by the conventional and solid-state synthesis methods were found to be like those of the individual compounds obtained by each method.

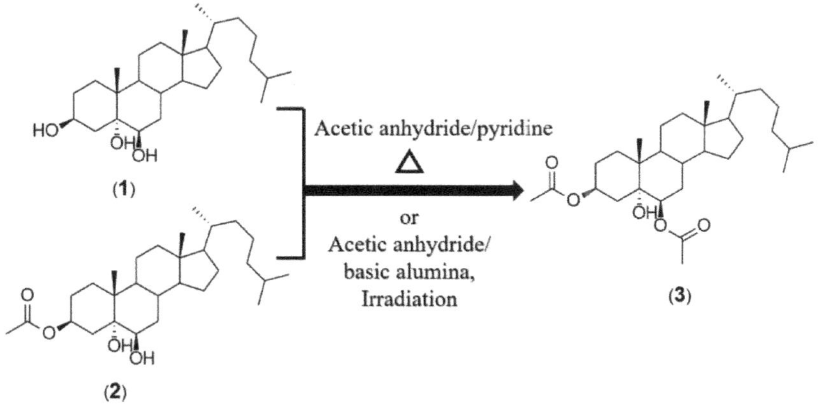

Scheme 1. Synthesis of 3β,6β-diethoxy-5α-cholestane-5-ol (**3**). Either cholestane-3β,5α,6β-triol (**1**) or 3β-acetoxy-5α-cholestane-5,6β-diol (**2**) can be used as the reactant, and the synthesis can be carried out using either conventional or solid-state methods.

Physical, spectral, and analytical data of the synthesized steroid (**3**):

M.p.: 167–168 °C (published m.p. 167 °C); Anal. calc. for $C_{31}H_{52}O_5$: C, 73.77; H, 10.38; Found: C, 73.82; H, 10.33%; IR (KBr, cm^{-1}): 3480 (tertiary alcohol, OH, asymmetric), 2931.6–2869.9 (-C–H, symmetric and asymmetric), 1741, 1736, 1266, 1238 for 2 × $-OCOCH_3$ (asymmetric and symmetric) 1031 cm^{-1} (C–O–C ether linkage, asymmetric); ^{1}HNMR (CDCl$_3$, 300 MHz); δ: 5. 37 (1H, dd, C6-αH, equatorial, J = 9.0, 6.0 Hz), 4.9 (1H, m, C3-aH, axial, W1/2 = 17 Hz, A/B ring function trans), 2.24 (1H, s, OH, exchangeable with D_2O), 2.07, 2.0 (2 × $-OCOCH_3$), 1.03 (s, C10-CH$_3$), 0.73 (s, C13-CH$_3$), 0.97–0.87 ppm for other side chain methyls; ^{13}C NMR (CDCl3, 100 MHz); δ: 171.0 −171.2 (2 × $-OC=O-$), 78.9 (C6), 75.9 (C5), 71.1 (C3), 57.1 (C14), 56.7 (C17), 44.6 (C9), 44.1 (C13), 43.1 (C10), 39.9 (C13), 39.4 (C24), 38.9 (C4), 36.5 (C20), 36.3 (C22), 35.1 (C1), 32.3 (C7), 30.5 (C8), 28.7 (C2), 27.9 (C25), 25.2–25.3 (C15 and C16), 24.1 (C23), 22.6–22.5 (C26/C27), 21.3 (C11), 21.2–20.9 (2 × CH3-COO-), 18.9 (C21), 16.2 (C19), and (C18) 13.3 ppm; UV λmax(methanol) 212 nm; HRMS (ESI): calcd. for $C_{31}H_{52}O_5$ [M + H]$^+$: 505.38; Found: 505.38. The IR and NMR spectra of compound **3**, shown in Figures S1–S3, are provided in the supplementary information accompanying this paper.

3.2. Single X-ray Crystallography and Computational Details

The diffraction data for crystals were obtained using the APEX2 diffractometer (Bruker–Noius, 2004). Subsequently, cell refinement and data reduction were performed using SAINT (Bruker, 2003) and the SAINT program, respectively. After making necessary corrections to the collected data, the title compound structure was determined using a direct method facilitated by the SHELXT-2015 software [55]. The structure was further refined using least squares with the help of the SHELXL-2015 software [56], which is included in the Olex2 package program [57]. The refinement process involves the anisotropic refinement of all non-hydrogen atoms. Subsequently, hydrogen atoms attached to non-carbon atoms were positioned based on the difference Fourier maps. The remaining hydrogen atoms were treated as rider atoms and refined using the isotropic displacement parameter. Using MERCURY software, diagrams and publication materials were produced [58].

3.3. Computational and Molecular Docking Studies

To examine the molecular structure of 3β,6β-diethoxy-5α-cholestan-5-ol, quantum chemical calculations were performed using the Gaussian 09 program [59]. The calculations were carried out in both vacuum and solvent phases using the 6-311G++(d,p) basis set and the B3LYP density functional [60,61]. The optimized structure is identified at the true minimum of the potential energy level, as confirmed by the actual value of the frequency derived at the same theoretical level. At the same theoretical level, the energy gaps of FMOs (HOMO and LUMO) of 3β,6β-diaoetoxy-5α-cholestane-5-ol are obtained. The effectiveness of long-range forces was proved using the Multiwfn program [62]. The calculation and analysis of a number of parameters related to these forces were made simpler by this application. Isosurface visualizations were also produced with the help of the VMD application to give the examined data a visual representation. To determine weak intermolecular interactions in the crystal, CrystalExplorer 21.5 [63] was used to study weak interactions between molecules in 3β,6β-diethoxy-5α-cholestan-5-ol crystals (3) creating Hirshfield surfaces and 2D fingerprints. Molecular docking is a method for predicting the extent to which small molecules bind to proteins. The Lamarckian Genetic Algorithm (LGA) operated in this study to perform an in-silico approach to the binding patterns of 3β, 6β-diacetoxy-5α-cholestan-5-ol with human serum albumin. The three-dimensional (3D) crystal structure of human serum albumin with PDB ID 1H9Z was used as a model of a receptor, and the crystallographic information file (CIF) of 3β, 6β-diacetoxy-5α-cholestan-5-ol was saved in PDF format and used as a ligand. Based on the reported literature regarding multiple active HSA sites [47], three different docking grid boxes were set up, each with a grid spacing of 0.375 and 60 × 60 × 60 points. For binding site I, the box centers were placed at x = 30.025, y = 9.580, z = 10.295; for binding site II, they were placed at x = 12.005, y = 8.737, z = 20.079; and for binding site III, they were placed at x = 33.049, y = 18.888, z = 35.8. The lowest energy structure docked at each active site was selected, and the Discovery Studio was employed to visualize the output and analyze hydrogen bonding and unbound interactions between the receptor and 3β,6β-diacetoxy-5α -cholestan-5-ol. Protein–ligand interaction profiles in multiple binding sites of the Human Serum Albumin (HSA) with steroid 3 were also obtained using the PLIP web tool [64].

3.4. HSA-Binding Experiments

3.4.1. HSA Sample Preparation

The concentration of human serum albumin (HSA) for a stock solution prepared by dissolving HSA in a 20 mM phosphate buffer solution at pH 7.4 was determined by measuring the absorbance of an HSA solution at 280 nm using a Perkin–Elmer–Lambda double beam UV–vis spectrophotometer. The extinction coefficient of HSA at 280 nm is 5.30. Sodium dodecyl sulfate-polyacrylamide gel electrophoresis (SDS-PAGE), which revealed a single band, was used to confirm the purity of HSA. To prepare different concentrations of the steroid, a stock solution of 2 mM 3β,6β-diacetoxy-5α-cholestan-5-ol in dimethyl sulfoxide (DMSO) was prepared and diluted in phosphate buffer. Blanks were made using 20 mM PB at pH 7.4 and 298 K. The pH of each solution was determined using an Orion-401-plus pH meter and an Orion glass electrode. The absorption spectra of HSA–steroid complexes were measured at increasing concentrations of the compound (5–50 μM). The concentration of HSA was held constant at 5 μM in this HSA binding study.

3.4.2. UV–Vis Absorption Titration

At a temperature of 298 K, the UV–vis absorption spectrum of HSA was examined. Measurements were performed with and without various concentrations of 3β,6β-diaoetoxy-5α-cholestan-5-ol. The experiment was carried out using quartz cuvettes with a 1 cm path length and a Peltier temperature controller on a Perkin–Elmer–Lambda double-beam UV–vis spectrophotometer. The concentration of human serum albumin was maintained at 5 micromolar in 20 mM sodium phosphate buffer, pH 7.4, and then treated with increasing concentrations of 3β,6β-diacetoxy-5α-cholestan-5-ol for titration. Appropriate values for negative controls were removed from HSA-3β,6β-diacetoxy-5α-cholestan-5-ol readings to eliminate sample absorbance.

3.4.3. Fluorescence Quenching Measurement

The intrinsic fluorescence of HSA was assessed using a Shimadzu fluorescence spectrophotometer RF-5301 equipped with a quartz cuvette and a 1 cm pathlength. The excitation wavelength was set at 295 nm, and emission spectra were recorded between 300 and 400 nm. The excitation and emission slit widths were both set to 5 nm. To explore fluorescence quenching experiments in more depth, The Stern–Volmer Equation (5) was used, as shown in references [65–67].

$$F0/F = Ksv\,[Q] + 1 = kq\tau 0\,[Q] + 1 \tag{5}$$

The effectiveness of a quencher's (Q) ability to quench fluorescence is gauged by the Stern–Volmer quenching constant (Ksv). It is determined by the average integrated fluorescence lifetime of the fluorophore ($\tau 0$), the molar concentration of the quencher, and the bimolecular rate constant (kq) of the quenching reaction. A tryptophan, $\tau 0$, equals roughly 10^{-9} s. The modified Stern-Volmer Equation (6) was used to calculate the binding constants (Kb) and number of binding sites (n) involved in the interaction of HSA samples. The free energy change (ΔG^0) of the process was determined by analyzing the fluorescence quenching data using Equation (6) and then plugging the results into Equation (7). These calculations facilitate the quantitative assessment of the binding constants (K_b) and number of binding sites (n) involved in the interaction of HSA samples.

$$\mathrm{Log}\,(F_0/F - 1) = \log K_b + n \log [Q] \tag{6}$$

where F_0 and F represent the fluorescence intensity in the absence and presence of the quencher (sample), K_b represents the binding constant, and n represents the number of binding sites, respectively.

$$\Delta G^0 = -RT\ln K_b \tag{7}$$

Using the above equation, the change in free energy (ΔG°) can be calculated from the binding constant (Kb) at temperature (T), where R is the universal gas constant $(1.987\ \mathrm{cal\ mol^{-1}K^{-1}})$.

3.5. In Vitro Measurement of Antioxidant Properties Using the DPPH Radical Scavenging Assay

The DPPH assay is a common method for determining a substance's antioxidant capacity. It assesses a substance's ability to reduce the stable free radical DPPH [68,69]. The method works by observing the reduction in alcoholic DPPH solutions in the presence of antioxidants that donate hydrogen atoms or electrons. Measure the hydrogen or electron-donating ability of the compound by observing the bleaching of a purple methanolic solution of 2,2-diphenyl-1-picrylhydrazyl (DPPH) using spectrophotometry. The percent inhibition of DPPH is a reliable indicator of the antioxidant capacity of a sample. The larger the inhibition percentage, the stronger the antioxidant activity. The DPPH assay provides a simple and convenient method for evaluating the antioxidant potential of different compounds in vitro. Of note, this assay specifically assessed the compound's ability to lower DPPH and did not provide a comprehensive measure of its overall antioxidant activity. In this study, the antioxidant activity of steroid 3 was assessed using the stable free radical 2,2-diphenyl-1-picrylhydrazyl (DPPH). Using the conventional DPPH assay, the sample's capacity to scavenge free radicals was evaluated. For the UV measurements, a freshly made solution of DPPH in methanol (6×10^{-5} M) was employed. Steroid 3 solutions were added to the DPPH solution in a 1:1 ratio and vortexed. The concentrations of the Steroid 3 solutions ranged from 6.25 to 100 µM. In other words, 1 mL of the steroid 3 solutions and 1 mL of the DPPH solution were combined. The reaction mixture was thoroughly vortexed and incubated for 30 min at room temperature in the dark. The standard used was ascorbic acid. At 517 nm, the inhibition percentage of the activity to scavenge DPPH radicals was determined using the following equation:

$$\mathrm{Inhibition\ (\%)} = [(A_0 - A)/A_0] \times 100$$

A plot of the percentage of DPPH inhibition against the concentration of the sample solutions was used to estimate the IC50 values (the concentration needed to scavenge 50% of the free radical). In this case, A_0 represents DPPH absorbance in the absence of the steroid, and A represents DPPH absorbance in the presence of the steroid.

3.6. Statistical Analysis

The data were analyzed using the mean and standard deviation (SD). The number of individual experiments is indicated by n.

4. Conclusions

A cholestane derivative was synthesized using an eco-friendly approach. Its binding to human serum albumin (HSA) was studied using fluorescence spectroscopy, UV–vis spectroscopy, molecular docking, and density functional theory. The results showed spontaneous binding ($\Delta G = -9.86$ kcal/mol) supported by negative binding scores in molecular docking. The HSA–ligand system exhibited static quenching, suggesting conformational changes in the protein upon binding. The molecule's solid-state packing is governed by dominant H$\cdots$H, C$\cdots$H, and O$\cdots$H interactions, while strong intramolecular interactions were found within the crystal structure through Reduced Density Gradients analysis. This is supported further by the frontier molecular orbital analysis, which revealed a large energy gap of 6.94 eV for the molecule. This indicates that the molecule is relatively stable and unlikely to decompose. The DPPH radical scavenging activity of the studied steroids was comparable to that of standard ascorbic acid. This indicates that the steroid may have antioxidant properties, but more research is needed to determine its mode of action and anticancer activity in vivo. Overall, our findings shed light on the mechanism of cholestane derivative binding to HSA as well as antioxidant analysis. The investigation into the interaction between cholesterol derivatives and human serum albumin (HSA) elucidates a fundamental framework conducive to the advancement of investigations in the domains of pharmaceutical delivery systems, agents for cholesterol reduction, cholesterol metabolism, and the integrity of peptide structures. These findings pave the way for potential future applications in research endeavors. These findings could also have implications for the development of new HSA-targeting therapeutic agents, as well as anticancer research.

Supplementary Materials: The following supporting information can be downloaded at: https://www.mdpi.com/article/10.3390/molecules28165942/s1, Figure S1: IR spectrum of 3, Figure S2: ^{1}H NMR spectrum of 3, Figure S3: ^{13}C NMR spectrum of 3, Table S1: Crystal data of titled compound 3. The supplementary crystallographic data has been deposited at the Cambridge Crystallographic Data Center, CCDC No. 901851 and may be obtained free of charge from http://www.ccdc.cam.ac.uk (accessed on 10 July 2023).

Funding: This research was not funded by any external agency.

Institutional Review Board Statement: Not applicable.

Informed Consent Statement: Not applicable.

Data Availability Statement: The findings are supported by experimental data. No further data are available.

Conflicts of Interest: The author declares no conflict of interest.

Sample Availability: Sample is available upon request.

References

1. Xie, J.; Zhang, S.-J.; Yue, G.G.-L.; Yu, L.-L.; Liu, H.; Ma, W.-Y.; Yan, H.; Ni, W.; Bik-San Lau, C.; Liu, H.-Y. Isolation, structural elucidation, and bioactivity of cholestane derivatives from *Ypsilandra thibetica*. *New J. Chem.* **2023**, *47*, 324–332. [CrossRef]
2. Ure, E.M.; Harris, L.D.; Cameron, S.A.; Weymouth-Wilson, A.; Furneaux, R.H.; Pitman, J.L.; Hinkley, S.F.; Luxenburger, A. Synthesis of 12β-methyl-18-nor-avicholic acid analogues as potential TGR5 agonists. *Org. Biomol. Chem.* **2022**, *20*, 3511–3527. [CrossRef] [PubMed]

3. Huang, Y.; Li, G.; Hong, C.; Zheng, X.; Yu, H.; Zhang, Y. Potential of Steroidal Alkaloids in Cancer: Perspective Insight into Structure–Activity Relationships. *Front. Oncol.* **2021**, *11*, 733369. [CrossRef] [PubMed]

4. Lin, C.-Y.; Huo, C.; Kuo, L.-K.; Hiipakka, R.A.; Jones, R.B.; Lin, H.-P.; Hung, Y.; Su, L.-C.; Tseng, J.-C.; Kuo, Y.-Y. Cholestane-3β, 5α, 6β-triol suppresses proliferation, migration, and invasion of human prostate cancer cells. *PLoS ONE* **2013**, *8*, e65734. [CrossRef] [PubMed]

5. Fan, P.; Wang, Y.; Lu, K.; Hong, Y.; Xu, M.; Han, X.; Liu, Y. Modeling maternal cholesterol exposure reveals a reduction of neural progenitor proliferation using human cerebral organoids. *Life Med.* **2023**, *2*, lnac034. [CrossRef]

6. German, C.A.; Liao, J.K. Understanding the molecular mechanisms of statin pleiotropic effects. *Arch. Toxicol.* **2023**, *97*, 1529–1545. [CrossRef]

7. Dolivo, D.M.; Reed, C.R.; Gargiulo, K.A.; Rodrigues, A.E.; Galiano, R.D.; Mustoe, T.A.; Hong, S.J. Anti-fibrotic effects of statin drugs: A review of evidence and mechanisms. *Biochem. Pharmacol.* **2023**, *214*, 115644. [CrossRef]

8. Andelova, K.; Bacova, B.S.; Sykora, M.; Hlivak, P.; Barancik, M.; Tribulova, N. Mechanisms underlying antiarrhythmic properties of cardioprotective agents impacting inflammation and oxidative stress. *Int. J. Mol. Sci.* **2022**, *23*, 1416. [CrossRef]

9. Wiggs, A.G.; Chandler, J.K.; Aktas, A.; Sumner, S.J.; Stewart, D.A. The effects of diet and exercise on endogenous estrogens and subsequent breast cancer risk in postmenopausal women. *Front. Endocrinol.* **2021**, *12*, 732255. [CrossRef]

10. McIver, L.A.; Siddique, M.S. *Atorvastatin*; Statpearls Publishing: Treasure Island, FL, USA, 2020.

11. Kim, T.-H.; Yu, G.-R.; Kim, H.; Kim, J.-E.; Lim, D.-W.; Park, W.-H. Network Pharmacological Analysis of a New Herbal Combination Targeting Hyperlipidemia and Efficacy Validation In Vitro. *Curr. Issues Mol. Biol.* **2023**, *45*, 1314–1332. [CrossRef]

12. Lewek, J.; Surma, S.; Banach, M. Statins and COVID-19 (Mechanism of Action, Effect on Prognosis). In *Cardiovascular Complications of COVID-19: Acute and Long-Term Impacts*; Springer: Berlin/Heidelberg, Germany, 2023; pp. 285–302.

13. Yu, D.; Liao, J.K. Emerging views of statin pleiotropy and cholesterol lowering. *Cardiovasc. Res.* **2022**, *118*, 413–423. [CrossRef]

14. Hariri, L.; Rehman, A. *Estradiol*; StatPearls Publishing: Treasure Island, FL, USA, 2020.

15. Geraci, A.; Calvani, R.; Ferri, E.; Marzetti, E.; Arosio, B.; Cesari, M. Sarcopenia and menopause: The role of estradiol. *Front. Endocrinol.* **2021**, *12*, 682012. [CrossRef] [PubMed]

16. Mishra, V.; Heath, R.J. Structural and biochemical features of human serum albumin essential for eukaryotic cell culture. *Int. J. Mol. Sci.* **2021**, *22*, 8411. [CrossRef]

17. Ma, J.; Yang, B.; Hu, X.; Gao, Y.; Qin, C. The binding mechanism of benzophenone-type UV filters and human serum albumin: The role of site, number, and type of functional group substitutions. *Environ. Pollut.* **2023**, *324*, 121342. [CrossRef]

18. Gallagher, A.; Kar, S.; Sepúlveda, M.S. Computational Modeling of Human Serum Albumin Binding of Per-and Polyfluoroalkyl Substances Employing QSAR, Read-Across, and Docking. *Molecules* **2023**, *28*, 5375. [CrossRef] [PubMed]

19. Yang, S.; Zhang, W.; Liu, Z.; Zhai, Z.; Hou, X.; Wang, P.; Ge, G.; Wang, F. Lysine reactivity profiling reveals molecular insights into human serum albumin–small-molecule drug interactions. *Anal. Bioanal. Chem.* **2021**, *413*, 7431–7440. [CrossRef]

20. Mariño-Ocampo, N.; Rodríguez, D.F.; Guerra Díaz, D.; Zúñiga-Núñez, D.; Duarte, Y.; Fuentealba, D.; Zacconi, F.C. Direct Oral FXa Inhibitors Binding to Human Serum Albumin: Spectroscopic, Calorimetric, and Computational Studies. *Int. J. Mol. Sci.* **2023**, *24*, 4900. [CrossRef] [PubMed]

21. Yang, F.; Zhang, Y.; Liang, H. Interactive association of drugs binding to human serum albumin. *Int. J. Mol. Sci.* **2014**, *15*, 3580–3595. [CrossRef]

22. Abdollahpour, N.; Soheili, V.; Saberi, M.R.; Chamani, J. Investigation of the interaction between human serum albumin and two drugs as binary and ternary systems. *Eur. J. Drug Metab. Pharmacokinet.* **2016**, *41*, 705–721. [CrossRef]

23. di Masi, A. Human serum albumin: From molecular aspects to biotechnological applications. *Int. J. Mol. Sci.* **2023**, *24*, 4081. [CrossRef]

24. Chadha, N.; Singh, D.; Milton, M.D.; Mishra, G.; Daniel, J.; Mishra, A.K.; Tiwari, A.K. Computational prediction of interaction and pharmacokinetics profile study for polyamino-polycarboxylic ligands on binding with human serum albumin. *New J. Chem.* **2020**, *44*, 2907–2918. [CrossRef]

25. Wang, J.; Cheng, J.-J.; Cheng, J.-H.; Liang, W. Characterization of the interactions of human serum albumin with carmine and amaranth using multi-spectroscopic techniques and molecular docking. *J. Food Meas. Charact.* **2022**, *16*, 4345–4354. [CrossRef]

26. Fan, J.; Gilmartin, K.; Octaviano, S.; Villar, F.; Remache, B.; Regan, J. Using human serum albumin binding affinities as a proactive strategy to affect the pharmacodynamics and pharmacokinetics of preclinical drug candidates. *ACS Pharmacol. Transl. Sci.* **2022**, *5*, 803–810. [CrossRef] [PubMed]

27. Alavi, F.S.; Ghadari, R.; Zahedi, M. Exploration of the binding properties of the human serum albumin sites with neurology drugs by docking and molecular dynamics simulation. *J. Iran. Chem. Soc.* **2017**, *14*, 19–35. [CrossRef]

28. Bratty, M.A. Spectroscopic and molecular docking studies for characterizing binding mechanism and conformational changes of human serum albumin upon interaction with Telmisartan. *Saudi Pharm. J. SPJ Off. Publ. Saudi Pharm. Soc.* **2020**, *28*, 729–736. [CrossRef] [PubMed]

29. Zhou, Z.; Hu, X.; Hong, X.; Zheng, J.; Liu, X.; Gong, D.; Zhang, G. Interaction characterizaton of 5-hydroxymethyl-2-furaldehyde with human serum albumin: Binding characteristics, conformational change and mechanism. *J. Mol. Liq.* **2020**, *297*, 111835. [CrossRef]

30. Pontremoli, C.; Barbero, N.; Viscardi, G.; Visentin, S. Insight into the interaction of inhaled corticosteroids with human serum albumin: A spectroscopic-based study. *J. Pharm. Anal.* **2018**, *8*, 37–44. [CrossRef]

31. Nda-Umar, U.I.; Ramli, I.B.; Muhamad, E.N.; Azri, N.; Amadi, U.F.; Taufiq-Yap, Y.H. Influence of heterogeneous catalysts and reaction parameters on the acetylation of glycerol to acetin: A review. *Appl. Sci.* **2020**, *10*, 7155. [CrossRef]
32. Castro, A.; Andrade, I.M.G.; Coelho, M.C.; da Costa, D.P.; Moreira, D.d.N.; Maia, R.A.; Lima, G.d.S.; dos Santos, G.F.; Vaz, B.G.; Militão, G.C.G. Multicomponent synthesis of spiro 1, 3, 4-thiadiazolines with anticancer activity by using deep eutectic solvent under microwave irradiation. *J. Heterocycl. Chem.* **2023**, *60*, 392–405. [CrossRef]
33. Pasricha, S.; Rangarajan, T. Green Acetylation of Primary Aromatic Amines. *Resonance* **2023**, *28*, 325–331. [CrossRef]
34. Palma, V.; Barba, D.; Cortese, M.; Martino, M.; Renda, S.; Meloni, E. Microwaves and heterogeneous catalysis: A review on selected catalytic processes. *Catalysts* **2020**, *10*, 246. [CrossRef]
35. Rajnikant, V.; Jasrotia, D.; Chand, B. Comparative crystallographic and hydrogen-bonding analysis of pregnane derivatives. *J. Chem. Crystallogr.* **2008**, *38*, 211–230. [CrossRef]
36. Pinto, R.; Ramos Silva, M.; Matos Beja, A.; Salvador, J.; Paixao, J. 6β-Acetamido-5α-hydroxycholestan-3β-yl acetate. *Acta Crystallogr. Sect. E Struct. Rep. Online* **2008**, *64*, o2303. [CrossRef] [PubMed]
37. Cremer, D.t.; Pople, J. General definition of ring puckering coordinates. *J. Am. Chem. Soc.* **1975**, *97*, 1354–1358. [CrossRef]
38. Huang, Y.; Rong, C.; Zhang, R.; Liu, S. Evaluating frontier orbital energy and HOMO/LUMO gap with descriptors from density functional reactivity theory. *J. Mol. Model.* **2017**, *23*, 3. [CrossRef]
39. Padrón, J.; Carrasco, R.; Pellon, R. Molecular descriptor based on a molar refractivity partition using Randic-type graph-theoretical invariant. *J. Pharm. Pharmaceut. Sci* **2002**, *5*, 258–266.
40. Alam, M.J.; Ahmad, S. Molecular structure, anharmonic vibrational analysis and electronic spectra of o-, m-, p-iodonitrobenzene using DFT calculations. *J. Mol. Struct.* **2014**, *1059*, 239–254. [CrossRef]
41. Guo, X.; Zhang, L.; Sun, X.; Han, X.; Guo, C.; Kang, P. Spectroscopic studies on the interaction between sodium ozagrel and bovine serum albumin. *J. Mol. Struct.* **2009**, *928*, 114–120. [CrossRef]
42. Gowda, J.I.; Nandibewoor, S.T. Binding and conformational changes of human serum albumin upon interaction with 4-aminoantipyrine studied by spectroscopic methods and cyclic voltammetry. *Spectrochim. Acta Part A Mol. Biomol. Spectrosc.* **2014**, *124*, 397–403. [CrossRef]
43. Zhang, J.; Gao, X.; Huang, J.; Wang, H. Probing the interaction between human serum albumin and 9-hydroxyphenanthrene: A spectroscopic and molecular docking study. *ACS Omega* **2020**, *5*, 16833–16840. [CrossRef]
44. Ross, P.D.; Subramanian, S. Thermodynamics of protein association reactions: Forces contributing to stability. *Biochemistry* **1981**, *20*, 3096–3102. [CrossRef] [PubMed]
45. Hentschel, L.; Hansen, J.; Egelhaaf, S.U.; Platten, F. The crystallization enthalpy and entropy of protein solutions: Microcalorimetry, van't Hoff determination and linearized Poisson–Boltzmann model of tetragonal lysozyme crystals. *Phys. Chem. Chem. Phys.* **2021**, *23*, 2686–2696. [CrossRef] [PubMed]
46. Jayaraj, A.; Schwanz, H.A.; Spencer, D.J.; Bhasin, S.; Hamilton, J.A.; Jayaram, B.; Goldman, A.L.; Krishna, M.; Krishnan, M.; Shah, A. Allosterically coupled multisite binding of testosterone to human serum albumin. *Endocrinology* **2021**, *162*, bqaa199. [CrossRef] [PubMed]
47. Shen, F.; Liu, Y.-X.; Li, S.-M.; Jiang, C.-K.; Wang, B.-F.; Xiong, Y.-H.; Mao, Z.-W.; Le, X.-Y. Synthesis, crystal structures, molecular docking and in vitro cytotoxicity studies of two new copper (II) complexes: Special emphasis on their binding to HSA. *New J. Chem.* **2017**, *41*, 12429–12441. [CrossRef]
48. Veeralakshmi, S.; Sabapathi, G.; Nehru, S.; Venuvanalingam, P.; Arunachalam, S. Surfactant–cobalt (III) complexes: The impact of hydrophobicity on interaction with HSA and DNA–insights from experimental and theoretical approach. *Colloids Surf. B Biointerfaces* **2017**, *153*, 85–94. [CrossRef]
49. Brighente, I.; Dias, M.; Verdi, L.; Pizzolatti, M. Antioxidant activity and total phenolic content of some Brazilian species. *Pharm. Biol.* **2007**, *45*, 156–161. [CrossRef]
50. Mahadevan, V.; Lundberg, W. Preparation of cholesterol esters of long-chain fatty acids and characterization of cholesteryl arachidonate. *J. Lipid Res.* **1962**, *3*, 106–110. [CrossRef]
51. Petersen, Q.R. Reductive Preparation of Oximes and the Selective Hydrolysis of their Acetates on Alumina. *Indiana Acad. Sci.* **1963**, *73*, 127–131.
52. Tsui, P.; Just, G. Some Reactions of 3 β-Mesyloxycholestane-5α, 6β-diol and Cholest-2-ene-5α, 6β-diol Acetates. *Can. J. Chem.* **1973**, *51*, 3502–3507. [CrossRef]
53. Lieberman, S.; Fukushima, D.K. Δ 5-Cholestene-3β, 4β, 7α-triol and the Inhibition of the Oxidation of Hydroxyl Groups by Vicinal Substituents. *J. Am. Chem. Soc.* **1950**, *72*, 5211–5218. [CrossRef]
54. Rowland, A.T. An Attempted Westphalen Rearrangement of a 5β-Hydroxy Steroid1. *J. Org. Chem.* **1964**, *29*, 222–224. [CrossRef]
55. Sheldrick, G.M. SHELXT–Integrated space-group and crystal-structure determination. *Acta Crystallogr. Sect. A Found. Adv.* **2015**, *71*, 3–8. [CrossRef]
56. Sheldrick, G.M. Crystal structure refinement with SHELXL. *Acta Crystallogr. Sect. C Struct. Chem.* **2015**, *71*, 3–8. [CrossRef] [PubMed]
57. Dolomanov, O.V.; Bourhis, L.J.; Gildea, R.J.; Howard, J.A.; Puschmann, H. OLEX2: A complete structure solution, refinement and analysis program. *J. Appl. Crystallogr.* **2009**, *42*, 339–341. [CrossRef]
58. Macrae, C.F.; Edgington, P.R.; McCabe, P.; Pidcock, E.; Shields, G.P.; Taylor, R.; Towler, M.; Streek, J. Mercury: Visualization and analysis of crystal structures. *J. Appl. Crystallogr.* **2006**, *39*, 453–457. [CrossRef]

59. Frisch, M.; Trucks, G.; Schlegel, H.; Scuseria, G.; Robb, M.; Cheeseman, J.; Scalmani, G.; Barone, V.; Mennucci, B.; Petersson, G. *Gaussian 09, Revision, D. 01*; Gaussian, Inc.: Wallingford, CT, USA, 2009. Available online: http://www.gaussian.com (accessed on 1 January 2014).
60. Becke, A.D. Densityfunctional thermochemistry. III. the role of exact exchange. *J. Chem. Phys.* **1993**, *98*, 5648–5652. [CrossRef]
61. Lee, C.; Yang, W.; Parr, R.G. Development of the Colle-Salvetti correlation-energy formula into a functional of the electron density. *Phys. Rev. B* **1988**, *37*, 785. [CrossRef]
62. Lu, T.; Chen, F. Multiwfn: A multifunctional wavefunction analyzer. *J. Comput. Chem.* **2012**, *33*, 580–592. [CrossRef]
63. Spackman, P.R.; Turner, M.J.; McKinnon, J.J.; Wolff, S.K.; Grimwood, D.J.; Jayatilaka, D.; Spackman, M.A. CrystalExplorer: A program for Hirshfeld surface analysis, visualization and quantitative analysis of molecular crystals. *J. Appl. Crystallogr.* **2021**, *54*, 1006–1011. [CrossRef]
64. Adasme, M.F.; Linnemann, K.L.; Bolz, S.N.; Kaiser, F.; Salentin, S.; Haupt, V.J.; Schroeder, M. PLIP 2021: Expanding the scope of the protein–ligand interaction profiler to DNA and RNA. *Nucleic Acids Res.* **2021**, *49*, W530–W534. [CrossRef]
65. Anand, U.; Jash, C.; Mukherjee, S. Spectroscopic probing of the microenvironment in a protein– surfactant assembly. *J. Phys. Chem. B* **2010**, *114*, 15839–15845. [CrossRef] [PubMed]
66. Wu, H.; Zhao, X.; Wang, P.; Dai, Z.; Zou, X. Electrochemical site marker competitive method for probing the binding site and binding mode between bovine serum albumin and alizarin red S. *Electrochim. Acta* **2011**, *56*, 4181–4187. [CrossRef]
67. Neamtu, S.; Mic, M.; Bogdan, M.; Turcu, I. The artifactual nature of stavudine binding to human serum albumin. A fluorescence quenching and isothermal titration calorimetry study. *J. Pharm. Biomed. Anal.* **2013**, *72*, 134–138. [CrossRef]
68. Katsube, T.; Tabata, H.; Ohta, Y.; Yamasaki, Y.; Anuurad, E.; Shiwaku, K.; Yamane, Y. Screening for antioxidant activity in edible plant products: Comparison of low-density lipoprotein oxidation assay, DPPH radical scavenging assay, and Folin– Ciocalteu assay. *J. Agric. Food Chem.* **2004**, *52*, 2391–2396. [CrossRef] [PubMed]
69. Ono, M.; Takamura, C.; Sugita, F.; Masuoka, C.; Yoshimitsu, H.; Ikeda, T.; Nohara, T. Two new steroid glycosides and a new sesquiterpenoid glycoside from the underground parts of Trillium kamtschaticum. *Chem. Pharm. Bull.* **2007**, *55*, 551–556. [CrossRef]

Review

Fascinating Furanosteroids and Their Pharmacological Profile

Valery M. Dembitsky †

Centre for Applied Research, Innovation and Entrepreneurship, Lethbridge College, 3000 College Drive South, Lethbridge, AB T1K 1L6, Canada; valery.dembitsky@lethbridgecollege.ca
† Current address: Bio-Geo-Chem Laboratories, 23615 El Toro Rd X, P.O. Box 049, Lake Forest, CA 92630, USA.

Abstract: This review article delves into the realm of furanosteroids and related isoprenoid lipids derived from diverse terrestrial and marine sources, exploring their wide array of biological activities and potential pharmacological applications. Fungi, fungal endophytes, plants, and various marine organisms, including sponges, corals, molluscs, and other invertebrates, have proven to be abundant reservoirs of these compounds. The biological activities exhibited by furanosteroids and related lipids encompass anticancer, cytotoxic effects against various cancer cell lines, antiviral, and antifungal effects. Notably, the discovery of exceptional compounds such as nakiterpiosin, malabaricol, dysideasterols, and cortistatins has revealed their potent anti-tuberculosis, antibacterial, and anti-hepatitis C attributes. These compounds also exhibit activity in inhibiting protein kinase C, phospholipase A2, and eliciting cytotoxicity against cancer cells. This comprehensive study emphasizes the significance of furanosteroids and related lipids as valuable natural products with promising therapeutic potential. The remarkable biodiversity found in both terrestrial and marine ecosystems offers an extensive resource for unearthing novel biologically active compounds, paving the way for future drug development and advancements in biomedical research. This review presents a compilation of data obtained from various studies conducted by different authors who employed the PASS software 9.1 to evaluate the biological activity of natural furanosteroids and compounds closely related to them. The utilization of the PASS software in this context offers valuable advantages, such as screening large chemical libraries, identifying compounds for subsequent experimental investigations, and gaining insights into potential biological activities based on their structural features. Nevertheless, it is crucial to emphasize that experimental validation remains indispensable for confirming the predicted activities.

Keywords: furanosteroids; isoprenoid lipids; fungi; fungal endophytes; plants; marine organisms

Citation: Dembitsky, V.M. Fascinating Furanosteroids and Their Pharmacological Profile. *Molecules* **2023**, *28*, 5669. https://doi.org/10.3390/molecules28155669

Academic Editors: Marina Savić, Erzsébet Mernyák, Jovana Ajdukovic and Suzana Jovanović-Šanta

Received: 27 June 2023
Revised: 21 July 2023
Accepted: 24 July 2023
Published: 26 July 2023

1. Introduction

Natural and synthesized organic compounds that contain a furan ring(s) exhibit diverse cardiovascular activity [1–3]. These compounds find extensive use as antibacterial, antiviral, anti-inflammatory, antifungal, anticancer, antihyperglycemic, analgesic, and anticonvulsant medications [1–6]. According to *ChemNetBase*, the number of natural compounds containing a furan ring exceeds 12,000 per molecule. Additionally, compounds with a 2,5-dihydrofuran ring surpass 12,200, while those with a 2,3-dihydrofuran ring exceed 1850. Moreover, compounds with a tetrahydrofuran ring account for over 47,570 compounds. Furthermore, steroids and related isoprenoid lipids that possess the additional ring(s) can be referred to as furanosteroids and their analogues and derivatives, comprising a collection of over 1000 compounds [7].

Furanosteroids represent a class of pentacyclic isoprenoid lipids that are synthesized by fungi and various other organisms. These compounds possess additional fused furan, dihydrofuran, or tetrahydrofuran ring(s) attached to their pentacyclic backbone [8–10].

Initially, furanosteroids were identified in mushrooms nearly 80 years ago, characterized by the presence of an extra furan ring connecting positions 4 and 6 of the steroid skeleton. However, the term "*furanosteroids*" has since been expanded to include any steroids and related compounds that contain the furan ring, 2,3-dihydrofuran ring, 2,5-dihydrofuran ring, and/or tetrahydrofuran ring (s). It is believed that the strained furan cycle contributes to the diverse biological activities exhibited by these compounds. Several metabolites belonging to this class of natural products have garnered significant attention in pharmacological research due to their potent anti-inflammatory and antibiotic properties, potential anti-proliferative activity, and ability to inhibit inositide-3-kinase [8,9,11–14].

This review is dedicated to furanosteroids and related isoprenoid lipids, their occurrence in fungi, plants, and marine organisms, and the exploration of their biological activity.

2. Furanosteroids Produced by Fungi and Fungal Endophytes

The furanosteroid scaffold (**1**) represents a highly oxygenated framework comprising a [5,6,6,6]-tetracycle (A-B-C-D rings), along with the presence of a furan ring, as depicted in Figure 1. Furanosteroids, exemplified by compounds **2–9**, **11**, **12**, **14**, and **15**, with their biological activities summarized in Table 1, encompass both the furan ring and steroids, viridin (**2**, 3D model is shown in Figure 2), **3–5**, and **15**.

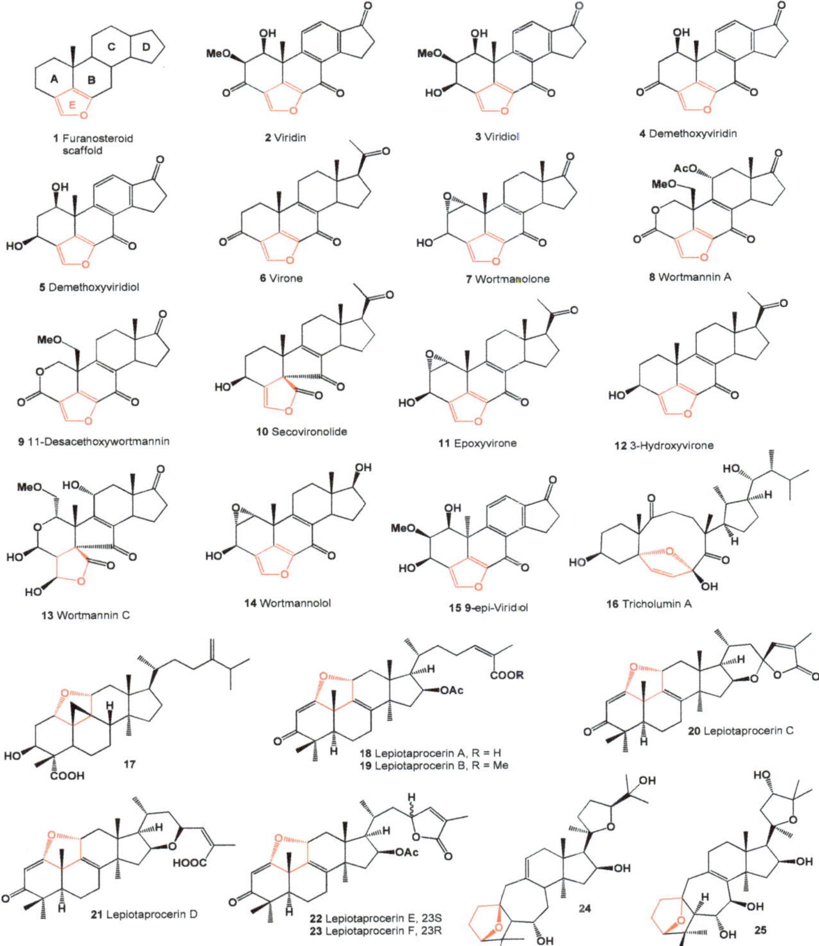

Figure 1. Furanosteroids produced by fungal species.

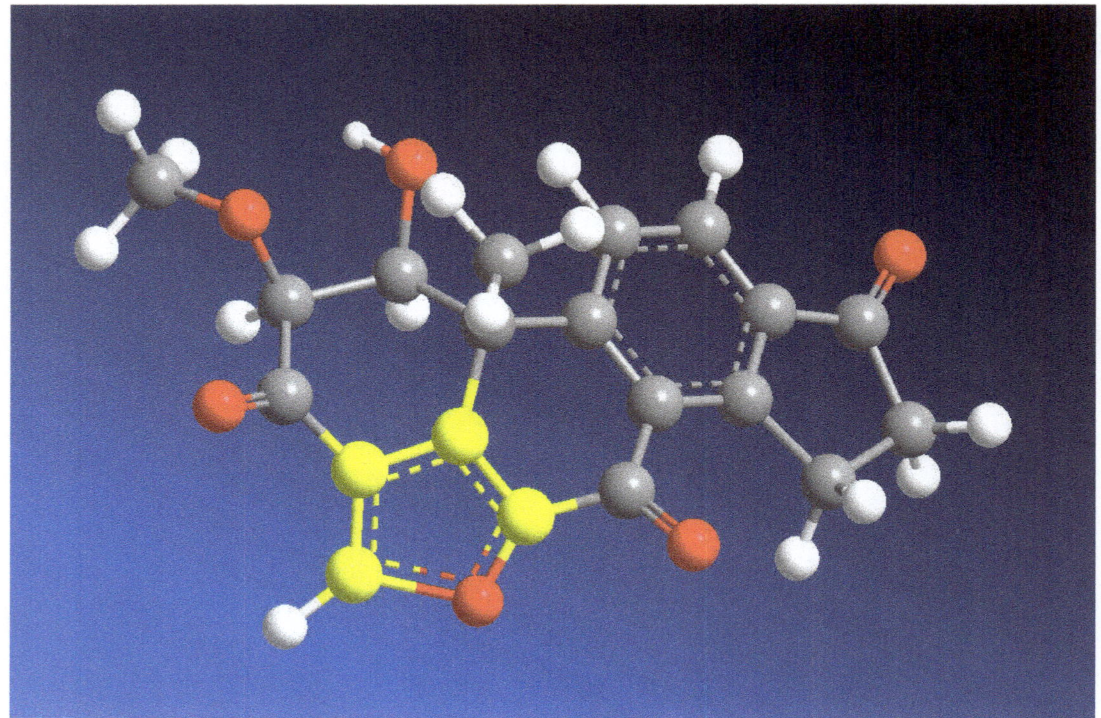

Figure 2. Three-dimensional structure of furanosteroid named viridin (**2**) and showing a wide range of antifungal and other biological activities. A distinctive characteristic of natural furanosteroids, as was believed in the 1950s of the last century, is the presence of an additional aromatic furan ring in the structure of the steroid. The aromatic furane ring is highlighted in yellow. The percentage of biological activities is shown in Figure 3. Viridin, an antibiotic produced by a pigment-forming strain of the common soil fungus *Trichoderma viride*, was first described by Brian and McGowan, and Brian, Curtis, Hemming and McGowan. Viridin is not antibacterial but is highly antifungal. Its activity against certain fungi is remarkably high. Germination of the spores of, for example, *Botrytis allii* is prevented by a concentration of 0.019 p.p.m. of α-viridin or 0.156 p.p.m. of β-viridin. Among synthetic fungicides, only the organo-mercurials are of the same order of activity. gray is carbon, white is hydrogen, and red is oxygen.

These steroids can be classified as aromatic steroids due to the presence of the C aromatic ring. Among them, the steroid known as secovironolide (**10**) contains an additional 2,3-dihydrofuran ring, while tricholumin A (**16**) is an example that includes a 2,5-dihydrofuran ring. Furthermore, the most commonly encountered steroids, **13** and **17–25**, feature a tetrahydrofuran ring, as illustrated in Figure 1.

The data presented in Tables 1–10 are taken from published data and obtained using the German computer software PASS (http://www.akosgmbh.de/mobile/pass.htm accessed on 1 April 2023). This program is in the public domain, and annually it is used by more than 26,000 scientists from around the world. The site of this program provides complete information on its use as well as the interpretation of the data obtained.

Table 1. Predicted biological activities of steroids derived from fungi (**2–25**).

No.	Dominated Biological Activity (Pa) *	Additional Predicted Activities (Pa) *
2	Antifungal (0.912) Anti-inflammatory (0.912) Antiprotozoal (Plasmodium) (0.794) Antibacterial (0.783)	Antineoplastic (0.909) Apoptosis agonist (0.828) Prostate disorders treatment (0.600) Proliferative diseases treatment (0.578)
3	Antineoplastic (0.802) Apoptosis agonist (0.647) Chemopreventive (0.573)	Anti-inflammatory (0.629) Antibacterial (0.547) Cytochrome P450 inhibitor (0.503)
4	Antineoplastic (0.779) Apoptosis agonist (0.766)	Antibacterial (0.718) Antifungal (0.709)
5	Antineoplastic (0.837) Apoptosis agonist (0.704)	Antibacterial (0.745)
6	Anti-infertility, female (0.789) Lipid metabolism regulator (0.789)	Antineoplastic (0.756) Apoptosis agonist (0.749)
7	Antineoplastic (0.913) Apoptosis agonist (0.907) Inositol-3-kinase inhibitor (0.898)	Lipid metabolism regulator (0.700) Hypolipemic (0.693) Antibacterial (0.642)
8	Inositol-4-kinase inhibitor (0.945) Apoptosis agonist (0.865) Inositol-3-kinase inhibitor (0.848)	Antifungal (0.876) Antibacterial (0.733)
9	Antineoplastic (0.934) Apoptosis agonist (0.865)	Antifungal (0.881) Antibacterial (0.766)
10	Inositol-4-kinase inhibitor (0.949) Inositol-3-kinase inhibitor (0.843) Monoamine oxidase inhibitor (0.837)	Apoptosis agonist (0.899) Antineoplastic (0.879)
11	Antineoplastic (0.924) Apoptosis agonist (0.833)	Antifungal (0.876) Antibacterial (0.733)
12	Antineoplastic (0.903) Apoptosis agonist (0.807) Prostate disorders treatment (0.800)	Lipid metabolism regulator (0.700) Hypolipemic (0.693)
13	Antineoplastic (0.934) Apoptosis agonist (0.907) Prostate disorders treatment (0.866)	Lipid metabolism regulator (0.897) Antifungal (0.881) Hypolipemic (0.793)
14	Antiproliferative (0.938) Antineoplastic (0.902) Apoptosis agonist (0.889)	Prostate disorders treatment (0.900) Proliferative diseases treatment (0.878)
15	Antiproliferative (0.929) Antineoplastic (0.922)	Prostate disorders treatment (0.876) Proliferative diseases treatment (0.844)
16	Antiproliferative (0.915) Apoptosis agonist (0.882)	Prostate disorders treatment (0.854) Proliferative diseases treatment (0.823)
17	Antineoplastic (0.877) Apoptosis agonist (0.856)	Ovulation inhibitor (0.698) Anti-inflammatory (0.682)
18	Antineoplastic (0.879) Apoptosis agonist (0.876)	Ovulation inhibitor (0.728) Anti-inflammatory (0.717)
19	Antineoplastic (0.865) Antiprotozoal (Plasmodium) (0.621)	Antifungal (0.779) Antibacterial (0.747)
20	Antineoplastic (0.888)	Antifungal (0.843)
21	Antineoplastic (0.910)	Apoptosis agonist (0.877)
22	Nitric oxide inhibitor (0.921)	Apoptosis agonist (0.879)

Table 1. *Cont.*

No.	Dominated Biological Activity (Pa) *	Additional Predicted Activities (Pa) *
23	Nitric oxide inhibitor (0.914) Antifungal (0.862)	Apoptosis agonist (0.867) Antineoplastic (0.844)
24	Antineoplastic (0.937)	Apoptosis agonist (0.892)
25	Antineoplastic (0.929)	Apoptosis agonist (0.887)

* Only activities with Pa > 0.5 are shown.

Table 2. Predicted biological activities of steroids derived from fungi (**26–47**).

No.	Dominated Biological Activity (Pa) *	Additional Predicted Activities (Pa) *
26	Antineoplastic (0.922) Apoptosis agonist (0.844)	Antifungal (0.887) Antibacterial (0.754)
27	Antineoplastic (0.929) Apoptosis agonist (0.839)	Antifungal (0.832) Antibacterial (0.782)
28	Antineoplastic (0.921) Apoptosis agonist (0.841)	Antifungal (0.833) Antibacterial (0.755)
29	Antineoplastic (0.889)	Antibacterial (0.766)
30	Antineoplastic (0.833)	Antibacterial (0.733)
31	Hepatoprotectant (0.860) Kidney function stimulant (0.719)	Muscle relaxant (0.599) Spasmolytic (0.587)
32	Hepatoprotectant (0.875) Kidney function stimulant (0.686)	Muscle relaxant (0.612) Spasmolytic (0.595)
33	Hepatoprotectant (0.839) Kidney function stimulant (0.719)	Muscle relaxant (0.782)
34	Antiprotozoal (Plasmodium) (0.718)	Anti-inflammatory (0.778)
35	Antiprotozoal (0.657)	Anti-inflammatory (0.766)
36	Antiprotozoal (Plasmodium) (0.947) Antiprotozoal (Leishmania) (0.888) Antiprotozoal (0.879)	Antineoplastic (0.866) Apoptosis agonist (0.795) Chemopreventive (0.722)
37	Antiprotozoal (Plasmodium) (0.949) Antiprotozoal (Leishmania) (0.868) Antiprotozoal (0.823)	Antineoplastic (0.859) Apoptosis agonist (0.791) Chemopreventive (0.717)
38	Antiprotozoal (Plasmodium) (0.927) Antiprotozoal (Leishmania) (0.914) Antiprotozoal (0.884)	Antineoplastic (0.843) Apoptosis agonist (0.747) Chemopreventive (0.698)
39	Antiprotozoal (Plasmodium) (0.912) Antiprotozoal (Leishmania) (0.886) Antiprotozoal (0.778)	Antineoplastic (0.876) Chemopreventive (0.773)
40	Antiprotozoal (Plasmodium) (0.955) Antiprotozoal (Leishmania) (0.948) Antiprotozoal (0.903)	Chemopreventive (0.903) Antineoplastic (0.822) Apoptosis agonist (0.747)
41	Antiprotozoal (0.822)	Anti-inflammatory (0.712)
42	Antibacterial (0.883) Antifungal (0.876)	Laxative (0.735) Anti-eczematic (0.717)
43	Antibacterial (0.907) Antifungal (0.865)	Laxative (0.722) Anti-eczematic (0.686)
44	Antibacterial (0.885) Antifungal (0.876)	Anti-inflammatory (0.712)

Table 2. *Cont.*

No.	Dominated Biological Activity (Pa) *	Additional Predicted Activities (Pa) *
45	Antibacterial (0.912) Antifungal (0.875)	Anti-eczematic (0.638) Anti-inflammatory (0.612)
46	Antibacterial (0.901) Antifungal (0.867)	Anti-eczematic (0.689)
47	Antibacterial (0.889) Antifungal (0.882)	Anti-eczematic (0.654)

* Only activities with Pa > 0.5 are shown.

Table 3. Predicted biological activities of steroids derived from fungi (**48–72**).

No.	Dominated Biological Activity (Pa) *	Additional Predicted Activities (Pa) *
48	Antibacterial (0.856) Antifungal (0.811)	Antiviral (0.578)
49	Nitric oxide inhibitor (0.906) Apoptosis agonist (0.667)	Antineoplastic (0.764)
50	Nitric oxide inhibitor (0.874) Apoptosis agonist (0.685)	Antineoplastic (0.785)
51	Plant growth inhibitor (0.876)	Antifungal (0.732)
52	Plant growth inhibitor (0.855)	Antifungal (0.689)
53	Antibacterial (0.769) Antifungal (0.716)	Antiviral (0.577)
54	Nitric oxide inhibitor (0.779)	Antibacterial (0.686)
55	Nitric oxide inhibitor (0.812)	Antibacterial (0.677)
56	Antineoplastic (0.755)	Anti-inflammatory (0.612)
57	Antineoplastic (0.734)	Antifungal (0.521)
58	Antibacterial (0.772) Antifungal (0.631)	Antiviral (0.589)
59	Antibacterial (0.856)	Antiviral (0.618)
60	Antifungal (0.789) Antibacterial (0.634)	Antiviral (0.611)
61	Antineoplastic (0.785)	Anti-inflammatory (0.622)
62	Cytoprotectant (0.758)	
63	Antibacterial (0.882)	Antifungal (0.668)
64	Antibacterial (0.872)	Antifungal (0.675)
65	Antibacterial (0.868)	Antifungal (0.712)
66	Antibacterial (0.856)	Antifungal (0.682)
67	Antimicrobial (0.902) Antibacterial (0.856)	Antifungal (0.754)
68	Antibacterial (0.911)	Antimicrobial (0.773)
69	Antiprotozoal (0.886) Antibacterial (0.847)	Antimicrobial (0.638)
70	Antineoplastic (0.772)	Antifungal (0.722)

Table 3. *Cont.*

No.	Dominated Biological Activity (Pa) *	Additional Predicted Activities (Pa) *
71	Antineoplastic (0.882) Chemopreventive (0.791)	Anti-inflammatory (0.634)
72	Antineoplastic (0.913) Chemopreventive (0.882) Apoptosis agonist (0.798)	Antiviral (0.832) Anti-inflammatory (0.711)

* Only activities with Pa > 0.5 are shown.

Table 4. Predicted biological activities of steroids derived from fungi (**73–101**).

No.	Dominated Biological Activity (Pa) *	Additional Predicted Activities (Pa) *
73	Antineoplastic (0.884)	Antifungal (0.621)
74	Antineoplastic (0.975) Apoptosis agonist (0.889) Chemopreventive (0.887)	Lipid metabolism regulator (0.858) Hypolipemic (0.677)
75	Antineoplastic (0.968) Chemopreventive (0.882) Apoptosis agonist (0.798)	Lipid metabolism regulator (0.897) Hypolipemic (0.693)
76	Antineoplastic (0.973) Apoptosis agonist (0.798) Chemopreventive (0.882)	Lipid metabolism regulator (0.872) Hypolipemic (0.654)
77	Antineoplastic (0.865)	Antifungal (0.633)
78	Antineoplastic (0.834)	Antifungal (0.611)
79	Antineoplastic (0.844)	Anti-inflammatory (0.666)
80	Antineoplastic (0.851)	Antifungal (0.597)
81	Antiproliferative (0.834)	Cytotoxic (0.658)
82	Antiproliferative (0.812)	Anti-inflammatory (0.645)
83	Antiproliferative (0.876) Antineoplastic (0.875)	Cytotoxic (0.745) Antifungal (0.677)
84	Antiproliferative (0.876)	Anti-inflammatory (0.668)
85	Antiproliferative (0.924) Cytotoxic (0.892)	Cytotoxic (0.821) Antifungal (0.734)
86	Antiproliferative (0.902) Cytotoxic (0.883)	Cytotoxic (0.833) Antifungal (0.680)
87	Antineoplastic (0.823)	Anti-inflammatory (0.619)
88	Antiproliferative (0.831)	Anti-inflammatory (0.644)
89	Antiproliferative (0.849)	Hypolipemic (0.623)
90	Antineoplastic (0.736)	Antifungal (0.567)
91	Antineoplastic (0.729)	Antifungal (0.564)
92	Antineoplastic (0.751)	Antifungal (0.592)
93	Antineoplastic (0.722)	Antimicrobial (0.612)
94	Antineoplastic (0.733)	Antimicrobial (0.622)
95	Antineoplastic (0.728)	Antimicrobial (0.641)
96	Antineoplastic (0.802)	Antimicrobial (0.632)
97	Allergic conjunctivitis treatment (0.687)	Antifungal (0.566)
98	Allergic conjunctivitis treatment (0.705)	Antifungal (0.587)

Table 4. *Cont.*

No.	Dominated Biological Activity (Pa) *	Additional Predicted Activities (Pa) *
99	Antineoplastic (0.902) Antiproliferative (0.883) Apoptosis agonist (0.870)	Cytotoxic (0.844) Anti-inflammatory (0.634)
100	Antibacterial (0.903)	Antimicrobial (0.734)
101	Antibacterial (0.868)	Antimicrobial (0.698)

* Only activities with Pa > 0.5 are shown.

Table 5. Predicted biological activities of steroids derived from plants (**102–127**).

No.	Dominated Biological Activity (Pa) *	Additional Predicted Activities (Pa) *
102	Antineoplastic (0.879) Chemopreventive (0.683) Apoptosis agonist (0.641)	Anti-inflammatory (0.840) Antifungal (0.671)
103	Antineoplastic (0.851) Chemopreventive (0.662) Apoptosis agonist (0.637)	Anti-inflammatory (0.790) Antifungal (0.653)
104	Analgesic (0.761) Antitussive (0.652)	Respiratory analeptic (0.744) Oxygen scavenger (0.651)
105	Antiprotozoal (Plasmodium) (0.869) Antiprotozoal (0.832)	Antineoplastic (0.798) Apoptosis agonist (0.795)
106	Antiprotozoal (Plasmodium) (0.939) Antiprotozoal (Leishmania) (0.891) Antiprotozoal (0.883)	Antineoplastic (0.874) Apoptosis agonist (0.765) Chemopreventive (0.703)
107	Antiprotozoal (Plasmodium) (0.954) Antiprotozoal (Leishmania) (0.912) Antiprotozoal (0.889)	Antineoplastic (0.911) Apoptosis agonist (0.893) Chemopreventive (0.791)
108	Antiprotozoal (Plasmodium) (0.941) Antiprotozoal (Leishmania) (0.866) Antiprotozoal (0.836)	Antineoplastic (0.876) Apoptosis agonist (0.795) Chemopreventive (0.722)
109	Antiprotozoal (Plasmodium) (0.940) Antiprotozoal (Leishmania) (0.848) Antiprotozoal (0.842)	Antineoplastic (0.868) Apoptosis agonist (0.794) Chemopreventive (0.713)
110	Antineoplastic (0.955) Apoptosis agonist (0.802)	Antiviral (0.818) Anti-inflammatory (0.721)
111	Antineoplastic (0.946) Apoptosis agonist (0.766)	Antiviral (0.832) Anti-inflammatory (0.734)
112	Antineoplastic (0.932) Apoptosis agonist (0.778)	Antiviral (0.822)
113	Antineoplastic (0.829) Cytochrome P450 inhibitor (0.691)	Anti-inflammatory (0.790) Antimitotic (0.592)
114	Antineoplastic (0.867) Cytochrome P450 inhibitor (0.748)	Anti-inflammatory (0.801) Antimitotic (0.622)
115	Antineoplastic (0.793)	Anti-inflammatory (0.715)
116	Antiprotozoal (Plasmodium) (0.707) Antiprotozoal (0.697)	Antineoplastic (0.707) Apoptosis agonist (0.570)
117	Antineoplastic (0.872) Apoptosis agonist (0.712)	Antiprotozoal (Plasmodium) (0.746) Antiprotozoal (0.703)
118	Antineoplastic (0.854) Apoptosis agonist (0.743)	Antiprotozoal (Plasmodium) (0.721) Antiprotozoal (0.693)

Table 5. *Cont.*

No.	Dominated Biological Activity (Pa) *	Additional Predicted Activities (Pa) *
119	Antineoplastic (0.850) Apoptosis agonist (0.691) Cytostatic (0.683)	Hepatoprotectant (0.767) Immunosuppressant (0.712) Anti-hypercholesterolemic (0.566)
120	Antineoplastic (0.855) Apoptosis agonist (0.697)	Hepatoprotectant (0.768) Immunosuppressant (0.718)
121	Antineoplastic (0.858) Apoptosis agonist (0.690)	Hepatoprotectant (0.753) Anti-hypercholesterolemic (0.584)
122	Antineoplastic (0.860) Apoptosis agonist (0.699)	Hepatoprotectant (0.755) Anti-hypercholesterolemic (0.616)
123	Antiviral (0.832)	Anti-inflammatory (0.721)
124	Antiviral (0.856)	Anti-inflammatory (0.734)
125	Antiviral (0.887) Antiviral (Arbovirus) (0.790)	Antiprotozoal (0.693)
126	Antiviral (0.877)	Anti-inflammatory (0.715)
127	Antiviral (0.839)	Anti-inflammatory (0.698)

* Only activities with Pa > 0.5 are shown.

Table 6. Biological activities of steroids derived from plants (**128–154**).

No.	Dominated Biological Activity (Pa) *	Additional Predicted Activities (Pa) *
128	Hepatoprotectant (0.784)	Immunosuppressant (0.734)
129	Hepatoprotectant (0.772)	Immunosuppressant (0.728)
130	Hepatoprotectant (0.777)	Immunosuppressant (0.729)
131	Antimicrobial (0.841)	Antibacterial (0.823)
132	Antineoplastic (0.823)	Antiprotozoal (0.821)
133	Antineoplastic (0.894) Chemopreventive (0.763)	Antiprotozoal (0.865)
134	Antineoplastic (0.902) Chemopreventive (0.883)	Antiprotozoal (0.872)
135	Antifungal (0.833)	Anti-inflammatory (0.790)
136	Antifungal (0.797)	Antimicrobial (0.671)
137	Antifungal (0.803)	Antimicrobial (0.666)
138	Antimicrobial (0.854)	Antifungal (0.687)
139	Antimicrobial (0.798)	Antifungal (0.693)
140	Cortisone reductase inhibitor (0.875)	Antineoplastic (0.825)
141	Cortisone reductase inhibitor (0.902)	Antineoplastic (0.887)
142	Cortisone reductase inhibitor (0.898)	Antineoplastic (0.879)
143	Anti-HIV-1 (0.894) Antiviral (0.839)	Anti-inflammatory (0.768)
144	Anti-HIV-1 (0.904) Antiviral (0.885)	Anti-inflammatory (0.754)
145	Prostate cancer treatment (0.928) Antineoplastic (0.911)	Antimicrobial (0.723)
146	Prostate cancer treatment (0.914) Antineoplastic (0.896)	Antimicrobial (0.741)

Table 6. *Cont.*

No.	Dominated Biological Activity (Pa) *	Additional Predicted Activities (Pa) *
147	Antineoplastic (0.907) Prostate cancer treatment (0.894)	Antimicrobial (0.654)
148	Antineoplastic (0.897) Prostate cancer treatment (0.887)	Antimicrobial (0.678)
149	Antineoplastic (0.838)	Antimicrobial (0.611)
150	Antineoplastic (0.864) Apoptosis agonist (0.723)	Antileukemic (0.822)
151	Antineoplastic (0.843) Apoptosis agonist (0.711)	Antifungal (0.733)
152	Antineoplastic (0.929) Antimetastatic (0.834)	Antileukemic (0.814)
153	Antineoplastic (0.924) Antimetastatic (0.876)	Antileukemic (0.754) Antimicrobial (0.629)
154	Antineoplastic (0.932) Antimetastatic (0.839)	Antileukemic (0.724) Antimicrobial (0.623)

* Only activities with Pa > 0.5 are shown.

Table 7. Predicted biological activities of steroids derived from plants (**155–186**).

No.	Dominated Biological Activity (Pa) *	Additional Predicted Activities (Pa) *
155	Antifungal (0.778)	Anti-inflammatory (0.709)
156	Antineoplastic (0.918) Prostate cancer treatment (0.915) Antimetastatic (0.734)	Antifungal (0.734) Antimicrobial (0.652)
157	Antineoplastic (0.923) Prostate cancer treatment (0.922) Antimetastatic (0.834)	Antifungal (0.736) Antimicrobial (0.655)
158	Antineoplastic (0.888) Apoptosis agonist (0.793)	Antifungal (0.713) Antimicrobial (0.654)
159	Antineoplastic (0.854) Apoptosis agonist (0.721)	Antifungal (0.743) Antimicrobial (0.664)
160	Anti-tuberculosis treatment (0.896)	Antibacterial (0.809)
161	Anti-tuberculosis treatment (0.903) Antimicrobial (0.745)	Antibacterial (0.811)
162	Anti-tuberculosis treatment (0.937) Antimicrobial (0.772)	Antibacterial (0.854)
163	Anti-tuberculosis treatment (0.875)	Antibacterial (0.754)
164	Anti-tuberculosis treatment (0.862)	Antibacterial (0.761)
165	Acetyl-cholinesterase inhibitor (0.931)	Antimetastatic (0.829)
166	Acetyl-cholinesterase inhibitor (0.942)	Antimetastatic (0.842)
167	DNA topoisomerase inhibitor (0.928) Antineoplastic (0.916) Antimetastatic (0.863)	Apoptosis agonist (0.834)
168	DNA topoisomerase inhibitor (0.917) Antineoplastic (0.922) Antimetastatic (0.821)	Apoptosis agonist (0.871)
169	Anti-inflammatory (0.909) Antiviral (0.844)	Antimicrobial (0.688)

Table 7. *Cont.*

No.	Dominated Biological Activity (Pa) *	Additional Predicted Activities (Pa) *
170	Anti-inflammatory (0.902) Antiviral (0.829)	Antimicrobial (0.679)
171	Antineoplastic (0.886)	Apoptosis agonist (0.756)
172	Antineoplastic (0.874)	Apoptosis agonist (0.771)
173	Antineoplastic (0.859)	Apoptosis agonist (0.754)
174	Antineoplastic (0.851)	Apoptosis agonist (0.764)
175	Antiproliferative (0.889)	Anti-inflammatory (0.654)
176	Antiproliferative (0.882)	Anti-inflammatory (0.659)
177	Antiproliferative (0.895)	Anti-inflammatory (0.663)
178	Anti-hepatitis C virus (0.912) Antiviral (0.898)	Antimicrobial (0.687)
179	Antineoplastic (0.873)	Apoptosis agonist (0.775)
180	Antineoplastic (0.869)	Apoptosis agonist (0.768)
181	Hepatocellular carcinoma inhibitor (0.897)	Hepatoprotectant (0.860)
182	Hepatocellular carcinoma inhibitor (0.906)	Hepatoprotectant (0.873)
183	Hepatocellular carcinoma inhibitor (0.911)	Hepatoprotectant (0.869)
184	Hepatocellular carcinoma inhibitor (0.889)	Hepatoprotectant (0.856)
185	Anti-inflammatory (0.832) Antiviral (0.754)	Antibacterial (0.785)
186	Anti-inflammatory (0.834) Antiviral (0.765)	Antibacterial (0.744)

* Only activities with Pa > 0.5 are shown.

Table 8. Predicted biological activities of steroids derived from marine sources (**187–221**).

No.	Dominated Biological Activity (Pa) *	Additional Predicted Activities (Pa) *
187	Antineoplastic (0.913) Antimetastatic (0.786)	Apoptosis agonist (0.619)
188	Antineoplastic (0.908) Antimetastatic (0.804)	Apoptosis agonist (0.634)
189	Antineoplastic (0.902) Antimetastatic (0.798)	Apoptosis agonist (0.624)
190	Cytoprotectant (0.768) Antineoplastic (0.723)	Antimicrobial (0.776)
191	Cytoprotectant (0.787)	Antimicrobial (0.685)
192	Cytoprotectant (0.766)	Antimicrobial (0.692)
193	Antineoplastic (0.900) Apoptosis agonist (0.836)	Antimetastatic (0.698)
194	Antineoplastic (0.889) Apoptosis agonist (0.821)	Antimetastatic (0.682)
195	Antineoplastic (0.872)	Antiprotozoal (0.765)
196	Antineoplastic (0.868)	Antiprotozoal (0.756)
197	Antiviral (0.776)	Anti-inflammatory (0.682)
198	Antiviral (0.749)	Anti-inflammatory (0.678)

Table 8. *Cont.*

No.	Dominated Biological Activity (Pa) *	Additional Predicted Activities (Pa) *
199	Antineoplastic (0.773)	Apoptosis agonist (0.653)
200	Antibacterial (0.747)	Antifungal (0.683)
201	Antibacterial (0.736)	Antifungal (0.651)
202	Apoptosis agonist (0.833)	Antineoplastic (0.711)
203	Apoptosis agonist (0.824)	Antineoplastic (0.683)
204	Antineoplastic (0.744)	Apoptosis agonist (0.621)
205	Antineoplastic (0.738)	Apoptosis agonist (0.619)
206	Antiallergic (0.728)	Anti-asthmatic (0.671)
207	Antiallergic (0.734)	Anti-asthmatic (0.698)
208	Antiallergic (0.742)	Anti-asthmatic (0.661)
209	Antiallergic (0.741)	Anti-asthmatic (0.676)
210	Antineoplastic (0.926)	Antimetastatic (0.855)
211	Antineoplastic (0.914)	Antimetastatic (0.836)
212	Antiproliferative (0.818)	Antineoplastic (0.654)
213	Antiproliferative (0.828)	Antineoplastic (0.662)
214	Antiproliferative (0.833)	Antineoplastic (0.659)
215	Antiproliferative (0.867)	Antineoplastic (0.678)
216	Apoptosis agonist (0.818) Antineoplastic (0.762)	Lipid metabolism regulator (0.638) Anti-inflammatory (0.691)
217	Antineoplastic (0.917)	Antimetastatic (0.743)
218	Antineoplastic (0.919)	Antimetastatic (0.767)
219	Antineoplastic (0.928) Apoptosis agonist (0.854)	Antimetastatic (0.711)
220	Antineoplastic (0.931) Apoptosis agonist (0.871)	Antimetastatic (0.721)
221	Antineoplastic (0.924)	Antimetastatic (0.734)

* Only activities with Pa > 0.5 are shown.

Table 9. Predicted biological activities of steroids derived from marine sources (**222–238**).

No.	Dominated Biological Activity (Pa) *	Additional Predicted Activities (Pa) *
222	Antiviral (0.788) Antiviral (Arbovirus) (0.750)	Anti-inflammatory (0.682)
223	Allergic conjunctivitis treatment (0.705)	Antifungal (0.637)
224	Antineoplastic (0.898) Apoptosis agonist (0.834)	Antimetastatic (0.723)
225	Antineoplastic (0.918) Apoptosis agonist (0.856)	Antimetastatic (0.743)
226	Cathepsin B inhibitor (0.889)	Apoptosis agonist (0.698)
227	Antifeedant (0.867)	Antifungal (0.703)
228	Protein kinase C inhibitor (0.921) Antineoplastic (0.834) Antimetastatic (0.748)	Antiviral (0.829) Autoimmune disorders treatment (0.619)
229	Antineoplastic (0.924) Apoptosis agonist (0.745)	Antifungal (0.698)

Table 9. *Cont.*

No.	Dominated Biological Activity (Pa) *	Additional Predicted Activities (Pa) *
230	Antineoplastic (0.931) Apoptosis agonist (0.805)	Anti-inflammatory (0.654)
231	Antineoplastic (0.918) Apoptosis agonist (0.764)	Anti-inflammatory (0.662)
232	Antineoplastic (0.932) Chemopreventive (0.738)	Apoptosis agonist (0.768)
233	Antineoplastic (0.929) Chemopreventive (0.787)	Apoptosis agonist (0.729)
234	Antineoplastic (0.943) Chemopreventive (0.818)	Apoptosis agonist (0.815)
235	Antimicrobial (0.886)	Antifungal (0.687)
236	Antineoplastic (0.943) Antimicrobial (0.922)	Apoptosis agonist (0.898) Antifungal (0.847)
237	Antibacterial (0.921)	Antifungal (0.727)
238	Antibacterial (0.765)	

* Only activities with Pa > 0.5 are shown.

Table 10. Predicted biological activities of steroids derived from marine sources (**239–266**).

No.	Dominated Biological Activity (Pa) *	Additional Predicted Activities (Pa) *
239	Antineoplastic (0.913) Apoptosis agonist (0.886)	Antimetastatic (0.829)
240	Antineoplastic (0.922) Chemopreventive (0.731) Anticarcinogenic (0.698)	Apoptosis agonist (0.876)
241	Antineoplastic (0.929) Chemopreventive (0.720) Anticarcinogenic (0.678)	Apoptosis agonist (0.854)
242	Antineoplastic (0.889) Anticarcinogenic (0.718)	Chemopreventive (0.683)
243	Antibacterial (0.881)	Antifungal (0.683)
244	Antibacterial (0.879)	Antifungal (0.627)
245	Antibacterial (0.902)	Antifungal (0.638)
246	Antineoplastic (0.938) Apoptosis agonist (0.834)	Chemopreventive (0.712)
247	Antineoplastic (0.886)	Apoptosis agonist (0.726)
248	Antineoplastic (0.887)	Apoptosis agonist (0.734)
249	Antineoplastic (0.889)	Apoptosis agonist (0.722)
250	Antineoplastic (0.900)	Apoptosis agonist (0.783)
251	Antineoplastic (0.911)	Apoptosis agonist (0.803)
252	Antineoplastic (0.935) Apoptosis agonist (0.814) Chemopreventive (0.756)	Antimetastatic (0.820)
253	Antineoplastic (0.929) Apoptosis agonist (0.821) Chemopreventive (0.744)	Antimetastatic (0.822)

Table 10. *Cont.*

No.	Dominated Biological Activity (Pa) *	Additional Predicted Activities (Pa) *
254	Phospholipase A2 inhibitor (0.923) Antibacterial (0.883)	Anti-inflammatory (0.689)
255	Phospholipase A2 inhibitor (0.827) Antibacterial (0.757)	Anti-inflammatory (0.683)
256	Antineoplastic (0.884) Apoptosis agonist (0.756)	Antimetastatic (0.688)
257	Antineoplastic (0.891) Apoptosis agonist (0.785)	Antimetastatic (0.679)
258	Antineoplastic (0.814)	Anti-inflammatory (0.661)
259	Antineoplastic (0.828)	Anti-inflammatory (0.658)
260	Antineoplastic (0.832)	Anti-inflammatory (0.650)
261	Antineoplastic (0.915) Antimetastatic (0.839)	Chemopreventive (0.672)
262	Antineoplastic (0.912) Antimetastatic (0.821)	Chemopreventive (0.669)
263	Antineoplastic (0.910) Apoptosis agonist (0.752)	Antimetastatic (0.712)
264	Antineoplastic (0.921) Apoptosis agonist (0.774)	Antimetastatic (0.729)
265	Antineoplastic (0.905) Apoptosis agonist (0.739)	Antimetastatic (0.704)
266	Antineoplastic (0.918) Apoptosis agonist (0.699)	Antimetastatic (0.687)

* Only activities with Pa > 0.5 are shown.

Viridin (**2**, activity is shown in Figure 3), a furano-steroidal antibiotic, represents the first identified member of the furanosteroid family. It was initially isolated in 1945 from a pigment-forming strain of the common soil fungus *Trichoderma viride* (as shown in Figure 4) [15,16]. Viridin, along with its derivatives **3**, **4**, **5**, **6**, and **11**, have demonstrated remarkable antifungal and antibacterial activities. Moreover, these compounds are known as potent inhibitors of the lipid kinase PI-3K [16]. Phosphatidylinositol 3-kinases (PI3Ks) are lipid kinases that play a central role in cell cycle regulation, apoptosis, DNA repair, aging, angiogenesis, cellular metabolism, and motility [16–18]. These enzymes catalyze the synthesis of specific members of the lipid family collectively known as "phosphoinositides". These PI3K products can in turn modulate the activation of many downstream proteins, ultimately regulating several cellular processes. Mammalian cells possess eight PI3Ks, which are grouped into three classes depending on their structure and substrate specificity [16–18].

More recently, viridin (**2**) and the phytotoxin viridiol (**3**) were discovered in liquid cultures produced by the fungus *Gliocladium virens* [19]. Additionally, demethoxyviridin (**4**) and demethoxyviridiol (**5**) were isolated for the first time from an unidentified fungus [20]. The ash dieback-causing fungus *Hymenoscyphus pseudoalbidus* yielded furanosteroids (**17** and **18**) along with known compounds like viridiol (**6**) and demethoxyviridiol (**5**) [14,20,21]. Epoxyvirone (**11**) was detected in *Talaromyces* species (as depicted in Figure 5) and produced by the marine sponge-associated fungus *Talaromyces stipitatus* KUFA 0207 [22,23]. Ding and co-workers [24] reported the isolation of wortmannolone (**7**), wortmannin (**8**), 11-desacetoxywortmannin-17β-ol (**9**), secovironolide (**10**, represented by the 3D graph in Figure 6), epoxyvirone (**11**), wortmannin C (**13**), and 3-dihydrovirone (**12**) from the culture broth of the endophytic fungus *Talaromyces wortmannii* LGT-4, which was derived from the Chinese medicinal plant *Tripterygium wilfordii*.

Furthermore, a furanosteroid (**10**), along with viridiol (**3**), and **8**, **11**, and **13** were isolated from the soil fungus *Trichoderma virens*. Notably, 9-*epi*-viridiol (**23**) and viridiol (**2**) exhibited cytotoxicity against HeLa and KB cells, with IC_{50} values of 19 and 50 µg/mL, respectively [25].

Wortmannin (**8**) was initially isolated in 1957 by Brian and his colleagues from the broth of *Penicillium wortmanni* and subsequently from other fungi such as *P. funiculosum*, *Talaromyces wortmannii*, *Fusarium oxysporum*, *P. radicum*, and *Talaromyces* sp. [17,26,27]. Wortmannin and its derivatives, including compounds **7**, **8**, **9**, **13**, and **23**, exhibit antiproliferative properties and phosphatidylinositol 3-kinase activity [28–31]. Two furanosteroids, wortmannolone (**7**) and wortmannolol (**14**), were isolated from the fungal endophyte *Talaromyces* sp., which was obtained from *Tripterygium wilfordii* [31].

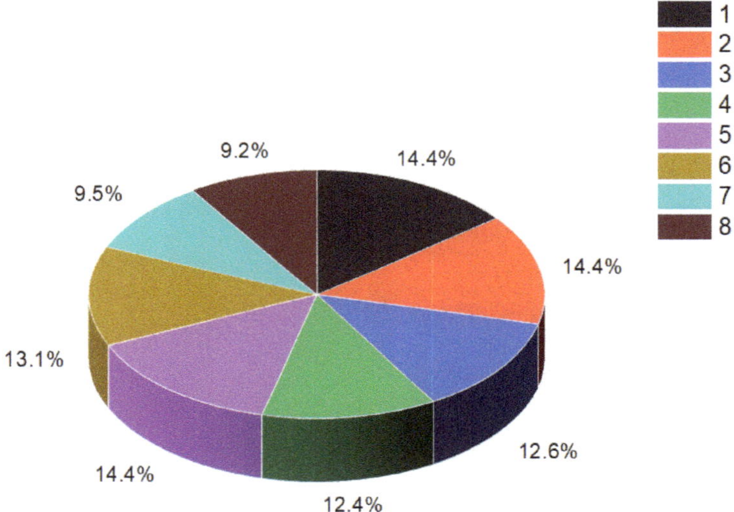

Figure 3. Percentage distribution of dominant and related biological activities using the widely recognized furanosteroid, viridin (**2**), which possesses unique pharmacological properties. The activities are indicated by the following numbers: 1. *Antifungal (14.4%)*; 2. *Anti-inflammatory (14.4%)*; 3. *Antiprotozoal (Plasmodium) (12.6%)*; 4. *Antibacterial (12.4%)*; 5. *Antineoplastic (14.4%)*; 6. *Apoptosis agonist (13.1%)*; 7. *Prostate disorders treatment (9.5%)*; and 8. *Proliferative diseases treatment (9.2%)*. The antifungal activity of the fungus *Trichoderma viride* has been observed to exhibit an inhibitory effect against various pathogens, such as *Fusarium solani*, *Rhizoctonia solani*, and *Sclerotium rolfsii*. Moreover, the extract obtained from raw mycelium has demonstrated notable antimicrobial activity, specifically displaying an antibacterial effect against *Bacillus subtilis*, *Escherichia coli*, and *Pseudomonas fluorescens*. Maximum antifungal effectiveness has also been recorded against *Candida albicans*, *Rhizoctonia solani*, *Pythium ultimum*, *Fusarium solani*, and *F. oxysporium*. The viridin metabolite present in the alcoholic mycelia extract of *Trichoderma viride* exerts antimicrobial, antifungal, and anticancer effects [14,32–34].

Another set of furanosteroids, namely 9-*epi*-viridiol (**15**) and viridiol (**3**), were isolated from *Trichoderma virens* [25]. Tricholumin A (**16**) possesses a unique carbon skeleton and is a highly transformed ergosterol derivative. It was isolated from the alga-endophytic fungus *Trichoderma asperellum*. Tricholumin A has demonstrated inhibitory effects against several pathogenic microbes, including *V. harveyi*, *V. splendidus*, and *Pseudoalteromonas citrea*. Additionally, it has displayed antifungal activity against *Glomerella cingulata* and has shown inhibition towards various marine phytoplankton species such as *Chattonella marina*, *Heterosigma akashiwo*, *Karlodinium veneficum*, and *Prorocentrum donghaiense* [35].

An unusual steroid compound, 1,11-epoxy-3-hydroxy-methylenecycloartan-28-oic acid (**17**), was isolated from an unidentified fungus [36]. Similar steroid structures have been identified in species belonging to the Agaricaceae family. Notably, a series of rare 1,11-epoxy lanostane-type triterpenoids called lepiotaprocerins A–F (**18–23**) were isolated from the fruiting bodies of the edible mushroom *Macrolepiota procera* collected in Poland. These compounds exhibited significant inhibitory effects on nitric oxide production, outperforming the positive control, L-NG-monomethyl arginine [37].

Figure 4. (**a**) *Trichoderma viride* is a fungus or mold that reproduces asexually through spores. The mycelium of this fungus can produce various enzymes, such as cellulases and chitinases. These enzymes are involved in the degradation of cellulose and chitin, respectively. *Trichoderma viride* commonly grows on wood, which predominantly consists of cellulose, as well as on fungi whose cell walls are primarily composed of chitin. It exhibits a parasitic nature towards the mycelium and fruiting bodies of other fungi, including cultivated ones. The fungus is known to cause "*Green mold fungus disease*" [38,39]. (**b**) *Hymenoscyphus pseudoalbidus*, also known as the Ash Wax Moth, is a saprotrophic fungus. It can be found growing on decaying trunks and branches of ash trees. The fruiting bodies of this fungus are hardly distinguishable from a macroscopic or microscopic perspective. It is commonly observed on the same substrate [40,41].

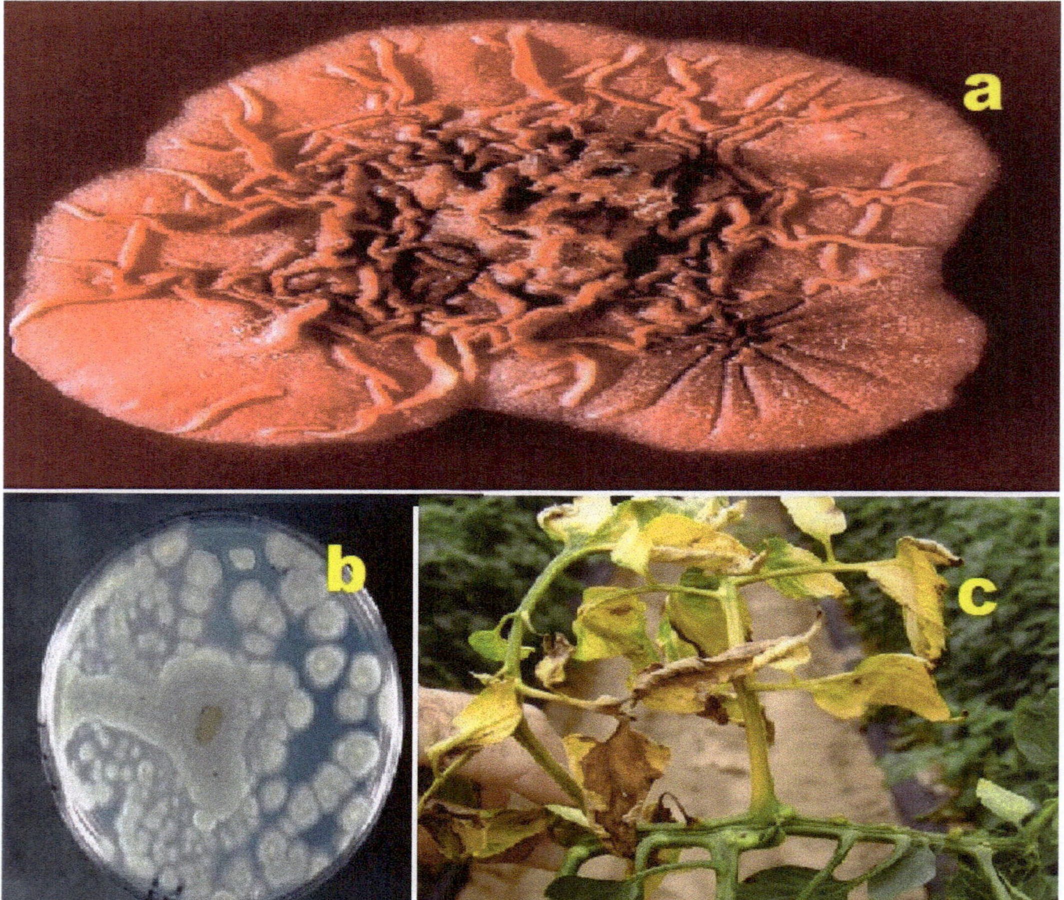

Figure 5. (**a**) *Talaromyces marneffei*: Talaromyces is a genus of fungi belonging to the family Trichocomaceae. It was first described in 1955 by the American mycologist Chester Ray Benjamin, who specialized in the taxonomy of fungal molds in the Eurotiales and Mucorales orders. *Talaromyces marneffei*, previously known as *Penicillium marneffei*, was identified in 1956. This fungus is endemic to Southeast Asia and is a significant cause of opportunistic infections in individuals with immunodeficiency associated with HIV / AIDS [42,43]. (**b**) *Penicillium funiculosum*: *P. funiculosum* is a plant pathogenic fungus that infects pineapple fruits. The disease it causes is referred to as fruit core rot, which is characterized by the browning or blackening and rotting of the fruit's core [44,45]. (**c**) *Fusarium oxysporum*: *F. oxysporum* is an ascomycete fungus belonging to the family Nectriaceae. *F. oxysporum* strains are commonly found in soil and can exist as saprophytes, deriving nutrients from decaying organic matter. They are capable of degrading lignin and complex carbohydrates present in soil residues [46,47].

Endophytic fungi originating from *Astragalus* species were found to catalyze the biotransformation of cycloastragenol, leading to the synthesis of an uncommon meroterpenoid compound **24** [48]. Additionally, through the utilization of an *Astragalus* endophyte called *Alternaria eureka* 1E1BL1, the biotransformation of cyclocephagenol, a novel cycloartane-type sapogenin with a tetrahydropyran unit, resulted in the formation of rare metabolites (**25–27**). The structures of these metabolites are depicted in Figure 7 [49].

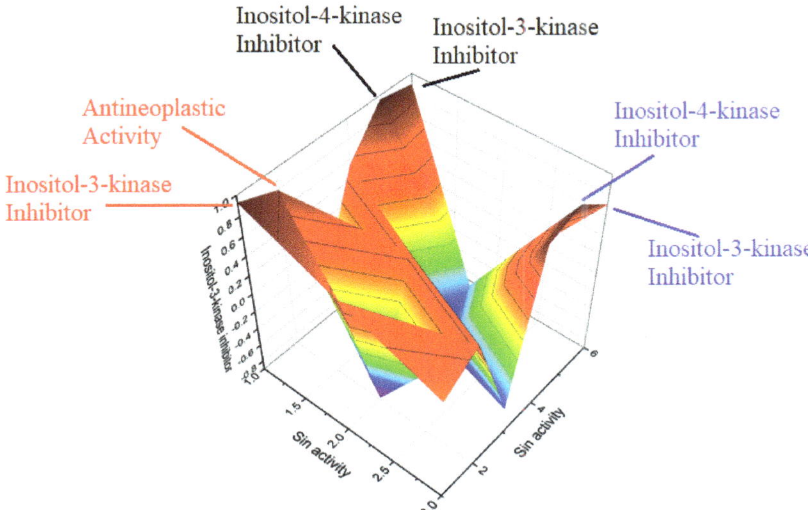

Figure 6. Three-dimensional graph illustrates the predicted and calculated activity as inositol-3-kinase inhibitors for steroids **7**, **8**, and **10** produced by the endophytic fungus *Talaromyces wortmannii*. The graph demonstrates a high confidence level of over 94% in the activity predictions. According to the data from Table 1, the dominant activity of steroids **7**, **8**, and **10** in the graph is shown in red. The steroid wortmanolone (**7**) is an antineoplastic and inositol-3-kinase inhibitor. The steroid wortmannin A (**8**) is an inositol-4-kinase inhibitor and inositol-3-kinase inhibitor of activity, and steroid secovironolide (**10**) is an inositol-4-kinase inhibitor and inositol-3-kinase inhibitor of activity. The activity text is in red for the steroid wortmanolone; in black is the activity of the steroid wortmannin A; and in blue is the activity of the steroid secovironolide. Red color—strong activity, blue color—no activity.

Various *Ganoderma* species, used in traditional Asian medicine to prevent and treat diseases [50], have yielded intriguing seco-steroids (**28–41**), as depicted in Table 2. Extracts from the wood-decay fungus *Ganoderma applanatum* revealed the discovery of ganoapplanic acid A (**28**), ganodapplanoic acids A (**29**), and B (**30**). These rearranged lanotane-type triterpenoids possess a 6/6/5/6-fused tetracyclic structure, featuring an uncommon C-13/C-15 oxygen bridge moiety [51,52]. Furthermore, *G. cochlear* yielded cochlates A and B (**31** and **32**), isomeric compounds with a 3,4-seco-9,10-seco-9,19-cyclo skeleton [53], while fornicatin A (**33**), a 3,4-seco-trinortriterpenoid, was first isolated from the fruiting bodies of *G. fornicatum* [54].

An intriguingly rearranged hexanorlanostane, known as cochlate C (**34**), was obtained from the fruiting bodies of *Ganoderma cochlear* [55]. Additionally, *Ganoderma tropicum* and *G. boninense*, two medicinal mushrooms, yielded a steroid (**35**) and a series of secolanostanes (**36–41**, 3D graph, see Figure 8) with anti-plasmodial activity [56–58]. Furthermore, the fungal strain *Emericella* sp. TJ29, an endophyte derived from the root of *Hypericum perforatum*, produced several extraordinary meroterpenoids named emeridones A–F (**42–47**) [59].

Trichocitrin (**48**), a diterpene, was extracted from the alga-endophytic fungus *Trichoderma citrinoviride*. Its structure is depicted in Figure 9, and the biological activity is outlined in Table 3. This compound represents the first furan-bearing fusicoccane diterpene derived from *Trichoderma* sp. and has shown inhibitory effects against *E. coli* [60]. The fungus *Aspergillus ustus*, isolated from the Mediterranean sponge *Suberites domuncula*, produced the sesterterpenoids ophiobolin-types **49** and **50** [61].

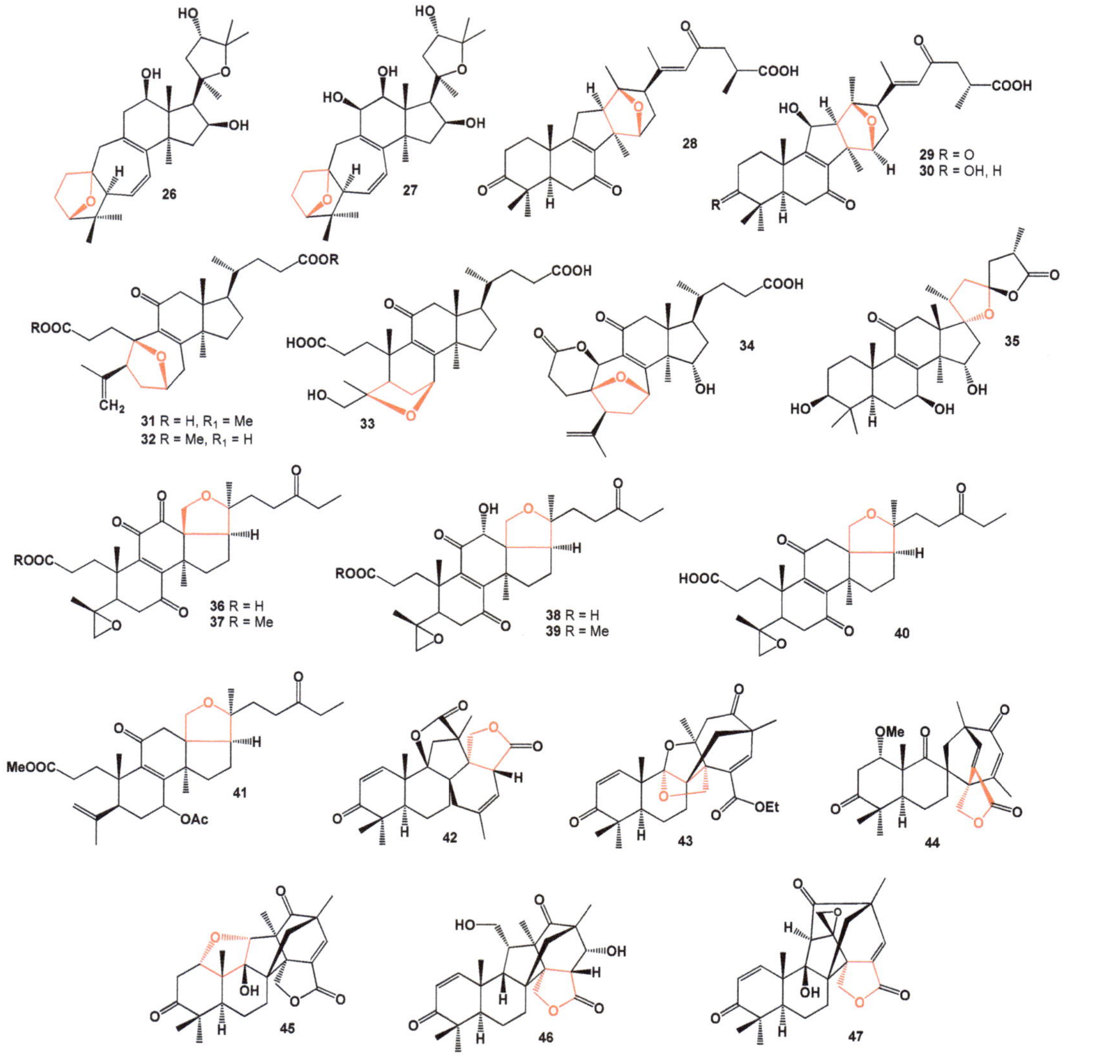

Figure 7. Furanosteroids produced by fungi and fungal endophytes.

Two phytotoxic sesterterpenoids, namely ophiobolin A lactone (**51**) and B (**52**), are synthesized by *Pseudomonas aeruginosa*, while compound **52** is also produced by the pathogenic fungi *Drechslera maydis* and *D. sorghicola* [62–64]. Additionally, the two mangrove fungi *Aspergillus terreus* H010 and *Lophiostoma bipolare* BCC25910 yielded compounds **53–55**; structures are shown in Figure 9, and a 3D graph is shown in Figure 10 [65,66].

Microbial transformation by the bacterial strain *Bacillus* sp. IMM-006 yielded two harziene-type diterpenoids, furanharzianones A (**56**) and B (**57**), featuring an unusual 4/7/5/6/5 ring system [67,68].

Dictyophora rubrovolvata, a saprophytic mushroom extensively cultivated in China, particularly in Guizhou Province, serves as a valuable source of two diterpenoids: 7,16,17-trihydroxy-19,6-kauranolide (**58**) and 2,7,11,14-tetrahydroxy-16-kauren-19,6-olide (**59**) [69]. Additionally, an *Acrostalagmus* fungus yielded LL-Z 1271a (**60**), a C16 terpenoid known for its antifungal properties [70,71]. Aspergilone A (**61**), an isopimarane compound, was isolated from a fungal species called *Epicoccum* [72], whereas xylarenolide (**62**), a pimarane derivative, has been identified in endophytic *Tubercularia* species [73] and certain *Xylaria* species [74,75].

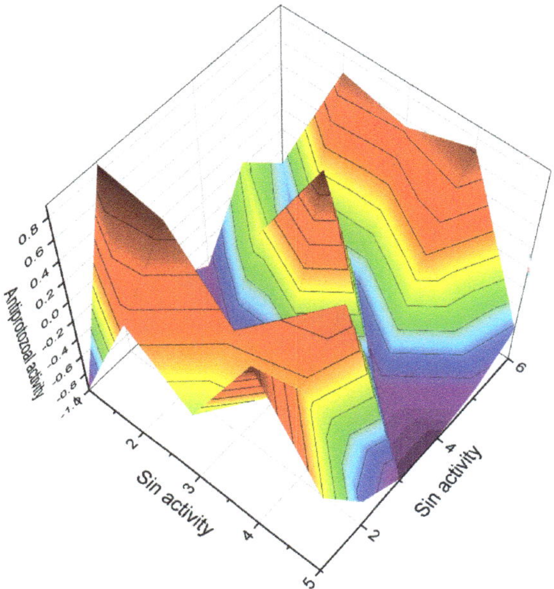

Figure 8. Three-dimensional graph illustrates the predicted and calculated antiprotozoal activity of steroids (**36**, **37**, **38**, **39**, and **40**) with a confidence level exceeding 95%. These steroids are produced by the medicinal mushroom *Ganoderma tropicum*.

From the solid culture of *Aspergillus flocculosus* 16D-1, two distinct steroids were isolated. The first one, asperflotone (**63**), is an unusual 8(14 → 15)-abeo-ergostane-type steroid. It possesses a unique ergosteroid structure characterized by a rearranged bicyclo [4.2.1]non-2-ene ring system, potentially resulting from α-ketol rearrangement during biosynthesis. Remarkably, both asperflotone and the second steroid, asperfloroid (**64**), exhibited inhibitory activity against IL-6 production in induced THP-1 cells [76]. In addition, the sponge-derived fungus *Aspergillus flocculosus* 16D-1 produced two other steroids, aspersecosteroids A (**65**) and B (**66**), which are 11(9→10)-abeo-5,10-secosteroids. Notably, both compounds demonstrated a potent inhibitory effect on the production of TNF-α and IL-6 [77].

Tricholumin A (**67**) was obtained from the alga-endophytic fungus *Trichoderma asperellum*. It retains cycle A, the final structural element of the original ergosterol, after undergoing deep oxidative transformations in the rest of the molecule, including the side chain fragment [35].

Xylarglycoside B (**68**), an antibacterial steroid, was isolated from the fungus *Xylaria* sp. KYJ-15, which was derived from the leaves of *Illigera celebica*. It displayed antibacterial activity against *Staphylococcus aureus* [78]. The endophytic fungus *Emericella variecolor* led to the isolation of emericellic acid (**69**), a meroterpenoid compound [79]. From the fruiting bodies of the mushroom *Stropharia rugosoannulata*, an unusual sterol with an unprecedented ether ring (**70**) was isolated. This mushroom is known as *saketsubatake* in Japanese and wine-cap *stropharia* in English [80]. Ergopyrone (**71**), an extraordinary styrylpyrone-fused ergosterol derivative, was isolated and structurally characterized from the mushroom *Gymnopilus orientispectabilis*. This steroid features a hexacyclic 6/5/6/6/6/5 skeleton that is formed via [3 + 2] cycloaddition between ergosterol and styrylpyrone precursors [81].

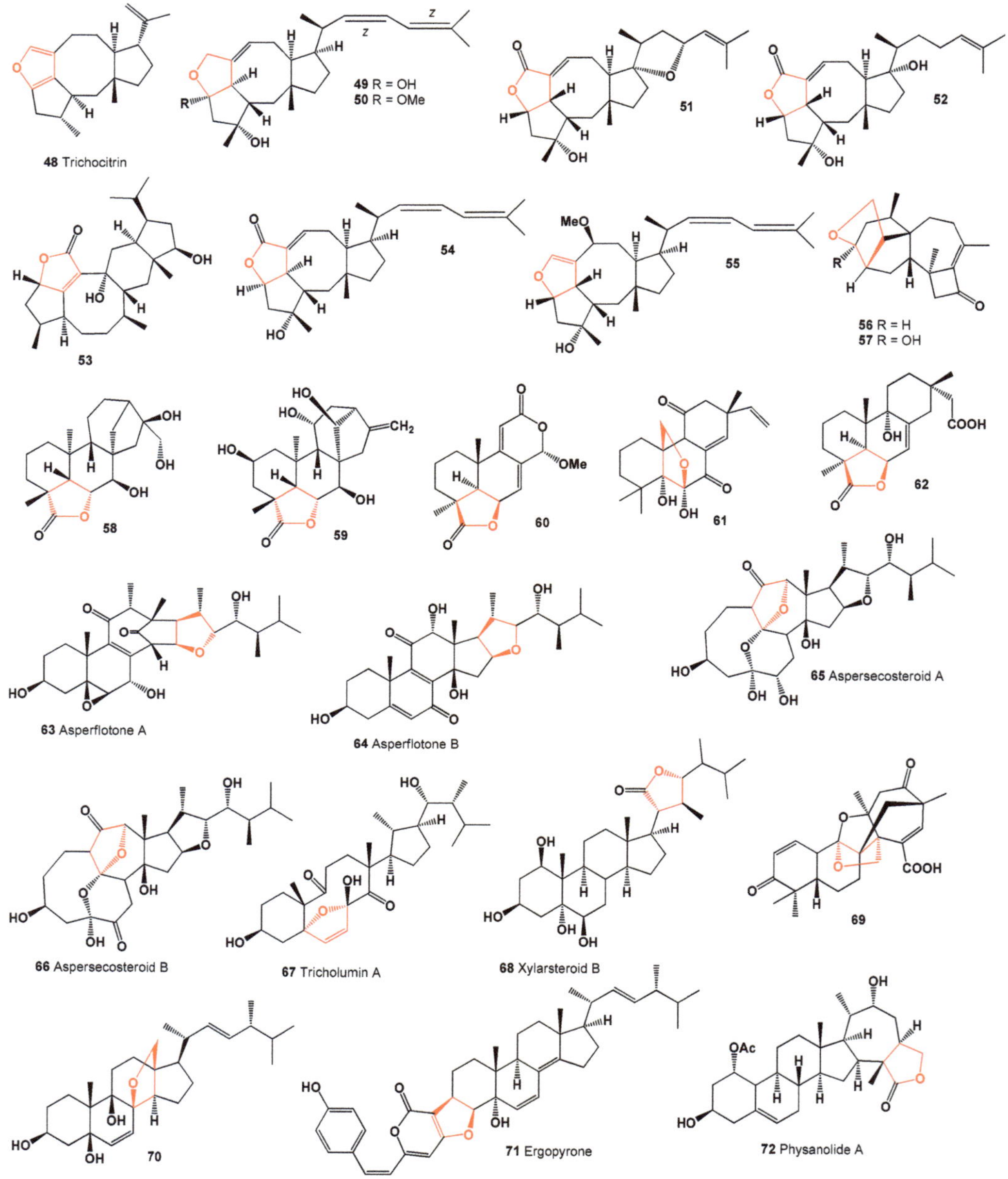

Figure 9. Steroids and isoprenoid lipids produced by fungi.

Physalis angulata yielded several C28 steroids of the withasteroid family, including physanolide A (**72**) with an unprecedented skeleton containing a seven-membered ring and various physalins (**73–80**), as shown in Figure 11. These compounds displayed biological activity, as outlined in Table 4. Physalins B, D, and F exhibited potent cytotoxicity against multiple tumor cell lines, including KB, A431, HCT-8, PC-3, and ZR751, with EC_{50} values below 0.4 μM [82,83]. Antheridiol (**81**), the fungal sex hormone, was isolated from *Achlya* *bisexualis*, a water mold [84]. Physangulide B (**82**), a steroid, was identified in the calyxes of *Physalis angulata*, featuring an additional tetrahydrofuran ring in its structure [85].

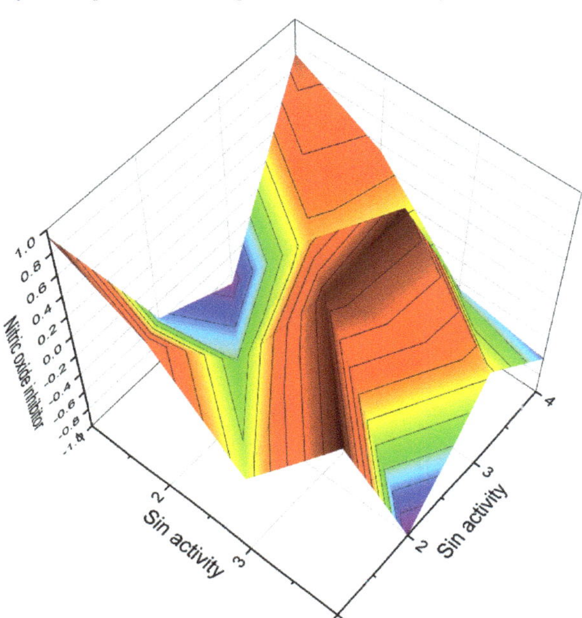

Figure 10. Three-dimensional graph illustrates the predicted and calculated activity of steroids (**49**, **50**, **54**, and **55**) as nitric oxide inhibitors, with a confidence level exceeding 90%.

Bioactive compounds **83–99**, known as withaphysalins, were isolated from *Physalis minima*. These compounds typically possess a hemiacetal or lactone linkage between C-18 and C-20, and they have demonstrated cytotoxic, antiproliferative, and anti-inflammatory activities [86,87].

Agaricus blazei (a picture of this fungus is shown in Figure 12), also known as '*Cogumelo do Sol*' in Brazil or '*Himematsutake*' in Japan, is a widely cultivated mushroom with medicinal uses.

It has been traditionally employed to treat various common ailments such as atherosclerosis, hepatitis, hyperlipidemia, diabetes, dermatitis, and cancer. The mushroom contains bioactive, highly oxygenated des-A-ergostane derivatives, including agariblazeispirols A (**90**) and B (**91**), as well as blazeispirols B (**92**), C (**93**), E (**94**), and F (**95**), which were isolated from cultured mycelia of *Agaricus blazei* [88,89]. Agariblazeispirols (**97–99**, 3D graph, see Figure 13) exhibited a moderate circumvention of drug resistance in mouse leukemia P388/VCR cells [89]. Additionally, blazeispirol A (**96**), featuring an unprecedented skeleton, has been isolated from the cultured mycelia of the same fungus [90].

A triterpenoid named irpexolidal (**97**) with an unprecedented carbon skeleton, along with its biogenetic-related compound irpexolide A (**98**) were isolated from the fruiting bodies of the medicinal fungus *Irpex lacteus* [91]. Study of the extract of the fruiting bodies of the mushroom *Leucopaxillus gentianeus*, allowed the isolation of minor cucurbitane triterpene, leucopaxillone B (**99**, the 3D graph is shown in Figure 13). The antiproliferative activity of the isolated triterpene was determined against the NCI-H460 human tumour cell line [92]. Simplifusidic acids B (**100**) and D (**101**), fusidane-type nortriterpenoids, were isolated from the marine-derived fungus *Simplicillium* sp. SCSIO 41513 [93].

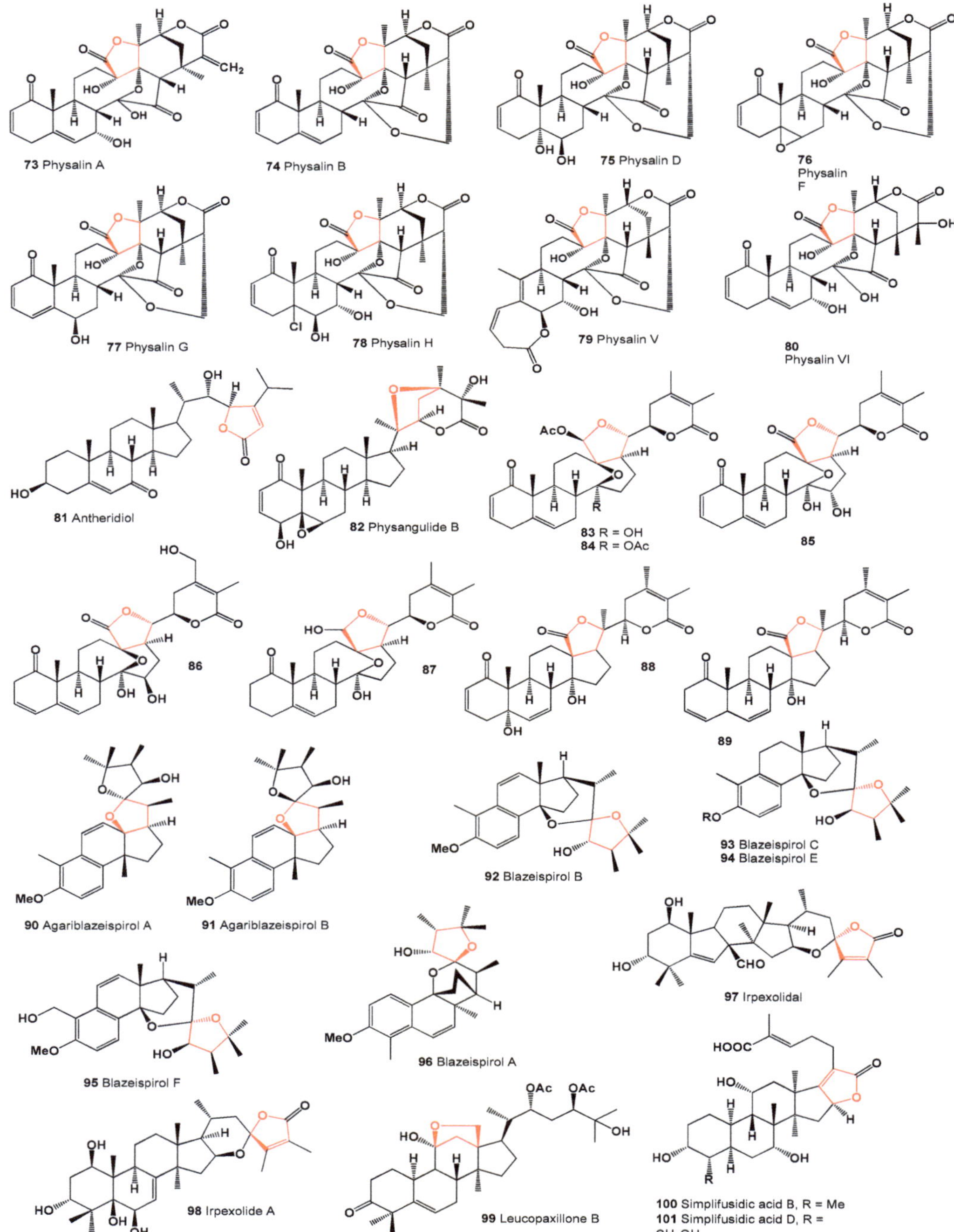

Figure 11. Steroids and meroterpenoids derived from fungal species.

Figure 12. (a) *Xylaria* sp.: This ascomycete fungus typically thrives on dead wood and is known for producing steroids (**68**). The genus *Xylaria* comprises fungal endophytes associated with both vascular and non-vascular plants, and it serves as a valuable source of bioactive secondary metabolites. These metabolites include sesquiterpenoids, esters, alcohols, terpenoids, cytochalasins, mellein, alkaloids, polyketides, and aromatic compounds. Some of these compounds have demonstrated potential activity as herbicides, fungicides, insecticides, antibacterials, antimalarials, antifungals, or α-glucosidase inhibitors [94]. (b) *Agaricus blazei*: This medicinal fungus contains secosteroids (**90–95**). With significant commercial value, *Agaricus blazei* offers a wide range of health benefits. The mushrooms are rich in biologically active substances such as polysaccharides, lipids, sterols, proteins, vitamins B, C, and D, as well as phenolic compounds. Polysaccharides from *A. blazei* have been shown to possess immunoregulatory, anti-inflammatory, hepatoprotective, and antitumor properties. Extracts from this fungus have been used to treat diabetes and bacterial infections, exhibiting anticarcinogenic and antimutagenic effects [95,96]. (c) *Irpex lacteus*: This medicinal fungus serves as a source of triterpenoids (**97** and **98**). *Irpex lacteus*, a white rot fungus, is widely employed in bioremediation and food biotechnology due to its exceptional lignin-degrading capabilities. The fungus produces various extracellular enzymes, including lignin peroxidase, laccase, glucose oxidase, proteases, and α-galactosidase, involved in oxidative, hydrolytic, and lignocellulose degradation processes. *Irpex lacteus* readily oxidizes steroids, triterpenoids, alkanes, and cyclic ketones [97,98].

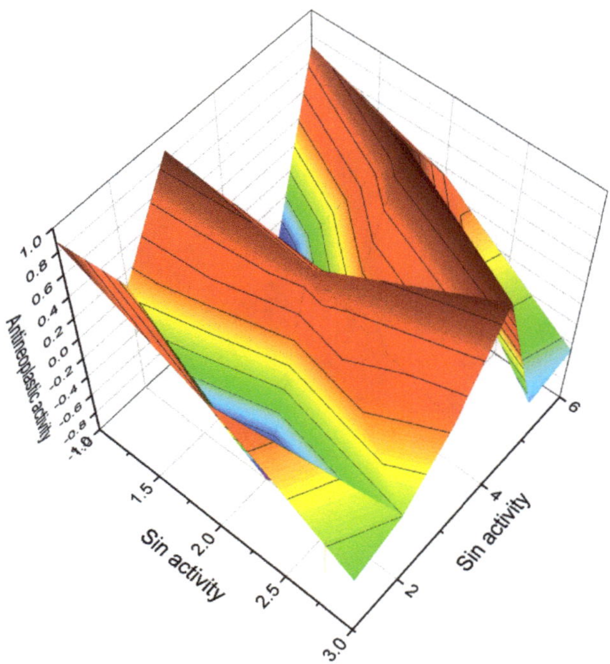

Figure 13. Three-dimensional graph illustrating the predicted and calculated antineoplastic activity of steroids (**74**, **75**, and **99**) with a confidence level exceeding 91%. These steroids are derived from meroterpenoids produced by the fungus *Physalis angulata*, while the steroid is sourced from the mushroom *Leucopaxillus gentianeus*.

3. Furanosteroids Derived from Plant Species

Furanosteroids derived from plant species have been extensively studied. Around 50 years ago, Indian scientists isolated a hexacyclic tetranortriterpenoid called vilasinin (**102**) from the green leaves of the neem tree (*Azadirachta indica*). Since then, numerous related metabolites have been discovered in various plant species [99]. 1,3-Diacetylvilasinin (**103**) has been reported from *Melia volkensii* [100], *Chisocheton paniculatus* [101], *Azadirachta indica* [102], as well as two African Turraea species, *Turraea holstii* and *T. parvifolia* [103]. Another compound, 1,3-diacetyl-12α-hydroxy-7-tigloylvilasinin (**104**), has been found in *Azadirachta indica* [104] and *Malleastrum antsingyense* (depicted in Figure 14) [105].

The leaves of *Trichilia gilgiana*, extracted with CH_2Cl_2-MeOH, yielded vilasinin-type limonoids known as rubescin H (**105**), gilgianin A (**106**), gilgianin B (**107**), TS3 (**108**), and trichirubine A (**109**). Furanosteroids **105**, **106**, and **107** demonstrated potent anti-plasmodial activity alongside significant cytotoxicity. Compounds **108** and **109** exhibited the highest anti-plasmodial activity, with IC_{50} values of 1.1 and 1.3 μM, respectively [106]. The structures of compounds **105**, **106**, and **107** are shown in Figure 15, and their biological activity is detailed in Table 5. Furthermore, a 3D graph illustrating the activity of compounds **108** and **109** is displayed in Figure 16.

Figure 14. (**a**) *Malleastrum*: *Malleastrum* is a genus comprising over 20 species within the Meliaceae family. Native to Madagascar, the Comoros, and Aldabra, plants in the Meliaceae family are known to contain limonoids, terpenoids, alkaloids, flavonoids, and phenolic compounds as their primary chemical constituents. Many species within this family exhibit cytotoxic, antimicrobial, or antimalarial activity [107]. (**b**) *Azadirachta indica* (Neem): *Azadirachta indica*, commonly known as Neem, belongs to the Meliaceae family. The leaves of Neem are widely used in Chinese, Ayurvedic, and Unani medicines, particularly in the Indian subcontinent. Neem leaves have demonstrated antibacterial, anthelmintic, antiviral, and anticancer properties, and most notably, they act as an immunomodulatory agent [108]. (**c**) *Annona squamosa* (Sugar Apple): *Annona squamosa*, also known as sugar apple, is a member of the Annonaceae family. It has been traditionally used in Indian, Thai, and American medicine. The leaves of sugar apple are commonly used as a decoction to treat dysentery and urinary tract infections [109].

Azadirachta indica, belonging to the family Meliaceae, is commonly known as neem or Indian lilac. It is utilized for its antimalarial, anti-inflammatory, antipyretic, antitumor, and anthelmintic properties. The leaves of *Azadirachta indica* contain tetranortriterpenoids such as **110** and **111** [110].

Sutherlandia frutescens, a plant from the Fabaceae family native to South Africa, is commonly referred to as Cancer bush. It is renowned for its multifunctional medicinal uses, and infusions and decoctions of the plant are widely employed in South Africa for treating cancer, inflammation, viral infections, and gastrointestinal diseases. An unusual cycloartane glycoside called sutherlandioside A (**112**) has been isolated from the water-methanol fraction of the plant [111].

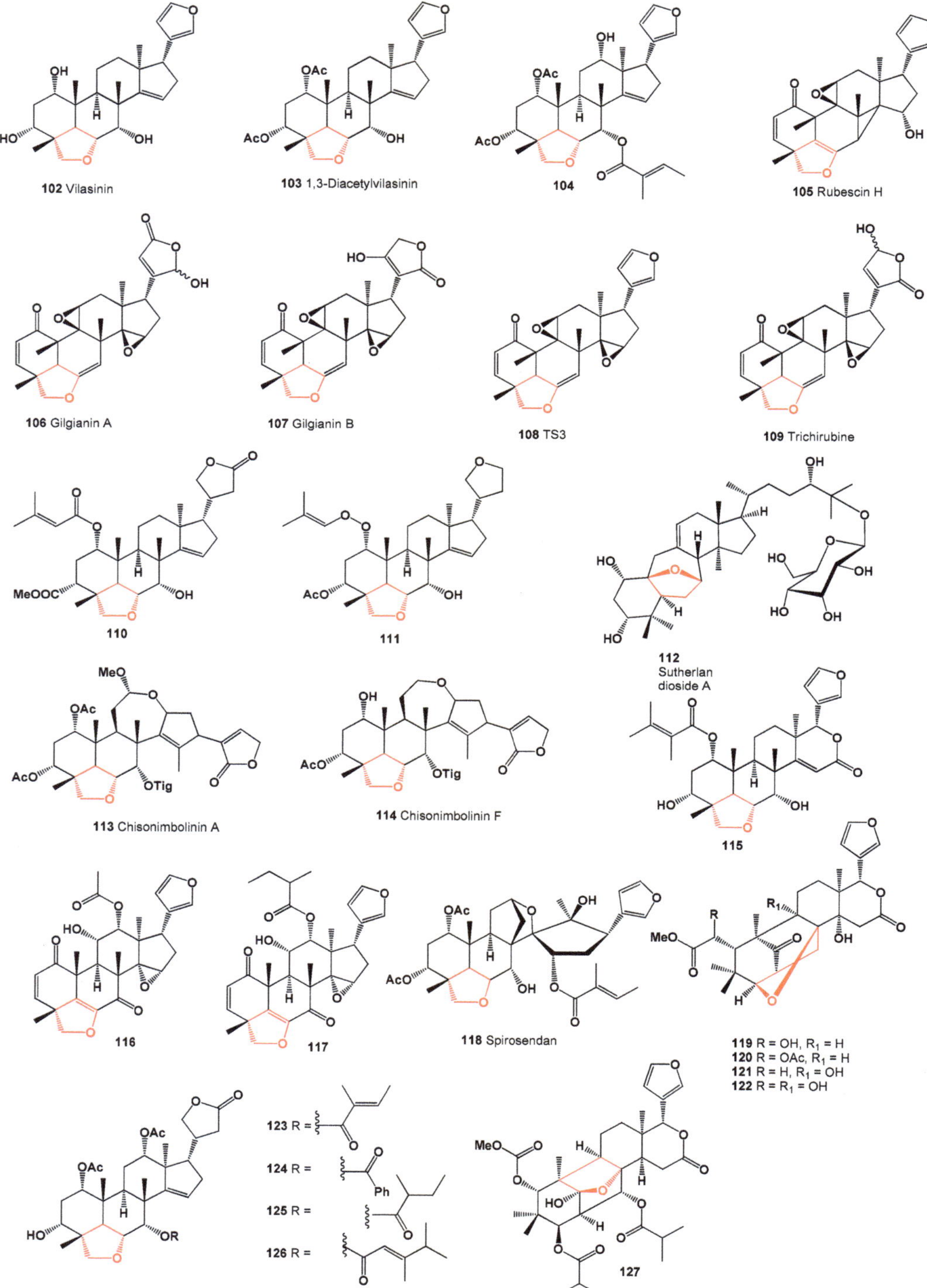

Figure 15. Steroids and meroterpenoids derived from plant species.

Chisocheton paniculatus twigs yielded two tetranortriterpenoids named chisonimbolinins A (**113**) and F (**114**) [112]. Furthermore, a triterpenoid **115** with insecticidal activity was isolated from a methanol extract of fresh leaves of *Azadirachta indica* [113].

Walsura cochinchinensis bark extract yielded two limonoids, walsucochinones B (**116**) and C (**117**). The ethyl acetate extract and walsucochinone C (**117**) displayed cytotoxic activity against MCF-7 human breast cancer cells [114]. *Melia toosendan* (Meliaceae) root bark provided spirosendan (**118**), a skeletal limonoid with a spiro-structure [115]. Additionally, an aqueous methanolic extract of Cedrela odorata leaves yielded four tetranortriterpenoids: cedrodorin (**119**), 6-acetoxycedrodorin (**120**), 6-deoxy-9*R*-hydroxycedrodorin (**121**), and 9*R*-hydroxycedrodorin (**122**) [116].

Dysoxylum gaudichaudianum, commonly known as ivory mahogany, yielded four tetranortriterpenoids named dysoxylins A–D (**123–126**), which exhibited potent antiviral activity against the respiratory syncytial virus [117]. *Xylocarpus rumphii* heartwood provided a triterpenoid derivative identified as xylorumphiins E (**127**) [118].

Stem bark extracts of *Khaya anthotheca* contained three limonoids: anthothecanolide (**128**), 3-O-acetylanthothecanolide (**129**), and 2,3-di-O-acetyl-anthothecanolide (**130**) [119]. The structure of compound **130** is depicted in Figure 17, and its biological activities are outlined in Table 6.

A tetranortriterpenoid called kokosanolide D (**131**) has been isolated from the methanol extract of fruit peels of *Lansium domesticum*, found in West Java, Indonesia [120]. Furthermore, an ethanol extract of a plant belonging to the *Malleastrum* genus yielded three steroids known as malleastrones A–C (**132**, **133**, and **134**), respectively. Compounds **133** and **134** exhibited antiproliferative activity against a range of cancer cell lines, with IC_{50} values ranging from 0.19 to 0.63 µM [121,122].

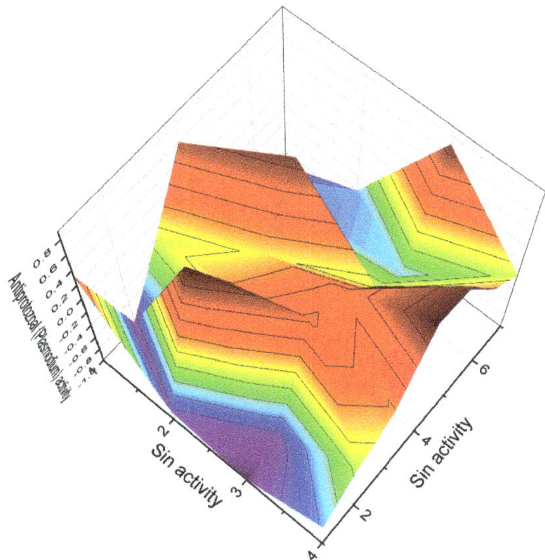

Figure 16. Three-dimensional graph illustrates the predicted and calculated antiprotozoal (*Plasmodium*) activity of the vilasinin-type limonoids (**106**, **107**, **108**, and **109**) with a confidence level exceeding 95%. These limonoids are derived from the tree *Trichilia gilgiana*, which is predominantly found in southern Nigeria and eastern Congo. In the Congo region, the extract of *Trichilia gilgiana* bark is utilized for its analgesic and stimulant properties. It has been traditionally used in traditional medicine to treat abdominal pain, chest pain, fever, and as a tonic. The juice of the young leaves is applied to circumcised wounds, while crushed leaves are added to drinking water for the treatment of respiratory diseases. Additionally, the juice of the leaves has demonstrated potent activity against the malarial *Plasmodium* [123–125].

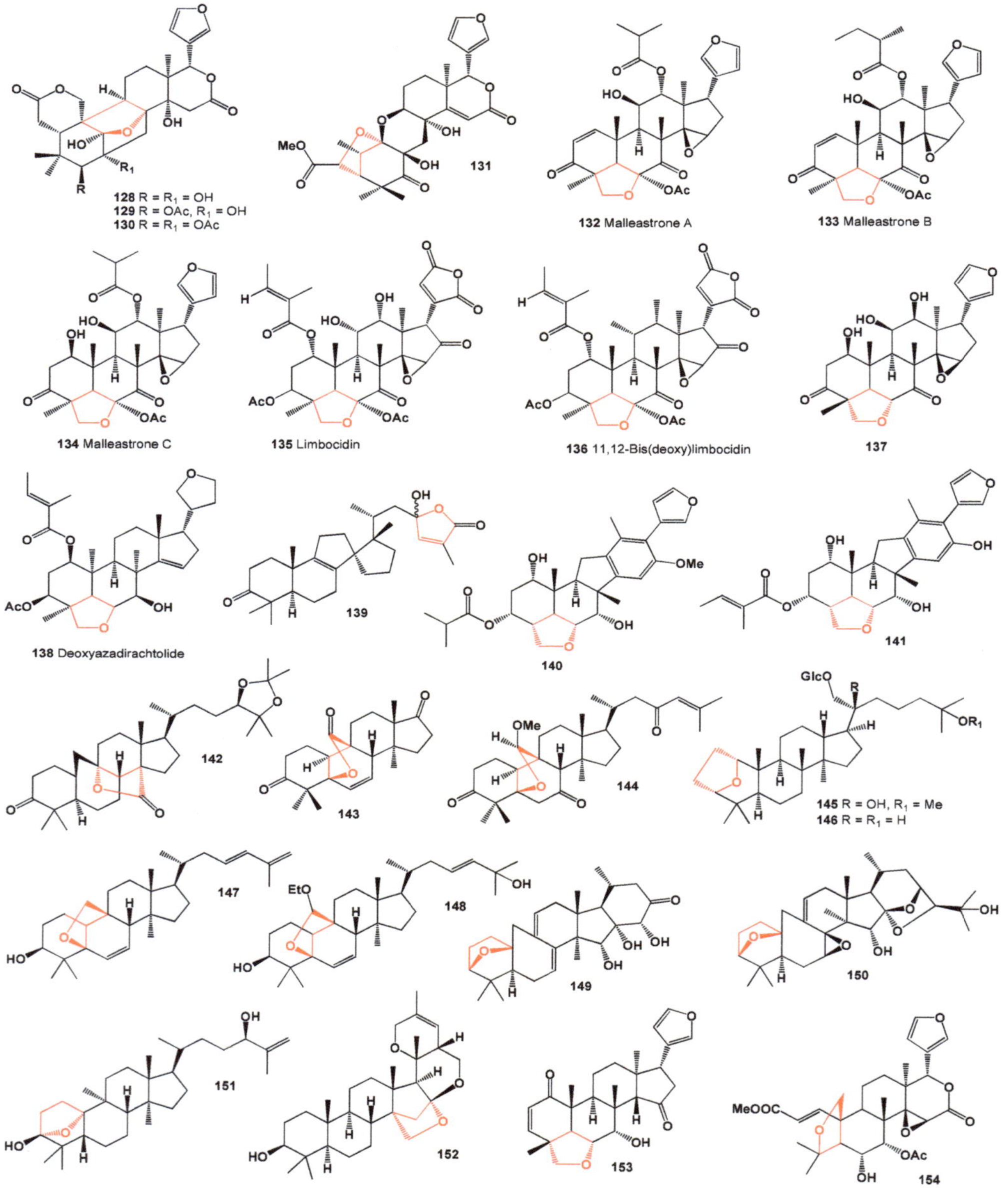

Figure 17. Bioactive Steroids and triterpenoids derived from plant species.

Limbocinin (**135**) and compound **136** were isolated from ethanolic extracts of neem seeds and the leaves of *Azadirachta indica* and *Annona squamosa* [126,127]. Both compounds exhibited antifungal activity [128].

Furanosteroids (**137** and **138**) have been isolated from plants belonging to the Meliaceae and Simaroubaceae families [129–132]. Furthermore, a rearranged lanostane (**139**) has been isolated from *Abies nephrolepis*, commonly known as Khingan fir [133].

Two limonoids with an aromatic ring D, walsucochinoids A (**140**) and B (**141**) were detected in the air-dried plant *Walsura cochinchinensis* (Figure 18 depicts a picture of this plant).

Figure 18. (**a**) *Walsura cochinchinensis*: This plant, belonging to the Meliaceae family, contains bioactive limonoids known as walsucochinoids A (**140**) and B (**141**). Medicinal plants from the *Walsura* genus, found in tropical areas of several Asian countries, are widely used in traditional medicine systems. Studies have identified over 200 compounds from ten species within this genus, including sesquiterpenoids, flavonoids, sterols, lignans, xanthones, and anthraquinones. Many of these compounds exhibit diverse properties such as cancer cell cytotoxicity, antimicrobial activity, antidiabetic effects, anti-inflammatory effects, antioxidant properties, antifeedant properties, antifertility effects, ichthyotoxic effects, and neuroprotective effects [134,135]. (**b**) *Gynostemma pentaphyllum*: Also known as Jiaogulan, *Gynostemma pentaphyllum* is a dioecious climbing vine from the Cucurbitaceae family. It is widely distributed in South and East Asia, as well as New Guinea. In Chinese medicine, it is commonly used to treat various conditions, including hepatitis, diabetes, and cardiovascular diseases. Extracts from *G. pentaphyllum* contain sterols, flavonoids, and polysaccharides that exhibit inhibitory activity against cancer cell proliferation. These extracts have demonstrated effects such as cell cycle arrest, apoptosis induction, inhibition of invasion and metastasis, inhibition of glycolysis, and immunomodulatory activity [136,137]. (**c**) *Momordica charantia*: The fruit of *M. charantia*, which contains anticancer steroids (**147 and 148**), has been consumed as food and used as medicine since ancient times. *M. charantia*, commonly known as bitter melon, holds a significant place in various systems of traditional medicine. It has been used to treat a wide range of conditions, including diabetes, abortive purposes, anthelmintic effects, contraception, dysmenorrhea, eczema, emmenagogue properties, antimalarial activity, lactagogue effects, gout, jaundice, abdominal pain, renal issues (stones), laxative effects, leprosy, leucorrhea, hemorrhoids, pneumonia, psoriasis, rheumatism, fever, and scabies [138,139].

From the leaves of *Caloncoba glauca*, a triterpenoid called caloncobalactone C (**142**) was isolated. It exhibited inhibitory activity against human and mouse 11β-hydroxysteroid dehydrogenase types 1 and 2 [140,141]. The 3D graph illustrating its structure is presented in Figure 19.

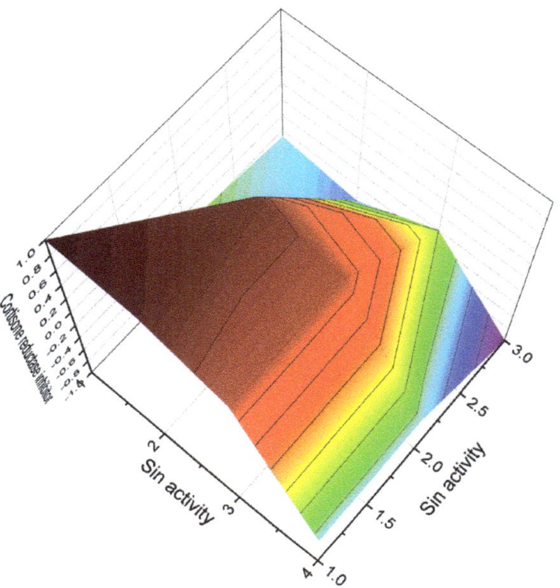

Figure 19. Three-dimensional graph illustrating the predicted and calculated activity of steroids (**140**, **141**, and **142**) as cortisone reductase inhibitors with over 90% confidence. These steroids include the limonoids with an aromatic ring D (**140** and **141**) found in the leaves of the *Walsura cochinchinensis* plant, as well as the triterpenoid (**142**) found in the leaves of *Caloncoba glauca*.

Within the EtOAc and n-BuOH extracts of the vines and leaves of *Momordica aurantia*, two steroids, octanorcucurbitacin (**143**) and kuguacin I (**144**), were discovered. These compounds demonstrated anti-HIV-1 activities in vitro [142]. Triterpenes gypensapogenin H (**145**) and I (**146**) were isolated from the hydrolyzate of the total saponin extract obtained from the dioecious herbaceous climber *Gynostemma pentaphyllum*, which is widely distributed in South and East Asia [143].

Two cucurbitane-type triterpenoids, namely (23*E*)-5β,19-epoxycucurbita-6,23,25-triene-3β-ol (**147**) and (19*R*,23*E*)-5β,19-epoxy-19-ethoxycucurbita-6,23-diene-3β,25-diol (**148**), have been isolated from the fruit of *Momordica charantia*. These compounds exhibited weak cytotoxic activity against cancer cell lines including MCF-7, HepG2, Du145, Colon205, and HL-60 [144].

The roots of Cimicifuga heracleifolia, a plant included in the Chinese Pharmacopoeia and used in traditional medicine in China for centuries, contain two unusual ring A cracking 9,19-cycloartane triterpenes (**149**) and (**150**) [145]. The dried fruit of Vitex negundo yielded the 3,10-epoxide (**151**) with antitumor activity [146].

From the seeds of *Hovenia trichocarpa*, two saponins named hoduloside XI and hoduloside XII were isolated. Both compounds shared the genin 20,26-epoxy-pseudojujubogenin (**152**) and displayed inhibitive activities against human cancer cell lines HL-60 and K562 [147]. In addition to these steroids, a furanosteroid named ceramicine J (**153**) was discovered in the hexane layer of *Chisocheton ceramicus* bark extract. This compound exhibited dose-dependent, moderate cytotoxicity against the HL-60 cell line [148]. Furthermore, a secosteroid called 6-O-deacetylseverinolide (**154**) was identified in the stem barks of *Atalantia buxifolia* extract [36].

The steroid inertogenin (**155**) is a compound containing a rare 7,15-tetrahydrofuran group. It is found in the leaves of *Strophanthus amboensis*, an erect deciduous shrub harvested for medical use in Southwest Africa [149]. The structure of inertogenin is depicted in Figure 20, and its biological activity is shown in Table 7. The leaves of *Toona ciliata* var. *yunnanensis* yielded several seco steroids: tooonayunnanins F (**156**), G (**157**), J (**158**), and K (**159**) [150].

Fritillaria pallidiflora bulbs are a source of unique jervinine-type alkaloids [151]. Peimissine (**160**), cycloparnine (**161**), and cycloposine (**162**) were isolated from the bulbs of *F. pallidiflora*, as well as the isosteroidal alkaloid yibeinone A (**163**) [152]. Puqienine F (**164**), a veratramine alkaloid with a 12,16-epoxy ring, was isolated from the bulbs of *Fritillaria puqiensis* [153]. A picture of *Fritillaria pallidiflora* is shown in Figure 21, and the 3D graph representing the activity of puqienine F (**164**) can be found in Figure 22.

From *Buxus hyrcana*, collected in Iran, two steroidal alkaloids were isolated: (+)-O6-buxafurandiene (**165**) and (+)-7-deoxy-O6-buxafurandiene (**166**). These compounds belong to the rare class of *Buxus* alkaloids with a tetrahydrofuran ring incorporated into their structures. Furthermore, they exhibited acetylcholinesterase enzyme inhibitory activity [154].

Steroidal alkaloids, solasodine (**167**), and tomatidine (**168**) were isolated from the aerial parts of *Solanum leucocarpum*, a plant belonging to the Solanaceae family. The collection was made at the regional natural park Ucumarí in Colombia. Both alkaloids have demonstrated biological activity. Solasodine exhibits DNA-damaging activity, while tomatidine displays activity through DNA topoisomerase II inhibition [155].

Two unique abeo-steroids, spirochensilides A (**169**) and B (**170**), were isolated from *Abies chensiensis*. These compounds represent the first example of triterpenoids with a distinctive 8,10-cyclo-9,10-seco and methyl-rearranged carbon skeleton [156]. *Abies faxoniana*, an endemic plant found in several provinces in China, serves as the source of lanostane and cycloartane derivatives A1/A2, **171–174**, which feature epimeric spiro-side chains [157].

Bungsteroid A (**175**), possessing an unreported carbon skeleton, was isolated from the pericarps of *Zanthoxylum bungeanum*. It represents a C34 steroid analogue with a unique 6/6/6/6/5-fused pentacyclic skeleton. This compound exhibited antiproliferative effects against HepG2, MCF-7, and HeLa cell lines, with IC_{50} values of 56.3, 64.2, and 74.2 µM, respectively [158].

The roots and stems of *Cyathula officinalis*, also known as *Cyathula Root* or *Radix Cyathula*, yielded two cyasterone stereoisomers: 28-*epi*-cyasterone (**176**) and 25-*epi*-28-*epi*-cyasterone (**177**) [159]. The biologically active plant steroid, 28-*epi*-cyasterone (**176**), has also been found in *Eriophyton wallchii* [160] and fronds of the fern *Microsorum scolopendria* [161].

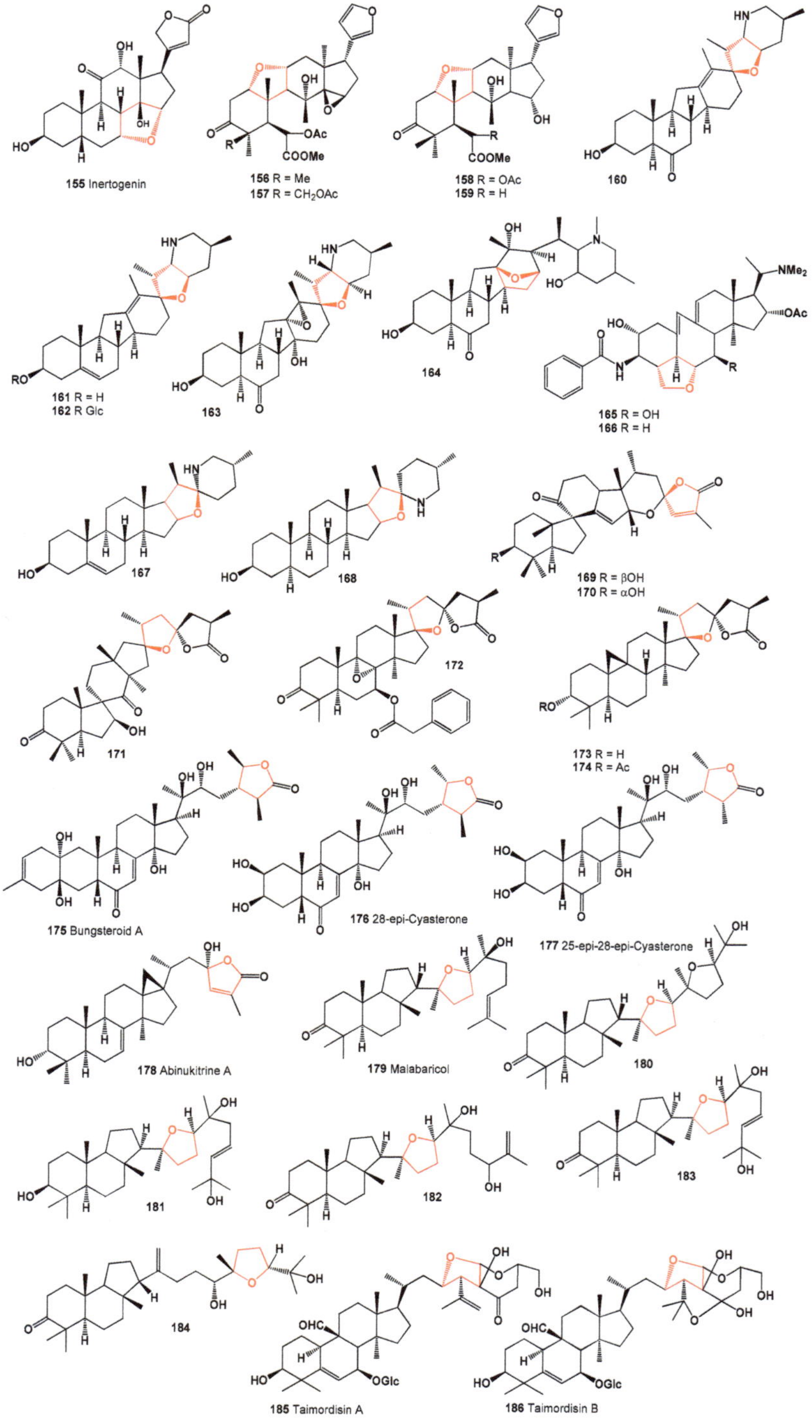

Figure 20. Steroids, di-, and triterpenoids derived from plant species.

Figure 21. (**a**) The plant *Fritillaria pallidiflora* is a source of unusual steroids (**160–164**). *F. pallidiflora*, also known as Siberian hazel grouse, is a species that was initially misnamed as it does not grow wild in Siberia. It was discovered in 1857 in the western regions of the Himalayas and Asia Minor. The plant is known for its medicinal uses, and an infusion of dried chopped onion from this plant is commonly used orally to treat cough, bronchitis, pneumonia, febrile illnesses, and abscesses [162,163]. Shaanxi fir (**b**) is a tree that grows in Gansu, Hubei, and Sichuan. *Abies chensiensis*, a species of Shaanxi fir, produces unusual triterpenoids (**169** and **170**). The essential oil of Shaanxi fir, as well as Siberian fir (*Abies sibirica*), has a pleasant, fresh pine aroma and contains bornyl acetate, which contributes to its soothing, balancing, and anti-inflammatory effects. The oil is used to relieve anger, promote contentment, alleviate intolerance in toxic relationships, promote self-connection, and foster a fearless attitude [164,165]. The pericarps of *Zanthoxylum bungeanum* (**c**) produce an unusual steroid (**175**). Extracts from *Z. bungeanum* fruit are widely used in the cosmetics industry to produce creams. Various parts of this plant, including the fruit, stems, leaves, and bark, have been utilized in local medical systems to treat fever, stomach pain, toothache, and inflammation [166,167].

Triterpenoids with medicinal properties have been identified from various plant sources. Abinukitrine A (**178**), a triterpenoid, was isolated from *Abies nukiangensis* extracts and exhibited a potent anti-hepatitis C virus (HCV) effect [62]. Malabaricol (**179**), another triterpenoid, was first reported by Indian chemists from the National Chemical Laboratory in 1967, isolated from the tropical tree *Ailanthus malabarica* [168]. Subsequently, malabaricol (**179**) and **180** were also found in the heartwood of *Ailanthus excelsa* [169]. Malabaricol and its derivatives [170] possess antibacterial effects, supporting their traditional use in folk medicine [171]. *Ailanthus triphysa* is a source of ailanthusins F (**181**) and G (**182**) [172], while two other derivatives, **183** and **184**, were discovered in the leaves of *Caloncoba echinata* [173].

Furthermore, triterpene glycosides called taimordisins A (**185**) and B (**186**) were isolated from the fresh fruits of Taiwanese *Momordica charantia* [174]. Although these compounds demonstrate favorable anti-inflammatory activity, they do not exhibit any anti-cancer properties and have been determined to be safe for use.

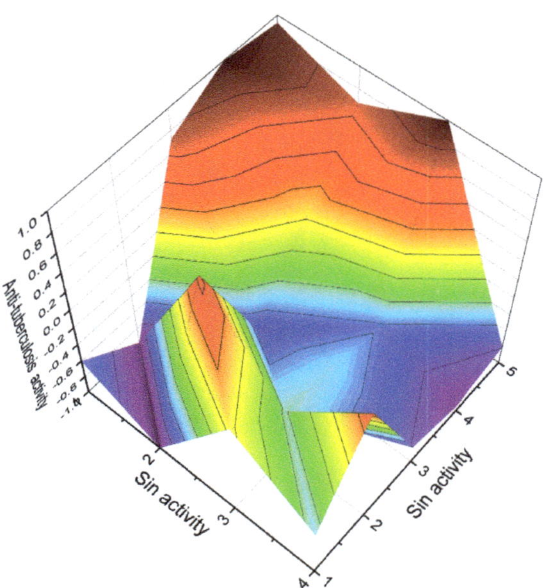

Figure 22. Three-dimensional graph illustrating the predicted and calculated anti-tuberculosis activity of steroids (**160**, **161**, **162**, **163**, and **164**) with a confidence level exceeding 90%. These steroids include the unusual steroidal jervinine-type alkaloids (**160–163**) derived from the bulbs of *Fritillaria pallidiflora*, as well as alkaloid (**164**) obtained from the bulbs of *Fritillaria puqiensis*. Notably, these compounds possess the rare property of exhibiting anti-tuberculosis activity. The genus *Fritillaria* has been widely used worldwide for medicinal and culinary purposes. For over 2000 years, decoctions made from the bulbs of various Fritillaria species have been utilized in traditional Chinese medicine to address diverse ailments such as asthma, pharyngitis, bronchitis, coughs, goiter, and hemoptysis. Additionally, it has been employed as an expectorant and antitussive agent [175,176].

4. Furanosteroids Derived from Marine Sources

Furanosteroids, a fascinating group of natural compounds, are derived from marine sources and are produced by various marine invertebrates, including cyanobacteria, fungal endophytes associated with algae, sponges, soft corals, or molluscs [2–8].

Merosterol A (**187**), a cyanobacterial cytotoxin found in *Scytonema* sp. PCC 10023, has been exhibited against HeLa cells with IC_{50} values of 1.8 µM [177]. In an active organic extract obtained from an Okinawan marine sponge of the genus *Dysidea* (depicted in Figure 23), two polyoxygenated steroids, dysideasterols F (**188**) and G (**189**), were identified. These compounds demonstrated a similar cytotoxic effect, with IC_{50} values of 0.15 and 0.3 µM, respectively, against human epidermoid carcinoma A431 cells [178]. Additionally, a toxic polyoxygenated steroid named cholest-6-en-11β,19-epoxy-3β,5α,8α,9α-tetrol (**190**) has been isolated from the sponge *Dysidea tupha* [179]. Furthermore, a sponge species, *Strongylophora* sp., yielded furano-pregnanes, namely 3,4-dihydroxypregna-5,17-diene-10,2-carbolactone (**191**, whose structure is shown in Figure 24) and 3,4-dihydroxypregna-5,15-dien-20-one-10,2-carbolactone (**192**) [180].

Nakiterpiosin (**193**) and nakiterpiosinone (**194**, whose structures are shown in Figure 24), which are halogenated and rearranged norsteroids, were isolated from the Okinawan marine sponge *Terpios hoshinota*. These compounds have demonstrated cytotoxicity against murine P388 leukemia cells [181].

Unusual steroids known as erylosides T (**195**) and U (**196**), derived from the sea sponge *Erylus goffrilleri*, contain novel genins [182]. Lanostanes (**197** and **198**), whose structures are depicted in Figure 24 and whose biological activity is presented in Table 8, were obtained from the marine sponge *Penares* sp., found in Vietnamese waters [183]. These discoveries highlight the diverse range of bioactive compounds originating from marine sources.

A spiroketal steroid, **199**, was obtained from a collection of *Gorgonella umbraculum* found in the Indian Ocean off the Tuticorin coast [184]. From a Japanese octocoral species, *Dendronephthya* sp., two secosteroids named isogosterones A (**200**) and B (**201**) were isolated. These compounds share common structures characterized as 12α-acetoxy-13,17-seco-cholesta-1,4-dien-3-ones with hemiacetal functionality. Isogosterones A and B exhibited inhibition of larval settlement in the barnacle *Balanus amphitrite*, with an EC$_{50}$ value of 0.2 μM [185].

Figure 23. (a) The marine sponge belonging to the genus *Dysidea* is the origin of dysideasterols F (**188**) and G (**189**). (b) The Okinawan marine sponge *Terpios hoshinota* produces the norsteroid nakiterpiosin (**193**). (c) The marine sponge *Lendenfeldia frondosa*, found in the Solomon Islands, contains epihomoscalaralactone IIA (**204**). (d) The Vietnamese marine sponge *Penares* sp. serves as the source of lanostanes (**197** and **198**). These diverse marine organisms contribute to the production of biologically active compounds with potential pharmacological applications.

Furthermore, two steroids were isolated from the marine sponge *Isis hippuris*, also known as sea bamboo, collected from the Andaman Islands, India. These compounds are 3,11-diacetylhippurin-1 (**202**) and 22-*epi*-hippuri-stanol (**203**) [186]. Additionally, the marine sponge *Lendenfeldia frondosa*, collected from the Solomon Islands, yielded epihomoscalaralactone IIA (**204**) [187], while another marine sponge, *Phyllospongia dendyi,* from the Indian Ocean contained homoscalaralactone I1 B (**205**) [188]. Bioactive scalaranes (**206–208**) have been identified in the bio-toxic extracts of the marine sponge *Hyrtios erecta* [189]. Phyllofolactone A (**209**), isolated from the marine sponge *Phyllospongia* (syn. *Carteriospongia*) *foliascens* found in the South China Sea, has demonstrated cytotoxicity against P-388 cells [190].

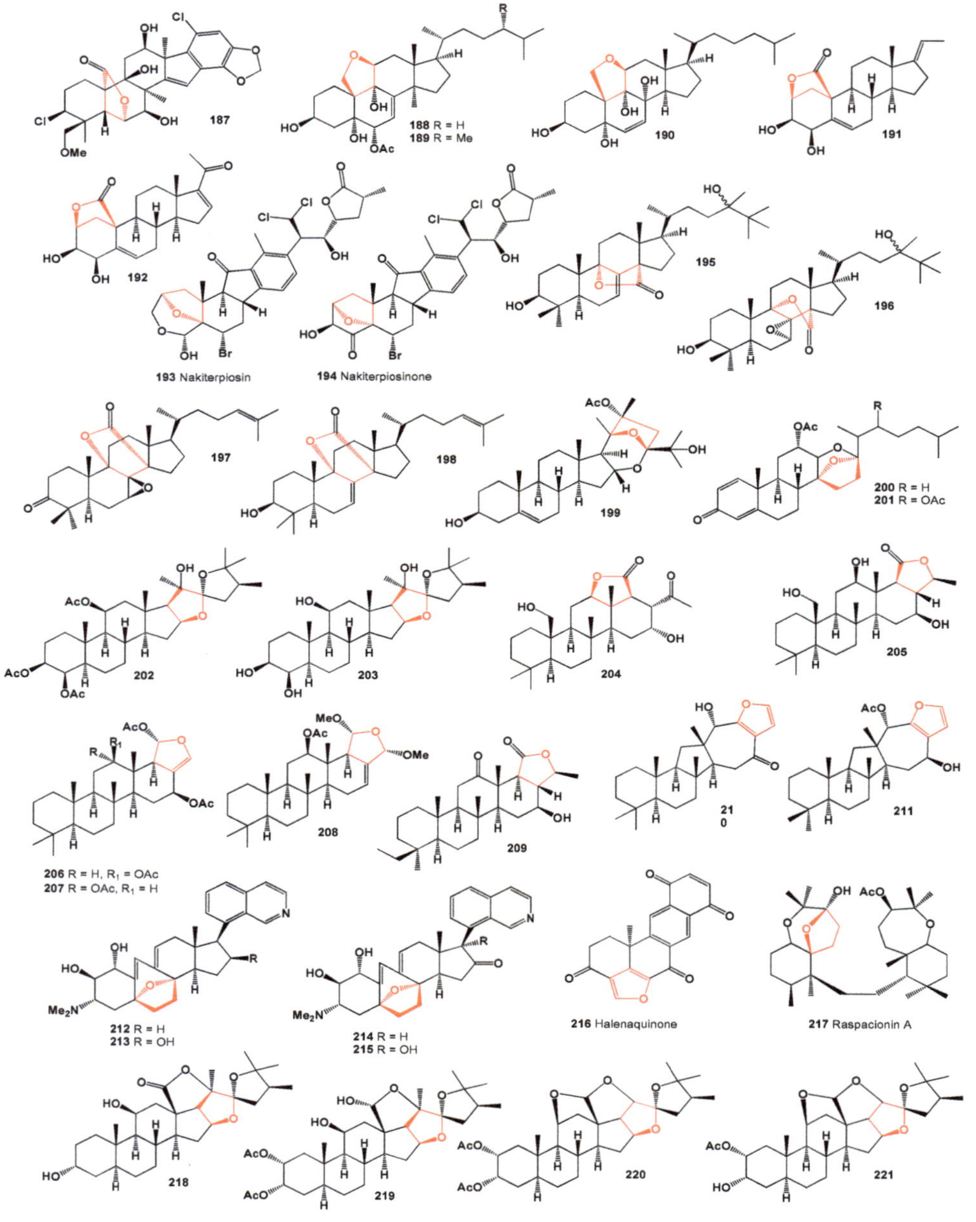

Figure 24. Furanosteroids and isoprenoid lipids derived from marine sources.

The marine sponge *Hippospongia* sp. collected from Taitung, Taiwan, serves as the source of a cytotoxic metabolite called hippospongide A (**210**) [191], as depicted in the 3D graph shown in Figure 25. Furthermore, the same compound, named salmahyrtisol A (**211**), has been isolated from the marine sponge *Hyrtios erecta* and exhibits significant cytotoxicity against murine leukemia (P-388), A-549, and HT-29 human cancer cells [192]. These findings underscore the potential of marine sponges as a valuable source of bioactive compounds with cytotoxic properties.

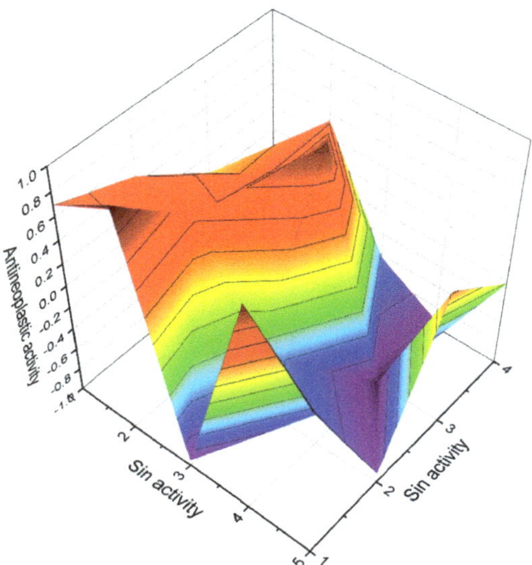

Figure 25. Three-dimensional graph illustrates the predicted and calculated antineoplastic activity of steroids (**210**, **218**, **219**, **220**, and **221**) with a confidence level exceeding 92%. These cytotoxic, highly polyoxygenated steroids are isolated from marine organisms inhabiting the waters surrounding Taiwan. Specifically, a metabolite (**210**) is found in the marine sponge *Hippospongia* sp., while the bamboo coral *Isis hippuris* contains compounds (**218–221**). These discoveries highlight the potential of marine organisms to provide valuable bioactive compounds with antineoplastic properties.

Steroidal anticancer alkaloids known as cortistatins A (**212**), B (**213**), C (**214**), and D (**215**) have been isolated from the marine sponge *Corticium simplex*. These alkaloids possess a unique 9 (10–19)-abeo-androstane and isoquinoline skeleton and have demonstrated the ability to inhibit the proliferation of human umbilical vein endothelial cells with high selectivity [193].

Metabolites containing the core structure of viridin have been extracted from marine invertebrates. Over 40 years ago, a research group led by Paul Scheuer at the University of Hawaii at Mānoa isolated demethoxyviridin (**4**) and its furano-quinone analogue called halenaquinone (**216**) from the marine sponge *Xestospongia exigua* [194]. Raspacionin A (**217**), a triterpenoid, was obtained from the red sponge *Raspaciona aculeata* found in the Mediterranean Sea. This compound exhibited cytotoxicity against the MCF-7 tumor cell line, with an IC_{50} value of 4 μM [195–197]. Highly polyoxygenated steroids (**218–221**) containing three or four additional tetrahydrofuran fragments were discovered in the bamboo coral *Isis hippuris*, collected from the Southeast coast of Taiwan. These steroids displayed cytotoxic activity against Hep G2, Hep 3B, A549, MCF-7, and MDAMB-231 cells [198]. Additionally, two steroids (**222** and **223**), whose structures are depicted in Figure 26 and whose biological activity is presented in Table 9, were isolated from the soft coral *Sarcophyton crassocaula* found in the Indian Ocean [199]. Furthermore, the cytotoxic compound sinubrasone B (**224**) was identified in the reef soft coral *Sinularia brassica* [200].

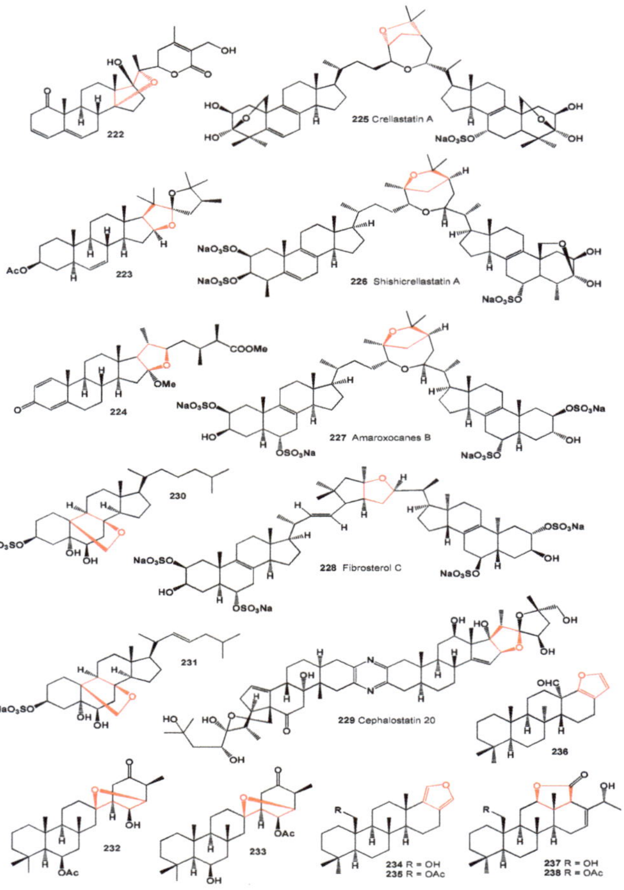

Figure 26. Furanosteroids, dimeric and isoprenoid lipids derived from marine sources.

Crellastatin A (**225**), a unique nonsymmetric dimeric steroid, was isolated from the marine sponge *Crella* sp. found on Vanuatu Island. This compound showcases an unprecedented connection through its side chains. Crellastatin A has been shown to possess significant in vitro cytotoxic activity against NSCLC-N6 cells, with an IC_{50} value of 0.5 µM [201]. This discovery underscores the potential of marine organisms to produce novel bioactive compounds with cytotoxic properties.

A similar dimeric steroid derivative, shishicrellastatin A (**226**), has been isolated from the marine sponge *Crella (Yvesia) spinulata*. This compound functions as a cathepsin B inhibitor, exhibiting an IC_{50} value of 8 µg/mL [202]. Another dimeric steroid, amaroxocane B (**227**), was discovered in the Caribbean coral reef sponge *Phorbas amaranthus* collected off Key Largo, Florida. It has shown effectiveness as an antifeedant [203]. Additionally, the same sterol dimer, hamigerol B (**227**), has been identified in the extract of the Mediterranean sponge *Hamigera hamigera* [204].

Furthermore, a sulfated sterol dimer named fibrosterol C (**228**), obtained from *Lissodendoryx (Acanthodoryx) fibrosa* collected in the Philippines, has been found to inhibit protein kinase Cζ with an IC_{50} value of 5.6 µM [205]. The 3D graph illustrating its activity is shown in Figure 27. Cephalostatin 20 (**229**), a member of the cephalostatin family known for its anticancer properties, has been isolated as a minor component of extracts from the marine worm *Cephalodiscus gilchristi* [206]. A picture of this sea sponge is displayed in Figure 28.

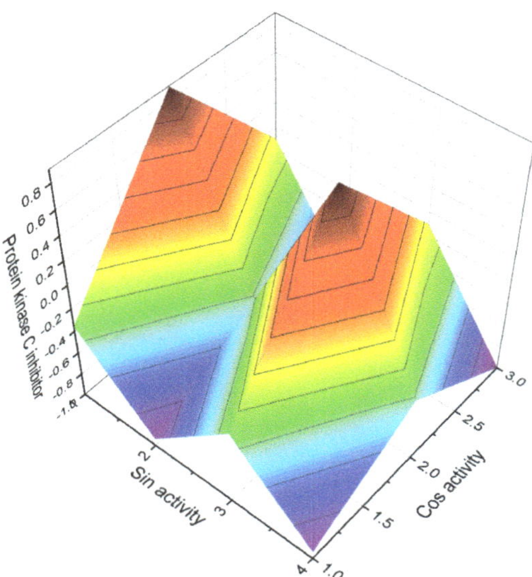

Figure 27. Three-dimensional graph presents the predicted and calculated activity of the steroid (**228**) with a confidence level exceeding 92%. This steroid is recognized as a protein kinase C inhibitor and exhibits potential as an anticancer and antiviral agent. It is a sulfated sterol dimer found in the marine sponge *Lissodendoryx (Acanthodoryx) fibrosa*, which belongs to the family Coelosphaeridae. This marine sponge is commonly found in the oceanic waters of the Philippines. The discovery of this compound highlights its promising bioactive properties derived from marine sources.

Figure 28. (**a**) The octocoral *Dendronephthya* sp. from the Sea of Japan produces secosteroids, namely isogosterones A (**200**) and B (**201**). (**b**) The marine sponge *Isis hippuris* serves as a source of steroids (**202**) and (**203**). (**c**) The marine sponge *Corticium simplex* produces steroidal alkaloids (**212–215**). (**d**) The marine worm *Cephalodiscus gilchristi* synthesizes a unique steroidal alkaloid (**229**). These diverse marine organisms contribute to the production of bioactive compounds with potential pharmacological applications.

An undescribed marine sponge from the genus *Euryspongia* serves as the source of sulfated steroids, namely eurysterols A (**230**) and B (**231**). Eurysterol A exhibits cytotoxicity against human colon carcinoma (HCT-116) cells, with an IC_{50} value of 0.3 μM. It also demonstrates antifungal activity against amphotericin-B-resistant *Candida albicans* [207]. Two sesterterpenoids, oxaspirosuberitenone (**232**) and isooxaspirosuberitenone (**233**), have been isolated from the marine sponge *Phorbas areolatus*. These compounds exhibit significant growth-inhibitory effects against A549, HepG2, HT-29, and MCF-7 tumor cell lines [208].

Furthermore, 12-dehydroxy-16-deacetoxy-22-hydroxyscalarafuran (**234**) and its corresponding acetate, 12-dehydroxy-16-deacetoxy-22-acetoxy-scalarafuran (**235**), were identified in the sponge *Smenospongia* sp. from Soheuksan Island (Korea). These compounds display antimicrobial activity and strong cytotoxicity against the human chronic myelogenous leukemia K562 cell line [209]. Additionally, coscinafuran (**236**) was detected in the MeOH fraction of the sponge *Coscinoderma mathewsi* [210].

Sednolide (**237**) and sednolide 22-acetate (**238**) were identified in extracts of the nudibranch *Chromodoris sedna*, collected in Baja California (Mexico). A picture of this mollusc is displayed in Figure 29. Sednolide (**237**) demonstrated growth inhibition of the marine bacterium *Vibrio anguillarum* at a concentration of 100 μg/disk [211].

Figure 29. (a) The reef soft coral *Sinularia brassica* is a producer of the cytotoxic steroid sinubrasone B (**224**). (b) The Caribbean coral reef sponge *Phorbas amaranthus* contains the dimeric steroid amaroxocane B (**227**). (c) The Mediterranean sponge *Hamigera hamigera* produces the same sterol dimer, hamigerol B (**227**). (d) The nudibranch *Chromodoris sedna* contains two sesterterpenoids, namely sednolide (**237**) and sednolide 22-acetate (**238**). These marine organisms contribute to the production of bioactive compounds with diverse pharmacological properties.

A rare and unusual aminosteroid called clionamine D (**239**), with its structure shown in Figure 30 and biological activity presented in Table 10, has been isolated from South African specimens of the sponge *Cliona celata*. This aminosteroid possesses a unique spiro bis-lactone side chain and exhibits cytotoxicity. It also modulates autophagy [212].

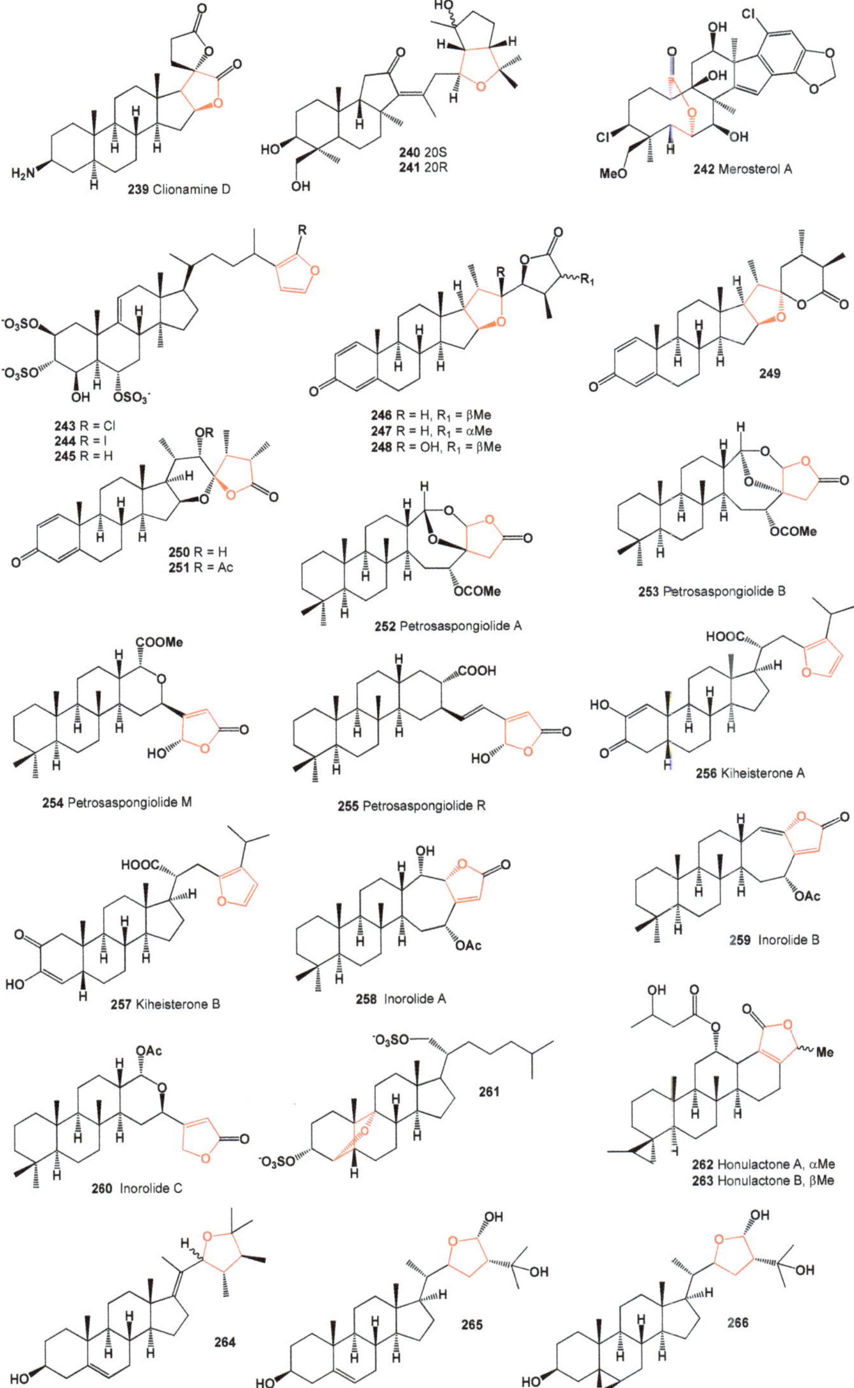

Figure 30. Furanosteroids and triterpenoids derived from marine sources.

Isomalabaricane-type triterpenoids, namely globostelletins P (**240**) and Q (**241**), have been isolated from the marine sponge *Rhabdastrella globostellata*. These triterpenoids, along with the CH_2Cl_2 fraction of the sponge, demonstrate inhibitory activities against various human tumor cell lines, including A549 (human lung adenocarcinoma), BGC-823 (human gastric carcinoma), HCT-8 (colonic carcinoma), Bel-7402 (human liver carcinoma), and A2780 (human ovarian carcinoma) [213].

The cyanobacterium *Scytonema* sp. from Bermuda serves as the source of merosterol A (**242**) [177]. Halogenated polar steroids, including chlorine-containing steroid sulfate (**243**), iodinated steroid **244**, and topsentiasterol sulfate D (**245**), have been isolated from the marine sponge *Topsentia* sp. [214,215]. Steroid (**243**) has been shown to effectively inhibit endo-1,3-β-D-glucanase from the marine mollusc *Spisula sachalinensis* [214].

A series of steroids known as sinubrasolides A–F (**246–251**), which belong to the class of withanolide-type steroids, were isolated from cultured specimens of *Sinularia brassica* from Taiwan [216]. Petrosaspongiolides A (**252**) and B (**253**) were the first cheilantane sesterterpene lactones to be isolated from a New Caledonian sponge initially assigned to the genus *Dactylospongia* but later reclassified as a new genus and species called *Petrosaspongia nigra* [217,218]. Additionally, other petrosasponiolides, namely M (**254**) and R (**255**), isolated from the New Caledonian marine sponge *Petrosaspongia nigra*, exhibited a γ-hydroxybutenolide moiety and a hemiacetal function [219,220]. The 3D graph illustrating the activity of compound **255** is shown in Figure 31.

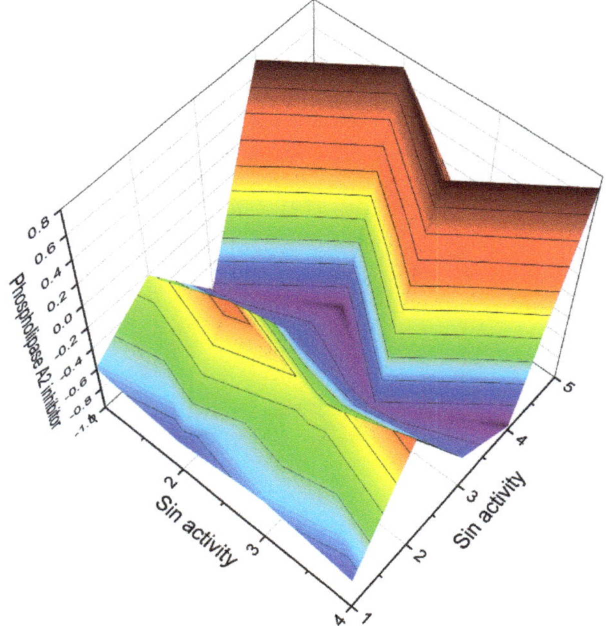

Figure 31. Three-dimensional graph illustrating the predicted and calculated activity of steroids (**254** and **255**) with a confidence level exceeding 90%. These steroids are recognized as phospholipase A2 inhibitors, showcasing their potential pharmacological activity.

Cytotoxic steroids named kiheisterones A (**256**) and B (**257**) have been isolated from a sponge of the order *Poecilosclerida* collected along the coast of the island of Maui (Hawaii). These sterols possess an α,β-disubstituted furan in the sidechain, a cis-fused A/B ring, a monoenolized α-di-ketone in the A ring, and a C-21 carboxyl group. Both steroids exhibit mild cytotoxicity against several human tumor cell lines, including A-549 lung carcinoma, HT-29 colon adenocarcinoma, and the P-388 murine lymphocytic leukemia cell line [220].

Furthermore, cytotoxic sesterterpenoids named inorolide A (**258**), B (**259**), and C (**260**) have been isolated from the Japanese nudibranch *Chromodoris inornata* (Chromodorididae) [221].

A unique 5β-steroid disulfate **261** with a distinctive 4α,9α-ether bridge has been isolated from the ophiuroid species *Ophiomastix annulosa* [222]. Furthermore, the marine sponge *Strepsichordaia aliena* from Indonesia has been found to contain the 20,24-bishomoscalarane sesterterpenes honulactones A (**262**) and B (**263**). Both sesterterpenes exhibit cytotoxicity against P-388, A-549, HT-29, and MEL-28 cancer cell lines, with an IC_{50} value of 0.1 μM [223].

A sarcosterol derivative, 22,25-epoxy-23,24-dimethylcholesta-5,17(20)-dien-3-ol (**264**), has been discovered in the soft coral *Sinularia mayi* [224]. Additionally, the soft coral *Lobophytum depressum* contains two steroids: lobophytosterol (**265**), which is a 22,28-epoxyergost-5-ene-3,25,28-triol, and 5,6-epoxy-lobophytosterol (**266**). A picture of this soft coral is displayed in Figure 32 [225,226]. Both steroids possess a double bond at positions 5,6. Lobophytosterol from the soft coral *Lobophytum laevigatum* has demonstrated cytotoxicity against A549 and HL-60 cell lines, with IC_{50} values of 4.5 and 5.6 μM, respectively [227].

Figure 32. (**a**) The ophiuroid species *Ophiomastix annulosa* serves as a holder of the steroid (**261**). (**b**) The soft coral *Lobophytum depressum* is a producer of steroids (**265**) and (**266**). (**c**) The marine sponge *Rhabdastrella globostellata* contains triterpenoids (**240** and **241**). (**d**) The marine mollusc *Spisula sachalinensis* contains the steroid (**243**). *S. sachalinensis* is a bivalve mollusc found in the Sea of Japan and the Sea of Okhotsk, inhabiting medium- to fine-grained sands at depths ranging from 0.5 to 10 m. It is the largest of the molluscs, reaching a length of approximately 130 mm and weighing around 500 g. This mollusc is considered a delicacy and is actively fished in Japan and Primorsky Krai (Russia) [228].

5. Conclusions

This comprehensive review examines the occurrence of furanosteroids and related isoprenoid lipids in various sources, including fungi, plants, and marine organisms. It provides an in-depth analysis of their biological activity, shedding light on their potential applications and significance in drug discovery. Furanosteroids and related isoprenoid lipids have garnered significant attention due to their diverse biological properties. This review aims to explore the occurrence of these compounds in fungi, plants, and marine organisms, as well as delve into their associated biological activity. By analyzing their medicinal and pharmacological potential, this review highlights the importance of furanosteroids as promising natural products for future research and development. The biological activity of furanosteroids and related isoprenoid lipids constitutes a crucial aspect of their investigation. This review focuses on the diverse range of activities exhibited by these compounds, including but not limited to anti-inflammatory, anticancer, and antimicrobial properties.

Funding: This research received no external funding.

Institutional Review Board Statement: Not applicable.

Informed Consent Statement: Not applicable.

Data Availability Statement: Not applicable.

Conflicts of Interest: The author declares that he has no known competing financial interests or personal relationships that could affect the work described in this article.

References

1. Peterson, L.A. Reactive metabolites in the biotransformation of molecules containing a furan ring. *Chem. Res. Toxicol.* **2013**, *26*, 6–25. [CrossRef] [PubMed]
2. Saeid, H.; Al Sayed, H.; Bader, M. A review on biological and medicinal significance of furan. *Alqalam J. Med. App. Sci.* **2023**, *6*, 44–58.
3. Tian, M.; Peng, Y.; Zheng, J. Metabolic activation, and hepatotoxicity of furan-containing compounds. *Drug Metabol. Disposit.* **2022**, *50*, 655–670. [CrossRef] [PubMed]
4. Banerjee, R.; Kumar, H.K.S.; Banerjee, M. Medicinal significance of furan derivatives: A Review. *Int. J. Rev. Life Sci.* **2012**, *2*, 7–16.
5. Alizadeh, M.; Jalal, M.; Hamed, K.; Saber, A. Recent updates on anti-inflammatory and antimicrobial effects of furan natural derivatives. *J. Inflammat. Res.* **2020**, *13*, 451–463. [CrossRef]
6. Mawlong, I.; Sujith Kumar, M.S.; Singh, D. Furan fatty acids: Their role in plant systems. *Phytochem. Rev.* **2016**, *15*, 121–127. [CrossRef]
7. Buckingham, J. *Dictionary of Natural Products*; CRC Press: Boca Raton, FL, USA, 2022.
8. Senapati, B.K. Recent progress in the synthesis of the furanosteroid family of natural products. *Org. Chem. Front.* **2021**, *8*, 2608–2642. [CrossRef]
9. Xue, D.; He, H.; Gao, S. Strategies for the total synthesis of the furanosteroids: Wortmannin and viridin. *Chem. Lett.* **2021**, *50*, 497–502. [CrossRef]
10. Hu, D.; Gao, Y.H.; Yao, X.S.; Gao, H. Recent advances in dissecting the demethylation reactions in natural product biosynthesis. *Curr. Opin. Chem. Biol.* **2020**, *59*, 47–53. [CrossRef]
11. Zhungietu, G.I.; Dorofeenko, G.N. Progress in the field of the chemistry of steroidal heterocycles. *Russ. Chem. Rev.* **1967**, *36*, 24–36. [CrossRef]
12. Hatti, K.S.; Muralitharan, R.L.; Kush, H.; Kush, A. Convenient database for neem secondary metabolites. *Bioinformation* **2014**, *10*, 314–315. [CrossRef] [PubMed]
13. Ui, M.; Okada, T.; Hazeki, K.; Hazeki, O. Wortmannin as a unique probe for an intracellular signalling protein, phosphoinositide 3-kinase. *Trends Biochem. Sci.* **1995**, *20*, 303–307. [CrossRef] [PubMed]
14. Hanson, J.R. The viridin family of steroidal antibiotics. *Nat. Prod. Rep.* **1995**, *12*, 381–384. [CrossRef] [PubMed]
15. Brian, P.W.; Curtis, P.J.; Hemming, H.G. The production of viridin by pigment-forming strains of *Trichoderma viride*. *Ann. Appl. Biol.* **1946**, *33*, 190–200. [CrossRef]
16. Brian, P.W.; McGowan, J.C. Viridin: A highly fungistatic substance produced by *Trichoderma viride*. *Nature* **1945**, *156*, 144–145. [CrossRef]
17. Brian, P.W.; Curtis, P.J.; Hemming, H.G.; Norris, G.L.F. Wortmannin, an antibiotic produced by *Penicillium wortmanni*. *Transact. Br. Mycol. Soc.* **1957**, *40*, 365–368. [CrossRef]

18. Viswanathan, K.; Ononye, S.N.; Cooper, H.D.; Kyle Hadden, M.; Anderson, A.C.; Wright, D.L. Viridin analogs derived from steroidal building blocks. *Bioorg. Med. Chem. Lett.* **2012**, *22*, 6919–6922. [CrossRef]

19. Jones, R.W.; Hancock, J.G. Conversion of viridin to viridiol by viridin-producing fungi. *Can. J. Microbiol.* **1987**, *33*, 963–966. [CrossRef]

20. Aldridge, D.C.; Turner, W.B.; Geddes, A.J.; Sheldrick, B. Demethoxyviridin and demethoxyviridiol: New fungal metabolites. *J. Chem. Soc. Perkin Trans. 1* **1975**, *10*, 943–945. [CrossRef]

21. Andersson, P.F.; Bengtsson, S.; Cleary, M.; Stenlid, J.; Broberg, A. Viridin-like steroids from *Hymenoscyphus pseudoalbidus*. *Phytochemistry* **2013**, *86*, 195–200. [CrossRef]

22. Zhai, M.M.; Li, J.; Jiang, C.X.; Shi, J.Y.P.; Di, D.L.; Crews, P. The bioactive secondary metabolites from Talaromyces species. *Nat. Prod. Bioprospect.* **2016**, *6*, 1. [CrossRef]

23. Noinart, J. Bioactive Secondary Metabolites from the Culture of the Marine Sponge-Associated Fungus *Talaromyces stipitatus* KUFA. Ph.D Thesis, University of Porto, Porto, Portugal, 2017.

24. Ding, H.-E.; Yang, Z.-D.; Sheng, L.; Zhou, S.-Y.; Li, S.; Yao, X.-J.; Zhi, K.-K.; Wang, Y.-G.; Zhang, F. Secovironolide, a novel furanosteroid scaffold with a fivemembered B ring from the endophytic fungus *Talaromyces wortmannii* LGT-4. *Tetrahedron Lett.* **2015**, *56*, 6754–6757. [CrossRef]

25. Phuwapraisirisan, P.; Rangsan, J.; Siripong, P.; Tip-Pyang, S. 9- epi -Viridiol, a novel cytotoxic furanosteroid from soil fungus *Trichoderma virens*. *Nat. Prod. Res.* **2006**, *20*, 14. [CrossRef] [PubMed]

26. Abbas, H.K.; Mirocha, C.J. Isolation and purification of a hemorrhagic factor (wortmannin) from *Fusarium oxysporum* (N17B). *Appl. Environ. Microbiol.* **1988**, *54*, 1268–1274. [CrossRef]

27. Singh, V.; Praveen, V.; Tripathi, D.; Haque, S.; Somvanshi, P. Isolation, characterization and antifungal docking studies of wortmannin isolated from *Penicillium radicum*. *Sci. Rep.* **2015**, *5*, 11948. [CrossRef]

28. Wipf, P.; Halter, R.J. Chemistry and biology of wortmannin. *Org. Biomol. Chem.* **2005**, *3*, 2053–2061. [CrossRef] [PubMed]

29. Smith, R.A.; Yuan, H.; Weissleder, R.; Cantley, L.C.; Josephson, L. A wortmannin-cetuximab as a double drug. *Bioconjug. Chem.* **2009**, *20*, 2185–2189. [CrossRef]

30. Kong, D.; Yamori, T. Phosphatidylinositol 3-kinase inhibitors: Promising drug candidates for cancer therapy. *Cancer Sci.* **2008**, *99*, 1734–1740. [CrossRef] [PubMed]

31. Morales-Oyervidesa, L.; Ruiz-Sánchez, J.P.; Oliveira, J.C. Biotechnological approaches for the production of natural colorants by Talaromyces/Penicillium: A review. *Biotechnol. Adv.* **2020**, *43*, 107601. [CrossRef] [PubMed]

32. Nagwa, E.; Kassem, A.H.; Hamed, M.A.; El-Fekya, A.M.; Elnaggard, M.A.A.; Mahmouda, K.; Alie, M.A. Isolation and characterization of the bioactive metabolites from the soil derived fungus *Trichoderma viride*. *Mycology* **2018**, *9*, 70–80.

33. Ghisalberti, E.L.; Sivasithamparam, K. Antifungal antibiotics produced by *Trichoderma* spp. *Soil Biol. Biochem.* **1991**, *23*, 1011–1020. [CrossRef]

34. Mukherjee, P.K.; Horwitz, B.A.; Kenerley, C.M. Secondary metabolism in Trichoderma—A genomic perspective. *Microbiology* **2012**, *158*, 35–45. [CrossRef] [PubMed]

35. Song, Y.-P.; Shi, Z.-Z.; Miao, F.-P.; Fang, S.-T.; Yin, X.-L.; Ji, N.-Y. Tricholumin A, a highly transformed ergosterol derivative from the alga endophytic fungus *Trichoderma asperellum*. *Org. Lett.* **2018**, *20*, 6306–6309. [CrossRef] [PubMed]

36. Hill, R.A.; Makin, H.L.J.; Kirk, D.N.; Murphy, G.M. *Dictionary of Steroids*, 1st ed.; Chapman and Hall: Boca Raton, FL, USA, 1991.

37. Chen, H.P.; Zhao, Z.Z.; Li, Z.H.; Huang, Y.; Zhang, S.B.; Tang, Y.; Yao, J.N.; Chen, L.; Isaka, M.; Feng, T.; et al. Anti-proliferative and anti-Inflammatory lanostane triterpenoids from the Polish edible mushroom *Macrolepiota procera*. *J. Agric. Food Chem.* **2018**, *66*, 3146–3154. [CrossRef]

38. Guzmán-Guzmán, P.; Kumar, A.; de los Santos-Villalobos, S.; Parra-Cota, F.I.; del Orozco-Mosqueda, C.; Fadiji, A.E.; Hyder, S.; Babalola, O.O.; Santoyo, G. Trichoderma species: Our best fungal allies in the biocontrol of plant diseases—A Review. *Plants* **2023**, *12*, 432. [CrossRef]

39. Sood, M.; Kapoor, D.; Kumar, V.; Sheteiwy, M.S.; Ramakrishnan, M.; Landi, M.; Araniti, F.; Sharma, A. Trichoderma: The "Secrets" of a multitalented biocontrol agent. *Plants* **2020**, *9*, 762. [CrossRef]

40. Gomdola, D.; Bhunjun, C.S.; Hyde, K.D.; Jeewon, R.; Pem, D.; Jayawardena, R.S. Ten important forest fungal pathogens: A review on their emergence and biology. *Mycosphere* **2022**, *13*, 612–671. [CrossRef]

41. Lorenc, F.; Samek, M. Pathogens threatening Czech Republic Forest ecosystems—A review. *Sylwan* **2021**, *165*, 853–871.

42. Benjamin, C.R. Ascocarps of Aspergillus and Penicillium. *Mycologia* **1955**, *47*, 669–687. [CrossRef]

43. Frisvad, J.C. Taxonomy, chemodiversity, and chemoconsistency of Aspergillus, Penicillium, and Talaromyces species. *Front. Microbiol.* **2015**, *5*, 773. [CrossRef]

44. Houbraken, J.; de Vries, R.P.; Samson, R.A. Modern taxonomy of biotechnologically important Aspergillus and Penicillium species. *Adv. Appl. Microbiol.* **2014**, *86*, 199–249. [PubMed]

45. Ramani, G.; Meera, B.; Vanitha, C.; Rao, M. Production, purification, and characterization of a β-glucosidase of *Penicillium funiculosum* NCL1. *Appl. Biochem. Biotechnol.* **2012**, *167*, 959–972. [CrossRef] [PubMed]

46. Fravel, D.; Olivain, C.; Alabouvette, C. *Fusarium oxysporum* and its biocontrol. *New Phytol.* **2003**, *157*, 493–502. [CrossRef]

47. Gordon, T.R.; Martyn, R.D. The evolutionary biology of *Fusarium oxysporum*. *Annu. Rev. Phytopathol.* **1997**, *35*, 111–128. [CrossRef]

48. Ekiz, G.; Yılmaz, S.; Yusufoglu, H.; Kırmızıbayrak, P.B.; Bedir, E. Microbial transformation of cycloastragenol and astragenol by endophytic fungi isolated from *Astragalus* species. *J. Nat. Prod.* **2019**, *82*, 2979–2985. [CrossRef]

49. Küçüksolak, M.; Üner, G.; Kırmızıbayrak, P.B.; Bedir, E. Neuroprotective metabolites via fungal biotransformation of a novel sapogenin, cyclocephagenol. *Sci. Rep.* **2022**, *12*, 18481. [CrossRef] [PubMed]

50. Galappaththi, M.C.A.; Patabendige, N.M.; Premarathne, B.M.; Hapuarachchi, K.K.; Tibpromma, S.; Dai, D.-Q.; Suwannarach, N.; Rapior, S.; Karunarathna, S.C. A Review of Ganoderma triterpenoids and their bioactivities. *Biomolecules* **2023**, *13*, 24. [CrossRef] [PubMed]

51. Su, H.-G.; Wang, Q.; Zhou, L.; Peng, X.-R.; Xiong, W.-Y.; Qiu, M.-H. Functional triterpenoids from medicinal fungi *Ganoderma applanatum*: A continuous search for antiadipogenic agents. *Bioorg. Chem.* **2021**, *112*, 104977. [CrossRef] [PubMed]

52. Li, L.; Peng, X.R.; Dong, J.R.; Lu, S.Y.; Li, X.N.; Zhou, L. Rearranged lanostane-type triterpenoids with anti-hepatic fibrosis activities from *Ganoderma applanatum*. *RSC Adv.* **2018**, *8*, 31287–31295. [CrossRef]

53. Peng, X.-R.; Liu, J.-Q.; Wang, C.-F.; Li, X.-Y.; Shu, Y.; Zhou, L.; Qiu, M.-H. Hepatoprotective effects of triterpenoids from *Ganoderma cochlear*. *J. Nat. Prod.* **2014**, *77*, 737–743. [CrossRef]

54. Niu, X.; Qiu, M.; Li, Z.; Lu, Y.; Cao, P.; Zheng, Q. Two novel 3,4-seco-trinorlanostane triterpenoids isolated from *Ganoderma fornicatum*. *Tetrahedron Lett.* **2004**, *45*, 2989–2993. [CrossRef]

55. Hill, R.A.; Connolly, J.D. Triterpenoids. *Nat. Prod. Rep.* **2020**, *37*, 962. [CrossRef]

56. Ma, K.; Li, L.; Bao, L.; He, L.; Sun, C.; Zhou, B.; Si, S.; Liu, H. Six new 3, 4-seco-27-norlanostane triterpenes from the medicinal mushroom *Ganoderma boninense* and their antiplasmodial activity and agonistic activity to LXRβ. *Tetrahedron* **2015**, *71*, 1808–1814. [CrossRef]

57. Zhang, S.-S.; Wang, Y.-G.; Ma, Q.-Y.; Huang, S.-Z.; Hu, L.-L.; Dai, H.-F.; Yu, Z.-F.; Zhao, Y.-X. Three new lanostanoids from the mushroom *Ganoderma tropicum*. *Molecules* **2015**, *20*, 3281–3289. [CrossRef]

58. Peng, X.; Liu, J.; Xia, J.; Wang, C.; Li, X.; Deng, Y.; Bao, N.; Zhang, Z.; Qiu, M. Lanostane triterpenoids from *Ganoderma hainanense*. *Phytochemistry* **2015**, *114*, 137–145. [CrossRef]

59. Li, Q.; Chen, C.; Cheng, L. Emeridones A–F, a series of 3,5-demethylorsellinic acid-based meroterpenoids with rearranged skeletons from an endophytic fungus *Emericella* sp. TJ29. *J. Org. Chem.* **2019**, *84*, 1534–1541. [CrossRef]

60. Liang, X.-R.; Miao, F.-P.; Song, Y.-P.; Guo, Z.-Y.; Ji, N.-Y. Trichocitrin, a new fusicoccane diterpene from the marine brown alga-endophytic fungus *Trichoderma citrinoviride* cf-27. *Nat. Prod. Res.* **2016**, *30*, 1605–1610. [CrossRef]

61. Liu, H.B.; Edrada-Ebel, R.A.; Ebel, R.; Wang, Y.; Schulz, B.; Draeger, S. Ophiobolin sesterterpenoids and pyrrolidine alkaloids from the sponge derived fungus *Aspergillus ustus*. *Helvetica Chim. Acta* **2011**, *94*, 623–631. [CrossRef]

62. Li, Y.L.; Xue, L.J.; Li, J.; Chen, L.M.; Wu, J.J.; Xu, Z.N.; Chen, Y.; Tiana, Y.P.; Yang, X.W. Abinukitrine A, a unique 17,18-cyclolanostane triterpenoid from *Abies nukiangensis*. *Org. Biomol. Chem.* **2019**, *17*, 2107–2109.

63. Sugawara, F.; Strobel, G.; Strange, R.N.; Siedow, J.N.; Van Duyne, G.D.; Clardy, J. Phytotoxins from the pathogenic fungi *Drechslera maydis* and *Drechslera sorghicola*. *Proc. Natl. Acad. Sci. USA* **1987**, *84*, 3081–3085. [CrossRef]

64. Li, E.; Clark, A.M.; Rotella, D.P.; Hufford, C.D. Microbial metabolites of ophiobolin A and antimicrobial evaluation of ophiobolins. *J. Nat. Prod.* **1995**, *58*, 74–81. [CrossRef]

65. Ebel, R. Terpenes from marine-derived fungi. *Mar. Drugs* **2010**, *8*, 2340–2368. [CrossRef]

66. Elissawy, A.M.; El-Shazly, M.; Ebada, S.S.; Singab, A.B.; Proksch, P. Bioactive terpenes from marine-derived fungi. *Mar. Drugs* **2015**, *13*, 1966–1992. [CrossRef]

67. Zhang, M.; Liu, J.; Chen, R.; Zhao, J.; Xie, K.; Chen, D.; Feng, K.; Dai, J. Two furanharzianones with 4/7/5/6/5 ring system from microbial transformation of harzianone. *Org. Lett.* **2017**, *19*, 1168–1171. [CrossRef]

68. Zhang, M.; Liu, J.; Chen, R.; Zhao, J.; Xie, K.; Chen, D.; Feng, K.; Dai, J. Microbial oxidation of harzianone by *Bacillus* sp. IMM-006. *Tetrahedron* **2017**, *73*, 7195–7199. [CrossRef]

69. Dong, H.; Zhou, C.; Li, X.; Gu, H.; Zhang, H.E.; Zhou, F.; Zhao, Z.; Fan, T. Ultraperformance liquid chromatography-quadrupole time-of-flight mass spectrometry based untargeted metabolomics to reveal the characteristics of *Dictyophora rubrovolvata* from different drying methods. *Front. Nutr.* **2022**, *9*, 1056598. [CrossRef]

70. Sato, M.; Kakisawa, H. Structures of three new C16 terpenoids from an Acrostalagmus fungus. *J. Chem. Soc. Perkin Trans. I* **1976**, *22*, 2407–2413. [CrossRef]

71. Kakisawa, H.; Sato, M.; Ruo, T.I.; Hayashi, T. Biosynthesis of a C16-terpenoid lactone, a plant growth regulator. *J. Chem. Soc. Chem. Commun.* **1973**, *9*, 802–803. [CrossRef]

72. Xia, X.; Zhang, J.; Zhang, Y.; Wei, F.; Liu, X. Pimarane diterpenes from the fungus *Epicoccum* sp. HS-1 associated with *Apostichopus japonicus*. *Bioorg. Med. Chem. Lett.* **2012**, *22*, 3017–3019. [CrossRef]

73. Li, Y.; Lu, C.; Hu, Z.; Huang, Y.; Shen, Y. Secondary metabolites of *Tubercularia* sp. TF5, an endophytic fungal strain of *Taxus mairei*. *Nat. Prod. Res.* **2009**, *23*, 70–76.

74. Shiono, Y.; Motoki, S.; Koseki, T.; Murayama, T.; Tojima, M.; Kimura, K.I. Isopimarane diterpene glycosides, apoptosis inducers, obtained from fruiting bodies of the ascomycete *Xylaria polymorpha*. *Phytochemistry* **2009**, *70*, 935–939. [CrossRef]

75. Li, Y.Y.; Hu, Z.Y.; Lu, C.H.; Shen, Y.M. Four new terpenoids from *Xylaria* sp. 101. *Helv. Chim. Acta* **2010**, *93*, 796–802. [CrossRef]

76. Gu, B.B.; Wu, W.; Jiao, F.R.; Jiao, W.J.; Li, L.; Sun, F.; Wang, S.P.; Yang, F.; Lin, H.W. Asperflotone, an 8(14→15)-abeo-ergostane from the sponge derived fungus *Aspergillus flocculosus* 16D-1. *J. Org. Chem.* **2019**, *84*, 300–306. [CrossRef]

77. Gu, B.-B.; Wu, W.; Jiao, F.-R.; Jiao, W.-h.; Li, L.; Sun, F.; Wang, S.-P.; Yang, F.; Lin, H.-W. Aspersecosteroids A and B, two 11(9→10)-abeo-5,10-secosteroids with a dioxatetraheterocyclic ring system from *Aspergillus flocculosus* 16D-1. *Org. Lett.* **2018**, *20*, 7957–7960. [CrossRef]

78. Dong, G.A.N.; Li, C.; Shu, Y.; Wu, J. Steroids and dihydroisocoumarin glycosides from *Xylaria* sp. by the one strain many compounds strategy and their bioactivities. *Chin. J. Nat. Med.* **2023**, *21*, 154–160.

79. Liangsakul, J.; Srisurichan, S.; Pornpakakul, S. Anthraquinone–steroids, evanthrasterol A and B, and a meroterpenoid, emericellic acid, from endophytic fungus, *Emericella variecolor*. *Steroids* **2016**, *106*, 78–85. [CrossRef]

80. Wu, J.; Suzuki, T.; Choi, J.H.; Yasuda, N.; Noguchi, K.; Hirai, H.; Kawagishi, H. An unusual sterol from the mushroom *Stropharia rugosoannulata*. *Tetrahedron Lett.* **2013**, *54*, 4900–4902. [CrossRef]

81. Lee, S.; Kim, C.S.; Yu, J.S.; Kang, H.; Yoo, M.J.; Youn, U.J.; Ryoo, R.; Bae, H.K.; Kim, K.H. Ergopyrone, a styrylpyrone-fused steroid with a hexacyclic 6/5/6/6/6/5 skeleton from a mushroom *Gymnopilus orientispectabilis*. *Org. Lett.* **2021**, *23*, 3315–3319. [CrossRef]

82. Kuo, P.C.; Kuo, T.H.; Damu, A.G.; Su, C.R.; Lee, E.J.; Wu, T.S.; Shu, R.; Chen, C.M.; Kenneth, F.; Bastow, T.-H.; et al. Physanolide A, a novel skeleton steroid, and other cytotoxic principles from *Physalis angulate*. *Org. Lett.* **2006**, *8*, 2953–2956. [CrossRef]

83. Sun, C.P.; Qiu, C.Y.; Zhao, F. Physalins V-IX, 16,24-cyclo-13,14-seco withanolides from *Physalis angulata* and their antiproliferative and anti-inflammatory activities. *Sci. Rep.* **2017**, *7*, 4057. [CrossRef]

84. Arsenault, G.P.; Biemann, K.; Barksdale, A.W.; McMorris, T.C. The structure of antheridiol, a sex hormone in *Achlya bisexualis*. *J. Am. Chem. Soc.* **1968**, *90*, 5635–5636. [CrossRef]

85. Maldonado, E.; Hurtado, N.E.; Pérez-Castorena, A.L.; Martínez, M. Cytotoxic 20,24-epoxy-withanolides from *Physalis angulata*. *Steroids* **2015**, *104*, 72–78. [CrossRef] [PubMed]

86. Meng, Q.; Fan, J.; Liu, Z.; Li, X.; Zhang, F.; Zhang, Y.; Sun, Y.; Li, L.; Xia Liu, X.; Hua, E. Cytotoxic withanolides from the whole herb of *Physalis angulata* L. *Molecules* **2019**, *24*, 1608. [CrossRef]

87. Sun, C.P.; Qiu, C.Y.; Yuan, T.; Nie, X.F.; Sun, H.X.; Zhang, Q.; Li, H.X.; Ding, L.Q.; Zhao, F.; Chen, L.X.; et al. Antiproliferative and anti-inflammatory withanolides from *Physalis angulata*. *J. Nat. Prod.* **2016**, *79*, 1586–1597. [CrossRef]

88. Hirotani, M.; Sai, K.; Hirotani, S.; Yoshikawa, T. Blazeispirols B, C, E and F, des-A-ergostane-type compounds, from the cultured mycelia of the fungus *Agaricus blazei*. *Phytochemistry* **2002**, *59*, 571–577. [CrossRef] [PubMed]

89. Hirotani, M.; Hirotani, S.; Takayanagi, H.; Komiyama, K.; Yoshikawa, T. Agariblazeispirols A and B, an unprecedented skeleton from the cultured mycelia of the fungus, *Agaricus blazei*. *Tetrahedron Lett.* **2003**, *44*, 7975–7979. [CrossRef]

90. Hirotani, M.; Hirotani, S.; Takayanagi, H.; Yoshikawa, T. Blazeispirol A, an unprecedented skeleton from the cultured mycelia of the fungus *Agaricus blazei*. *Tetrahedron Lett.* **1999**, *40*, 329–332. [CrossRef]

91. Tang, Y.; Zhao, Z.Z.; Hu, K.; Feng, T.; Li, Z.H.; Chen, H.P.; Liu, J.K. Irpexolidal Represents a class of triterpenoid from the fruiting bodies of the medicinal fungus *Irpex lacteus*. *J. Org. Chem.* **2019**, *84*, 1845–1852. [CrossRef]

92. Clericuzio, M.; Tabasso, S.; Bianco, M.A.; Pratesi, G.; Beretta, G.; Tinelli, S.; Zunino, F.; Vidari, G. Cucurbitane triterpenes from the fruiting bodies and cultivated mycelia of *Leucopaxillus gentianeus*. *J. Nat. Prod.* **2006**, *69*, 1796–1799. [CrossRef]

93. Cheng, X.; Liang, X.; Yao, F.H.; Liu, X.B.; Qi, S.H. Fusidane-type antibiotics from the marine-derived fungus *Simplicillium* sp. SCSIO 41513. *J. Nat. Prod.* **2021**, *84*, 2945–2952. [CrossRef]

94. Macías-Rubalcava, M.L.; Sánchez-Fernández, R.E. Secondary metabolites of endophytic Xylaria species with potential applications in medicine and agriculture. *World J. Microbiol. Biotechnol.* **2017**, *33*, 15. [CrossRef] [PubMed]

95. Huang, K.; El-Seedi, H.R.; Xu, B. Critical review on chemical compositions and health-promoting effects of mushroom *Agaricus blazei* Murill. *Curr. Res. Food Sci.* **2022**, *5*, 2190–2203. [CrossRef] [PubMed]

96. Bertollo, A.G.; Mingoti, M.E.D.; Plissari, M.E.; Betti, G.; Junior, W.A.R.; Luzardo, A.R.; Ignácio, Z.M. *Agaricus blazei* Murrill mushroom: A review on the prevention and treatment of cancer. *Pharmacol. Res. Mod. Chin. Med.* **2022**, *2*, 100032. [CrossRef]

97. Dong, X.M.; Song, X.H.; Liu, K.B.; Dong, C.H. Prospect and current research status of medicinal fungus *Irpex lacteus*. *Mycosystema* **2017**, *36*, 28–34.

98. Mezule, L.; Civzele, A. Bioprospecting white-rot basidiomycete Irpex lacteus for improved extraction of lignocellulose-degrading enzymes and their further application. *J. Fungi* **2020**, *6*, 256. [CrossRef]

99. Pachapurkar, R.V.; Kornule, P.M.; Narayana, C.R. A new hexacyclic tetranortriterpenoid. *Chem. Lett.* **1974**, *3*, 357–358. [CrossRef]

100. Rogers, L.L.; Zeng, L.; Kozlowski, J.F.; Shimada, H.; Alali, F.Q.; Johnson, H.A.; McLaughlin, J.L. New bioactive triterpenoids from *Melia volkensii*. *J. Nat. Prod.* **1998**, *61*, 64. [CrossRef]

101. Connolly, J.D.; Labbé, C.; Rycroft, D.S.; Taylor, D.A.H. Tetranortriterpenoids and related compounds. Part 22. New apotirucailol derivatives and tetranortriterpenoids from the wood and seeds of *Chisocheton paniculatus* (Meliaceae). *J. Chem. Soc. Perkin Trans.* **1979**, *21*, 2959–2964. [CrossRef]

102. Kraus, W.; Cramer, R.; Sawitzki, G. Tetranortriterpenoids from the seeds of *Azadirachta indica*. *Phytochemistry* **1981**, *20*, 117–120. [CrossRef]

103. Mulholland, D.A.; Monkhe, T.V.; Coombes, P.H.; Rajab, M.S. Limonoids from *Turraea holstii* and *Turraea floribunda*. *Phytochemistry* **1998**, *49*, 2585. [CrossRef]

104. Kumar, C.S.S.R.; Srinivas, M.; Yakkundi, S. Limonoids from the seeds of *Azadirachta indica*. *Phytochemistry* **1996**, *43*, 451. [CrossRef]

105. Coombesa, P.H.; Mulhollanda, D.A.; Randrianarivelojosia, M. Vilasinin limonoids from *Malleastrum antsingyense* J.F. Leroy (Meliodeae: Meliaceae). *Biochem. Syst. Ecol.* **2008**, *36*, 74–76. [CrossRef]

106. Kowa, T.K.; Jansen, O.; Ledoux, A.; Mamede, L.; Wabo, H.K.; Tchinda, A.T.; Genta-Jouve, G.; Frédérich, M. Bioassay-guided isolation of vilasinin–type limonoids and phenyl alkene from the leaves of *Trichilia gilgiana* and their antiplasmodial activities. *Nat. Prod. Res.* **2022**, *36*, 5039–5047. [CrossRef] [PubMed]

107. Yadav, R.; Pednekar, A.; Avalaskar, A.; Rathi, M. A comprehensive review on Meliaceae family. *World J. Pharm. Sci.* **2015**, *3*, 1572–1577.

108. Islas, J.F.; Acosta, E.; Buentello, Z.G.; Delgado-Gallegos, J.L.; Moreno-Treviño, M.G.; Escalante, B.; Moreno-Cueva, J.E. An overview of Neem (*Azadirachta indica*) and its potential impact on health. *J. Function. Foods* **2020**, *74*, 104171. [CrossRef]

109. Ma, C.; Chen, Y.; Chen, J.; Li, X.; Chen, Y. A Review on *Annona squamosa* L.: Phytochemicals and biological activities. *Am. J. Chin. Med.* **2017**, *45*, 1–32. [CrossRef] [PubMed]

110. Ragasa, C.Y.; Nacpil, Z.D.; Natividad, G.M.; Tada, M.; Coll, J.C.; Rideout, J.A. Tetranortriterpenoids from *Azadirachta indica*. *Phytochemistry* **1997**, *46*, 555–558. [CrossRef]

111. Fu, X.; Li, X.C.; Smillie, T.J.; Carvalho, P.; Mabusela, W.; Syce, J.; Johnson, Q.; Folk, W.; Avery, M.A.; Khan, I.A. Cycloartane glycosides from *Sutherlandia frutescens*. *J. Nat. Prod.* **2008**, *71*, 1749–1753. [CrossRef]

112. Yang, M.H.; Wang, J.S.; Luo, J.G.; Wang, X.B.; Kong, L.Y. Tetranortriterpenoids from *Chisocheton paniculatus*. *J. Nat. Prod.* **2009**, *72*, 2014–2018. [CrossRef]

113. Siddiqui, B.S.; Afshana, F.; Ghiasuddina, F.; Faizi, S.; Naqvi, S.N.H.; Tariq, R.M. Two insecticidal tetranortriterpenoids from *Azadirachta indica*. *Phytochemistry* **2000**, *53*, 371–376. [CrossRef]

114. Trinh, B.T.D.; Nguyen, H.D.; Nguyen, H.T.; Pham, P.D.; Ngo, N.T.N. Cytotoxic limonoids from the bark of *Walsura cochinchinensis*. *Fitoterapia* **2019**, *133*, 75–79. [CrossRef] [PubMed]

115. Munehiro, N.; Bo, Z.J.; Kenjiro, T.; Hideo, N. Spirosendan, a novel spiro limonoid from *Melia toosendan*. *Chem. Lett.* **1998**, *27*, 1279–1280.

116. Veitch, N.C.; Wright, G.A.; Stevenson, P.C. Four new tetranortriterpenoids from *Cedrela odorata* associated with leaf rejection by *Exopthalmus jekelianus*. *J. Nat. Prod.* **1999**, *62*, 1260–1263. [CrossRef] [PubMed]

117. Chen, J.L.; Kernan, M.R.; Jolad, S.D.; Stoddart, C.A.; Bogan, M.; Cooper, R. Dysoxylins A–D, tetranortriterpenoids with potent anti-RSV activity from *Dysoxylum gaudichaudianum*. *J. Nat. Prod.* **2007**, *70*, 312–315. [CrossRef]

118. Waratchareeyakul, W.; Hellemann, E.; Gil, R.R.; Chantrapromma, K.; Langat, M.K.; Mulholland, D.A. Application of residual dipolar couplings and selective quantitative NOE to establish the structures of tetranortriterpenoids from *Xylocarpus rumphii*. *J. Nat. Prod.* **2017**, *80*, 391–402. [CrossRef]

119. Tchimene, M.K.; Tane, P.; Ngamga, D.; Connolly, J.D.; Farrugia, L.J. Four tetranortriterpenoids from the stem bark of *Khaya anthotheca*. *Phytochemistry* **2005**, *66*, 1088–1093. [CrossRef]

120. Fauzi, F.M.; Meilanie, S.R.; Zulfikar; Farabi, K.; Herlina, T.; Al Anshori, J.; Mayanti, T. Kokosanolide D: A New tetranortriterpenoid from fruit peels of *Lansium domesticum* Corr. cv *Kokossan*. *Molbank* **2021**, *2021*, M1232. [CrossRef]

121. Murphy, B.T. Isolation, and Structure Elucidation of Antiproliferative Natural Products from Madagascar. Ph.D. Thesis, Virginia Polytechnic Institute and State University, Blacksburg, VA, USA, 2007.

122. Murphy, B.T.; Brodie, P.; Slebodnick, C.; Miller, J.S.; Birkinshaw, C.; Randrianjanaka, L.M.; Andriantsiferana, R.; Rasamison, V.E.; Dyke, K.T.; Suh, E.M.; et al. Antiproliferative limonoids of a *Malleastrum* sp. from the Madagascar rainforest. *J. Nat. Prod.* **2008**, *71*, 325–329. [CrossRef]

123. Krief, S.; Martin, M.T.; Grellier, P.; Kasenene, J.; Sevenet, T. Novel antimalarial compounds isolated in a survey of self-medicative behavior of wild chimpanzees in Uganda. *Antimicrob. Agents Chemother.* **2004**, *48*, 3196–3199. [CrossRef]

124. Batista, R.; de Jesus Silva Júnior, A.; Braga de Oliveira, A. Plant-derived antimalarial agents: New leads and efficient phytomedicines. Part II. Non-alkaloidal natural products. *Molecules* **2009**, *14*, 3037–3072. [CrossRef]

125. Curcino Vieira, I.J.; da Silva Terra, W.; dos Santos Gonçalves, M.; Braz-Filho, R. Secondary metabolites of the genus Trichilia: Contribution to the chemistry of Meliaceae family. *Am. J. Anal. Chem.* **2014**, *5*, 42542. [CrossRef]

126. Siddiqui, S.; Ghiasuddin, S.; Siddiqui, B.S.; Faizi, S. Tetranortriterpenoids and steroidal glycosides from the seeds of *Azadirachta indica* A. Juss. *Pak. J. Sci. Indust. Res.* **1989**, *32*, 435.

127. Schwikkard, S.L. Extractives from the Meliaceae and Simaroubaceae of Madagascar. Ph.D Thesis, University of Natal, Durban, South Africa, 1997.

128. Tan, Q.G.; Luo, X.D. Meliaceous limonoids: Chemistry and biological activities. *Chem. Rev.* **2011**, *111*, 7437–7522. [CrossRef]

129. Banerji, R.; Misra, G.; Nigam, S.K. On the triterpenes of *Azadirachta indica* (*Melia azadirachta*). *Fitoterapia* **1977**, *48*, 166.

130. Banerji, B.; Nigam, S.K. Wood constituents of Meliaceae: A review. *Fitoterapia* **1984**, *55*, 3.

131. Akhila, A.; Rani, K. Chemistry of the neem tree (*Azadirachta indica* A. Juss.). In *Fortschritte der Chemie Organischer Naturstoffe/Progress in the Chemistry of Organic Natural Products*; Herz, W., Falk, H., Kirby, G.W., Moore, R.E., Tamm, C., Eds.; Springer: Vienna, Austria, 1999.

132. Gao, Q.; Sun, J.; Xun, H.; Yao, X.; Wang, J.; Tang, F. A new azadirachta from the crude extracts of neem (*Azadirachta indica* A. Juss) seeds. *Nat. Prod. Res.* **2017**, *31*, 1739–1746. [CrossRef]

133. Ou-Yang, D.-W.; Wu, L.; Li, Y.-L.; Yang, P.-M.; Kong, D.-Y.; Yang, X.-W.; Zhang, W.-D. Miscellaneous terpenoid constituents of *Abies nephrolepis* and their moderate cytotoxic activities. *Phytochemistry* **2011**, *72*, 2197–2204. [CrossRef]

134. Yang, S.P.; Yue, J.M. Discovery of structurally diverse and bioactive compounds from plant resources in China. *Acta Pharmacol. Sin.* **2012**, *33*, 1147–1158. [CrossRef]

135. Son, N.T. The genus Walsura: A rich resource of bioactive limonoids, triterpenoids, and other types of compounds. In *Progress in the Chemistry of Organic Natural Products 118*; Kinghorn, A.D., Falk, H., Gibbons, S., Asakawa, Y., Liu, J.K., Dirsch, V.M., Eds.; Springer: Cham, Switzerland, 2022.

136. Li, Y.; Lin, W.; Huang, J. Anti-cancer effects of *Gynostemma pentaphyllum* (Thunb.) Makino (Jiaogulan). *Chin. Med.* **2016**, *11*, 43. [CrossRef]
137. Razmovski-Naumovski, V.; Huang, T.H.W.; Tran, V.H.; Li, G.Q.; Duke, C.C.; Roufogalis, B.D. Chemistry and pharmacology of *Gynostemma pentaphyllum*. *Phytochem. Rev.* **2005**, *4*, 197–219. [CrossRef]
138. Grover, J.K.; Yadav, S.P. Pharmacological actions and potential uses of *Momordica charantia*: A review. *J. Ethnopharmacol.* **2004**, *93*, 123–132. [CrossRef]
139. Jia, S.; Shen, M.; Zhang, F.; Xie, J. Recent Advances in *Momordica charantia*: Functional components and biological activities. *Int. J. Mol. Sci.* **2017**, *18*, 2555. [CrossRef]
140. Han, M.-L.; Zhang, H.; Yang, S.-P.; Yue, J.-M. Walsucochinoids A and B: New rearranged limonoids from *Walsura cochinchinensis*. *Org. Lett.* **2012**, *14*, 486–489. [CrossRef]
141. Simo Mpetga, J.D.; He, H.-P.; Hao, X.-J.; Leng, Y.; Tane, P. Further cycloartane and friedelane triterpenoids from the leaves of *Caloncoba glauca*. *Phytochem. Lett.* **2014**, *7*, 52–56. [CrossRef]
142. Chen, J.-C.; Liu, W.-Q.; Lu, L.; Qiu, M.H.; Zheng, Y.T.; Yang, L.M. Kuguacins F-S, cucurbitane triterpenoids from *Momordica charantia*. *Phytochemistry* **2009**, *70*, 133–140. [CrossRef]
143. Zhang, X.-S.; Cao, J.-Q.; Zhao, C.; Wang, X.-d.; Wu, X.-j.; Zhao, Y.-Q. Novel dammarane-type triterpenes isolated from hydrolyzate of total *Gynostemma pentaphyllum* saponins *Bioorg. Med. Chem. Lett.* **2015**, *25*, 3095–3099. [CrossRef]
144. Cao, J.-Q.; Zhang, Y.; Cui, J.-M.; Zhao, Y.-Q. Two new cucurbitane triterpenoids from *Momordica charantia* L. *Chin. Chem. Lett.* **2011**, *22*, 583–586. [CrossRef]
145. Wang, W.-H.; Nian, Y.; He, Y.-J.; Wan, L.-S.; Bao, N.-M.; Zhu, G.-L.; Wang, F.; Qiu, M.-H. New cycloartane triterpenes from the aerial parts of *Cimicifuga heracleifolia*. *Tetrahedron* **2015**, *71*, 8018–8025. [CrossRef]
146. Huang, D.; Qing, S.; Zeng, G.; Wang, Y.; Guo, H.; Tan, J.; Zhou, Y. Lipophilic components from fructus *Viticis Negundo* and their antitumor activities. *Fitoterapia* **2013**, *86*, 144–148. [CrossRef]
147. Zhou, Y.; Yang, J.; Peng, L.; Li, Y.; Chen, W. Two novel saponins of 20, 26-epoxy derivatives of pseudojujubogenin from the seeds of *Hovenia trichocarpa*. *Fitoterapia* **2013**, *87*, 65–68. [CrossRef]
148. Wong, C.P.; Shimada, M.; Nugroho, A.E.; Hirasawa, Y.; Kaneda, T.; Hadi, A.H.A.; Osamu, S.; Morita, H. Ceramicines J–L, new limonoids from *Chisocheton ceramicus*. *J. Nat. Med.* **2012**, *66*, 566–570. [CrossRef]
149. Renkonen, O.; Schindler, O.; Reichstein, T. Die Konstitution von Sinogenin. Glykoside und Aglykone. 181. Mitteilung. *Croatica Chem. Acta* **1957**, *29*, 239–245.
150. Liu, J.-Q.; Wang, C.-F.; Li, Y.; Chen, J.-C.; Zhou, L.; Qiu, M.-H. Limonoids from the leaves of *Toona ciliata* var. *yunnanensis*. *Phytochemistry* **2012**, *76*, 141–149. [CrossRef]
151. Xu, D.-M.; Huang, E.-X.; Wang, S.-Q.; Wen, X.-G.; Wu, X.-Y. Studies on the chemical constituents of *Fritillaria pallidiflora* Schrenk. *J. Integr. Plant Biol.* **1990**, *32*, 789–793.
152. Li, Y.; Yili, A.; Li, J.; Muhamat, A.; Aisa, H.A. New isosteroidal alkaloids with tracheal relaxant effect from the bulbs of *Fritillaria pallidiflora* Schrenk. *Bioorg. Med. Chem. Lett.* **2016**, *26*, 1983–1987. [CrossRef]
153. Li, H.J.; Jiang, Y.; Ping Li, W.-C. Puqienine F, a novel veratramine alkaloid from the bulbs of *Fritillaria puqiensis*. *Chem. Pharm. Bull.* **2006**, *54*, 722–724. [CrossRef]
154. Babar, Z.U.; Ata, A.; Meshkatalsadat, M.H. New bioactive steroidal alkaloids from *Buxus hyrcana*. *Steroids* **2006**, *71*, 1045–1051. [CrossRef]
155. Niño, J.; Correa, Y.M.; Mosquera, O.M. Biological activities of steroidal alkaloids isolated from *Solanum leucocarpum*. *Pharm. Biol.* **2009**, *47*, 255–259. [CrossRef]
156. Zhao, Q.-Q.; Song, Q.-Y.; Jiang, K.; Li, G.-D.; Wei, W.-J.; Li, Y.; Gao, K. Spirochensilides A and B, two new rearranged triterpenoids from *Abies chensiensis*. *Org. Lett.* **2015**, *17*, 2760–2763. [CrossRef]
157. Wang, G.-W.; Lv, C.; Fang, X.; Tian, X.-H.; Ye, J.; Li, H.-L.; Shan, L.; Shen, Y.-H.; Zhang, W.-D. Eight pairs of epimeric triterpenoids involving a characteristic spiro-E/F ring from *Abies faxoniana*. *J. Nat. Prod.* **2015**, *78*, 50–60. [CrossRef]
158. Meng, X.H.; Chai, T.; Shi, Y.P.; Yang, J.L. Bungsteroid A: One Unusual C34 Pentacyclic Steroid Analogue from *Zanthoxylum bungeanum* Maxim. *J. Org. Chem.* **2020**, *85*, 10806–10812. [CrossRef]
159. Okuzumi, K.; Hara, N.; Uekusa, H.; Fujimoto, Y. Structure elucidation of cyasterone stereoisomers isolated from *Cyathula officinalis*. *Org. Biomol. Chem.* **2005**, *3*, 1227–1232. [CrossRef]
160. Yi, J.; Luo, Y.; Li, B.; Zhang, G. Phytoecdysteroids and glycoceramides from *Eriophyton wallchii*. *Steroids* **2004**, *69*, 809–815. [CrossRef]
161. Snogan, E.; Vahirua-Lechat, I.; Ho, R.; Bertho, G.; Girault, J.P.; Ortiga, S.; Maria, A.; Lafont, R. Ecdysteroids from the medicinal fern *Microsorum scolopendria* (Burm. f.). *Phytochem. Anal.* **2007**, *18*, 441–450. [CrossRef]
162. Hao, D.-C.; Gu, X.-J.; Xiao, P.-G.; Peng, Y. Phytochemical, and biological research of Fritillaria medicinal resources. *Chin. J. Nat. Med.* **2013**, *11*, 0330–0344. [CrossRef]
163. Wang, Y.; Hou, H.; Ren, Q. Natural drug sources for respiratory diseases from Fritillaria: Chemical and biological analyses. *Chin. Med.* **2021**, *16*, 40. [CrossRef] [PubMed]
164. Rushforth, K.D. Notes on Chinese silver firs 2. *Notes R. Bot. Gard. Edinb.* **1984**, *41*, 535–540.
165. Dai, J.; Han, R.; Xu, Y.; Li, N.; Wang, J.; Dan, W. Recent progress of antibacterial natural products: Future antibiotics candidates. *Bioorg. Chem.* **2020**, *101*, 103922. [CrossRef]

166. Zhang, M.; Wang, J.; Zhu, L.; Li, T.; Jiang, W.; Zhou, J.; Peng, W.; Wu, C. *Zanthoxylum bungeanum* Maxim. (Rutaceae): A systematic review of its traditional uses, botany, phytochemistry, pharmacology, pharmacokinetics, and toxicology. *Int. J. Mol. Sci.* **2017**, *18*, 2172. [CrossRef]

167. Bao, Y.; Yang, L.; Fu, Q.; Fu, Y.; Tian, Q.; Wang, C.; Huang, Q. The current situation of *Zanthoxylum bungeanum* industry and the research and application prospect. A review. *Fitoterapia* **2023**, *164*, 105380. [CrossRef]

168. Chawla, A.; Dev, S. A new class of triterpenoids from *Ailanthus malabarica* DC derivatives of malabaricane. *Tetrahedron Lett.* **1967**, *48*, 4837–4843. [CrossRef]

169. Srinivas, P.V.; Rao, R.R.; Rao, J.M. Two new tetracyclic triterpenes from the heartwood of *Ailanthus excelsa* Roxb. *Chem. Biodivers.* **2006**, *3*, 930–934. [CrossRef]

170. Ebada, S.S.; Lin, W.; Proksc, P. Bioactive sesterterpenes and triterpenes from marine sponges: Occurrence and pharmacological significance. *Mar. Drugs* **2010**, *8*, 313–346. [CrossRef]

171. Khare, C.P. *Indian Medicinal Plants*; Springer: Berlin/Heidelberg, Germany, 2007; ISBN 978-0-387-70637-5.

172. Thongnest, S.; Boonsombat, J.; Prawat, H.; Mahidol, C.; Ruchirawat, S. Ailanthusins A-G and nor-lupane triterpenoids from *Ailanthus triphysa*. *Phytochemistry* **2017**, *134*, 98–105. [CrossRef] [PubMed]

173. Ziegler, H.L.; Christensen, J.; Olsen, C.E.; Sittie, A.A.; Jaroszewski, J.W. New dammarane and malabaricane triterpenes from *Caloncoba echinata*. *J. Nat. Prod.* **2002**, *65*, 1764–1768. [CrossRef]

174. Liaw, C.C.; Lo, I.W.; Lin, Y.C.; Huang, H.T.; Zhang, L.J.; Hsiao, P.C. Four cucurbitane glycosides taimordisins A–D with novel furopyranone skeletons isolated from the fruits of *Momordica charantia*. *Food Chem.* **2022**, *14*, 100286. [CrossRef] [PubMed]

175. Rashid, I.; Yaqoob, U. Traditional uses, phytochemistry and pharmacology of genus Fritillaria—A review. *Bull. Natl. Res. Cent.* **2021**, *45*, 124. [CrossRef]

176. Xu, Y.-H.; Dai, G.-M.; Tang, S.H.; Yang, H.-J.; Sun, Z.-G. Changes of anti-tuberculosis herbs formula during past three decades in contrast to ancient ones. *Chin. J. Integr. Med.* **2021**, *27*, 388–393. [CrossRef]

177. Moosmann, P.; Ueoka, R.; Grauso, L.; Mangoni, A.; Morinaka, B.I.; Gugger, M.; Piel, J. Cyanobacterial ent-sterol-like natural products from a deviated ubiquinone pathway. *Angew. Chem. Int. Ed.* **2017**, *56*, 4987–4990. [CrossRef]

178. Govindam, S.V.S.; Choi, B.-K.; Yoshioka, Y.; Kanamoto, A.; Fujiwara, T.; Okamoto, T.; Ojika, M. Novel cytotoxic polyoxygenated steroids from an Okinawan sponge *Dysidea* sp. *Biosci. Biotechnol. Biochem.* **2012**, *76*, 999–1002. [CrossRef]

179. Braekman, J.C.; Daloze, D.; Moussiaux, B.; Vandervyver, G.; Riccio, R. Cholest-6-EN-11β,19-epoxy-3β,5α,8α,9α-tetrol, a novel polyoxygenated steroid from the sponge *Dysidea tupha*. *Bull. Soc. Chim. Belg.* **1988**, *97*, 293. [CrossRef]

180. Corgiat, J.M.; Scheuer, P.J.; Rios Steiner, J.L.; Clardy, J. Three pregnane-10, 2-carbolactones from a sponge, *Strongylophora* sp. *Tetrahedron* **1993**, *49*, 1557. [CrossRef]

181. Teruya, T.; Nakagawa, S.; Koyama, T.; Suenaga, K.; Kita, M.; Uemura, D. Nakiterpiosin, a novel cytotoxic C-nor-D-homosteroid from the Okinawan sponge *Terpios hoshinota*. *Tetrahedron Lett.* **2003**, *44*, 5171. [CrossRef]

182. Afiyatullov, S.Sh.; Kalinovsky, A.I.; Antonov, A.S.; Ponomarenko, L.P.; Dmitrenok, P.S.; Aminin, D.L.; Krasokhin, V.B.; Nosova, V.M.; Kisin, A.V. Isolation and structures of erylosides from the Carriben sponge *Erylus goffrilleri*. *J. Nat. Prod.* **2007**, *70*, 1871–1877. [CrossRef]

183. Kolesnikova, S.A.; Lyakhova, E.G.; Kalinovsky, A.I.; Pushilin, M.A.; Afiyatullov, S.S.; Yurchenko, E.A.; Dyshlovoy, S.A.; Minh, C.V.; Stonik, V.A. Isolation, structures, and biological activities of triterpenoids from a marine sponge *Penares* sp. *J. Nat. Prod.* **2013**, *76*, 1746–1752. [CrossRef]

184. Anjaneyulu, A.S.R.; Rao, V.L.; Sastry, V.G. A new spiroketal steroid from *Gorgonella umbraculum*. *Nat. Prod. Res.* **2003**, *17*, 149. [CrossRef]

185. Tomono, Y.; Hirota, H.; Fusetani, N. Isogosterones A–D, antifouling 13,17-secosteroids from an octocoral *Dendronephthya* sp. *J. Org. Chem.* **1999**, *64*, 2272. [CrossRef]

186. Rao, C.B.; Ramana, K.V.; Rao, D.V.; Fahy, E.; Faulkner, D.J. Metabolites of the gorgonian *Isis hippuris* from India. *J. Nat. Prod.* **1988**, *51*, 954. [CrossRef]

187. Rao, C.B.; Kalidindi, R.S.H.S.N.; Trimurtulu, G.; Rao, D.V. Metabolites of porifera, part III. New 24-methylscalaranes from *Phyllospongia dendyi* of the Indian ocean. *J. Nat. Prod.* **1991**, *54*, 364. [CrossRef]

188. Alvi, K.A.; Crews, P. Homoscalarane sesterterpenes from *Lendenfeldia frondosa*. *J. Nat. Prod.* **1992**, *55*, 859. [CrossRef]

189. Crews, P.; Bescansa, P. Sesterterpenes from a common marine sponge, *Hyrtios erecta*. *J. Nat. Prod.* **1986**, *49*, 1041. [CrossRef] [PubMed]

190. Zeng, L.; Fu, X.; Su, J.; Pordesimo, E.O.; Traeger, S.C.; Schmitz, F.J. Novel bishomoscalarane sesterterpenes from the sponge *Phyllospongia foliascens*. *J. Nat. Prod.* **1991**, *54*, 421. [CrossRef]

191. Chang, Y.-C.; Tseng, S.-W.; Liu, L.-L.; Chou, Y.; Ho, Y.-S.; Lu, M.-C.; Su, J.-H. Cytotoxic sesterterpenoids from a sponge *Hippospongia* sp. *Mar. Drugs* **2012**, *10*, 987–997. [CrossRef]

192. Youssef, D.T.; Yamaki, R.K.; Kelly, M.; Scheuer, P.J. Salmahyrtisol A, a novel cytotoxic sesterterpene from the Red Sea sponge *Hyrtios erecta*. *J. Nat. Prod.* **2002**, *65*, 2–6. [CrossRef]

193. Aoki, S.; Watanabe, Y.; Sanagawa, M.; Setiawan, A.; Kotoku, N.; Kobayashi, M. Cortistatins A, B, C, and D, anti-angiogenic steroidal alkaloids, from the marine sponge *Corticium simplex*. *J. Am. Chem. Soc.* **2006**, *128*, 3148–3149. [CrossRef] [PubMed]

194. Roll, D.M.; Scheuer, P.J.; Matsumoto, G.K.; Clardy, J. Halenaquinone, a pentacyclic polyketide from a marine sponge. *J. Am. Chem. Soc.* **1983**, *105*, 6177–6178. [CrossRef]

195. Cimino, G.; Madaio, A.; Trivellone, E. Minor triterpenoids from the Mediterranean sponge, *Raspaciona aculeata*. *J. Nat. Prod.* **1994**, *57*, 784–790. [CrossRef]

196. Cimino, G.; Crispino, A.; Madaio, A.; Trivellone, E. Raspacionin B, a further triterpenoid from the mediterranean sponge *Raspaciona aculeata*. *J. Nat. Prod.* **1993**, *56*, 534–538. [CrossRef]

197. Cimino, G.; Epifanio, R.D.A.; Madaio, A.; Puliti, R.; Trivellone, E. Absolute stereochemistry of raspacionin, the main triterpenoid from the marine sponge *Raspaciona aculeata*. *J. Nat. Prod.* **1993**, *56*, 1622–1626. [CrossRef]

198. Chao, C.H.; Huang, L.F.; Yang, Y.L.; Su, J.H.; Wang, G.H.; Chiang, M.Y.; Wu, Y.C.; Dai, C.F.; Sheu, J.H. Polyoxygenated steroids from the Gorgonian *Isis hippuris*. *J. Nat. Prod.* **2005**, *68*, 880–885. [CrossRef]

199. Anjaneyulu, A.S.R.; Krishna Murthy, M.V.R.; Gowri, P.M. Novel epoxy steroids from the Indian ocean soft coral *Sarcophyton crassocaule*. *J. Nat. Prod.* **2000**, *63*, 112–118. [CrossRef] [PubMed]

200. Huang, C.Y.; Su, J.S.; Liaw, C.C.; Sung, P.J.; Chiang, P.L.; Hwang, T.L.; Chang-Feng Dai, J.-H.; Sheu, H. Bioactive steroids with methyl ester group in the side chain from a reef soft coral *Sinularia brassica* cultured in a tank. *Mar. Drugs* **2017**, *15*, 280. [CrossRef] [PubMed]

201. D'Auria, M.V.; Giannini, C.; Zampella, A.; Minale, L.; Debitus, C.; Roussakis, C. Crellastatin A: A cytotoxic bis-steroid sulfate from the Vanuatu marine sponge *Crella* sp. *J. Org. Chem.* **1998**, *63*, 7382–7388. [CrossRef] [PubMed]

202. Murayama, S.; Imae, Y.; Takada, K.; Kikuchi, J.; Nakao, Y.; van Soest, R.W.M.; Okada, S.; Matsunaga, S. Shishicrellastatins, inhibitors of cathepsin B, from the marine sponge *Crella (Yvesia) spinulata*. *Bioorg. Med. Chem.* **2011**, *19*, 6594–6598. [CrossRef] [PubMed]

203. Morinaka, B.I.; Pawlik, J.R.; Molinski, T.F. Amaroxocanes A and B: Sulfated dimeric sterols defend the Caribbean coral reef sponge *Phorbas amaranthus* from fish predators. *J. Nat. Prod.* **2009**, *72*, 259–264. [CrossRef]

204. Cheng, J.F.; Lee, J.S.; Sun, F.; Jares-Erijman, E.A.; Cross, S.; Rinehart, K.L. Hamigerols A and B, unprecedented polysulfate sterol dimers from the Mediterranean sponge *Hamigera hamigera*. *J. Nat. Prod.* **2007**, *70*, 1195–1199. [CrossRef]

205. Whitson, E.L.; Bugni, T.S.; Chockalingam, P.S.; Conception, G.P.; Feng, X.; Jin, G. Fibrosterol sulfates from the Philippine sponge *Lissodendoryx (Acanthodoryx) fibrosa*: Sterol dimers that inhibit PKC. *J. Org. Chem.* **2009**, *74*, 5902–5908. [CrossRef]

206. Pettit, G.R.; Xu, J.-P.; Chapuis, J.-C.; Melody, N. The cephalostatins. Isolation, structure, and cancer cell growth inhibition of cephalostatin 20. *J. Nat. Prod.* **2015**, *78*, 1446–1450. [CrossRef]

207. Boonlarppradab, C.; Faulkner, D.J. Eurysterols A and B, cytotoxic and antifungal steroidal sulfates from a marine sponge of the genus *Euryspongia*. *J. Nat. Prod.* **2007**, *70*, 846–848. [CrossRef]

208. Solanki, H.; Angulo-Preckler, C.; Calabro, K.; Kaur, N.; Lasserre, P.; Cautain, B.; de la Cruz, M.; Reyes, F.; Avila, C.; Thomas, O.P. Suberitane sesterterpenoids from the Antarctic sponge *Phorbas areolatus* (Thiele, 1905) *Tetrahedron Lett.* **2018**, *59*, 3353–3356. [CrossRef]

209. Song, J.; Jeong, W.; Wang, N.; Lee, H.-S.; Sim, C.J.; Oh, K.-B.; Shin, J. Scalarane sesterterpenes from the sponge *Smenospongia* sp. *J. Nat. Prod.* **2008**, *71*, 1866–1871. [CrossRef] [PubMed]

210. Kimura, J.; Hyosu, M. Two new sesterterpenes from the marine sponge, *Coscinoderma mathewsi*. *Chem. Lett.* **1999**, *52*, 61–62. [CrossRef]

211. Hochlowski, J.E.; Faulkner, D.J.; Bass, L.S.; Clardy, J. Metabolites of the dorid nudibranch *Chromodoris sedna*. *J. Org. Chem.* **1983**, *48*, 1738–1740. [CrossRef]

212. Keyzers, R.A.; Daoust, J.; Davies-Coleman, M.T.; Van Soest, R.; Balgi, A.; Donohue, E.; Roberge, M.; Andersen, R.J. Autophagy-modulating aminosteroids isolated from the sponge *Cliona celata*. *Org. Lett.* **2008**, *10*, 259–262. [CrossRef]

213. Li, J.; Zhu, H.; Ren, J.; Deng, Z.; de Voogd, N.J.; Proksch, P.; Lin, W. Globostelletins JeS, isomalabaricanes with unusual cyclopentane sidechains from the marine sponge *Rhabdastrella globostellata*. *Tetrahedron* **2012**, *68*, 559–565. [CrossRef]

214. Guzii, A.G.; Makarieva, T.N.; Denisenko, V.A.; Dmitrenok, P.S.; Burtseva, Y.V.; Krasokhin, V.B.; Stonik, V.A. Topsentiasterol sulfates with novel iodinated and chlorinated side chains from the marine sponge *Topsentia* sp. *Tetrahedron Lett.* **2008**, *49*, 7191–7193. [CrossRef]

215. Fusetani, N.; Takahashi, M.; Matsunaga, S. Topsentiasterol sulfates, antimicrobial sterol sulfates possessing novel side chains, from a marine sponge, *Topsentia* sp. *Tetrahedron* **1994**, *50*, 7765–7770. [CrossRef]

216. Huang, C.-Y.; Liaw, C.-C.; Chen, B.-W.; Chen, P.-C.; Su, J.-H. Withanolide-based steroids from the cultured soft coral *Sinularia brassica*. *J. Nat. Prod.* **2013**, *76*, 1902–1908. [CrossRef]

217. Lal, A.R.; Cambie, R.C.; Rickard, C.E.F.; Bergquist, P.R. Sesterterpene lactones from a sponge species of the genus *Dactylospongia*. *Tetrahedron Lett.* **1994**, *35*, 2603–2606. [CrossRef]

218. Cambie, C.R.; Lal, A.R.; Rickard, C.E.F. A sesterterpene lactone from *Petrosaspongia nigra* sp. nov. *Acta Cryst.* **1996**, *C52*, 709–711. [CrossRef]

219. Randazzo, A.; Debitus, C.; Minale, L.; Pastor, P.G.; Alcaraz, M.J.; Payá, M.; Gomez-Paloma, L. Petrosaspongiolodes M–R: New potent and selective phosphoplipase A2 inhibitors from the New Caledonian marine sponge *Petrosaspongia nigra*. *J. Nat. Prod.* **1998**, *61*, 571–575. [CrossRef] [PubMed]

220. Carney, J.R.; Yoshida, W.Y.; Scheuer, P.J. Kiheisterones, new cytotoxic steroids from a Maui sponge. *J. Org. Chem.* **1992**, *57*, 6637–6640. [CrossRef]

221. Miyamoto, T.; Sakamoto, K.; Amano, H.; Higuchi, R.; Komori, T.; Sasaki, T. Three new cytotoxic sesterterpenoids, inorolide A, B, and C from the nudibranch *Chromodoris inornata*. *Tetrahedron Lett.* **1992**, *33*, 5811–5814. [CrossRef]

222. D'Auria, M.V.; Paloma, L.G.; Minale, L.; Riccio, R.; Zampella, A. Isolation, structure characterization and conformational analysis of a unique 4α,9α-epoxysteroid sulphate from the okinawan ophiuroid *Ophiomastix annulosa*. *Tetrahedron Lett.* **1992**, *33*, 4641–4644. [CrossRef]
223. Jiménez, J.I.; Yoshida, W.Y.; Scheuer, P.J.; Lobkovsky, E.; Clardy, J.; Kelly, M. Honulactones: New bishomoscalarane sesterterpenes from the Indonesian sponge *Strepsichordaia aliena*. *J. Org. Chem.* **2000**, *65*, 6837–6840. [CrossRef] [PubMed]
224. Kobayashi, M. Marine terpenes and terpenoids. IV. Isolation of new cembranoid and secocembranoid lactones from the soft coral *Sinularia mayi*. *Chem. Pharm. Bull.* **1988**, *36*, 488–494. [CrossRef]
225. Carmely, S.; Kashman, Y. Isolation and structure elucidation of lobophytosterol, depresosterol and three other closely related sterols: Five new C28 polyoxygenated sterols from the red sea soft coral *Lobophytum depressum*. *Tetrahedron* **1981**, *37*, 2397–2403. [CrossRef]
226. Kashman, V.; Carmely, S. Four novel C28 sterols from *Lobophytum depressum*. *Tetrahedron Lett.* **1980**, *21*, 4939–4942. [CrossRef]
227. Kim, S.K.; van Ta, Q. Bioactive sterols from marine resources and their potential benefits for human health. *Adv. Food Nutrit. Res.* **2012**, *65*, 261–268.
228. Donets, M.M.; Tsygankov, V.Y. Trace elements in commercial marine organisms from the Russian part of the Northwest Pacific (2010–2018). *Environ. Chem. Lett.* **2019**, *17*, 1727–1740. [CrossRef]

MDPI

St. Alban-Anlage 66

4052 Basel

Switzerland

www.mdpi.com

Molecules Editorial Office

E-mail: molecules@mdpi.com

www.mdpi.com/journal/molecules